Separations for Biotechnology 3

Papers presented at the Third International Symposium on 'Separations for Biotechnology' organised by the Biotechnology Group and Solvent Extraction and Ion Exchange Group of the Society of Chemical Industry, held on 12–15 September 1994, at The University of Reading.

Organising Committee

Professor D. L. Pyle (*Chairman*)	*University of Reading*
Dr J. A. Asenjo	*University of Reading*
Dr B. S. Baines	*Glaxo Group Research*
Dr J. R. Birch	*Celltech Biologics plc*
Dr G. Chapman	*Bio-products Laboratory*
Dr H. A. Chase	*University of Cambridge*
Dr J. B. Chaudhuri	*University of Bath*
Dr A. Lyddiatt	*University of Birmingham*
Dr T. de V. Naylor	*EPSRC Separations Coordinator*
Dr A. Thomson	*Independent Consultant*
Mr M. S. Verrall	*SmithKline Beecham Pharmaceuticals*
Mr S. P. Vranch	*Jacobs Engineering*
Mrs A. Potter	*SCI Conference Manager*

Referees

Dr F. Addo-Yobo

Professor A. Andrews

Dr B. Andrews

Dr T. Arnot

Mr M. Ashley

Professor D. Brown

Professor C. Bucke

Dr J. Bonerjea

Dr F. J. Davis

Professor P. Dunnill

Dr R. Field

Dr C. Hill

Professor J. Howell

Professor M. Hoare

Dr J. Hubble

Dr N. Kirby

Dr D. Leak

Dr J. C. Lee

Dr M. Lewis

Dr A. Livingston

Dr P. Luckham

Dr G. Lye

Dr K. Niranjan

Dr R. Rastall

Mr A. J. Reynolds

Professor N. Robson

Professor M. Streat

Dr D. Stuckey

Dr C. J. Skidmore

Dr N. Titchener-Hooker

Dr J. Varley

Professor L. Weatherley

Dr M. Winkler

Separations for Biotechnology 3

Edited by

D. L. Pyle

*Biotechnology and Biochemical Engineering Group, Department of
Food Science and Technology, The University of Reading, Reading,
UK*

Sep/ae
CHEM

Special Publication No. 158

ISBN 0-85186-724 3

A catalogue record of this book is available from the British Library

Published by The Royal Society of Chemistry,
Thomas Graham House, Science Park, Cambridge
CB4 4WF

Printed in Great Britain by Bookcraft (Bath) Ltd

Preface

TP248
.25
S47 I58
1994
CHEM

Bioseparations is the set of techniques by which the products of fermentations or biotransformations are recovered for subsequent use. The importance of recovery in high yield of pure biological products for process efficiency and economics as well as for product safety and quality cannot be overstated. Many of the technical problems posed in bioprocessing are new ones. A satisfactory solution must combine high yields and productivity with high levels of purity; it must also be predictable, reproducible and controllable. It is difficult to achieve these objectives at present, for a number of reasons. One is that the starting point of the downstream processing line is typically a dilute, but complex, impure and poorly characterised broth, often currently requiring several purification stages; secondly, many products are not robust, making the task of recovery difficult and one for which existing process technologies are not directly appropriate; thirdly, there are large gaps in fundamental understanding in areas of science and engineering relevant to separations technology for biomolecules; finally, extreme demands are imposed on product purity by regulatory requirements.

This book includes the papers - oral and poster - presented at the Third International Conference on Separations for Biotechnology. It includes papers on a very wide range of approaches and areas of biotechnology, reflecting the multidisciplinary nature of the problems. I believe that it contains many new and interesting contributions to the science, engineering and implementation of bioseparations technology, and that it will be of equal interest to researchers and to practising industrial biotechnologists.

The proceedings show that, whilst there are as yet few generally applicable solutions, and fewer obviously optimal ones, nonetheless, as a result of R&D, the range of available technologies is rapidly increasing. Research is also throwing up other exciting possibilities, although there still remain many problems for research and development. One of the most interesting features is the recognition that a variety of approaches may all have a contribution to make. This is certainly an area where multidisciplinary efforts are needed, and where molecular biology and process engineering must go hand in hand.

Whilst quality can never be absolutely assured, all the papers presented here have been refereed by experts. Although the conference itself was structured around a series of themes, I felt that to continue this into the book would have been to impose a rather artificial structure, and as a result these proceedings are presented alphabetically by author. The limited number of papers should, however, ensure that it will be rather easier for the reader to find their way around the text than to solve a typical sequencing problem.

I must acknowledge the help of the SCI and its staff and also of many individuals in putting this book together. I am very grateful to the conference organising committee for all their work, which was rewarded by a larger than average refereeing load. I am also very grateful to all those colleagues, acknowledged elsewhere, who helped with refereeing; such unsung and unpaid work is vital for the development of science and engineering. I wish to thank Amanda Wright and Chris Pyle for their help with editing and with the index and, last but by no means least, Jean Davis for all her efforts in handling the reams of paperwork associated with this volume.

Leo Pyle

Contents

Extraction Kinetics of Amino Acids to AOT Water-in-oil Microemulsion

M. Adachi, R. Nishita, and M. Harada

INSTITUTE OF ATOMIC ENERGY, KYOTO UNIVERSITY, UJI 611, JAPAN

1. INTRODUCTION

The kinetics of the extraction with microemulsions(ME), which is important for separating biologically active substances, is very complicated because reorganization of ME droplets proceeds simultaneously with the extraction. Although several models have been proposed for elucidating the kinetic mechanism[1-8], the extraction rates have not yet been elucidated for the effects of the charged state of solute, the location of solute entrapment in ME droplets, the direction of extraction and the diameter of a ME droplet.

In this paper, we present the time evolution of the concentration profile of Trp for Trp transfer near an AOT–CCl₄/brine solution interface, from which the rate processes concerned with the transfer can be directly elucidated. The kinetic behaviour in the Lewis cell was examined for the extraction of Trp, I⁻ and K⁺ between the aqueous phase and AOT in n–heptane solutions. We measured the rates of the forward and backward extractions of Trp in zwitterionic and anionic states. These solutes are suitable for investigating the effects of electrical charge and the location of solute taken up in a ME droplet on the extraction rate. In both systems, large interfacial resistance was observed. We propose a model for interpreting the interfacial resistance.

2. EXPERIMENTAL

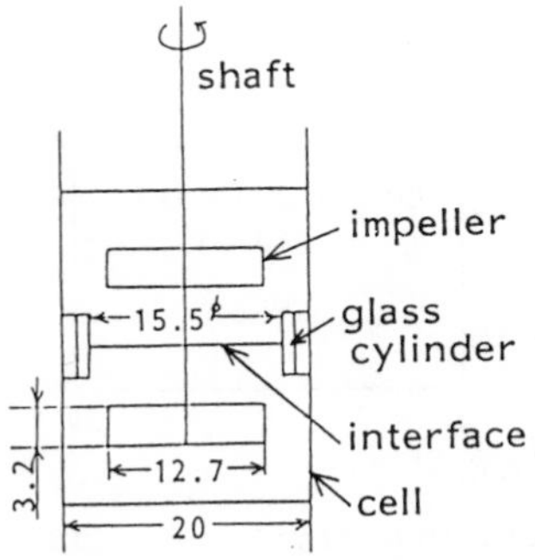

Fig. 1 Lewis-type cell.

Materials AOT provided by Tokyo Kasei Co. Ltd. was used without further purification. Trp, sodium chloride, sodium iodide and potassium chloride of reagent grade, and n–heptane, carbon tetrachloride of spectrophotometric reagent–grade were used as supplied.

Method A Lewis–type cell(Fig.1) was used to measure the extraction rate. The ME phase was prepared by dissolving a desired amount of AOT in n–heptane, and then by equilibrating it with the brine solution in order to saturate the

ME phase with water. The two phases of equal volume were brought into contact with each other and then stirred at desired rates. The hydrodynamic condition of each phase is slightly different from each other. The concentrations of Trp in both phases were determined by a spectrophotometer (280nm) in situ. The concentrations of I^- and K^+ in the solutions sampled were measured by a spectroscopic method (Shimadzu UV-200) at 226 nm and an inductively coupled argon plasma emission spectrophotometer (Jourrel Ash, ICAP 500), respectively. The water extracted in the ME phase was analysed by the Karl-Fisher method.

We also measured the time-evolution of the Trp-concentration profiles in a static diffusion cell in the AOT-CCl$_4$ microemulsion/NaCl aqueous solution system with the help of the position-scanning spectrophotometer[9]. The geometry of the diffusion cell is 4 cm in height and 1 mm in thickness.

We measured the concentration profile of AOT in AOT-ME droplets/ n-heptane solution with the aid of the position-scanning spectrophotometer at 250 nm, using the flow junction cell. The diffusion coefficient for the ME droplets in n-heptane was evaluated from the concentration profile for AOT.

3. RESULT AND DISCUSSION

Rate Processes concerned with Extraction of Trp by AOT microemulsion

The extraction of solute in the ME/aqueous solution system is generally a complicated process, because the reorganization of the droplets proceeds upon solute extraction. We first elucidate which rate processes play a key role in the solute transfer.

The time evolution of the concentration distribution of Trp in the static diffusion cell is shown in Fig.2. In this experiment, the Trp in the organic phase, which was prepared by bringing the AOT-ME in CCl$_4$ into contact with 1M brine solution containing Trp, was transferred to the 1M NaCl aqueous solution. The abscissa in Fig.2 represents the distances x from the interface normalized by $\sqrt{t}$ following Boltzmann's method[10]. These concentration distributions away from the interface are uniquely determined by the

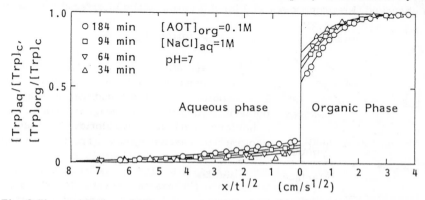

<u>Fig. 2</u> Time evolution of the concentration distribution of tryptophan during extraction in the static diffusion cell. [Trp]$_o$:feed concentration.

Boltzmann's variable $x/\sqrt{t}$ irrespective of the elapsed time. However, the concentrations near the interface strongly depend on time. This experimental result suggests that the rate–determining processes concerned with Trp transfer are the diffusion processes in both phases and the interfacial process. The governing equation is expressed by:

$$\partial C_J / \partial t = D_J\, \partial^2 C_J / \partial x^2, \quad \text{when } x \neq 0 \quad (J = \text{org. and aq.}) \qquad (1)$$

and,

$$-D_J\, \partial C_J / \partial x = k_i (C_{org} - K_d C_{aq}), \quad \text{at the interface } x = 0 \qquad (2)$$

with the initial and boundary conditions, $C_{org} = C_{org,o}$ and $C_{aq} = 0$ at $x = \pm\infty$ and $t = 0$. The solid curves in the figure are the values calculated from Eqs.(1) and (2) with $D_{org} = 1 \times 10^{-6}$ cm²/s, $D_{aq} = 1 \times 10^{-5}$ cm²/s, $k_i = 7 \times 10^{-6}$ cm/s, and $K_d = 0.3$. The calculated curves agree well with the observed values, indicating that the interfacial process plays a key role in the Trp transfer.

The rate of extraction J in the Lewis cell is then expressed by,

$$J = -(V_{org}/A)(dC_{org}/dt) = k_{org,ov}(K_d C_{aq} - C_{org}) \qquad (3)$$

$$1/k_{org,ov} = K_d/k_{aq} + 1/k_i + 1/k_{org} \qquad (4)$$

Since $V_{org} = V_{aq}$ in this work, the solution to Eq.(3) is expressed by,

$$\ln\{(C_{org,eq} - C_{org,o})/(C_{org,eq} - C_{org})\} = k_{org,ov}(A/V_{org})(1 + K_d)t \qquad (5)$$

where $C_{org,eq}$ is the value at $t = \infty$. In AOT(0.1M)–CCl₄/brine(1M) system, Trp was extracted in the Lewis cell at the stirring speed 200rpm. The time course of the extraction satisfies Eq.(5) as shown in Fig.3. The $k_{org,ov}$ value obtained is 9×10^{-6} cm/s.

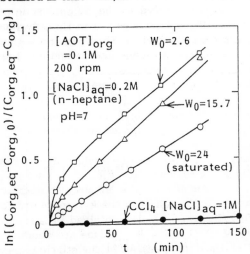

<u>Fig. 3</u> An illustration of relation between left hand side of Eq.(3) and t.
○△□ n–heptane system, ● CCl₄–system

The mass transfer coefficient k_{org}, which was obtained by extracting phenol dissolved in n–heptane(free from AOT) to 0.1M NaOH aqueous solution, is shown in Fig.4. k_{org} increases in proportion to $N^{0.71}$. The k_{org} value for ME droplets can be evaluated from the above mass transfer coefficient by taking into account the difference in the diffusion coefficients as mentioned later. The $k_{org,ov}$ value for the Trp extraction is much smaller than the mass transfer coefficient k_{org} evaluated, suggesting the existence of the rate–determining interfacial resistance. The $k_{org,ov}$ value almost agrees with the k_i

value obtained in the static diffusion cell.

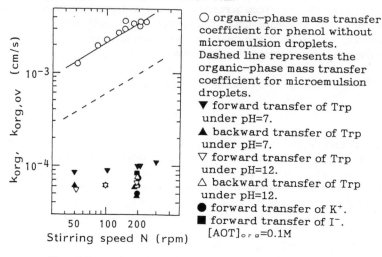

○ organic–phase mass transfer coefficient for phenol without microemulsion droplets. Dashed line represents the organic–phase mass transfer coefficient for microemulsion droplets.
▼ forward transfer of Trp under pH=7.
▲ backward transfer of Trp under pH=7.
▽ forward transfer of Trp under pH=12.
△ backward transfer of Trp under pH=12.
● forward transfer of K^+.
■ forward transfer of I^-.
$[AOT]_{org}$=0.1M

Fig. 4 Organic–phase mass transfer coefficients.

Extraction Rate of Trp in AOT/n–heptane/brine System

In this section, we selected n–heptane as a solvent instead of CCl_4 which penetrates into the hydrocarbon tails of AOT–ME droplet[11]. The electrical charge of Trp can be easily regulated. Trp is mainly entrapped in the water pool under pH=12 at $[NaCl]_{aq}$=0.2M, whereas under pH=7, Trp is entrapped in the interfacial zone of the ME droplets[12]. The partition coefficient of Trp between AOT w/o ME phase and NaCl aqueous phase strongly depends on pH.

The time course of the Trp extraction obtained in the Lewis cell under $[NaCl]_{aq}$ = 0.2M, $[AOT]_{org}$ = 0.1M, pH=7 are shown in Fig.3. When the organic phase is saturated with water(W_0=24), the time course satisfies Eq.(3), whereas in smaller W_0–cases, the extraction rates are initially much larger than that at the saturated condition. This rate enhancement arises from the water co–extraction. Taking this into account, the experiments in this work were performed under the water–saturated conditions. The obtained $k_{org,ov}$ values for Trp in zwitterionic and anionic charged states are shown in Fig.4 for both forward and backward extractions.

The k_{org} value for the ME phase with W_0=24 can be evaluated from that for the phenol transfer. The diffusion coefficient for the ME–droplet D_{ME} was determined to be 2.0×10^{-6} cm^2/s for W_0=24(the ME droplet diameter =12nm). The k_{org} for the ME phase is estimated from the k_{org} obtained for phenol transfer multiplying by $(D_{ME}/D_{phenol})^{1/2}$. The obtained k_{org} for the ME droplets is the dashed line in Fig.4. The $k_{org,ov}$ values are about one order lower than the dashed line, and are independent of the stirring speed N. From this result, the Trp transfer is controlled by the interfacial rate process as in the case of the CCl_4 solvent system.

The rates of the extraction of the ions K^+ and I^- were also measured

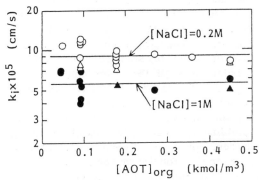

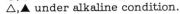

Fig. 5 Dependence of k_i on $[AOT]_{org}$
for Trp transfer.
○,● under neutral pH condition.
△,▲ under alkaline condition.

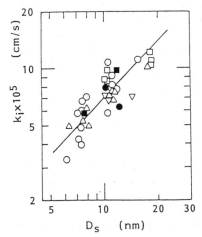

Fig. 6 k_i versus D_s.
○ forward transfer of Trp
under pH=7.
□ backward transfer of Trp
under pH=7.
△ forward transfer of Trp
under pH=12.
▽ backward transfer of Trp
under pH=12.
● forward transfer of K^+.
■ forward transfer of I^-.

under the same conditions as those in Fig.3, as shown in Fig.4. The observed $k_{org,ov}$ values situate around those for Trp. From these results, the interfacial rate processes appear to be insensitive to the location of solute entrapment in the ME-droplets, the solute charge states, and the direction of solute transfer. However, the solvent species strongly affects the interfacial rate process.

In order to elucidate the interfacial rate process, we measured k_i over wide ranges of $[AOT]_{org}$ and $[NaCl]_{aq}$. The observed k_i values are independent of $[AOT]_{org}$ up to $[AOT]_{org}$=0.45M as shown in Fig.5. Since K_d is proportional to $[AOT]_{org}$ or to the number density of ME droplets, the solute flux J should become proportional to $[AOT]_{org}$, i.e., to the number density of the ME droplets. The model of the rate–determining ME droplet formation at the interface by thermal fluctuations[2-4] is invalid, because the process of the droplet formation should be independent of $[AOT]_{org}$.

The k_i values are affected by $[NaCl]_{aq}$, as clearly seen in Fig.5. $[NaCl]_{aq}$ affects the microemulsion structure through the diameter of the ME droplets D_s. Then, $[NaCl]_{aq}$ would affect the k_i through D_s. The k_i values observed by altering $[NaCl]_{aq}$ are plotted against D_s in Fig.6 for the AOT/n-heptane/ brine solution. D_s was determined by the small angle x-ray scattering method. Though the k_i-values have some scatter, a linear dependence of k_i on D_s is obtained.

Model of Interfacial Rate Process

In the preceding section, the interfacial rate process is not concerned with the droplet formation at the interface. Then, the rate process must involve the processes of the collision/fusion of the ME droplets at the interface(see Fig.7). The proposed mechanism for the forward mass transfer is as follows:(a) The ME droplets existing in the first coordination shell near

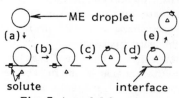

Fig. 7 A model for the
interfacial processes.

the interface can collide and fuse at the interface. (b) Occasionally the ME droplets fuse with the surfactant layer of the interface, and then the wall of a droplet opens. (c) Solute diffuses into the droplet through the open hole. (d) The open hole is closed. (e) The ME droplet is detached from the interface and returns to the organic phase containing the solubilized solute. The back transfer process proceeds through the reverse route to that for the forward transfer.

The experimental results indicate that the rate–determining process is independent of the charged state, the location of solute existing in the ME droplet and the direction of extraction. Thus, the solute–diffusing process (c) in Fig.7 should not be the rate–determining step. The duration of the open hole would be long enough for the process (c) to attain an equilibrium. The processes (a) and (e) are not rate–determining because the frequency for the ME droplet to collide with the interface is very high, and the adsorption of ME droplets on the interface is neglected (see Fig.5). Therefore, the process (b), i. e., the fusion process between the ME droplet and the surfactant layer of liquid/liquid interface is the rate–determining step.

Based on this conclusion, we formulated the interfacial rate process and finally obtained the following expression:

$$J = k_r D_a (K_d C_{aq,i} - C_{org,i}), \quad k_i = k_r D_a \qquad (6)$$

The k_r–value is determined from Fig.6 to be 60 to 100 s^{-1} for the n–heptane system. The k_r value obtained in this experiment is lower than the value reported by Albery et al.[1] for the release of H$^+$ from ME phase to KBr aqueous phase.

The effect of solvent species on k_i is strong, suggesting that solvent species strongly affects the fusion process. From the result for CCl$_4$, the k_r value for CCl$_4$ decreases to around 12 s^{-1}.

4 CONCLUSION

Experimental results on the extraction rate between an AOT w/o microemulsion saturated with water and a coexisting aqueous brine solution were presented. The rate processes concerned with the tryptophan extraction were elucidated from the non–stirring extraction experiment, and the interfacial rate process was found to play an essential role in the extraction. The effects of several factors on the interfacial resistance were examined for tryptophan, K$^+$ and I$^-$ extractions in AOT/n–heptane/NaCl aqueous solution system. The interfacial rate process is controlled by the fusion of microemulsion droplet with the surfactant layer of the liquid/liquid interface. An expression for the rate of the interfacial process was formulated. The rate constant for the fusion process was obtained as 60 to 100 s^{-1}

for the n–heptane system. Organic solvent has a great effect on the fusion rate constant.

Nomenclature

A: interfacial area (cm^2),
C_J: concentration in J–phase (mol/cm^3)
D_J: diffusion coefficient in J–phase (cm^2/s)
D_a: diameter of ME droplet (nm)
D_{ME}: diffusion coefficient for ME droplet (cm^2/s)
$[I]_J$: concentration of I species in J–phase (kmol/m^3)
K_d: partition coefficient (–)
k_f: fusion rate constant between ME droplet and liquid/liquid interface (1/s)
k_i: rate constant for interfacial process (cm/s), N: stirring speed (rpm)
k_J: mass transfer coefficient of J–phase (cm/s), t: time (s), (min)
V_J: volume of J–phase (cm^3)
W_0: [H$_2$O]$_{org}$/[AOT]$_{org}$
x: distance (cm)
subscript
  0: initial value, i: interface, ov: overall
abbreviation
AOT: sodium bis(2–ethylhexyl)sulfosuccinate
ME:microemulsion
Trp: tryptophan

References

1. W.J.Albery, R.A.Chouhery, N.Z.Atay and, B.H.Robinson J.Chem. Soc. Faraday Trans. 1,1987,83,2407
2. P.Plucinski and W.Nitsch, Ber. Bunsenges. Phys. Chem.,1989,93,994
3. W.Nitsch, and P.Plucinski, J.Colloid. Int. Sci.,1990,136, 338
4. P.Plucinski and W.Nitsch, Solvent Extraction ISEC'90(ed.T.Sekine), 1992, P847
5. S.R.Dungan, T.Bausch, T.A.Hatton, P.Plucinski, W.Nitsch. J.Colloid Int. Sci., 1991, 145, 33
6. T.E.Bausch,P.K.Plucinski and W.Nitsch,J.Colloid Int. Sci.,1992,150, 226
7. P.Plucinski and W.Nitsch,J.Phys.Chem.,1993,97,8983
8. P.Plucinski,W.Nitsch,H.Reitmeir and J.Solano–Bauer,"Solvent Extraction in The Process Industries", Proceedings of ISEC'93,1993, p1064
9. W.Eguchi,M.Harada,M.Adachi,M.Tanigaki and K.Kondo, J.Chem. Eng. Japan, 1984,17,472
10. M.Harada,T.Imamura,K.Fujiyoshi and W.Eguchi,J.Chem.Eng.Japan, 1975,8, 233
11. A.Shioi,M.Harada and M.Tanabe,J.Phys.Chem.,1993,97,8281
12. M.Adachi, M.Harada, A.Shioi, and Y.Sato, J.Phys. Chem.,1991, 95,7925

Effect of Salt Type on Partition of Cytochrome c between AOT Water-in-oil Microemulsion and Coexisting Aqueous Phase

M. Adachi and M. Harada

INSTITUTE OF ATOMIC ENERGY, KYOTO UNIVERSITY, UJI 611, JAPAN

1 INTRODUCTION

It is well known that the partition coefficient of proteins between an AOT w/o microemulsion(ME) phase and a coexisting aqueous phase depends significantly upon the type of salt in the aqueous phase[1,2]. The effect of salt type on partitioning of proteins, however, has not been well understood.

Bruno et al.[1] investigated this effect for cytochrome c(cytc) in AOT/iso-octane/brine system based on the 'water shell model'. As long as cytc was used as a model protein, the extraction mechanism of cytc to ME phase does not follow the 'water shell model, in terms of a sole dependence on electrostatic interactions between the protein and the reverse micelle, for solubilization to occur'. Cytc adsorbs AOT molecules on its surface, and the change of cytc–surface from native hydrophilic to hydrophobic states by adsorbing AOT provides the most important driving force for cytc to be solubilized in ME phase[3].

In this paper, we elucidate the salt type effect on cytc extraction in AOT/n-heptane/brine system from the view point of the above extraction mechanism. First, we present the experimental results on the effect of cation on partitioning of cytc using NaCl, KCl and $BaCl_2$ as the salt. We measured the concentration of monomeric AOT in the aqueous phase $[(AOT)_1]_{aq}$, which is one of the most important factor for determination of the adsorption state of cytc by AOT. A good correlation between partition coefficient of cytc and $[(AOT)_1]_{aq}$ is shown. We also present the experimental results on the effect of anion using NaCl, $NaClO_4$ and NaI as the salt. The relation between partition of cytc and the order of Hofmeister series is discussed.

2 EXPERIMENTAL

Materials Cytochrome c(cytc) from horse heart, type III, was purchased from Sigma. AOT provided by Tokyo Kasei Co. Ltd. was used without further purification. NaCl, KCl, $BaCl_2$, $NaClO_4$ and NaI of reagent grade and n-heptane of spectrophotometric reagent grade obtained from Nacalai Tesque were used as supplied.

Method Measurements of the precipitation behaviour of cytc in the aqueous phase were carried out as follows: the aqueous solution, which contained desired amounts of cytc, AOT and salt were kept at 25 ℃ in a test tube for about 2 weeks. Cytc deposited at the bottom of the tube as precipitate. The concentrations of cytc and AOT remaining in the supernatant solution were measured. Experiments on the partitioning of cytc in AOT/n-heptane/ brine system were performed by bringing the AOT in n-heptane solution into contact with the same volume of the aqueous phase, which contained the desired amounts of cytc and salt. It took about 2 weeks for system to attain an equilibrium state.

The concentration of cytc was determined using an inductively coupled argon plasma emission spectrophotometer(Jourrel Ash,ICAP-500) by detect- ing haem iron and was also determined by ultraviolet and visible light spectrometer[4,5] (Shimadzu UV-200). The concentration of AOT was measured by a spectroscopic method with bis[2-[(5-chloro-2-pyridyl)oxo]-5 (diethylamino)-phenolate]cobalt(III) chloride[6]. Circular dichroism(CD) of cytc was measured(Jasco, J-500), and the size of ME droplet was measured by the small angle X-ray scattering(SAXS) method. Water in the organic phase was analysed by the Karl-Fisher Method.

3 RESULT AND DISCUSSION

Effect of Cation on Partitioning of Cytochrome c

Dependence of the distribution coefficient D of cytc on the salt concentrations is shown in Fig.1, where D is defined as $[cytc]_{org}/[cytc]_{aq}$. $D/[AOT]_{org}$ is taken as the ordinate because D was proportional to $[AOT]_{org}$.

In case of NaCl, cytc was completely extracted in the range $[NaCl]_{aq} <$ 0.5M, and D could not be determined. $D/[AOT]_{org}$ decreased steeply in the $[NaCl]_{aq}$ range from 0.5M to 2M. In the case of KCl, cytc was extracted completely in the range $[KCL]_{aq} < 0.2M$, and $D/[AOT]_{org}$ decreased in the range from 0.2M to 0.4M. These experimental results agree well with those reported by Bruno et al.[1] and Hatton[2]. When BaCl_2 was used as the salt, partition behaviour could be measured even at very low salt concentration because the ME of Winsor II type was stable even in the low salt concentration range. A bell- shaped curve was obtained for $D/[AOT]_{org}$ versus the salt concentration. $D/[AOT]_{org}$ de- creased in the high $[BaCl_2]_{aq}$ range from 0.05M to 0.2M and also decreased in the low range from 0.02M to 0.005M, and cytc was not extracted in the range, $[BaCl_2]_{aq} <$ 0.005M.

In order to understand the characteristic partitioning behav- iour obtained in BaCl_2 system, the precipitation behaviour of cytc in the aqueous single phase together with the detailed partitioning behaviour in the binary phase system are shown versus $[BaCl_2]_{aq}$ in Fig.2 and Fig.3, respectively. The precipitation behaviour was mea-

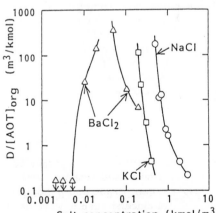

Fig.1 Effect of cation on distribution coefficient of cytochrome c. pH=7, $[AOT]_{org}=0.1M,[cytc]_o=0.5$ to $1\times10^{-4}M$

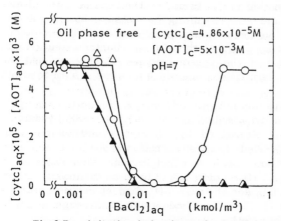

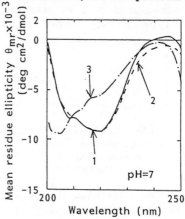

sured under the condition of feed concentration ratio ρ = $[AOT]_c/[cytc]_c$ = 103. Cytc dissolved completely in $[BaCl_2]_{aq}$ <0.003M. With increasing $[BaCl_2]_{aq}$, cytc began to precipitate and almost completely precipitated in 0.02M<$[BaCl_2]_{aq}$ <0.05M. With further increase in $[BaCl_2]_{aq}$, cytc began to dissolve in the aqueous phase again and dissolved completely in $[BaCl_2]_{aq}$>0.2M. The concentration of AOT dissolved in the supernatant solution $[AOT]_{aq}$ decreased monotonously with increasing

Fig.2 Precipitation behaviour of cytochrome c in BaCl₂ aqueous solution.
O[cytc]ₐ q, △[AOT]ₐ q, ▲[AOT]ₐ q(cytc free)

$[BaCl_2]_{aq}$ in both with- and without-cytc systems. The value of $[AOT]_{aq}$ decreased to 10^{-5}M in order of magnitude in the range, $[BaCl_2]_{aq}$>0.1M.

The complete precipitation region in Fig.2 agrees quite well with the complete extraction region in Fig.3. Good correspondence is also found between complete dissolution region in Fig.2 and the low D-region in Fig.3. Fig.4 shows the circular dichroism spectra of cytc in the aqueous phase sampled in the binary-phase experiment shown in Fig.3. The CD spectrum at $[BaCl_2]_{aq}$=0.002M is quite different from that of native cytc in water and agrees with the spectrum of denatured cytc (see Fig.7 in ref.3), which adsorbs many AOT molecules mainly by hydrophobic interaction and becomes hydrophilic with negative charge. On the other hand, the CD spectrum for

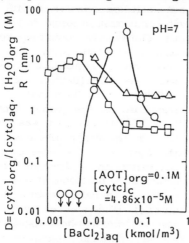

Fig.3 Partitioning of cytochrome c in AOT/n-heptane/BaCl₂ aqueous solution system. OD, □[H₂O]ₒ ᵣ ₉, △R: radius of ME droplet

Fig.4 Circular dichroism spectra of cytochrome c.
1: native state
2,3: aqueous sample in AOT/ n-heptane/BaCl₂ aq. soln. system of Fig.3
2:[BaCl₂]=0.1M,3:[BaCl₂]=0.002M

$[BaCl_2]_{aq} = 0.1M$ is same as that of native state, indicating that cytc adsorbs almost no AOT.

We can understand the partitioning and precipitation behaviour of cytc in $BaCl_2$ system from the results of the CD spectra, the AOT concentration dissolved in $BaCl_2$ aqueous solution obtained in the precipitation experiments and the results obtained for NaCl system described in reference 3. $[AOT]_{aq}$ is very low when $[BaCl_2]_{aq} > 0.2M$, and cytc stays mainly in native hydrophilic state. Thus, cytc remains completely dissolved in the aqueous phase in the precipitation experiment, and $D/[AOT]_{org}$ is low. With decreasing $[BaCl_2]_{aq}$, $[AOT]_{aq}$ increases, and cytc begins to adsorb AOT molecules mainly by electrostatic interaction, making the cytc surface hydrophobic, and cytc becomes amphiphilic. Thus, cytc begins to precipitate in the absence of the organic phase, and $D/[AOT]_{org}$ increases steeply because amphiphilic cytc is stably solubilized in the interfacial layer of ME droplet. With further decrease in $[BaCl_2]_{aq}$, $[AOT]_{aq}$ increases steeply. In the range $[BaCl_2]_{aq} < 0.003M$, AOT dissolves easily in the aqueous phase, and cytc adsorbs many AOT molecules mainly by the hydrophobic interaction under the excess AOT condition as shown by the CD spectrum. Since the adsorption of many AOT molecules on cytc through the hydrophobic interaction makes cytc surface hydrophilic and also makes cytc negatively charged state, cytc dissolves completely in the aqueous phase in the precipitation experiment and can not be extracted in the ME phase. Therefore, the transformation from native hydrophilic cytc surface to hydrophobic one due to the adsorption of AOT through the electrostatic interaction is the most important factor for cytc extraction to ME phase in $BaCl_2$ system.

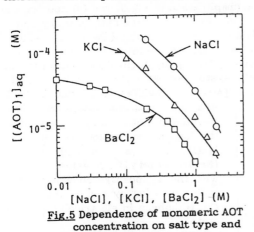

<u>Fig.5</u> Dependence of monomeric AOT concentration on salt type and salt concentration. pH=7

We can conclude that the adsorption of AOT on cytc through the electrostatic interaction is most important factor for cytc extraction regardless of salt type. Let us formulate the extraction process of cytc to the ME phase based on the above conclusion. Adsorption of AOT molecules is closely concerned with the activity of monomeric AOT in the extraction system. As described in the previous paper[3], $[AOT]_{aq}$ increases with increasing $[AOT]_{org}$ because of the formation of micelles in the aqueous phase. Then, we can determine the monomeric AOT concentration $[(AOT)_i]_{aq}$ from the constant $[AOT]_{aq}$ value observed in the range $10^{-4}M < [AOT]_{org} < 3 \times 10^{-4}M$. The determined $[(AOT)_i]_{aq}$ values for NaCl, KCl and $BaCl_2$ are shown in Fig.5. The formation of the extractable species of cytc is given as,

$$(cytc)_{o,aq} + n(AOT)_{1,aq} \rightleftharpoons (cytc(AOT)_n)_{aq},$$

$$K_n = [cytc(AOT)_n]_{aq} / ([(cytc)_o]_{aq}[(AOT)_i]_{aq}{}^n) \qquad (1)$$

Here, $(cytc)_{o.aq}$ represents a cytc molecule without adsorption of AOT, and $(cytc(AOT)_n)_{aq}$ the complex. Extractable species $(cytc(AOT)_n)_{aq}$ is an amphiphilic molecule and is solubilized in the interfacial zone of ME droplet. Then, the partition coefficient of $(cytc(AOT)_n)_{aq}$ between the ME phase and the aqueous phase is considered to be proportional to interfacial area of ME droplets, when the concentration of $(cytc(AOT)_n)_{aq}$ is low. Since the oc-cupied area by an AOT molecule S_{AOT} at the interface of ME droplet is constant regardless of the curvature of the ME droplet, the interfacial area of ME droplets per unit volume is given as $[AOT]_{org}S_{AOT}$. The partition of $(cytc(AOT)_n)_{aq}$ between two phases is given as,

$$(cytc(AOT)_n)_{aq} \rightleftharpoons (cytc(AOT)_n)_{org}$$

$$[cytc(AOT)_n]_{org}/[cytc(AOT)_n]_{aq} \equiv P[AOT]_{org}S_{AOT} \quad (2).$$

From Eqs.(1) and (2), the distribution coefficient D is given by,

$$D/[AOT]_{org} = [cytc(AOT)_n]_{org}/([AOT]_{org}([(cytc)_o]_{aq}+[cytc(AOT)_n]_{aq}))$$

$$= PS_{AOT}K_n[(AOT)_1]_{aq}{}^n/(1+K_n[(AOT)_1]_{aq}{}^n) \quad (3).$$

Cytc dissolved in the aqueous phase in the binary-phase extraction system was found to be in native state from the spectroscopic measurements,i.e., UV, fluorescence and CD spectra, indicating that cytc in the aqueous phase adsorbs no AOT. This indicates that $[(cytc)_o]_{aq}$ is much higher than $[cytc(AOT)_n]_{aq}$. Thus, Eq.(3) can be simplified to

$$D/[AOT]_{org} = PS_{AOT}K_n[(AOT)_1]_{aq} \quad (4).$$

$D/[AOT]_{org}$ was plotted versus $[(AOT)_1]$ in Fig.6. The variation in $D/[AOT]_{org}$ with $[(AOT)_1]_{aq}$ in NaCl system almost agree with that in KCl system, whereas in BaCl$_2$ system, $D/[AOT]_{org}$ deviates from that in NaCl system. From the slope in Fig.6 n-value is obtained as 8, except in the low concentration region of $[(AOT)_1]_{aq}$, i.e., in the region $[NaCl]_{aq}>1.5M$. The n-value 8 agrees with the net positive charge of cytc, indicating that cytc is solubilized in the ME phase in electrically neutral state.

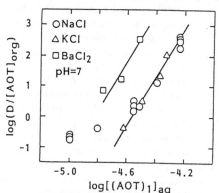

Fig.6 Dependence of partition coefficient of cytc on $[(AOT)_1]_{aq}$

Since AOT molecules are mainly adsorbed through the electrostatic interaction, the variation in the activity coefficient of AOT-anion and amino group of lysine residue in cytc due to both salt concentration and salt type must be taken into account for the determination of K_n value. Although the variation in the activity coefficient for the ions on the protein surface is not easy, the variation in the activity coefficients for a univalent electrolyte in NaCl or KCl aqueous solution has been formulated[7] by Guggenheim and by Scatchard. In the present study, uniunivalent electrolyte is $R-NH_3^+$ and SO_3^--R'. The activity coefficients estimated by the above method for these electrolytes vary only by 10% in the range, 0.5M< $[NaCl]_{aq}$ <1M and 0.2M< $[KCl]_{aq}$ <0.4M. The variations in salt concentration and salt type also affect P-value through the salting out effect for the hydrophobic groups. Following Tanford[8], the salting out effect can be neglected when the concentration of a uniunivalent salt is about 0.3M or less. Thus, the salting out effect is estimated to be small for our experimental conditions, except in high $[NaCl]_{aq}$ range. Another characteristic of cations is the specific interaction with AOT-anion, which causes the change in the diameter of ME droplet and in the phase diagram. Especially, the specific interaction of Ba^{2+} ion is strong as shown above in Fig.1 and Fig.3, and may cause the deviation of $D/[AOT]_{org}$ from NaCl system in Fig.6.

Effect of Anion on Partitioning of Cytochrome c

The distribution coefficient of cytc was measured in AOT/n-heptane/ brine system using NaCl, $NaClO_4$ and NaI as the salt. The variation in anion species has no effect on $[H_2O]_{org}$ and on the radius of ME droplet R as shown in Fig.7. The D-value varies in the order $D(NaCl) > D(NaClO_4) > D(NaI)$ at the same salt concentration. The effect of the anion species follows the Hofmeister series[9]. The direct interaction between singly charged anions and cations increases in the reverse order of the Hofmeister series[10]. D is lowest for I^-, which lowers the adsorption of AOT molecules on cytc by interacting directly with the positively charged lysine residue and decreases the distribution coefficient of cytc.

The effect of salt type on the precipitation behaviour of cytc by AOT in the aqueous single phase system was measured in order to confirm the above

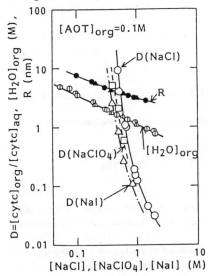

Fig.7 Effect of anion on partition coefficient of cytochrome c. pH=7, $[cytc]_o$=0.5 to 1×10^{-4}M

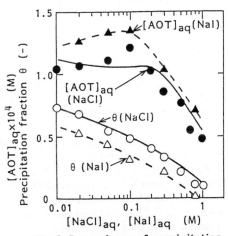

Fig.8 Dependence of precipitation behaviour of cytc on anion type. ρ=8.23, $[AOT]_c$=4×10^{-4}M, pH=7 Solid curves for NaCl system are calculated results using Eqs.(2) and (3) in reference 3.

findings. Fig.8 shows the effect of the anion species on the precipitation at fixed feed concentration ratio $\rho = 8.23$. The fraction of precipitation θ, which is defined as the amounts of cytc precipitated divided by that of cytc fed to the system, takes low values for NaI in comparison with that for NaCl, and $[AOT]_{aq}$ became higher for NaI than that for NaCl, indicating that I^- effectively suppress the AOT adsorption in comparison with Cl^-. Therefore, we can conclude that the effect of anion on the distribution coefficient of cytc is attributed to the preventive effect of anion on the adsorption of AOT-anion.

4 CONCLUSION

The effect of salt type on the distribution coefficient of cytc in AOT/n-heptane/brine system was investigated. The effect of cation was mainly attributed to the variation in the concentration of monomeric AOT in the aqueous phase, which affected the formation of extractable cytc by controlling the adsorption of AOT molecules on cytochrome c. The effect of anion is due to the preventive effect for the AOT adsorption through the direct interaction with the positively charged lysine residues.

NOMENCLATURE

$(AOT)_1$: monomeric AOT
D: distribution coefficient$=[I]_{org}/[I]_{aq}$ $(-)$
$[I]_J$: concentration of I species in J-phase $(kmol/m^3)$
ρ: $[AOT]_o/[cytc]_o$ $(-)$
subscript c: feed concentration
abbreviation
AOT: sodium bis(2-ethylhexyl)sulfosuccinate
CD: circular dichroism
cytc: cytochrome c
ME: microemulsion

REFERENCE

1. P.Bruno,M.Caselli,P.L.Luisi,M.Maestro and A.Traini,J.Phys.Chem.,1990, 94,5908
2. T.A.Hatton,'Surfactant-based SeparationProcess' Marcel Dekker,New York,1989,chap.3
3. M.Adachi and M.Harada,J.Phys.Chem.,1993,97,3631
4. P.Brochette,C.Petit and M.P.Pileni,J.Phys.Chem.,1988,92,3505
5. Y.P.Myer,Biochemistry,1968,7,765
6. S.Taniguchi and K.Goto,Talanta,1980,27,289
7. H.S.Harned and R.A.Robinson,'The International Encyclopedia of Physical Chemistry and Chemical Physics Topic 15.vol.2 Multicomponent Electrolyte Solutions.',Pergamon Press,1968
8. C.Tanford,'The Hydrophobic Effect: Formation of Micelles and Biological Membranes' 2nd.Ed.John Wiley & Sons,New York,1980
9. A.Watanabe,'Kaimen Denki Gensho',Kyoritsu Shuppan,1972
10. Y.Goto,Seibutsu Butsuri(Biophysics),1991,31,122

Analysis of Size Exclusion Chromatographic Properties of Proteins

Alessandra Adrover and Diego Barba[a]

DIPARTIMENTO DI INGEGNERIA CHIMICA, UNIVERSITÀ DI ROMA "LA SAPIENZA", VIA EUDOSSIANA 18, 00184 ROMA, ITALY

[a] DIPARTIMENTO DI INGEGNERIA CHIMICA, UNIVERSITÀ DELL'AQUILA MONTE LUCO DI ROIO, 67040 L'AQUILA, ITALY

1 INTRODUCTION

In biotechnological applications, size-exclusion chromatographic columns (SEC columns; also known as gel-filtration chromatographic columns) have become useful separation units for the purification of biological mixtures and especially for the separation of proteins. The mechanism of separation in SEC is related almost exclusively to the geometrically hindered transport of solute molecules in the pore network and in the gel beads, whose average pore radius is of the same order of magnitude as the radius of the solute molecules[1]. For separation purposes, a SEC column can be characterized by specifying the dependence of the retention time on the characteristics of the porous packing and on the size of the solute molecules. Moreover, the analysis of resolution properties and column performances depends on the behaviour of dispersion as a function of solute velocity. In other words,the response of chromatographic columns for analytical and separation purposes can be characterized by the behaviour of retention time and of dispersion. In practical terms this means the evaluation and prediction of the first two moments of the outlet concentration distribution.

In this work, we develop the analysis of the functional dependence of the retention time t_r on biomolecule size for a wide class of globular proteins, showing how a SEC column can be characterized by means of four parameters expressing the geometrical constraints upon solute molecules induced by the pore-network structure. A first application is the prediction (from retention time measurements) of the globular biomolecule radius and the study of separation capacity. The dispersion properties of globular proteins in the usual operating conditions of preparatory analyses are also briefly discussed.

2 EXPERIMENTAL DETAILS

The columns used in performing experiments were: Bio-Sil TSK 250 (packing G 3000 SW), average particle size 10μm, average pore radius $< r >= 125$Å, void fraction $\varphi_o = 0.408$ and Bio-Sil TSK 125 (packing G 2000 SW), average particle size 10μm, average pore radius $< r >= 62.5$Å, void fraction $\varphi_o = 0.375$, from Toyo-Soda (column length $L = 60$ cm; column section $S = 0.44$ cm^2). The elution solutions were 0.1 M Na$_2$SO$_4$, 0.1 M NaH$_2$PO$_4$ and 0.02 % Sodium Azide (w/v) for the TSK 250 column; 0.05 M Na$_2$SO$_4$, 0.05 M NaH$_2$PO$_4$ and 0.02 % Sodium Azide for the TSK 125 column. The eluents were adjusted to pH 6.8 by using a NaOH solution. The processing apparatus consisted of a twin-headed reciprocating pump, Water mod 510, a selectable wavelength U.V. detector (481 Lambda

Max) and a Rheodyne injector, model 7125, purchased from Millipore U.K. The protein concentration varied from 0.1 to 20 mg/ml and the injection volume from 20 to 40 μl. All the experiments were performed at 25 °C. The pH and the ionic strength were chosen (as suggested by Bio-Rad) in order to avoid any interaction (electrically, hydrophilically, etc) between the molecules considered with the gel. The biomolecules considered were: bovine thyroglobulin (669 kDa), horse spleen apoferritin (443 kDa), bovine γ-globulin (160 kDa), mouse IgG (150 kDa), hen's egg ovalbumin (44 kDa), horse myoglobin (17 kDa) and cyanocobalamin (1.35 kDa). The flow rate varied from 0.1 to 1.0 ml/min.

3 ACCESSIBILITY FACTOR AND RETENTION TIME

In SEC, the overall transport properties of solute molecules can be described by means of the accessibility factor $G(\lambda)$, representing the ratio between the effective diffusion coefficient of solute molecules in column D and the corresponding Stokes-Einstein diffusivity[1] D_E

$$G(\lambda) = D/D_E , \quad \lambda = r_p/ < r > \tag{1}$$

which, in the absence of adsorption phenomena or field effects (e.g. dipole or hydrophobic/hydrophilic interaction between solute molecules and the pore matrix), depends exclusively on the ratio λ between the radius of the solute molecule r_p and the average pore radius $< r >$. Various theoretical and/or semi-empirical models have been proposed for $G(\lambda)$ which are all valid for solute molecules of spherical shape and are derived from simple assumptions for the geometry of the pores, assumed e.g. to be of cylindrical $(G(\lambda) = (1 - \lambda)^2)$ or conical shape $(G(\lambda) = (1 - \lambda)^3)$, or from semi-empirical analysis (see the Anderson-Quinn model[2]). However, it should be noted that most of these expressions for $G(\lambda)$ come from the analysis of hindered diffusion in membrane technology, while there is a fundamental topological difference between membrane transport and SEC. In the latter, the structure of the pore-network is extremely complex and the functional form of $G(\lambda)$ is significantly influenced by topological effects related to disorder. It will be shown below that the influence of the complex topological pore-network structure can be encompassed in the expression of $G(\lambda)$ in terms of a single scaling exponent which depends specifically on the properties of the SEC-packing. In SEC, the mean retention time t_r of a solute molecule can be expressed in the form

$$t_r = L\varphi/v_o , \quad \varphi = \varphi_o + (1 - \varphi_o)\chi G(\lambda) , \tag{2}$$

where v_o is the mean elution velocity, $(v_o = F_o/S$, where F_o is the volumetric elution flow rate and S the column section), φ_o the void fraction of the column and χ the internal void fraction. The parameter φ is the *equivalent void fraction* (encompassing both internal porosity and hindering geometric constraints expressed by $G(\lambda)$) which a solute molecule experiences in the column.

Figure 1A shows the behaviour of t_r vs $1/v_o$ for the TSK 125 column. The slope of the corresponding lines is equal to $L\varphi$. As an initial rough approximation we may regard the void fraction φ_o as equal to the value of φ evaluated from the retention time of thyroglobulin, as indicated by Bio-Rad Laboratories in their test chromatogram, which yields φ_o=0.408 (TSK 125), φ_o=0.375 (TSK 250). Information on protein radii was obtained from the collected data given by Felgenhauer[3] in his analysis of mass-radius scaling. The least square fitting of Felgenhauer's data yields $r_p = AM^\beta$, where r_p is expressed in Å and M in Dalton. The values of the fitting parameters are A=0.52074, and β=0.375. By assuming the Felgenhauer data correlation as the mass-radius scaling for globular proteins, the dependence of the normalized parameter $\varphi^* = (\varphi - \varphi_o)/(1 - \varphi_o)$ as a function of $1 - \lambda$ is readily obtained from the data of figure 1A and from eqs. (1)-(2). The values of φ deriving

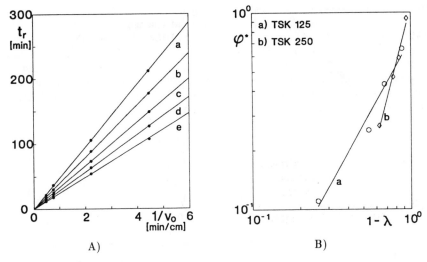

Figure 1: *A) Retention time t_r vs $1/v_o$ for TSK 125 column: a) cyanocobalamin; b) myoglobin; c) ovalbumin; d) bovine-globulins; e) thyroglobulin ; B) φ^* vs $1 - \lambda$ for TSK 125 and TSK 250, for a void fraction chosen from the retention time of thyroglobulin.*

from the experimental data of the two columns are shown, in a log-log plot, in figure 1B and suggest a power law correlation of the form

$$\varphi = \chi G(\lambda) = \chi(1 - \lambda)^\alpha . \tag{3}$$

The values for the parameters α and χ obtained from figure 1B are $\alpha=3.06$, $\chi=1.00$ (TSK 250) and $\alpha=1.76$, $\chi=0.728$ (TSK 125). The values of the exponent α found experimentally indicate clearly that the usual correlations adopted in hindered diffusion cannot provide an accurate description of the behaviour of the accessibility factor as a function of λ.

An interesting consequence of the analysis outlined above is the quantitative prediction of the radii of globular proteins from retention-time measurements. This aspect is closely related to the separation capacity of SEC columns and to their resolution. If φ_o, χ, $< r >$ and α are known, the solute radius can be predicted from the equation

$$r_p = < r > [1 - (\varphi^*/\chi)^{1/\alpha}] , \tag{4}$$

where φ^* is the normalized void fraction of the solute molecule (see eq. (2) and figure 1B). Figure 2A shows a comprehensive review of the prediction of molecular radii by means of SEC columns as obtained from eq.(4) as well as comparison with the results of Felgenhauer and with data from Tyn and Gusek[4], obtained by the measurement of protein diffusivities. In addition to the above-mentioned proteins, predictions for the radius of human IgG, mouse IgG, and alcohol dehydrogenase have also been reported. As can be seen, the evaluation of mass-radius scaling from retention time is sufficiently accurate as compared to other estimates, with the exception of the data for cyanocobalamin ($M=1350$ Dalton) obtained from TSK 250, for which the effect of the microporosity of the gel-matrix could explain the underestimation of the radius. On the other hand, in TSK 250 cyanocobalamin exhibits a value of λ of about 0.06, for which the peak resolution factor attains a significantly low value (figure 2B). The error range found in the estimation of r_p from SEC measurements can be estimated as 10% of the radius of the protein.

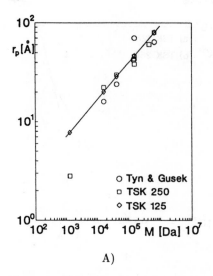

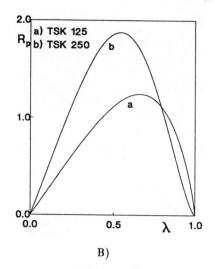

A)   B)

Figure 2: A) Review plot of the experimental results for the prediction of protein radii for SEC experiments. The dashed line represents the correlation of Felgenhauer data; B) Peak resolution factor R_p vs λ.

We may therefore conclude that, given a proper characterization of the columns, size exclusion chromatography is a fairly satisfactory experimental procedure for the evaluation of molecular sizes as compared with other transport techniques (e.g. those which make use of the Stokes-Einstein relation between diffusivity and molecular radius).

4 PEAK RESOLUTION

Given the accessibility factor, from eqs. (2) and (3) it is possible to establish the protein range (expressed in terms of molecular weight or radius) within which the column optimally separates two different peaks (peak resolution). A measurement of the peak resolution can be obtained by considering the relative influence of solute mass in the change of the elution volume (or equivalently of the retention time). Given the average pore radius and the Felgenhauer correlation parameters the peak resolution factor R_p can be expressed as

$$R_p(M) = \frac{M}{V_e} \frac{dV_e(M)}{dM} = R_p(\lambda) = \frac{\lambda}{\beta\varphi} \frac{d\varphi(\lambda)}{d\lambda} = \frac{\alpha\lambda}{\beta\varphi} \frac{\varphi - \varphi_o}{1 - \lambda} , \qquad (5)$$

where the accessibility factor is expressed by means of a power law behaviour, (eq.(3)). For $\lambda=0$ and $\lambda=1$, $R_p=0$ and the peak resolution factor has a unimodal shape as a function of λ (figure 2B). Therefore, there exists an optimal value of λ, λ_{opt}, for which R_p is maximum and the column exhibits the greatest peak resolution capacity. For the two columns considered $\lambda_{opt} = 0.65 \pm 0.03$ (TSK 125) and $\lambda_{opt} = 0.536 \pm 0.015$ (TSK 250). The value of λ_{opt} is indicative of the region in which a SEC column can operate in optimal conditions for analytical purposes, e.g. for the estimation of protein radii from chromatographic data. For separation purposes, however, it would be better to introduce a resolution factor R_s which takes into account the information related to peak dispersion and to its dependence on molecular weight[5].

5 COLUMN CHARACTERIZATION

The analysis reported in the previous section can be regarded as a first-order approximation for a satisfactory characterization of chromatographic columns. In fact, the most approximate aspect lies in taking the value of φ of thyroglobulin as the void factor of the column, since there is no a priori evidence that this protein exhibits total exclusion.

Let φ_M be the value of φ for the largest protein considered (thyroglobulin in our experiments), and λ_M the ratio of its radius to the mean pore radius. If λ_M is less than unity the protein does not exhibit total exclusion and φ_M should be regarded as an upper bound for φ_o. An iterative optimization procedure may be used to estimate the proper void fraction of the column. Let $\varphi_o^{(1)} = \varphi_M$ be the first approximation for φ_o. From SEC data on retention times of the other test proteins, a preliminary estimate for $\alpha^{(1)}$ and $\chi^{(1)}$ can be obtained by fitting the data of φ^* of the test proteins with the power law relation (3). From this approximation a new value of φ_o can be obtained from the iterative procedure :

$$\varphi_o^{(n+1)} = \frac{\varphi_M - \chi^{(n)}(1 - \lambda_M)^{\alpha^{(n)}}}{1 - \chi^{(n)}(1 - \lambda_M)^{\alpha^{(n)}}} \tag{6}$$

and the limit of this procedure furnishes an estimate of φ_o, χ and α; for the two columns considerd φ_o=0.408, χ =0.728, α=1.76 for TSK 125, and φ_o=0.316, χ=1.00, α=2.44, for TSK 250 . Comparison with the first approximation of the previous section reveals that thyroglobulin exhibits complete exclusion for TSK 125, while the optimized φ_o for TSK 250 is much less than the φ value of thyroglobulin.

Figure 3A shows the comparison of our experimental results with the values of $G(\lambda)$ predicted by the classical models mentioned above, clearly indicating that there is no hope of obtaining an a priori characterization of SEC columns based exclusively on oversimplified models of hindered transport. This statement derives from the observation that most of the correlations adopted in hindered transport have been developed in relation to membrane experiments, where the structure of the pore network is fairly regular and can be described by an ensemble of parallel non-interconnected pores with smooth pore-size distribution.

The experimental literature on chromatography often presents calibration curves of the elution volume V_e vs. molecular weight in the form $V_e = K \log M$, which implies that φ scales linearly with the logarithm of the molecular weight ($\varphi = V_e/V_t$, V_t being the total volume of the column). Figure 3B shows the behaviour of φ with M as derived from eq. (2) with the value of the parameters α, χ and φ_o corresponding to those of TSK 250. It can be seen that a logarithmic regression fits these data quite well, even if the mechanism of hindered transport is described by the power law relation, eq.(3). This observation explains why a logarithmic calibration curve can be drawn for most SEC from experimental observations.

Comparison of the analysis based on the accessibility factor and the calibration curves usually adopted in SEC reveals the fundamental difference between the two methods. Even if $G(\lambda)$ has been characterized by means of a fitting exponent α at the level of macroscopic description, $G(\lambda)$ is a physically measurable function whose functional form ultimately depends on the geometry and topology of the pore network. In the recent literature it has been shown that the fine structure of silica gels constituting the packing of SEC columns exhibits fractal properties which can be correlated to the pore-size distribution function. On the basis of this result it is possible to argue that theoretical models based on percolation theory and on related models of disorder could be developed in order to predict the functional form of $G(\lambda)$ analytically[6]. However, this kind of modeling calls for detailed experimental characterization of the porous structure of the gel particle at the microscopic level and goes beyond the limits of a macroscopic characterization of SEC columns.

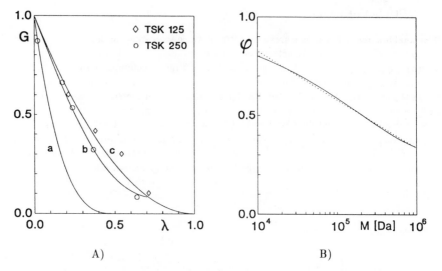

A)   B)

Figure 3: *A) Comparison of the experimental data for $G(\lambda)$ with various literature models for the accessibilty factor: a) cylindrical pore model; b) Anderson and Queen; c) conical pore model; B) Calibration curve of φ vs the molecular weight M for TSK 250. The dashed line is the logarithmic interpolating curve $log(M)$.*

6 DISCUSSION

A clear picture of resolution capability and separation performances cannot be obtained without considering dynamical features related to the dispersion properties of biomolecules. The dispersion characteristics of globular proteins for the fluiddynamical regimes adopted in HPLC indicate that the dispersion σ^2 scales with the second power of the retention time

$$\sigma^2(t_r) \sim t_r^2 , \tag{7}$$

which implies that the macroscopic dispersion coefficient D_a is a linear function of the mean velocity v_o. This means that the dispersion of a biomolecule is characterized by a unique dispersion Peclet number $Pe_d = L^2/(t_r D_a)$ constant for each flow regime. For the proteins considered, $10^2 < Pe_d < 10^5$. Nevertheless, in the case of smaller pore radii and very complex pore-network topology (as in the case of TSK 125) the scaling of σ^2 for smaller biomolecules may deviate from the simple scaling law (eq. 7), i.e. $\sigma^2(t_r) \sim t_r^\gamma$, with $1 < \gamma < 2$. A theoretical analysis of solute transport at the microscopic scales as a stochastic process (i.e. as a Langevin equation) leads to the description of dispersion by means of the exit-time equation associated with the corresponding Langevin equation of motion[7]. The predictions of this model are consistent with the dispersion law (eq. (7)), while the deviation from eq.(7) may be related to the effect of a complex topological pore-network. Even in this case, the leading role played by the disordered nature of the porous packing suggests that, in an engineering perspective, it is to be expected that the improvement of separation processes based on size-exclusion effects will also come from the design of porous packing so as to optimize the steric interactions between biomolecules and the pore network topology and geometry. A more detailed analysis of dispersion properties can be found in Adrover et al.[8].

Acknowledgments

The authors thank Lia Mosca for her help in performing experiments.

References

1. W. M. Deen, *AIChE J.*, 1987,**33**, 1409.

2. J.L. Anderson and J.A. Quinn, *Biophys. J.*,1974, **14**, 130.

3. K. Felgenhauer, *Hoppe-Seyler's Z. Physiol. Chem.*,1974, **355**, 1281.

4. M.T. Tyn and T.W. Gusek, *Biot. & Bioeng.*,1990, **35**, 327.

5. M. Giona and D. Spera, *Design and Control of Size-Exclusion Chromatographic Columns as Separation Units*, this Conference, 1994.

6. M. Giona, A. Adrover and D. Spera, *Chem. Biochem. Eng. Q.*, 1993, **7(4)**, 199.

7. M. Giona and A. Adrover, *Chem. Engng. Sci.*,1993,**48(11)**, 1933.

8. M. Giona, A. Adrover, D. Barba and D. Spera, *Chem. Engng. Sci.*, 1994,**49(4)**, 541.

Polymers for Selective Adsorption, Applications of Molecular Imprinting to Biotechnology

C. Alexander, C. R. Smith, E. N. Vulfson, and M. J. Whitcombe

BBSRC INSTITUTE OF FOOD RESEARCH, READING LABORATORY, EARLEY GATE, WHITEKNIGHTS ROAD, READING RG6 2EF, UK

ABSTRACT

This paper reviews the concept of molecular imprinting, the techniques involved and the scope of applications of imprinted polymers to separations in biotechnology. New methods of imprinting, suitable for compounds bearing spatially separated hydroxyl groups and for bacterial cells are briefly discussed.

1. INTRODUCTION

Molecular recognition is crucial to the selectivity of biocatalysts and the high specificity displayed by antibodies and other biological receptors. It may also be a factor determining reagent selectivity in organic synthesis and in chemical separations. Awareness of this fact has led to the expenditure of considerable effort towards the synthesis of chemical structures with the ability to discriminate molecular species. The development of artificial receptors of this type, combining the versatility of chemical interactions with the precision of biological recognition, may provide solutions to wide ranging problems associated with biotransformations, food manufacture, waste treatment, environmental analysis, etc. Indeed, if the preparation of "tailor-made" matrixes for individual separations becomes a possibility, they will be extremely suited to, for example, selective product recovery from dilute fermentation streams and cell culture media. Similarly, such highly specific adsorbents could find applications in the removal of particularly toxic components from waste or of undesirable elements in foods. The aim of this paper is to briefly review the concepts and methods currently used for the preparation of chemical structures with introduced recognition elements.

A familiar concept to readers of biochemical textbooks is the lock and key analogy, (**Figure 1**) as a representation of the process of molecular recognition. The "key" represents a molecule of a substrate or ligand, the "lock" is an enzyme or antibody and the "keyhole" is the recognition site. On a molecular scale the "fit" is determined by contributions from hydrogen-bonding, ionic interactions, hydrophobicity, favourable steric interactions and, in some cases, the formation of covalent bonds. We will extend the same analogy to describe the two, as yet most successful, chemical strategies for the synthesis of artificial recognition sites.

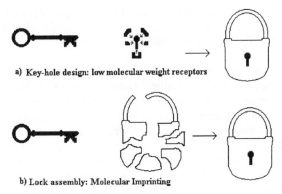

a) Key-hole design: low molecular weight receptors

b) Lock assembly: Molecular Imprinting

Figure 1. Artificial molecular recognition viewed as the lock and key analogy.

The first approach relies on the *de novo* design of a "keyhole" capable of recognising the target molecule. This is achieved by assembling complementary groups in a very specific fashion (for examples, see the reviews by Rebek[1] and Lehn[2]). It relies heavily on modelling molecular interactions and conformational factors and gives rise to some very elegant molecular design, usually requiring demanding and lengthy synthetic procedures. This "keyhole assembly" or low molecular weight receptor can then be introduced into a "lock", if required.

An alternative approach, known as molecular imprinting, makes use of self-assembly, allowing the complementary parts of the "keyhole" to find their proper place around the functional regions of the "key", assembling the "lock" in the process, (**Figure 1b**). In chemical terms this methodology is based on polymerization carried out in the presence of a specific print molecule or template, (the key) which forms a complex with some of the constituent monomers. The recognition sites are then created by removal of the template, leaving its replica (the keyhole) within the polymer structure (the lock). An alternative way of visualizing this process is the cinematic cliché of pressing a key into wax or soap with the intention of making a duplicate. Although this approach, essentially based on the self-assembly of inexpensive monomers around the template molecule, can sometimes be overlooked as not being particularly "sexy" science, it does allow rather precise reproduction of recognition sites and is deserving of more attention as an up and coming new technology.

One may wonder to what extent the above approaches can provide a viable alternative to Nature's designs such as enzymes and antibodies. In other words to what extent artificial receptors are inferior to natural ones. In fact the amino acid receptor of Galán *et al.*[3] shows enantioselective binding of amino acids with aromatic side chains with a selectivity similar to, or even exceeding that of enzymes, such as chymotrypsin or subtilisin. In a further example due to Vlatakis *et al.*,[4] a polymer containing methacrylic acid was imprinted with theophylline. Competitive binding experiments showed that this polymer exhibited cross-reactivities towards structural analogues of the template, of a remarkably similar nature to those of polyclonal antibodies raised against the same ligand. Both of these systems perform well and offer some significant advantages in terms of robustness and solvent and temperature tolerance over proteins. With the potential for large scale separations and economic manufacture of custom adsorbents, molecular imprinting is the more attractive option, and it is this on which we will concentrate.

The preparation of molecularly imprinted polymers is often illustrated as in **Figure 2**. The functional groups within a template molecule are the points of attachment for polymerizable monomers bearing complementary functionalities (**Figure 2a**). This "template monomer" (**Figure 2b**) is then polymerized in a mixture of monomers containing a high percentage of cross-linker (**Figure 2c**). Polymerization may be photochemically or thermally initiated, and solvent is included to act as "porogen", to provide the required morphology of the final polymer. Physical treatment such as grinding may be necessary to control the particle size of the material. The template is then removed by repeatedly washing the polymer with solvent, or by hydrolysis, depending on the nature of its attachment to the polymer. This leaves a "cavity" or binding site within the rigid polymer structure with complementary shape and functionality to the template (**Figure 2d**).

There are two methods of forming the template monomer, which gives rise to two methods of molecular imprinting. The first is **covalent imprinting**, pioneered by Wulff,[5] where the polymerizable species are attached to the template by covalent bonds. These covalent bonds are formed by reversible condensation processes which allow re-binding in exactly the same manner in the imprinted polymer binding site. The major drawback of the current state of covalent imprinting is the limited range of functional groups for which compatible protocols exist. For instance, the method of molecular imprinting monosaccharides, developed by Wulff and co-workers, relies on the formation of cyclic boronate esters, and hence on 1,2- and 1,3-diol functionalities being present in the template. Covalent methods have also been developed for amines,[6] aldehydes[7] and ketones.[8]

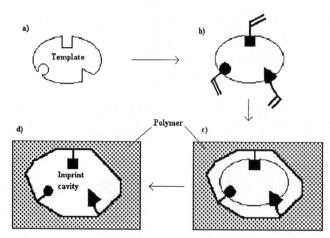

Figure 2. Schematic diagram of the molecular imprinting process.

An alternative imprinting method, developed by Mosbach and co-workers, does not rely on the formation of covalent bonds to arrange the polymerizable monomer units around the template, but on hydrogen-bonding, dipole-dipole and ionic interactions. This method is **non-covalent imprinting** and is generally complementary, in terms of the range of functional groups for which it is effective, to the covalent method. Imprinted polymers, prepared by non-covalent imprinting methods, have recently been reported as HPLC stationary phases capable of baseline separation of racemic mixtures of structurally related beta-blockers,[9] an important class of chiral drugs. This followed earlier reports of imprinted polymer-stationary phases capable of separating racemates of, for example, amino acid derivatives[10] and monosaccharides.[11]

The technique of molecular imprinting is now a well established means of introducing specific recognition sites in highly cross-linked polymeric matrixes. The scope of applications for molecularly imprinted polymers is clearly not limited to the preparation of chromatographic materials for the separation of racemates, as a number of recent papers demonstrate. Initial studies have been reported in areas of synthetic chemistry,[12] catalysis,[13] chemical sensor design,[14] and as a substitute for antibodies in immunoassays.[4] The review by Mosbach[15] gives an excellent overview of the current state of this technology. However several general problems need to be addressed in order to further develop the methodology of molecular imprinting as a powerful tool in the area of biotechnology. For example new general strategies are required for the generation of recognition sites capable of binding a wide range of ligands not necessarily carrying multiple functional groups. It is also of interest to extend the scope of imprinting techniques to encompass high molecular weight molecules, or even microorganisms. Our own efforts have been directed at these issues in general, as well as towards some specific applications with relevance to foods. Some new methods developed in the course of our recent studies are described below.

2. IMPRINTING METHODS FOR STEROL TEMPLATES

Both methods of molecular imprinting outlined above have generally been applied to highly-functionalized templates. Our own interest in the field of molecular imprinting is to develop methods suitable for poorly functionalized molecules and to extend the range of functional groups which may be imprinted. We have chosen sterols as a model system for our studies, as a class of compounds occurring widely in nature, with important implications in diet, health and pharmacology, but which are largely hydrocarbon. The application of the existing methodologies of imprinting to templates containing single hydroxyl groups presents two problems: First there is no adjacent functional group to assist in the formation of cyclic boronate esters, as in covalent imprinting, and secondly the formation of hydrogen bonds to a single hydroxyl is unlikely to be sufficient to provide a good templating effect in non-covalent imprinting. Two of our current projects address these problems and will be described in more detail in the accompanying[16,17] papers, but the general principles are outlined below.

2.1. Irreversible Covalent Imprinting

This method consists of elements from both of the above imprinting strategies, combined by new synthetic procedures, to produce novel imprinted polymers. The template monomers are prepared by covalent attachment of the polymerizable groups, but subsequent hydrolysis of the polymer results in an imprint with functionality which produces a non-covalent binding. This is exemplified by the imprinting of cholesterol using the template monomer, cholesteryl (4-vinyl)phenyl carbonate (**1**). This monomer, after copolymerization with ethyleneglycol dimethacrylate, can be cleaved with the loss of CO_2, to release a fragment derived from cholesterol, and generate a phenolic hydroxyl within the polymer binding site, suitably positioned to hydrogen-bond with cholesterol. The range of solvents in which cholesterol binds to the polymer is somewhat restricted, nevertheless a very large number of the theoretically available sites are active and binding capacities of $\geq 100\mu$moles per gram of polymer have been obtained. Some of the characteristics of these polymers are presented in more detail in the accompanying paper.[16]

2.2. Reversible Covalent Imprinting of Monoalcohols

The principle drawback of the method of Wulff and co-workers is the requirement for adjacent functionality such that cyclic esters of 4-vinylphenyl boronic acid (**2**) can be prepared. In order to overcome these limitations, we have prepared the modified polymerizable boronophthalide (**3**), which forms a 1:1 adduct with monohydroxy compounds such as sterols, yet can be used at higher ratios to imprint poly-hydroxy species. This greatly extends the range of compounds to which reversible covalent imprinting can be applied. The preparation of imprinted polymers based on this compound, and their applicability to analysis and bioseparation are discussed more fully in the accompanying paper.[17]

3. "IMPRINTING" OF WHOLE (BACTERIAL) CELLS

The chemistry of molecular imprinting generally involves the self-assembly of monomer components around the template, followed by a polymerization reaction which conserves the shape and orientation of the "key". In principle, this key could be of any size and shape, and thus we decided to test whether imprinting techniques could be applied to effect the separation of complex macromolecules and even whole cells. In particular the separation and concentration of bacteria from foods, to facilitate rapid analysis, was envisaged using cells as giant templates around which an imprinting polymerization could be carried out.

The approach we have adopted employs interfacial polymerization to help direct self-assembly of monomer components and the bacterial cell wall. A water-soluble polymer is stirred with a suspension of the cells to be "imprinted" and then added to an organic phase containing a crosslinking agent, dispersed in aqueous buffer. The bacteria partition between the organic droplets and the bulk aqueous phase, and the gelled polymer forms at the interface concurrently. This is shown schematically in **Figure 4**.

We have carried out these polymerizations at pH 7.8-10.5 to form hollow microcapsules with bacteria at the outer surface. The encapsulated organic phase, which

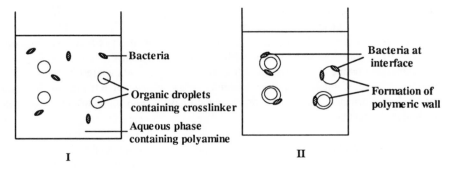

Polymer capsules are UV cured after stage II, and bacteria can either be removed at this stage to leave an "imprint", or can be further treated, as shown below, to enhance specificity.

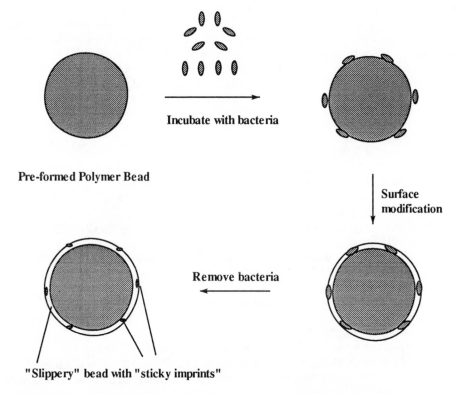

Figure 4. The preparation of bacteria-specific beads by the imprinting method.

also contained a photopolymerizable mixture and initiator, was then irradiated to cross-link the capsule interior, providing beads strong enough to withstand centrifugation and sterilisation. The bacteria were removed from the surface of the beads at this stage by physical shear. Alternatively the beads, with bacteria still attached, were subjected to a further chemical treatment. This rendered the areas of surface not covered by the bacteria template incapable of interacting with bacterial cells. Subsequent removal of the cells yielded polymer beads with bacterial "prints" which have shown some species-selectivity in binding experiments.

4. CONCLUSIONS

Molecular imprinting allows one to design and synthesize polymeric adsorbents with specific binding properties, using cheap and readily available starting materials. The new methodologies described in this paper substantially extend the range of compounds amenable to molecular imprinting. Imprinted polymers are robust and capable of repeated recycling with negligible loss of activity. This makes them ideally suited as custom-built recovery systems for the removal of contaminants for waste streams and the efficient recovery of the products of fermentations or biotransformations from complex mixtures. Coupled with other technologies, imprinted polymers may have applications in sensors for continuous monitoring of reactants in large-scale chemical reactors. In the long term, imprinting methods may be successfully applied to the design of efficient new catalysts with solvent and temperature stabilities superior to biocatalysts. Other applications to problems in the fields of chemistry and biochemistry are limited only by the imagination and the ability to recognise that imprinted polymers hold the key to the solution.

5. REFERENCES

1. J. Rebek Jr., Accnts. Chem. Res., 1990, 23, 399.
2. J-M. Lehn, Angew. Chem. Int. Ed. Engl., 1990, 29, 1304.
3. A. Galán, D. Andreu, A.M. Echavarren, P. Prados and J. de Mendoza, J. Am. Chem. Soc., 1992, 114, 1511.
4. G. Vlatakis, L.I. Andersson, R. Müller and K. Mosbach, Nature, 1993, 361, 645.
5. G. Wulff, Trends Biotech., 1993, 11, 85.
6. G. Wulff, W. Best and A. Akelah, Reactive Polym., 1984, 2, 167.
7. G. Wulff, B. Heide and G. Helfmeier, J. Am. Chem. Soc., 1986, 108, 1089.
8. K.J. Shea and D.Y. Sasaki, J. Am. Chem. Soc., 1991, 113, 4109.
9. L. Fischer, R. Müller, B. Ekberg and K. Mosbach, J. Am. Chem. Soc., 1991, 113, 9358.
10. B. Sellergren, Makromol. Chem., 1989, 190, 2703.
11. G. Wulff and S. Schauhoff, J. Org. Chem., 1991, 56, 395.
12. S.E. Byström, A. Börje and B. Akermark, J. Am. Chem. Soc., 1993, 115, 2081.
13. R. Müller, L.I. Andersson, and K. Mosbach, Makromol. Chem., Rapid Commun., 1993, 14, 637.
14. E. Hedborg, F. Winquist, I. Lundström, L.I. Andersson and K. Mosbach, Sensors and Actuators A, 1993, 37-38, 796.
15. K. Mosbach, Trends Biochem. Sci., 1994, 19, 9.
16. M.J. Whitcombe, M.E. Rodriquez and E.N. Vulfson, These proceedings, page 565.
17. C.R. Smith, M.J. Whitcombe and E.N. Vulfson, These proceedings, page 482.

The Effect of Protein Characteristics on Their Extraction in Reversed Micelle Systems

B. A. Andrews

BIOTECHNOLOGY AND BIOCHEMICAL ENGINEERING GROUP, DEPARTMENT OF FOOD SCIENCE AND TECHNOLOGY, THE UNIVERSITY OF READING, PO BOX 226, READING RG6 2AP, UK

The partitioning of proteins between various aqueous phases and a reversed micelle phase consisting of an anionic surfactant, AOT, in isooctane has been investigated. The influence of aqueous phase pH, ionic strength and ion species on the uptake of water and proteins into the reversed micellar phase as well as the influence of physicochemical characteristics of the protein have been investigated. Micelle size (characterized by the parameter w_0) was found to be constant with changes in pH but to vary depending on the type on ions present in the system. Protein solubilization in the reversed micelle phase appears to be dependent on the net charge and charge distribution of the protein. Ionic strength affects both protein solubilization and w_0.

1 INTRODUCTION

Reversed micelle systems have been the subject of extensive study in recent years as a technique for the extraction and purification of proteins[1-3]. They have great potential for the liquid-liquid extraction of biomolecules under moderate conditions. Partitioning of proteins is dependent upon many factors which are directly related to the interaction between properties inherent to the system itself and those pertaining to the protein under investigation. Fundamental studies of the factors which determine separation of biomolecules are necessary to establish correlations between the physico-chemical properties of the proteins and the reversed micelle system. These factors include pH, ionic strength and type of ions present in the system, type of surfactant and organic solvent used and physico-chemical properties of the protein such as isoelectric point (pI), hydrophobicity, size, charge density and charge distribution.

Reversed micelle systems formed from sulfosuccinic acid(2-ethylhexyl)ester, sodium salt (AOT), isooctane and water are commonly used. AOT is an anionic surfactant which will form micelles in apolar solvents without a cosurfactant. The polar head groups of the surfactant molecules are directed towards the interior of the micelles and form a polar core which can solubilize water. The lipophilic chains are exposed to the solvent. The molar water to surfactant ratio ($w_0 = [H_2O]/[AOT]$) is used to characterize micelle size. In AOT/isooctane systems the micelles are spherical, nanometer sized particles (with diameters ranging from 10 to 200 Å) which are thermodynamically stable. Water content of a micellar system and phase ratio both have a strong influence on protein solubilization and subsequent function.

The aqueous phase pH in which the protein is dissolved directly affects the charge on the protein: at pH values below its pI a protein will have a net positive charge and at a pH above its pI it will have a net negative charge. The net charge of a protein and the distribution of the charges on the surface of the protein will affect the interaction with the surfactant headgroups and hence protein transfer into the micelles. The aim of this paper is to study the effects of ionic strength and ion type of the aqueous phase on micelle size and to investigate the effects of protein characteristics such as net charge on their solubility in reversed micelle systems.

2 MATERIALS AND METHODS

2.1 Proteins

Ribonuclease-A (RA, type III-A from bovine pancreas) and soybean trypsin inhibitor (SBTI, type II-S) were purchased from Sigma Chemical Co. Ltd., thaumatin (Tha, 100% pure Talin, human food grade) was obtained from Four-F Nutrition, Northallerton, North Yorkshire, UK.

2.2 Chemicals

Sulfosuccinic acid bis(2-ethylhexyl)ester, sodium salt (AOT) was obtained from Sigma and spectrophotometric grade isooctane (2,2,4-trimethylpentane), from Aldrich Chemical Co. Inc. Both were used as supplied. All other chemicals used were of AnalaR grade.

2.3 Partition Experiments

Protein transfer experiments were done in 25ml stoppered flasks at room temperature (ca. 21°C). The organic phase was composed of 50mM AOT in isooctane. The aqueous phase consisted of NaCl, KCl, KBr or $MgCl_2$ at different concentrations and pHs. The pH was adjusted by the addition of concentrated acid or base, buffers were not used because over such a wide range of pH values several different buffers would have been necessary involving the addition of a mixture of ions to the system. The pH given is the aqueous phase pH measured after phase equilibrium. Protein was dissolved in the aqueous phase at a concentration of 0.25 mg/ml. Equal volumes of organic and aqueous solutions were mixed and stirred for 5 minutes which was found to be ample time for equilibration. Phase separation was aided by centrifugation at 2000 rpm for 5 min. Samples were then assayed for protein and water content. Each data point includes the results from at least 4 samples with a deviation of +/- 8%.

2.4 Assays

Protein in both phases was assayed by absorbance at 280 nm and 310 nm in a Pharmacia Ultrospec III spectrophotometer, standard curves were prepared for each phase with each protein. Water was measured by Karl Fischer titration using a Mettler DL37 KF coulometer.

3 RESULTS AND DISCUSSION

Proteins usually exist as charged molecules, at pH values equal to their isoelectric point (pI) the net charge is zero (the number of negative charges is equal to the number of positive charges), at pH values above their pI proteins have a net negative charge and at pH values below the pI they have a net positive charge. In a reversed micelle system containing AOT as surfactant the head groups are negatively charged and hence the proteins will interact with these headgroups, and hence be solubilized into the micelles, when they have a net positive charge, i.e. at pH values below their pI. The range of pH values over which protein solubilization occurs is dependent on the titration curve of the protein (change in net charge of the protein molecule with pH). If a protein has a steep titration curve it indicates that the net charge of the protein changes rapidly with changes in pH, conversely if the titration curve is flat throughout much of the pH range used it indicates that the net charge on the protein changes little as the pH changes.

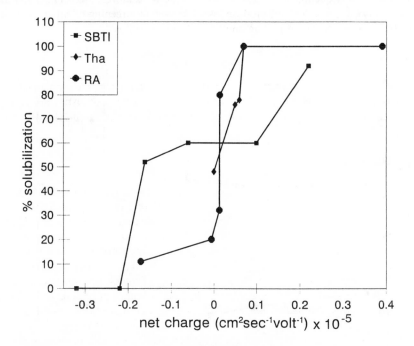

Figure 1 *Influence of net charge on solubilization in reversed micelles.*
SBTI - soybean trypsin inhibitor, Tha - thaumatin, RA - ribonuclease A

Figure 1 shows the solubilization of proteins in reversed micelles as a function of the net charge of the protein molecules (calculated from their titration curves). For the proteins thaumatin and ribonuclease the percentage solubilized in the micelle phase increases very rapidly with very small changes in net charge. With trypsin inhibitor solubilization varies little with net charge except at extreme values, it is expected that when the net charge is positive the interaction with the surfactant head groups is favourable and solubilization will occur. Conversely, when the net charge is negative interactions are not favourable and no protein should be solubilized. It is clear from figure 1 that solubilization of trypsin inhibitor does occur when the net charge is negative, this implies that another factor (such as charge distribution) is also influencing partition. If the protein molecules have an asymmetric distribution of charges (i.e. the charges are concentrated in specific regions of the molecule, creating highly charged patches) the protein-reversed micelle interaction could be different from those of a protein with symmetrically distributed charges.

Figure 2 shows the influence of pH on w_0 in two reversed micelle systems; one containing KBr in the aqueous phase and the other containing $MgCl_2$. w_0 is a measure of the water content of the reversed micelles and hence their size. From figure 2 it is clear that there is almost no change in w_0 with pH. However, there is an important difference in the value of w_0 depending on the type of salt present in the aqueous phase. The system containing $MgCl_2$ results in much larger micelles than the system with KBr. This may be a consequence of the relative sizes of the K^+ and Mg^{2+} ions (1.33 and 0.66 Å respectively), the smaller ions produce less electrostatic screening of the micelles and hence allow more transfer of protein and water to the micelles[3,4,5].

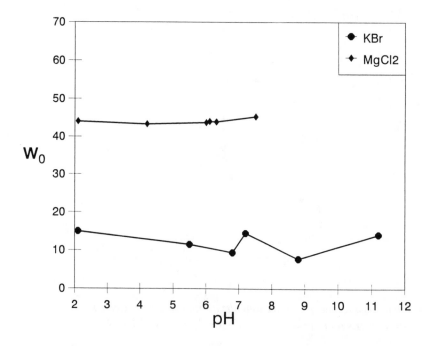

Figure 2 *Effect of pH on w_0 (molar water to surfactant ratio, $[H_2O]/[AOT]$).*

The ionic strength and type of ions will influence w_0. Figure 3 shows the w_0 values for reversed micelle systems containing NaCl of increasing ionic strength. Three proteins were solubilized into the reversed micelle phases, ribonuclease A, thaumatin and soybean trypsin inhibitor. In all cases the w_0 value decreases with increasing ionic strength of NaCl. As the ionic strength of NaCl increases the ions present form an electrostatic shield around the micelles making interaction of water and protein molecules with the charged surfactant headgroups more electrostatically unfavourable. Differences in the curves are only evident at low ionic strengths (< 0.3) and are very similar at higher ionic strengths. Trypsin inhibitor and thaumatin have much higher molecular weights than ribonuclease (28,000 Da, 24,500 Da and 13,500 Da respectively) and possibly more water enters the micelles with these proteins to accommodate their larger sizes. With ribonuclease A there is very little change in micelle size in this range of ionic strengths.

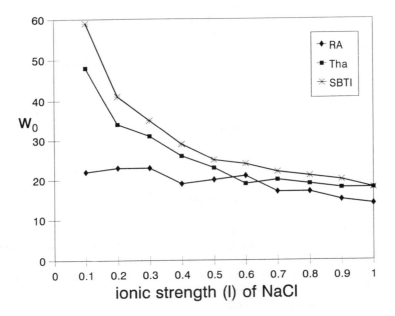

Figure 3 *Effect of ionic strength of NaCl on w_0 with three proteins.*

Figure 4 shows the influence of increasing ionic strength on w_0 values in four reversed micelle systems with three different salts in the aqueous phases and containing two proteins. From this figure it is evident that the type of ions present in the system are important. In the systems with $MgCl_2$ the w_0 values are much greater than with KBr and NaCl. Although the concentration of each salt was the same, the ionic strength of $MgCl_2$ is greater as the Mg^{2+} ion is divalent. It appears that, as in figure 2, the relative sizes of the ions has a large influence on water uptake by the micelles ($Mg^{2+} < Na^+ < K^+$).

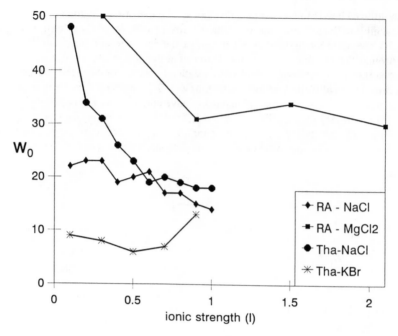

Figure 4 *Effect of ionic strength of three salts on w_0.*

The solubilization of protein into the reversed micellar phase is also influenced by the ionic strength and type of ions present in the system. Figure 5 shows the effect of increasing ionic strength of three salts ($MgCl_2$, NaCl and KCl) on the solubilization of ribonuclease. $MgCl_2$ allowed protein transfer in the range of ionic strengths from 0.3 to 2.7, significantly higher than both KCl and NaCl. The Mg^{2+} ion has an atomic radius of 0.66 Å and is thus smaller than both K^+ and Na^+, this result supports the idea that smaller ions produce less screening of the micelles and thus allow more protein transfer. Marcozzi et al.[2] have studied the factors affecting the forward and backward transfer of α-chymotrypsin in AOT/isooctane systems. They used four salts, KCl, NaCl, LiCl and $CaCl_2$ and found that protein transfer occurred at lowest ionic strengths with KCl followed by $CaCl_2$, NaCl and LiCl. The atomic radii of these ions decrease in the order K^+ > Ca^{2+} > Na^+ > Li^+ indicating that the size of the ions may also be important in this system.

4 CONCLUSIONS

Net charge and charge distribution of proteins can have a strong effect on their solubilization in reversed micelle systems.

pH influences protein charge depending on the isoelectric point and titration curve and thus influences partition.

pH has little effect on w_0.

Increasing ionic strength tends to decrease both w_0 and the amount of protein solubilized.

The ion species present in the system affect w_0 and protein solubilization. Smaller ions produce less electrostatic screening and allow more protein transfer and larger w_0.

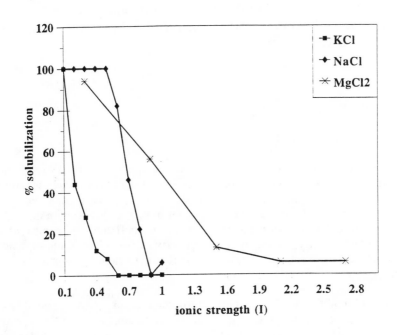

Figure 5 *Effect of ionic strength on solubilization of ribonuclease A, pH 5.*

5 REFERENCES

1. P.L. Luisi, <u>Angewandte Chemie</u>, 1985, <u>24</u>, 439.
2. G. Marcozzi, N. Correa, Luisi, P.L. and M. Caselli, <u>Biotechnol. Bioeng.</u>, 1991, <u>38</u>, 1239.
3. S.F. Matzke, A.L. Creagh, C.A. Haynes, J.M. Prausnitz and H.W. Blanch, <u>Biotechnol. Bioeng.</u>, 1992, <u>40</u>, 91.
4. B.A. Andrews, D.L. Pyle and J.A. Asenjo, <u>Biotechnol. Bioeng.</u>, 1994, <u>43</u>, 1052.
5. B.A. Andrews and K. Haywood, <u>J. Chromatography</u>, 1994, <u>668</u>, 55.

Effect of Complex Formation in Organic Acid Extraction by Supported Liquid Membranes

G. Aroca and D. L. Pyle

BIOTECHNOLOGY AND BIOCHEMICAL ENGINEERING GROUP, DEPARTMENT OF FOOD SCIENCE AND TECHNOLOGY, THE UNIVERSITY OF READING, READING RG6 2AP, UK

1 INTRODUCTION

The separation of bioproducts generally starts with a very diluted stream, implying that the aims will always include concentration and selectivity. When denaturation is not a constraint, solvent extraction has been useful in concentrating and separating polar and non-polar biomolecules, such as antibiotics. Concentration and selectivity can both be enhanced: increasing the partition coefficient by choice of diluent (solvent plus modifier), changing some property of the solute in the organic phase, i.e. by ion pairing, or relaxing the thermodynamic constraint by using a reagent in the stripping side.

Liquid Membranes (LM) are particularly suitable for recovery from dilute feeds, and have been proposed by several authors[1,2] as an alternative in the separation of biomolecules with low molecular weight. In fact there are several advantages of the different LM systems (emulsion, supported, contained, etc.) over traditional solvent extraction processes, such as: process intensification (extraction and stripping are carried out simultaneously) by increasing the overall effective partition coefficient, improved mass transfer kinetics by increasing the transfer area without high agitation cost, and solvent retention, in the cases of supported or contained systems.

A supported liquid membrane is made by soaking the pores of a solid support, usually a microfiltration membrane, with an organic phase (see Figure 1); transport takes place across this phase. When an extracting agent or carrier is added, the process is called facilitated.

The levels of recovery and selectivity achieved in the facilitated liquid membrane extraction of bioproducts such as organic or amino acids depend on various factors.

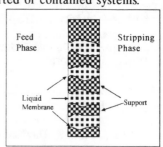

Figure 1: Supported Liquid Membrane

These include the chemistry (stoichiometry, reaction rates and equilibria) of the different compounds involved and the transfer kinetics. In the case of mixtures they also depend on the competition between the various species involved for the carrier.

2 THEORY

2.1 Organic acid extraction by SLM

When organic acids are separated using supported liquid membranes (SLM), the principal resistance, i.e. the rate limiting step, is always the rate of transport of the acid/carrier complex across the membrane[3]. It is thus very important to be able to relate the complex concentration to the extraction conditions. The simplest model for complex formation between an organic acid and a tertiary amine involves a single reversible ion-pairing reaction. However, depending on the type of organic acid and the polarity of the solvent, the organic phase is capable of taking up acid in excess of the stoichiometric requirement for neutralization of the amine base[4]. Then the chemical equilibria involved can be written as a set of reactions between **p** acid molecules and **q** molecules of amine to give various **{p,q}** complexes[5,6]:

$$p\,H_nA_{(w)} + q\,R_3N_{(o)} \rightleftharpoons (R_3N)_q(H_nA)_{p(o)} \tag{1}$$

where subscripts **(w)** and **(o)** refer to the aqueous and organic phases and H_nA is the undissociated acid. Thus we must also include the appropriate dissociation equilibria:

$$H_nA_{(w)} \rightleftharpoons n\,H^+_{(w)} + A^{-n}_{(w)} \tag{2}$$

The acid/carrier reactions have equilibrium constants:

$$K_{p,q} = \frac{[(R_3N)_q(H_nA)_{p(o)}]}{[H_nA_{(w)}]^p\,[R_3N_{(o)}]^q} \tag{3}$$

which must be solved simultaneously with the dissociation equilibria to give the interfacial concentrations of the complex $C_{p,q}$.

Assuming that the activity coefficients for the species in the extracting system are proportional to their concentrations, with the stoichiometric models outlined above it is possible to find the best fit values for the equilibrium constant(s) including the effects of the activity coefficients, using a non-linear least squares optimization procedure to satisfy the mass balances for the acid and amine at the organic phase in an extracting system. The general expressions for these balances are:

$$C_{HA(o)} = \sum_{i=1}^{p}\sum_{j=1}^{q} i K_{p,q} C^i_{HA(w)} C^j_{R_3N(o)}$$

$$C_{R_3N(o)} = C^o_{R_3N(o)} - \sum_{i=1}^{p}\sum_{j=1}^{q} j K_{p,q} C^i_{HA(w)} C^j_{R_3N(o)} \tag{5}$$

where c_{HA} is the concentration of the undissociated acid, c_{R3N} is the concentration of free amine in the organic phase and $c_{R3N}°$ is the initial concentration of the amine.

The stoichiometries of extraction of citric acid and lactic acid with trialkylamines are very different. Citric acid forms {1,q} {acid,amine} complexes; the formation of {1,1} and {1,2} complexes is reported[7] when extracted in various solvents. However lactic acid forms {p,1} {acid,amine} complexes; the formation of {1,1}, {2,1} and {3,1} complexes have been reported[5].

The facilitated extraction of organic acids by SLM, using trialkylamines as carriers, has been studied by several authors without taking proper account of the type or number of the {acid,amine} complexes formed: for example in studies of the extraction of citric acid with trialkylamines Friesen *et al.* (1991)[8] considered only the formation of one {1,1} complex, whilst Basu and Sirkar (1991,1992)[9, 10] consider the formation of a {1,3} complex. The same has happened with the studies on the extraction of lactic acid by liquid membranes, where extraction has been studied principally in emulsion liquid membrane systems[11].

In order to study the effect of the formation of multi-complexes, we postulate a chemical model for the extraction of citric acid using Na_2CO_3 as stripping agent, which assumes the formation of two complexes, and, in the case of lactic acid, the formation of three complexes. These are shown in summary form in Figures 2 and 3. The carbonate ions are counter-transported across the membrane and released in the internal phase, affecting the dissociation equilibrium at the feed phase.

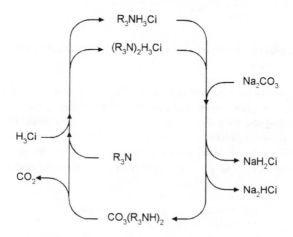

Figure 2: *Cycle of citric acid extraction by SLM using TOA as carrier and Na_2CO_3 as stripping agent.*

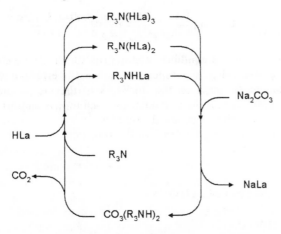

Figure 3: *Cycle of Lactic acid extraction by SLM using TOA as carrier and Na_2CO_3 as stripping agent.*

2.2 Mass Transfer

The following assumptions are made: (a) because the acid is very sparingly soluble in the organic phase, only the flux of complexes is considered, (b) the reactions at the interfaces are very fast, (c) the diffusion coefficients are the same for all the {acid,amine} complexes, (d) the acid concentration is the same in the bulk feed and interface and (e) with excess stripping reagent the strip side concentrations of the complexes are zero. The compounds involved in the acid transport must satisfy the continuity equation for one dimensional steady state transport across the membrane[12], and according to Fick's law.

$$-J_a = D_{ac} \sum_{i=1}^{p} \sum_{j=1}^{q} i \frac{dc_{a_i c_j}}{dx} \tag{6}$$

where J_a is the acid flux and $c_{a,c}$ are the concentrations of the complexes formed. From this expression and knowing the equilibrium and the stoichiometric relations between the different compounds involved in the transport, it is possible to develop analytical expressions relating the concentration of the undissociated acid (c_a) in the feed phase with time (t) (Details in publication in preparation). In the general case where there is only a {1,1} {acid,amine} complex the expression is:

$$k_{1,1}(c_{ao}-c_a) + \frac{1}{2}\ln\frac{c_{ao}}{c_a} = \frac{k_{1,1}ADc_{co}}{2LV}t \tag{7}$$

For lactic acid, assuming that $K_{1,3}$ is very small (see Results section) and therefore that the {1,1} and {2,1} complexes are formed, we have:

$$k_{1,1}(c_{ao}-c_a) + \ln\frac{c_{ao}}{c_a} + \frac{k_{1,1}-2k_{1,2}}{2k_{1,2}}\ln\frac{(1+2k_{2,1}c_{ao})}{(1+2k_{2,1}c_a)} = \frac{ADk_{1,1}c_{co}}{LV}t \tag{8}$$

Finally for citric acid, assuming {1,1} and {1,2} complex formation:

$$\ln\frac{c_{ao}}{c_a} + (1+2k_{1,2}c_{co})\ln[\frac{1+k_{1,2}c_{co}+2k_{1,1}k_{1,2}c_c c_{ao}}{1+k_{1,2}c_{co}+2k_{1,1}k_{1,2}c_c c_a}] = \frac{AD(1+k_{1,2}c_{co})k_{1,1}c_c}{LV}t \qquad (9)$$

where: $k_{p,q}$ are the respective equilibrium constants for the $\{p,q\}$ complexes ($k_{1,1} = K_{1,1}$, $k_{1,2} = K_{1,2}/k_{1,1}$ and $k_{2,1} = K_{2,1}/k_{1,1}$ respectively), A is the effective area of transfer, D is the diffusion coefficient, L is the thickness of the membrane and V is the volume of the feed phase, c_{co} is the initial carrier (amine) concentration, and c_c is the carrier (amine) concentration at time t. These expressions were used to evaluate the effect of the formation of the complexes in the transport of organic acids through a supported liquid membrane.

3 MATERIALS AND METHODS

3.1 Equilibrium studies

Equal volumes of organic and aqueous phase were added to stoppered flasks and equilibrated at 20 °C for 12 h. Then the phases were separated by centrifugation (3000 rpm, 2 h). Citric acid and lactic acid concentrations in the aqueous phase were determined enzymatically (Boehringer Mannheim GmbH). The organic acid in the organic phase was calculated by difference. The solvent used was xylene (BDH), the extractant agent used was trioctylamine (TOA, Sigma 98 %). All the reagents used are reagent grade.

3.2 Extractions by Supported liquid membrane

The experiments were performed in a cell consisting of two chambers provided with agitation, separated by a poly-propylene microfiltration membrane (Celgard 2500, Celanese Co., Charlotte, N.C., USA) soaked with the organic phase (xylene / 20 % TOA). The dimensions of the cell are: effective membrane area A: 17,5 cm², membrane thickness L: 0.0025 mm and reservoir volume V: 125 cm³. The agitation was maintained at 150 rpm to avoid mass transfer limitations at the interfaces on the feed and stripping sides[3]. The initial concentration of organic acid was 0.25 [M] and the initial concentration of Na_2CO_3 was 0.25 [M] in both cases. A volume of 0.5 ml was taken from the feed phase at intervals of times to analyse the acid concentration.

4 RESULTS

4.1 Determination of the Constants of Equilibrium

Non-linear regression on the equilibrium data obtained was performed using the Marquart algorithm with the 'Non-Lin' Procedure using SAS software (SAS Institute Inc., USA). The values for the equilibrium constants obtained for citric acid were $K_{1,1} = 25.7$ (1/mol) and $K_{1,2} = 100.5$ (l^2/mol^2). For lactic acid $K_{1,1} = 0.72$ (1/mol), $K_{2,1} = 2.26$ (l^2/mol^2) and $K_{3,1} = 0.369$ (l^3/mol^3).

Figures 4 and 5 show the equilibrium behaviour between citric acid and lactic acid respectively, with TOA (20 %) in xylene. The continuous lines represent the predicted values assuming two complex formation -{1,1} and {1,2}- in the case of citric acid, and the formation of three complexes -{1,1},{2,1}, and {3,1}- in the case of lactic acid.

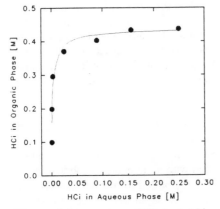

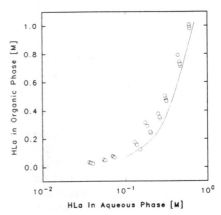

Figure 4: *Extraction of Citric Acid by TOA (20%) in Xylene.*

Figure 5: *Extraction of Lactic Acid by TOA (20%) in Xylene.*

4.2 Supported Liquid Membrane Extractions

Figures 5 and 6 show the variation with time of the acid concentration on the feed side of the supported liquid membrane and the predicted pattern assuming formation of one {acid,amine} complex (Eq.7, dotted line) and two complexes (Eq.8 and Eq.9, solid lines) respectively. The values of the diffusion coefficient, D_{ac}, used in the simulation were estimated from the Wilke-Chang correlation, for the trioctylammoniumcitrate complex $D_{ac} = 4.5 \times 10^7$ (cm^2/s) and for trioctyl-ammoniumlactate complex $D_{ac} = 9.4 \times 10^7$ (cm^2/s)[11]. The different symbols in the figures corresponds to different experiments made at the same conditions.

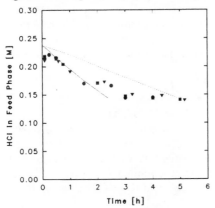

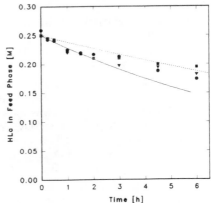

Figure 6: *Citric Acid extraction by SLM.*

Figure 7: *Lactic Acid extraction by SLM.*

5 DISCUSSION AND CONCLUSIONS

The good agreement between the experimental equilibrium data and the equilibrium model confirms the importance of correctly identifying the complexes in the corresponding systems.

As figure 5 and 6 show, quantitative models that consider the formation of more than one complex (i.e. Eq. 8 and 9), give a better agreement with the experimental data. The agreement is good over the early stages of transfer: subsequent deviation from the model may be due to irreversible interactions between the amine and carbonate cause loss of activity.

The rate of extraction of citric acid is greater than that of lactic acid, because of the higher value of the constant of equilibrium for citric acid extraction. In the case of Lactic Acid all the complexes are formed at the interface, while in the case of citric acid the complexes, other than the {1,1} type, may be formed within the liquid membrane phase, this would change the potential for the diffusion of the different species, which in this case could improve citric acid transport.

Current work aims to develop theory for other extraction systems with single and mixed acids, including systems where external mass transfer is important.

REFERENCES

1. R. Marr and A. Kopp, *Int. J. Chem. Engng.*, 1982, **22**, 44.
2. J. J. Pellegrino and R. D. Noble, *TIBTECH*, 1990, **8**, 216.
3. T. Sirman, D. L. Pyle and A. S. Grandison, Ed. D. L. Pyle 'Separations for Biotechnology', Elsevier Applied Science, London, 1991, p. 325.
4. A. S. Kertes and C. J. King, *Biotechnol. Bioengng.*, 1986, **28**, 269.
5. J. A. Tamada, A. S. Kertes and C. J. King, *Ind. Eng. Chem. Res.*, 1990, **29**, 1319.
6. V. Bizek, J. Horacek, R. Rericha and M. Kousova, *Ind. Eng. Chem. Res.*, 1992, <u>**31**</u>, 1554.
7. V. Bizek, J. Horacek, M. Kousova, A. Heyberger and J. Prochazka, *Chem. Eng. Sci.*, 1992, **47**, 1433.
8. D. T. Friesen, W. C. Babcock, D. T. Brose and A. R. Chambers, *J.Membrane Sci.*, 1991, **56**, 127.
9. R. Basu and K. K. Sirkar, *AIChE J.*, 1991, **37**, 383.
10. R. Basu and K. K. Sirkar, *Solv. Extr. & Ion Exch.*, 1992, **10**, 119.
11. J. Chaudhuri and D. L. Pyle, *Chem. Eng. Sci.*, 1992, **47**, 41.
12. E. L. Cussler, 'DIFFUSION:Mass Transfer in Fluid Systems', Cambridge University Press, Cambridge, 1984.

Cellulose-binding Domains as Affinity Tags for Protein Purification

Z. Assouline, J. B. Coutinho, N. R. Gilkes, J. M. Greenwood, D. G. Kilburn, K. D. Le, R. C. Miller, Jr., E. Ong, C. Ramirez, and R. A. J. Warren

DEPARTMENT OF MICROBIOLOGY AND IMMUNOLOGY AND THE PROTEIN ENGINEERING NETWORK OF CENTRES OF EXCELLENCE, THE UNIVERSITY OF BRITISH COLUMBIA, VANCOUVER, BC, CANADA V6T 1Z3

1 INTRODUCTION

The cellulose-binding domains (CBDs) from the cellulases of the bacterium *Cellulomonas fimi* are 100-150 amino acids long[1]. The CBDs function independently of their cognate catalytic domains[2], and have association constants (K_as) for crystalline cellulose of about 10^5 M, roughly comparable to the affinities of antibodies for antigens[3]. The CBDs differ in their affinities for different types of cellulose: CBD_{CenA} and CBD_{Cex} from endoglucanase A and an exoglucanase/xylanase, respectively, bind to both crystalline and amorphous cellulose; CBD_{CenC} from endoglucanase C binds to amorphous but not to crystalline cellulose.[1] They retain their binding properties when fused to heterologous proteins which makes them useful affinity tags for protein purification and/or immobilization.[4-6]

2 CBD FUSION PROTEINS

The CBDs facilitate purification of the *C. fimi* cellulases by affinity chromatography on cellulose (Table 1). Various CBD-containing fusion proteins have been constructed to analyse the utility of the CBDs in downstream processing (Table 2). The fusions are designed to test the CBDs as affinity tags for the purification of target proteins and for the removal of proteases used in the release of the CBDs from the target polypeptides.

Table 1 *Purification of endoglucanase A (CenA) from Escherichia coli*

Purification Step	Total Protein (mg)	Total Activity (u)	Specific Activity (u/mg)	Recovery (%)	Purification factor
Cell extract	19,800	11,090	0.56		
Streptomycin sulphate supernatant	15,444	11,189	0.725	100	1.3
Pooled fractions after elution from cellulose with 6M GnHCl and diafiltration	43.3	5,460	126	49	225

Gn HCl: guanidine hydrochloride

2.1 CBDs in protein purification

Gene fusions encoding a desired protein fusion are constructed by standard methods of *in vitro* gene manipulation. A CBD can be fused to the N- or C-terminus of a target polypeptide. Such gene fusions have been expressed in *Escherichia coli, Streptomyces lividans, Saccharomyces cerevisiae* and cultured mammalian cells. If the fusion protein contains the leader peptide as well as the CBD of a cellulase, it may be exported to the periplasm of *E. coli*.

Table 2 *CBD-containing fusion proteins*

Fusion protein	Fusion partner	Agent for desorption from cellulose	Reference
Abg-CBD$_{Cex}$	β-glucosidase	water	5
FXa-CBD$_{Cex}$	factor Xa	water	9
ProtA-CBD$_{Cex}$	protein A	GnHCl	12
CBD$_{CenA}$-PhoA	alkaline phosphatase	GnHCl	7
Sta-CBDCex	streptavidin	not tested	11

GnHCl: guanidine hydrochloride

Overexpression of the gene encoding such a fusion can result in the leakage of significant quantities of the fusion protein into the culture medium.[7] This greatly simplifies purification of the fusion protein because it obviates the need for cell rupture.

The fusion proteins can be purified extensively in a single step by affinity chromatography on cellulose (Table 3 and 4). They are adsorbed to cellulose in cell-free culture medium or in cell extracts in 50 mM buffer, depending on the cellular location of the fusion protein. Non-specifically adsorbed proteins are removed by washing with 50 mM buffer-NaCl. The fusion proteins are desorbed with either distilled water or 6M guanidinium hydrochloride (GnHCl) depending on the combination of CBD and target protein (Table 1). A protein fused to CBD_{CenA} can be separated from a protein fused to CBD_{CenC} by adsorption of the former to crystalline cellulose. The CBD_{CenC} fusion is then recovered with amorphous cellulose.[6]

In the event that the target protein in a fusion which can be desorbed from cellullose only with GnHCl is irreversibly denatured during purification, mutant CBDs are being sought which will desorb under milder, non-denaturing conditions.

Table 3 *Purification of CBD-PhoA from E. coli culture supernatant*

Purification Step	Total Activity (u)	Total Protein (mg)	Specific Activity (u/mg)	Yield (%)	Purification Factor
Culture supernatant	600	81	7.4	200	
H_2O elution,[a] before UF[b]	374	19	20	62	2.7
H_2O elution, UF retenate[c]	365	14	26	60	3.5

[a]Protein in 90 mL of supernatant was passed through a column of 100 mL bed volume containing CF1 cellulose (Sigma). After washing with 50 mM Tris-HCl, pH 7.5, with and then without 1M NaCl, bound proteins were eluted with a 200 mL linear gradient from the same buffer to distilled water.
[b]After ultrafiltration (Amicon PM30 membrane) and buffer reconstitution.
[c]After final centrifugation

2.2 Recovery of target proteins

If the fusion protein is constructed with a sequence sensitive to a specific protease, such as factor Xa or Kex2, between the CBD and the target protein, the CBD is released from the target with the protease, then removed with cellulose. Alternatively, the target protein is hydrolysed with the protease while still adsorbed to cellulose.[8] This is advantageous if the fusion protein requires GnHCl for desorption and is irreversibly denatured by the treatment.

Table 4 *Purification of Abg-CBC$_{Cex}$ from cell extracts of E. coli*

Purification Step	Total Activity (u)	Total Protein (mg)	Specific Activity (u/mg)	Yield (%)	Purification Factor
Cell extract (soluble proteins)	18,987	18,467	1.02	100	
Treated extract[a]	14,755	12,126	1.22	78	1.2
H$_2$O elution[b] and ultrafiltration[c]	9,045	74	119.6	48	117

[a]Extract after treatment with streptomycin sulphate.
[b]Proteins in the extract from the cells in a 60 L culture were passed through a column of 800 mL bed volume containing CF1 cellulose. After washing with 50 mM phosphate, pH 7.0, with and then without 1 M NaCl, bound proteins were eluted with a 2.05 L concave gradient of the same buffer to distilled water.
[c]Amicon PM10 membrane.

2.3 CBDs and proteases

In general, target proteins are released efficiently from fusion proteins with proteases such as factor Xa. Although the CBD is removable with cellulose, the target protein is now contaminated with the protease. Several types of fusion have been constructed to circumvent this problem. FXa-CBD$_{Cex}$ is active in solution and when adsorbed to cellulose.[9] This means that the CBD can be released from a fusion protein in solution, then both the CBD and FXa-CBD$_{Cex}$ removed from the target protein with cellulose. FXa-CBD$_{Cex}$ immobilized on cellulose can also be used to release other types of affinity tag from fusion proteins. FXaCH6, which has a hexahistidine sequence at its C-terminus[10], is used to release the target protein from a CBD fusion protein adsorbed to cellulose. $FXaCH_6$ is removed from the target protein with Ni^{2+}-NTA agarose. Sta-CBD_{Cex} will bind biotinylated proteins to cellulose.[11] This obviates the need for constructing gene fusions encoding CBD-protease fusions, or polyhistidine-protease fusions, providing the protease retains activity when biotinylated.

3   CONCLUSIONS

CBDs are useful affinity tags for protein purification. The affinity matrix, cellulose, is cheap, available in a variety of forms, and does not need modification prior to use.

Acknowledgements

This research was supported by the Natural Sciences and Engineering Research Council of Canada, the Medical Research Council of Canada and Ciba Geigy (Canada) Ltd.

References

1.   J.B. Coutinho, N.R. Gilkes, R.A.J. Warren, D.G. Kilburn and R.C. Miller, Jr., *Mol. Microbiol.*, 1992, **6**, 1243.
2.   N.R. Gilkes, R.A.J. Warren, R.C. Miller, Jr. and D.G. Kilburn, *J. Biol. Chem.*, 1988, **263**, 10401.
3.   N.R. Gilkes, E. Jervis, B. Henrissat, B. Tekant, R.C. Miller, Jr., R.A.J. Warren and D.G. Kilburn, *J. Biol. Chem.*, 1992, **267**, 6743.
4.   J. Greenwood, N.R. Gilkes, D.G. Kilburn, R.C. Miller, Jr. and R.A.J. Warren, *FEBS Lett.*, 1989, **244**, 127.
5.   E. Ong, N.R. Gilkes, R.A.J. Warren, R.C. Miller, Jr. and D.G. Kilburn, *Bio/Technology*, 1989, **7**, 604.
6.   J.B. Coutinho, N.R. Gilkes, D.G. Kilburn, R.A.J. Warren and R.C. Miller, Jr. *FEMS Microbiol. Lett.* 1993, **113**, 211.
7.   Greenwood, J.M. Ph.D. Thesis, University of British Columbia, 1993.
8.   P.G. Seeboth, R.A.J. Warren and J. Heim. *Appl. Microbiol. Biotechnol.* 1992, **37**, 621.
9.   Z. Assouline, H. Shen, D.G. Kilburn and R.A.J. Warren. *Prot. Engng.* 1993, **6**, 787.
10. Assouline, Z., unpublished observations.
11. K.D. Le, N.R. Gilkes, D.G. Kilburn, R.C. Miller, Jr., J.N. Saddler and R.A.J. Warren. *Enzyme Microb. Technol.* In press.
12. C. Ramirez, J. Fung, R.C. Miller, Jr., D.G. Kilburn and R.A.J. Warren. *Bio/Technology* 1993, **11**, 1570.

Purification of a Monoclonal IgG1 Antibody to Interleukin 5 and Preparation of its $F(ab)_2$ Fragments

Surjit K. Bains

DEPARTMENT OF NATURAL PRODUCTS DISCOVERY, GLAXO RESEARCH AND DEVELOPMENT LTD., GREENFORD ROAD, GREENFORD, MIDDLESEX UB6 0HE, UK

1. ABSTRACT

Interleukin 5 (IL-5) is a cytokine protein involved in the regulation of immunity and allergic inflammation.

Protein G Sepharose was used to affinity purify gram quantities of neutralising monoclonal IgG1 antibody (Mab 7) to Interleukin 5 from tissue culture fluid in a single step. Bioactive $F(ab)_2$ fragments of this antibody were generated by papain cleavage. Further purification of this cleaved fraction will be described.

2. INTRODUCTION

Enzymatic hydrolysis of antibodies has long been used to study their structure and function, more recently Fab or $F(ab)_2$ fragments obtained from monoclonal antibodies have been used for medical purposes (Rea and Ultee, 1993). Originally pepsin was used to produce $F(ab)_2$ fragments from murine IgG1 antibodies but gave low yields and often inactive fragments. More recently papain has been used in the absence of thiols to prepare active $F(ab)_2$ fragments with high yields from IgG1 murine antibodies (Parham *et al.*, 1982, Kurkela *et al.*, 1988). This report describes the use of these methods to generate gram quantities of $F(ab)_2$ fragments from monoclonal antibody Mab 7 (IgG1 isotype) active against recombinant human IL-5, for use in immunoassay and neutralisation studies.

3. MATERIALS AND METHODS

3.1 Monoclonal Antibodies

Mab 7 cells were cultured in a Hollow Fibre Bioreactor (Cell Pharm Quantum Biosystems) in medium supplemented with 2% Approsera (Advanced Protein Products - depleted IgG serum substitute).

3.2 Estimation of Antibody and Bovine IgG

A mouse IgG1 and bovine IgG low level radial immunodiffusion kits (R.I.D.) (Binding Site) were used to measure the amount of antibody and bovine IgG present in the culture tissue fluid and purified samples. The kits were set up according to the manufacturer's instructions.

3.3 Large Scale Purification of Antibody

4 litres of cell culture fluid containing Mab 7 at a concentration of 2.5mg/ml was loaded onto a 70 x 90mm Protein G Sepharose (Pharmacia) column (previously equilibrated with 20mM sodium phosphate pH 7.0) at a flow rate of 40ml/min. The column was eluted with 100mM glycine pH 2.5, and the pooled antibody fraction was neutralised immediately with 1M Tris-HCl pH 9.0. The antibody solution was dialysed thoroughly against 100mM sodium acetate, 3mM EDTA pH 5.5.

3.4 Papain Digestion

This was carried out using a slight modification of the method described by Parham *et al.*, (1982). 17mg of papain purified by covalent chromatography (Brocklehurst *et al.*, 1974) (kindly supplied by Professor K Brocklehurst) was incubated with 8ml of 50mM Tris, 3mM EDTA, 50mM cysteine pH 7.8 at 37°C for 1 h. Excess cysteine was removed by gel filtration on 4 PD10 Sephadex G25 columns (Pharmacia) with 100mM sodium acetate, 3mM EDTA pH 5.5. This yielded 8mg of fully activated papain, which was added to the antibody at an enzyme:antibody mass ratio of approximately 1:500 and then incubated for 18 h at 37°C. The reaction was stopped by addition of iodoacetamide to a final concentration of 30mM.

3.5 Purification of Papain Digested Fragments

The total digest was loaded onto a 70 x 90mm Protein A Sepharose (Pharmacia) column in 3M NaCl, 0.05M Tris pH 8.9 at a flow rate of 30ml/min, and washed with the binding buffer. The unbound fraction was collected and concentrated using a stirred cell concentrator, (Amicon) to 150ml. 50ml of the concentrate was loaded at a time onto a gel filtration 980 x 50mm Superdex 200 (Pharmacia) column at 10ml/min and eluted with PBS pH 7.2 buffer.

3.6 Analytical Studies

SDS PAGE of non-reduced and reduced proteins were used to assess purity of the antibody and its fragments.

SDS PAGE was carried out on a Phastsystem (Pharmacia) with Phast Gel gradient 10-15% and protein bands were visualised by silver stain. Protein concentrations were measured by absorbance of the solution at 280nm.

50

3.7 Direct Binding Enzyme Linked Immunosorbent Assay (ELISA) and B13 Cell Proliferation Assay

This was carried out using the method described by McNamee *et al.*, (1991).

4. RESULTS

A total of 10 grams of anti human IL-5 antibody was present in the starting culture fluid (as measured by R.I.D.). 5 grams was recovered on Protein G Sepharose at a concentration of 7mg/ml. Only trace levels of bovine IgG (less than 50mg/ml) was detected by R.I.D. Purity was estimated by SDS PAGE to be about 90% (Figure 1).

When the antibody was dialysed against digestion buffer it yielded 4275mg of purified antibody. Cleavage of the antibody with freshly activated thiol-free papain gave rise to F(ab)₂ fragments (100kD) and Fc fragment (26kD) (Figure 1). The F(ab)₂ fragments were purified on Protein A Sepharose to remove the Fc and any undigested antibody. As expected, the F(ab)₂ fragments did not bind to the column whereas the Fc fragment was retained. The F(ab)₂ pool was concentrated and further purified by gel filtration to remove any residual papain (Figure 1). The recovery of pure F(ab)₂ fragments was 2100mg. The F(ab)₂ migrated as a 100kD band in the non reducing gel (Figure 1) and a 25kD doublet in the reducing gel (Figure 2), a characteristic pattern for such fragments.

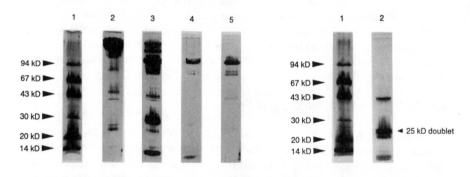

Figure 1 Silver Stained SDS-Page (Non Reduced)

Lane 1: Low molecular weight markers (Pharmacia)
Lane 2: Purified antibody
Lane 3: Papain hydrolysed fraction
Lane 4: Unbound fraction from Protein A Sepharose
Lane 5: Purified F(ab)₂ fragments separated on Superdex 200

Figure 2 Silver Stained SDS-Page (Reduced)

Lane 1: Low molecular weight markers (Pharmacia)
Lane 2: Doublet at 25kD

The purified F(ab)₂ fragments showed a reduced activity compared to the whole antibody in the IL-5 ELISA (data not presented) and 10 fold less activity in the B13 bioassay (Figure 3) (McNamee *et al.*, 1991).

Figure 3

B13 Bioassay Testing Activity of Purified Mab 7 and F(ab)₂ Fragments

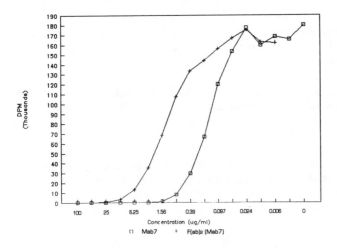

5. DISCUSSION

The aim of this work was to produce gram quantities of pure and active F(ab)₂ fragments for use in immunoassays and biological activity neutralisation studies. By using active papain in the absence of thiols, followed by a two step purification method, a recovery of 75% (gram equivalents) F(ab)₂ fragments was achieved from the pre-digestion stage. The recovered material was estimated to be 90% pure by SDS gel analysis and biologically active both in the IL-5 ELISA and in the B13 Bioassay.

Acknowledgements

I am grateful to Peter Boulton and Chris Roberts for growing the cell line, Paul Hissey and Dev Baines for their help and advice and to Dilynia Fattah for carrying out the bioassay.

References

1. Brocklehurst, K., Carlsson, J., Kierstan, M.P.J. and Crook, E.M. Covalent Chromatography by Thiol-Disulfide Interchange. Methods Enzymol. 34B (1974) 531-544.

2. Coligan J.E., Kruisbeek A.M., Margulies D.H., Shevach E.M. and Strober W., Current Protocols in Immunology Volume 1 (1992) 287-288.

3. Kurkela R., Vuolas L. and Vihko P., Preparation of F(ab)$_2$ fragments from monoclonal mouse IgG1 suitable for use in radioimaging. Journal of Immunological Methods 110 (1988) 229-236.

4. McNamee L.A., Fattah D.I., Baker T.J., Bains S.K. and Hissey P.H., Production, characterisation and use of monoclonal antibodies to human interleukin 5 in an enzyme-linked immunosorbent assay. Journal of Immunological Methods 141 (1991) 81-88.

5. Parham P., Androlewicz M.J., Brodsky F.M., Holmes N.J. and Ways J.P., Monoclonal antibodies: Purification fragmentation and application to structural and functional studies of class I MHC antigens, Journal of Immunological Methods 53 (1982) 133-173.

6. Rea D.W. and Ultee M.E., A novel method for controlling the pepsin digestion of antibodies, Journal of Immunological Methods 157 (1993) 165-173.

Validation of a Primary Capture Process for Production of Human Therapeutic Monoclonal Antibodies

R. M. Baker, A.-M. Brady, B. S. Combridge, L. J. Ejim, S. L. Kingsland, D. A. Lloyd, and P. L. Roberts

RESEARCH AND DEVELOPMENT DEPARTMENT, BIO PRODUCTS LABORATORY, DAGGER LANE, ELSTREE, HERTS WD6 3BX, UK

1 ABSTRACT

This validation study was carried out to determine the maximum number of purification cycles for which Protein G Sepharose FF (Pharmacia Biotech) can be used for the manufacture of therapeutic monoclonal antibodies.

A hollow fibre high density culture system has been used to grow two cell lines, one of which produces a human IgG1 monoclonal antibody (MAb) and the other a human IgG3 MAb. A three-stage chromatographic system has been developed for their purification. A number of parameters were studied during repeated runs of the primary capture column which is packed with Protein G Sepharose FF. The dynamic binding capacity of the gel was measured at regular intervals. Viral reduction, DNA clearance and Protein G leakage levels were also measured. Yield of IgG was measured for every run.

The yield of IgG1 showed no significant decrease after 55 runs. Recovery of IgG3 was not so well sustained, with the yield dropping by 16% after 60 runs. The dynamic capacity of the gel changed over the runs with earlier breakthrough of IgG in fractions of the unbound material. Protein G levels measured in the eluate were less than 3µg/mg IgG and did not change after 55 repeat runs. DNA clearances measured at the beginning and end of the lifetime studies were consistent at between 3.1 and 4.4 logs. Viral reductions were similarly consistent over the series with acid labile viruses giving a mean reduction greater than 5 logs.

2 MATERIALS AND METHODS

2.1 Chromatography System

The chromatography system used for validation work was a scaled-down version of the pilot scale system preserving the dynamic parameters of the full-scale system. A Biopilot[TM] was used with IMV8 valves modified as previously reported.[1] The system was controlled by an LCC 500 controller which was programmed to load tissue culture supernatant and collect Protein G Sepharose FF (Protein G Sepharose) eluate automatically. Sanitisation, equilibration into ethanol and equilibration into running buffers were controlled by separate programmes. 4 ml gel was packed into a 1cm diameter column to give a bed height of approximately 5cm (0.8cm² cross sectional area).

The tissue culture harvest was produced in a hollow fibre high density culture system and was collected in discrete volumes into sterile bottles. For these validation experiments, a uniform starting material was produced by pooling harvest into large-volumes with an IgG concentration of 70µg/ml for IgG3 and 150µg/ml for IgG1.

The purification scheme consisted of loading the tissue culture harvest without pre-treatment (except 0.2µm filtration) and then washing the loaded Protein G Sepharose column with the following series of buffers; phosphate buffered saline, pH6.8, high ionic strength acetate buffer, pH5.3 and low ionic strength acetate buffer, pH5.3. Elution was with citric acid, pH2.8 for IgG1 and pH3.0 for IgG3. As the IgG was eluted from the column it was diluted with acetate buffer.

The system was sanitised between each purification run using 6M guanidine chloride and re-equilibrated into running buffers. For periods of storage longer than 72 hours, the system was equilibrated into 20% ethanol.

In order to be able to obtain meaningful DNA results, the system was operated using the best aseptic techniques possible.

2.2 Validation Studies

Model viruses used in virus-reduction studies were Sindbis, polio-1, vaccinia virus and Herpes simplex-1. The latter was included as a high titre model for Epstein- Barr virus which had been used in cell line production. The viruses were propagated in cell-culture and added to the column load material at a dilution of 1/20. The virus infectivity titre of the load material and eluted product were determined by plaque assay and used to calculate the total virus in each fraction and thus the log virus-reduction value for the chromatographic process.

2.3 Assays

Total DNA was measured using the Threshold^TM assay (Molecular Devices Inc.). This is an immunoligand assay the endpoint of which is detected by a silicon sensor. Intra-assay variation on a standard internal control was 7% over 12 assays.

Protein G was measured by ELISA. Immunopurified chicken anti-Protein G (OEM Concepts Inc.) was used. The standard was a Protein G preparation obtained from Pharmacia Biotech. Polyclonal human IgG was added to samples and standards and samples were assayed with and without a Protein G spike. Plates were developed with chicken anti-Protein G HRP and OPD.

Samples were assayed for IgG by ELISA using anti-human whole IgG antisera (Seralab). Standards and controls were the appropriate purified MAb.

3 RESULTS AND DISCUSSION

3.1 Chromatograms

The shapes of the chromatograms were as expected for all runs except for those where the tissue culture supernatant was spiked with virus. The virus preparation contains many cell culture proteins as well as virus particles and some of these may have bound to the gel and eluted later in the purification. Nevertheless, this change in peak heights did not affect the collection of peaks and so the system remains a valid model of the full-scale purification system.

3.2 Virus reduction

The results of virus-reduction studies are given in **Table 1**. The chromatographic process was very effective in reducing the level of Herpes Simplex virus and sindbis virus. In part this reduction is due to acid inactivation during elution of the antibody from the column. However, the reduction value demonstrated with polio virus was exclusively due to physical virus removal because this virus is acid-resistant. Virus-reduction was similar with both monoclonal antibodies. In addition, virus-reduction was not influenced by whether early or late runs in the life-time of the column had been tested.

Table 1 *Virus-reduction during Protein G Sepharose Column Chromatography.*

Antibody	Virus-reduction (log)[a]			
	Herpes	Sindbis	Polio-1	Vaccinia
IgG3	>6.0 (4.0-6.9)	5.4 (4.9-5.5)	2.9 (2.8-2.9)	nd
IgG1	6.7 (6.6-7.1)	5.1 (4.9-5.2)	3.5 (3.2-3.6)	1.7 (1.6-1.8)

[a] - Results given as the mean from 6 - 9 IgG3 runs or 2 - 3 IgG1 runs, range in brackets.
nd - not determined.

3.3 Protein G leakage

IgG3 acid eluates show much higher levels of Protein G leakage than IgG1 eluates **(Table 2)**. The IgG1 results are similar to the published data for leakage of Protein G or Protein A in the purification of MAbs[2,3], whereas the IgG3 results are considerably higher. In our hands, measured Protein G levels in IgG3 eluate samples increased with storage at 4°C. This effect is not seen in samples with very low IgG levels and samples from later stages in the purification scheme show levels of Protein G similar to those reported for IgG1 here. The levels in purified IgG3 are stable with storage. It is probable that the high levels seen here are an artifact of storage.

Table 2. *Level of ligand in eluate from Protein G Sepharose FF.*

	Run No.	Protein G conc. (ng/ml)	Protein G level (ng/mg IgG)
IgG3	3	956	2600
	25	497	1500
	50	458	1370
IgG1	1	<31.2	<16.2
	2	40.4	27.2
	25	<31.2	<25.2
	50	<31.2	<33.8

The levels of Protein G in the eluates did not show any trend with run number but in the unbound fraction **(Figure 1)** the IgG3 in particular shows a very clear decrease with increasing use of the gel. This implies that there is a population of immobilised Protein G which is loosely bound and is removed from the gel in early runs. Although IgG1 is more tightly bound to the gel and is eluted at a lower pH than IgG3, the results from IgG1 runs do not show such a clear trend. This difference could be due to the use of different batches of Protein G Sepharose in the two columns.

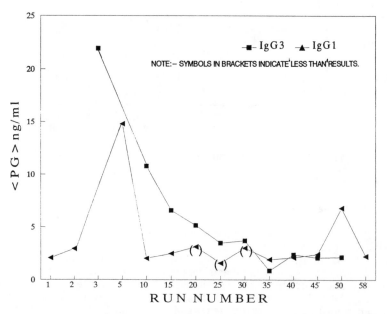

Figure 1. *Protein G levels in unbound fraction.*

3.4 Yield of IgG

The mean recovery of IgG1 from Protein G Sepharose was 91.8% over 59 runs. A graph of recovery with run number **(Figure 2)** shows no significant decrease in yield. The recovery of IgG3 was slightly lower with a mean of 76.9% over 63 runs. A graph of recovery with run number **(Figure 3)** shows the decrease in yield.

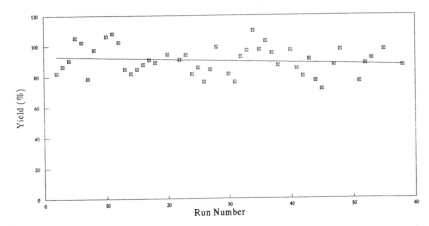

Figure 2. *Recovery of IgG1 from successive purifications on Protein G Sepharose FF.*

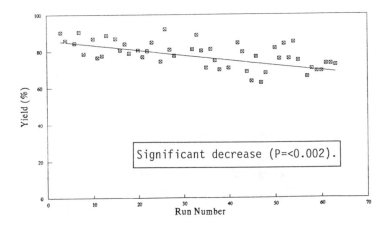

Figure 3. *Recovery of IgG3 from successive purifications on Protein G Sepharose FF.*

The dynamic capacity of the gel was measured by overloading the column approximately every 10th cycle and assaying the IgG recovery in fractions of the unbound material. The results for IgG1 are shown in **Figure 4.** Runs late in the series show a lower capacity for IgG as the amount loaded before significant breakthrough is seen is less. However, after 59 runs the total unbound IgG at a loading of 12mg/ml was 2%. The results for IgG3 are shown in **Figure 5.** Again the capacity for IgG decreases with run number with a total breakthrough of 6% at a loading of 8mg/ml in Run 49. These results show an acceptable gel performance throughout the validation series.

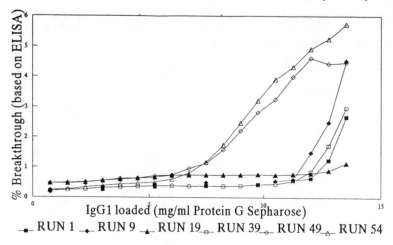

Figure 4. *Change in capacity of Protein G Sepharose FF for IgG1 with re-use.*

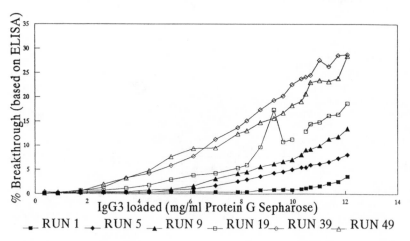

Figure 5. *Change in capacity of Protein G Sepharose FF for IgG3 with re-use.*

3.5 DNA clearance

The results for DNA clearance obtained in the series using the IgG3 harvest are shown in **Table 3**. They show that equivalent clearance of DNA is obtained during runs at different stages in the validation series. Variabilities in DNA loaded probably reflect differences we have seen in tissue culture supernatants produced at different stages in the continuous cell culture.

Table 3. *DNA clearance during purification of IgG3.*

Run No.	DNA loaded (μg)	DNA in product (ng)	(ng/mg IgG)	Log clearance
3	46.0	10.70	0.44	3.60
10	93.0	10.95	0.46	3.90
25	38.7	4.26	0.19	3.95
50	24.9	7.80	0.31	3.50

The results for IgG1 purification **(Table 4)** are similar to those above. Results were not obtained for runs later in this series.

Table 4. *DNA clearance during purification of IgG1.*

Run No.	DNA loaded (μg)	DNA in product (ng)	(ng/mg IgG)	Log clearance
2	34.0	1.4	0.44	4.40
11	41.6	6.5	0.13	3.80
26	17.6	13.3	0.39	3.10

4 CONCLUSIONS

This validation study demonstrates only the clearances over the first step of the purification scheme for these monoclonal antibodies. We consider that greater than 5 logs reduction of virus on a single step in a purification scheme shows effective clearance over that step. Viral clearance here is by combined acid inactivation and physical removal and the acid-labile viruses tested showed its effectiveness.

The DNA levels measured at this intermediate stage are such that clearance measured during subsequent steps in the purification would reduce them in the final product to the FDA recommended detection limit of <10 pg/dose. The measured Protein G levels in the IgG peak would give not more than 90ng in the final formulation. Again, further clearance would be expected during later stages of purification.

Over an economically acceptable gel life-time, this primary capture column consistently produces an eluate suitable for further purification for therapeutic use.

REFERENCES

(1) G.E. Chapman, P. Matejtschuk, J.E. More and P. Pilling, *Sep. for Biotech.*, 1990, **2**, 601.
(2) M. Godfrey, P. Kwasowski, R. Clift and V. Marks, *J. Immun. Methods*, 1993, **160**, 97.
(3) R. Francis, J. Bonnerjea and C. Hill, *Sep. for Biotech.*, 1990, **2**, 491.

Isolation Strategy for Identifying Bioactive Compounds from Fermentation Broth

Rossella Bortolo, Nunzio Andriollo, Emanuele Cauchi, and Giorgio Cassani[a]

ENICHEM ISTITUTO G. DONEGANI, VIA FAUSER 4, NOVARA, ITALY

[a]ENICHEM AUGUSTA, VIA REALI 4, PADERNO DUGNANO, MILANO, ITALY

During a screening programme for the identification of new fungicides, insecticides and herbicides from microbial origin, 5300 microorganism strains were isolated from soil samples collected in different world sides. The fermentation of microorganisms endowed of interesting biological activity led to the isolation of 40 bioactive compounds, which are listed in Table 1.

Table 1 Isolated bioactive compounds

Class

macrolide	Concanamycin A, Concanamycin B, Leucanicidin Macrotetrolydes complex, Bafilomycin A1, Bafilomycin C1 Neocopiamycin, Copiamycin, Antimycin complex, Oligomycin
macrocycle	Desertomycin, Desertomycin D*, Scopafungin, AB006 A*, AB006 B*
polyene	Flavofungin, Fungichromin, AB011 A*, AB011 B*, AB021 A*, AB021 B*, AB023 A*, AB023 B*
anthracycline	Cosmomycin A, Cosmomycin B, Cosmomycins complex, Ditrisarubicin.
nucleoside	Blasticidin S, 5 - Hydroxymethyl - Blasticidin S, Dealanyl - Ascamycin, Aristeromycin.
others	Conglobactin, Cycloheximide, Nigericin, Piericidins complex, Nocardamine, AB041*, Pseudobactin, Pseudobactin B*

*new antibiotics

The isolation strategy is outlined in Figure 1.

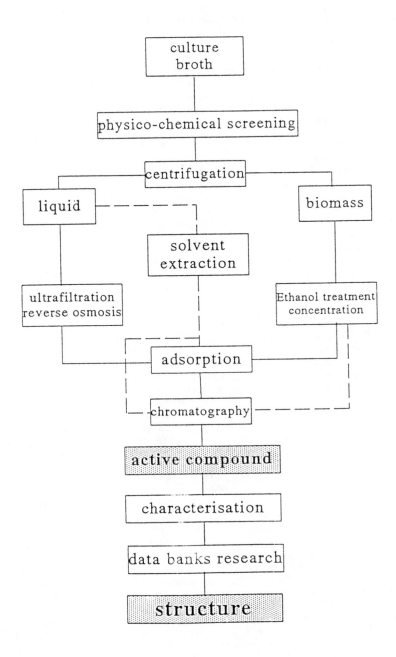

Figure 1 General isolation strategy

Fermentation broths contained by weight approximately 80-95% water, 1-6% biomass, 0-5% nutrient residues (sugars, salts, starch etc.), 1-5% macromolecular compounds (proteins, polysaccharides) and only 0.001-0.03% (1-30 mg/L) active compounds.

Since the goal of the research programme was to discover new antibiotics, the major task was to determine at the earliest possible stage the novelty of the compound and decide if it was worthy to carry on the purification procedure: for this purpose, the compound had to be isolated in a form suitable for identification by analytical methods and therefore as pure as possible.

Since the only recognizable characteristic of the unknown antibiotic was its biological activity, each step of purification had to be verified by biological activity tests; to avoid solvents and salts interferences, that could have questioned the validity of the test, a very laborious sample treatment was often necessary: the definition of a standard biological activity test was the first key-step of an isolation programme of new active compounds.

The second step was to determine the physicochemical characteristics of the active compound (stability towards pH, temperature and light, solubility, solvent extractability and other) by means of preliminary studies on small aliquots of broth. Such a physicochemical screening was decisive to plan out the purification strategy. For example, if the active compound was water soluble, the broth was clarified by ultrafiltration (UF): besides removing undesirable macromolecular compounds, such a cleaning step significantly extended the life of the columns used for the following chromatographic purification and improved the reproducibility of the process. However, this was not a general rule: for example, it was impracticable to clarify the broth of the polyene antibiotics AB021a and AB021b[1] (mw=1139 and 1165 respectively) with UF techniques because, unlike other similar compounds, these two antibiotics formed micellar aggregates which were retained by UF membranes even with a molecular weight cut off of 35 kDa.

The final purification of the active compounds was usually achieved by chromatography, which in all cases provided the purity required for the investigation of the antibiotic structure by the usual analytical techniques, such as Mass spectrometry, NMR, UV, IR spectroscopy and elemental analysis.

Various chromatographic techniques were used, (normal and reverse phase, ion pairing, GPC, SEC, ion exchange) depending on the physicochemical characteristics of the active compound, drawn from the physicochemical screening. Initially, the chromatographic conditions where assessed with a trial and error method, by using small home-made columns, commercially available SPE columns or TLC plates. The results were scaled up to preparative scale and were optimized by HPLC. At this stage, a photodiode array detector was a very useful tool to identify an active compound even in complex mixtures: in fact, certain classes of substances, such as the polyenes, have a characteristic spectrum and could be identified in an early stage of the purification process.

Once isolated, the antibiotics were characterized and on the basis of the resulting preliminary data (molecular weight, minimum formula, functional groups, short sequences of structure), the on-line International Data Banks were queried. Whenever the results of the query were not satisfactory, a thorough characterisation of the structure was carried out by means of MS and NMR techniques, as in the case of the new antibiotics AB023a and AB023b.[2]

Examples

The first example[3] concerns a very efficient recovery method that led to the complete resolution of Pseudobactin from Pseudobactin B, whose succinic acid moiety is replaced by a succinamide group (Figure 2).

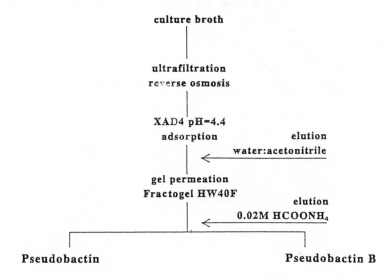

Figure 2 Structure of Pseudobactin and Pseudobactin B
        Z = -NH$_2$ Pseudobactin
        Z = -OH   Pseudobactin B

Pseudobactin B is a new siderophore, never described before, produced together with Pseudobactin by the microorganism Pseudomonas PD30. Pseudobactin was isolated by Teintze et al. (1981)[4], but the described method required many purification steps and provided the complex of Pseudobactin with Fe(III), which subsequently needed to be removed[5].
In our case, a new simple method for the recovery and the separation of Pseudobactin and Pseudobactin B from fermentation broth is described (Figure 3): the key-step was the gel permeation chromatography, carried out on Fractogel HW40 (F) and eluted with 0.02 M ammonium formate.

culture broth

|

ultrafiltration
reverse osmosis

|

XAD4 pH=4.4
adsorption   **elution**
                    ←   **water:acetonitrile**

gel permeation
Fractogel HW40F
                                  elution
                    ←   **0.02M HCOONH$_4$**

Pseudobactin   **Pseudobactin B**

Figure 3 Isolation procedure of Pseudobactin and Pseudobactin B.

The isolation of two structurally similar nucleoside antibiotics,
Blasticidin S and 5-Hydroxymethyl-Blasticidin S, is reported in the
second example[6]. The two antibiotics only differ in a hydroxymethyl group
on the cytosine ring; they were produced by the same microorganism and
showed a similar biological activity (Figure 4).

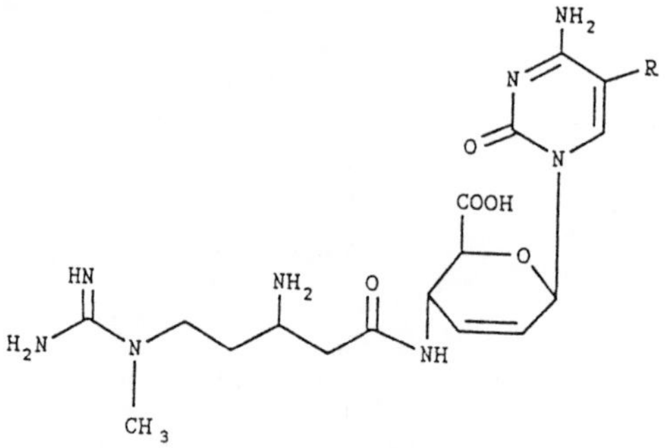

Figure 4 Structure of Blasticidin S and 5-Hydroxymethyl-Blasticidin S.
            R = H   Blasticidin S
            R = CH₂OH   5-Hydroxymethyl-Blasticidin S

Althought the first tentative approach to the purification of these
compounds from the culture broth was rather complex, including reverse
phase chromatography, adsorption, ion exchange column and a desalting
step (Figure 5.a), it was not possible to obtain a pure compound.
Nevertheless, when a HPLC peak could be correlated with the biological
activity, the isolation procedure could be optimized: in particular, HPLC
analysis showed that the active peak was constituted of two overlapped
peaks, that could be easily resolved by modifying the pH. A very quick
procedure of purification was planned: the ion exchange step was replaced
with a reverse phase column eluted at pH=3 and the complete resolution of
the two antibiotics was obtained (Figure 5.b).
The isolation of both Blasticidin S and 5-Hydroxymethyl-Blasticidin S was
reported by Larsen et al. (1989)[7], but the separation of the mixture into
individual antibiotics was accomplished only by using semi-preparative
HPLC. In our example, Blasticidin S and 5-Hydroxymethyl-Blasticidin S
could be easily separated in large scale.

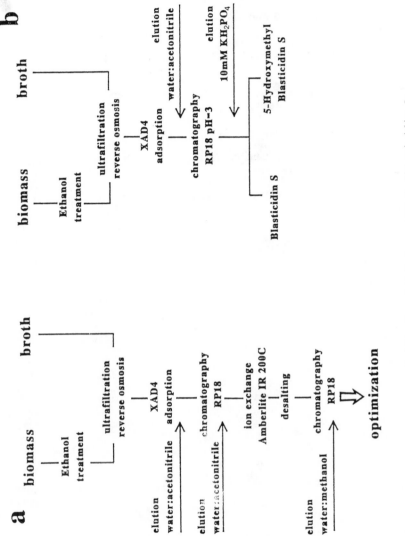

Figure 5 Isolation of Blasticidin S and 5-Hydroxymethyl-Blasticidin S:
a) preliminary purification procedure
b) optimized purification procedure

REFERENCES

1. D.Cidaria, N.Andriollo, G.Cassani, E.Crestani, S.Spera, C.Garavaglia, G.Pirali, G.Confalonieri, EP 371509, 1989

2. R. Bortolo, S. Spera, G. Guglielmetti, G. Cassani, J. Antibiotics, 1993, 46, 255-264.

3. N.Andriollo, A.Guarini, G.Cassani, J. Agric. Food Chem., 1992, 40, 1245-1248.

4. M. Teinze, M. B. Mossain, C. L. Barnes, J. Leong, D. Van der Helm, Biochemistry, 1981, 20, 6446-6457.

5. J. M. Meyer, M. A. Abdallah, J.Gen.Microbiol., 1978, 107, 319-328.

6. A.Scacchi, R.Bortolo, G.Cassani, G.Pirali, E.Nielsen J.Plant Growth Regul., 1992, 11, 39-46

7. S.H. Larsen, D.M. Berry, J.W .Paschal, J.M.Gillian J.Antibiotics, 1989, 470-471.

Engineering and Production in Bacterial Thermo-inducible Systems of Chimeric β-Galactosidase Fusion Proteins

A. Benito, J. L. Corchero, P. Vila, and A. Villaverde

INSTITUT DE BIOLOGIA FONAMENTAL AND DEPARTAMENT DE GENÈTICA I MICROBIOLOGIA, UNIVERSITAT AUTÒNOMA DE BARCELONA, E-08193 BELLATERA, BARCELONA, SPAIN

1 INTRODUCTION

Mammalian and viral proteins are often toxic for bacterial cells when produced in such recombinant expression systems. This can result in failures during scaled–up production, involving both plasmid loss and cell death. However, experience in bacterial cell growth and metabolism largely recommends their use for small, medium and large scale production whenever possible[1]. Moreover, plasmid development in expression vectors during the last decade has provided a wide spectrum of production strategies to be assayed for each particular protein[2]. To overcome toxicity of these proteins, regulable promoters are used to maintain the product concentration below the critical threshold. In addition, the obtaining recombinant fusion proteins also reduce the undesirable toxic effects. The presence of homologous domains makes them better tolerated by cells and folding is more correct[3,4]; as a result they become more resistant to bacterial proteases. On the other hand, the presence of an extra domain permits protein engineering to gain new properties, which could be useful during the production and downstream processes. The most desirable qualities conferred by the homologous tag to the product are the ability of being monitored and quantified during fermentation and the possibility of a single step purification by an universal protocol, irrespective of the nature and sequence of the heterologous part in the chimeric protein[5,6].

One of the most widely used homologous tags is the *Escherichia coli* β-galactosidase. In this paper, we present our results on the engineering of the *lacZ* gene to be expressed in a thermo–inducible, tightly controlled bacterial system. We also evaluate the ability of the resulting mutant proteins to be used as monitoring and purification tags for both recombinant peptides and complete proteins from foot–and–mouth disease virus (FMDV).

2 CHOOSING THE EXPRESSION VECTOR

A temperature inducible expression strategy was chosen to produce mutant β-galactosidases. We selected the plasmid pJLA602[7], because the combination of p_L and p_R lambda promoters placed in tandem permits a strong transcription of the downstream cloned genes. The presence of the *cI857* gene encoding a thermo– sensitive repressor for the promoters enables this vector to be used in any *E. coli* strain. A *lacZ* gene amplified by PCR was cloned in the polylinker of pJLA602, resulting in 2 new plasmids expressing *lacZ* under the control of lambda promoters: pJLACZ[8] and pJCO46[9]. These vectors only differ in some restriction sites at the 5' extreme of *lacZ* gene, and direct the synthesis of the enzyme with similar efficacy after heat induction to 42ºC (**Figure 1**).

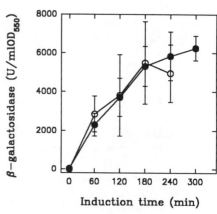

Figure 1. β–Galactosidase activity in MC1061/pJLACZ (empty circles) and MC1061/pJCO46 (filled circles) cultures after induction. Cells were grown in LB medium at 28ºC and at 250 rpm until the cultures reached an OD_{550} of about 0.4. Then, they were transferred to a water bath previously warmed to 42ºC, and further incubated at the same rpm. Enzymatic activity was calculated by Miller's method after chloroform– mediated cell lysis. Results are the mean and standard deviation of 3 independent experiments.

The sharpness of response of this system to temperature control was explored in MC1061/pJLACZ strain. The use of a single step continuous culture[10] allowed the study of expression levels at different temperatures and at a constant dilution rate of 0.1 h⁻¹. By assaying induction temperatures from 30 to 42ºC, increasing values of enzymatic activity were obtained (**Figure 2**). These data revealed an exponential behaviour of CI857–controlled gene expression, and a very efficient repression at temperatures below 32ºC. The constancy of the expression levels at each temperature and the strength of gene expression at 42ºC, makes this system very appropriate for further production experiments. Moreover, plasmid loss was undetected in MC1061/pJLA602 after 200

Figure 2. β–Galactosidase activity in a MC1061/pJLACZ continuous culture after induction at increasing temperatures. Cells were growing in CAM9 medium with glucose and ampicillin, at a dilution rate of 0.1 h⁻¹. More details about the experiments are given elsewhere[10].

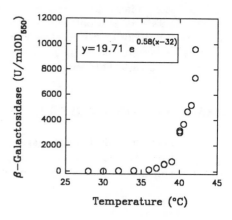

$$y = 19.71 \, e^{0.58(x-32)}$$

generations without ampicillin, at both 28 and 37ºC. In MC1061/pJCO46 cells, a nearly constant, moderate appearance rate of plasmid-free cells (pfc) relative to total viable cells (cfu) was estimated to be of 0.0041 pfc/cfu/generation at 28ºC and of 0.0057 pfc/cfu/generation at 37ºC, during about 150 generations.

3 PROPERTIES OF CHIMERIC VP1-β-GALACTOSIDASE PROTEINS

Several constructs, derivatives of pJLAC and pJCO46, were designed to produce the bacterial enzyme carrying the immunodominant region of foot–and–mouth disease virus (FMDV) VP1 protein, the site A[11]. In the pAB1 product[12], 55 amino acids of VP1 (containing the immunodominant site A) were linked to the amino terminus of β–galactosidase. This fusion protein was studied to determine its specific enzymatic activity. By comparing several samples of induced cell extracts, a linear dependence between enzymatic activity and amounts of fusion protein was found with a correlation coefficient of 0.989. This enabled us to reliably monitor the production of the protein by Miller's manual method, and encouraged us to develop a flow injection analysis (FIA) strategy applied to the on–line monitoring of β–galactosidase fusion proteins[12,13,14]. Moreover, in our hands, commercial β–galactosidase immuno–affinity columns from Promega™ worked with similar efficacy when comparing recombinant β–galactosidase produced by pJLACZ and the pAB1 product. The estimated binding ability of the column was 1.8 mg/ml bead for β–galactosidase and 1.55 mg/ml bead for AB1. The inactivation of both proteins after elution according the protocol given by the manufacturer was also similar and represented about 65 % of the total units for β–galactosidase and 60% for AB1.

Subsequently, we decided to investigate the ability of the β–galactosidase tag to accept large protein domains at both extremes, or small peptides in inner regions. Predicted surface–exposed inner regions of the enzyme were assayed for their ability to carry a 27 amino acid stretch reproducing the site A of FMDV, without a complete loss of enzymatic activity. Among the 6 regions explored we identified one, involving amino acids 275 and 276 of the enzyme, which was able to maintain activity after the insertion. The recombinant plasmid encoding for such a protein, pM275VP1, was introduced in MC1061 cells and the resulting protein further characterized, revealing a reduced thermal stability provoked by the insert **(Table 1)**.

Table 1. Relative enzymatic activity (%) after incubation at different temperatures during 60 min.

	β–Galactosidase	M275VP1
28ºC	100	100
42ºC	95.2	53.8
48ºC	91.7	40.8

On the other hand, the entire FMDV VP1 protein (of about 23 kDa) was joined to the native enzyme alternatively by its amino and carboxy termini[15]. All the encoding plasmids (pJVP1LAC, pJLACVP1 and pJV2LAC respectively) produced active proteins. Features of all these products regarding their structural and biotechnological properties are summarized in **Table 2**.

Table 2. Properties of several chimeric FMDV-β-galactosidase fusion proteins.

	β-Gal	AB1	M275VP1	VP1LAC	LACVP1	V2LAC
Molecular mass of monomers (kDa)	116	122	118	139	139	162
Specific activity (U/μg)	340	357	140	360	1040	ND
Recognition by SB10[a]	–	+	+	+	+	–
Recognition by anti-β-galactosidase serum	+	+	+	+	+	–
Recognition by anti-β-galactosidase MAb[b]	+	+	+	+	+	+/–
Yield in batch[c](U/ml)	10,725	2,625	6,475	4,420	10,200	850
Yield in fed-batch (U/ml)[c]	90,687	ND	46,575	ND	ND	ND

[a] Reactivity in Western blot analysis. Symbols are + (clear, specific bands), – absence of specific bands and +/– (weak, but specific bands)

[b] Anti-β-galactosidase monoclonal antibody was obtained from Boehringer Mannheim

[c] Maximum levels obtained in any of the performed experiments

ND Not studied

4 DISCUSSION

β-Galactosidase fusion proteins carrying the complete FMDV VP1 protein or segments of it, have been designed, constructed and produced in an *E. coli* thermo-inducible system. The stability of the vector and the sharpness of gene expression control by CI857 repressor are appropriate for laboratory-scale production experiments. Both ends of the enzyme are able to accept the complete viral protein without dramatic loss of specific activity. In the case of the LACVP1 protein, underestimation of the amounts of recombinant protein in Western blot analysis could have caused an apparently enhanced specific activity **(Table 2)**. On the contrary, several attempts to insert a peptide reproducing a VP1 immunodominant stretch inside the enzyme failed to provide active proteins, with the exception of the insertion located between position 275 and 276 of the enzyme. However, this chimeric protein has a reduced specific activity **(Table 2)** and it is more sensitive to high temperatures **(Table 1)**. The permissive identified site seems to be placed in a dispensable region, as suggested from comparative sequence analysis[16,17]. In any of these cases, toxicity of the products or extensive degradation after induction have been detected, and a correspondence between protein concentration and enzymatic units has been always confirmed. The identification of this permissive site opens a possibility for further engineering of β–galactosidase to be employed as a substrate in analytical assays. The insertion of large foreign peptides in this internal exposed site of the carrier protein could be used to detect and quantify biological properties of enzymes, antibodies and other active molecules in the way done by Baum and coworkers to analyze HIV protease activity[18].

On the other hand, the alternative use of both termini of the enzyme can provide more flexibility to solve problems related with the folding and the activity of the heterologous region in the fusion protein. However, the protein containing two copies of complete VP1 (V2LAC) is produced at very low amounts (**Table 2**). The presence of correct both *VP1* nucleotide sequences in the vector have been verified, but we have been unable to detect the product using anti–VP1 monoclonal antibodies. Moreover, some plasmid loss has been evidenced in MC1061/pJV2LAC cultures even in batch experiments in the presence of the selective antibiotic (not shown). Altogether, these results suggest that this product (or the encoding plasmid) is unstable in *E. coli* cells. Therefore, though the increase in molar ratio between heterologous and homologous parts could be desired regarding yield in a production process, we have renounced the 'both–side' strategy for further constructions of chimeric β–galactosidases.

ACKNOWLEDGEMENTS

We thank X. Carbonell for critical reading of the manuscript. This work has been supported by grant BIO92–0503 from CICYT, Spain. A. Benito and J.L. Corchero are recipients of predoctoral fellowships from MEC, Spain.

REFERENCES

1.   A.R. Shatzman, *Cur. Op. Biotech.*, 1993, *4*, 517.
2.   P.O. Olins and S.C. Lee, *Cur. Op. Biotech.*, 1993, *4*, 520.
3.   L. Strandberg and S.–O. Enfors, *Appl. Environ. Microbiol.*, 1991, *57*, 1669.
4.   J. Buchner, I. Pastan and U. Brinkmann, *Anal. Biochem.*, 1992, *205*, 263.
5.   R. Sherwood, *TIBTECH.*, 1991, *9*, 1.
6.   H.M. Sassenfeld, *TIBTECH.*, 1990, *8*, 88.
7.   B. Schauder, H. Blöcker, R. Frank and J.E.G. McCarthy, *Gene*, 1987, *52*, 279.
8.   A. Benito, M. Vidal and A. Villaverde, *J. Biotechnol.*, 1993, *29*, 299.
9.   P. Vila, J. L. Corchero, A. Benito and A. Villaverde, submitted for publication.
10.   A. Villaverde, A. Benito, E. Viaplana and R. Cubarsí, *Appl. Environ. Microbiol*, 1993, *59*, 3485.
11.   M.G. Mateu, M.A. Martínez, L. Capucci, D. Andreu, E. Giralt, F. Sobrino, E. Brocchi and E. Domingo, *J. Gen. Virol.*, 1990, *71*, 629.
12.   A. Benito, F. Valero, F.J. Lafuente, M. Vidal, J. Cairó, C. Solà and A. Villaverde, *Enzyme Microb. Technol.*, 1993, *15*, 66.

13.   J. Cairó, M. Vidal, A. Villaverde, F. Valero, F.J. Lafuente and C. Solà, *Biotech. Tech.*, 1991, *5,* 389.

14.   F. Valero, F.J. Lafuente, C. Solà, A. Benito, M. Vidal, J. Cairó and A. Villaverde, *Biotech. Tech.* 1992, *6,* 213.

15.   J.L. Corchero, A. Benito and A. Villaverde, submitted for publication.

16.   M. Hood, A.V. Fowler and I. Zabin, *Proc. Natl. Acad. Sci. USA*, 1978, *75,* 113.

17.   O. Poch, H. L'Hôte, V. Dallery, F. Debeaux, R. Fleer and R. Sodoyer, *Gene,* 1992, *118,* 55.

18.   E.Z. Baum, G.A. Bebernitz and Y. Gluzman, *Proc. Natl. Acad. Sci. USA*, 1990, *87,* 10023.

Isolation of Human α-Atrial Natriuretic Factor (α-hANF) from Recombinant Fusion Protein

V. Berzins,[a] V. Baumanis, A. Skangals, I. Mandrika, and I. Jansone

[a]DEPARTMENT OF BIOCHEMISTRY AND MOLECULAR BIOLOGY; BIOMEDICAL RESEARCH AND STUDY CENTRE, UNIVERSITY OF LATVIA, KRONVALDA 4, LVI842 RIGA, LATVIA

INTRODUCTION

The small cardiac peptide hormone - α-human atrial natriuretic factor (α-hANF) is involved in the homeostasis of blood pressure, fluid volume and vascular function[1]. Therefore, the predominant circulating form of α-hANF a 28-amino-acid oligopeptide is expected to be useful in clinical treatment of hypertension and congestive heart failure as a hypotensive and natriuretic factor.

In order to facilitate the production of human α-atrial natriuretic factor for clinical studies, one way is to use recombinant DNA techniques to express the α-hANF gene in Escherichia coli. For minimizing the intracellular proteolytic degradation of short peptide hormone, the α-hANF has been synthesized as a fusion protein either with the protective polypeptide Cd-LH[2] or with CAT[3] or with β-galactosidase[4] and then purified fusion protein was subjected to cleavage in vitro by specific endopeptidases. We have used this approach for construction of vector, expression of hybrid gene and purification of recombinant fusion protein containing bacteriophage fr 129-amino-acid coat protein linked to α-hANF via tripeptide Ile-Asp-Lys sequence designed as a specific cleavage site for Lys-C endopeptidase digestion[5].

Here we describe the cleavage conditions for the release of a α-hANF from the fusion protein together with the subsequent purification and characterization of the target peptide.

RESULTS

Production of phage fr coat-α-hANF fusion protein.

As an expression vector we chose recombinant plasmid pFAN-15, which carries the hybrid of bacteriophage fr coat protein and α-hANF structural genes placed under the control of the inducible E. coli tryptophan promoter and effective translation initiation region (TIR) of phage fr coat protein (**Fig. 1**). The selection of phage fr coat protein as a protector polypeptide of a α-hANF (sequence hybrid see **Fig. 2**) was influenced by the following considerations:
- expression vector pFAN-15 is derivative of plasmid pFRC 8-1-33 which overproduced fr coat protein in E. coli. Newly synthesized fr coat protein is very stable in host cells and may serve as efficient carrier.
- we can easily detect the fusion protein during manipulation by Western immunoblotting and by immunoprecipitation using a fr coat protein polyclonal antisera.

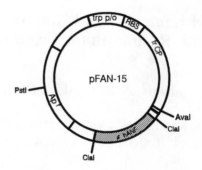

Figure 1. *Restriction site and function map of expression vector pFAN-15.*

1 MASNFEEFVL VDNGGTGDVK VAPSNFANGV AEWISSNSRS QAYKVTCSVR
51 QSSANNRKYT VKVEVPKVAT QVQGGVELPV AAWRSYMNME LTIPVFATND
101 DCALIVKALQ GTFKTGNPIA TAIAANSGIY IDKSLRRSSC FGGRMDRIGA
151 QSGLGCNSFR Y

Figure 2. *Amino-acid sequence for a fr coat-α-hANF hybrid protein. Linker sequence - Lys-C proteolytic cleavage site is underlined.*

The phage fr coat protein-α-hANF fusion protein was purified to near homogeneity by ion exchange chromatography on DEAE-, CM-cellulose, QAE Sephadex A25 columns and selective precipitation at pH9.0[5].
Nevertheless a considerable degree of purification can be achieved already at the CM-cellulose step (See **Table 1** and lane 5 at **Fig. 3**) and fusion protein of this purity was routinely used as substrate in the subsequent cleavage and α-hANF purification steps.

Table 1. *Purification of recombinant fr coat-α-hANF fusion protein.*

Purification stage	Total protein (mg)[a]	Fusion protein content		Yield %
		as % of total protein[b,c]	(mg)	
Lysate of cells (10 g wet weight)	826	18	148.7	100.0
DEAE-cellulose (flow through)	421	26	109.5	73.6
CM-cellulose (fraction 2)	109	90	98.1	66.0
Sephadex G100 (pooled fraction)	92	96	88.3	59.4

a Protein concentration was determined by Coomassie Blue G250 binding assay.
b,c Fusion protein contents were estimated by double radial imunodiffusion and by scanning of Coomassie Blue - stained SDS-polyacrylamide gels, respectively.

Lys-C endopeptidase cleavage.

The fusion protein from CM-cellulose column (fraction 2 in 75 mM ammonium acetate, pH5.0) comprising fr coat-α-hANF protein was dialysed against 1 mM acetic acid, pH4.0, for 20 hours at 4 °C. Approximately 1.2 mg of fusion protein in 2.4 ml of reaction volume was mixed with Lys-C endopeptidase (Boehringer Mannheim) at an enzyme to substrate ratio 1:240 (w/w) and incubated at 20 °C:
a - for 25 min at pH5.2 in 0.5 mM acetic acid, 20% acetonitrile, 1 mM Tris and 0.2 mM EDTA;

b - for 20 min at pH5.2 (conditions (a)) and then for 120 min at pH8.5 in 26 mM
    Tris-HCl, 1.2 mM EDTA;
c - for 120 min at pH8.5 in 25 mM Tris-HCl, 1 mM EDTA.

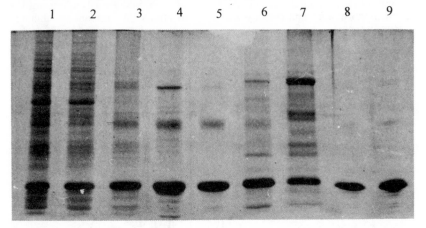

Figure 3. *SDS containing 12%-23% gradient polyacrylamide gel electrophoretic
analysis of fr coat-α-hANF fusion protein purification. Lane: 1 - solubilised E.
coli K802/pFAN-15 cells; 2 - lysate of cells; 3 - DEAE-cellulose fraction (flow
through); 4-7 - fractions 1, 2, 3 and 4 from CM-cellulose; 8, 9 - pooled fraction
from Sephadex G100.*

The cleavage reaction was stopped by addition of 1 vol of 2%
SDS-2% β-mercaptoethanol solution and boiling for 5 min or by addition of TFA
to final concentration 0.2% (v/v). The reaction mixtures were analysed by
SDS-PAAG electrophoresis and α-hANF was resolved from other cleavage
reaction peptides on a reverse phase HPLC.

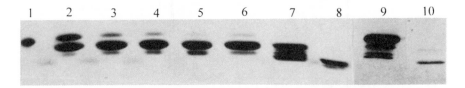

Figure 4. *SDS-PAAG electrophoretic analysis of fusion protein cleavage with
Lys-C endopeptidase.*

The SDS-gradient PAAG (12%-23%) in **Fig. 4** shows a representative time course
for the digestion of recombinant fusion protein with Lys-C endopeptidase, an
enzyme which specially cleaves on the C-terminal side of lysine residues. Lane 1
represents the zero time point while lanes 2 - 6 represent samples taken after 1, 5,
10, 20 and 25 min incubation at conditions (a); lanes 7, 8 show the effect of Lys-C
endopeptidase action on fusion protein at conditions (b) for 21 and 140 min; in
lanes 9 and 10 are samples from precipitated and soluble fractions of fusion
protein digested at conditions (c) for 120 min, respectively. At zero time, the
17.3 kDa band of recombinant fusion protein is the only prominent band observed.
Under given conditions (a) of cleavage (20% acetonitrile in reaction mixture,
pH5.2) which was not optimal for efficient Lys-C endopeptidase action, but

prevents the hydrophobic fusion protein from aggregation, the band of starting material was very quickly reduced and a new major band of about 14 kDa polypeptide was been seen and short peptides, including α-hANF using Laemli conditions of electrophoresis are difficult to observe. Simultaneously aliquots of the reaction mixture were analysed by HPLC to determine the amount of liberated

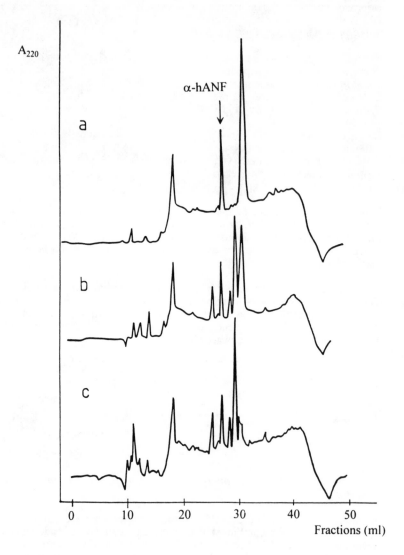

Figure 5. *Reverse phase HPLC separation of Lys-C endopeptidase cleaved fr coat-α-hANF fusion protein.*
a - 30 μg fusion protein digested for 25 min at pH5.2 under conditions (a);
b - 30 μg fusion protein digested for 20 min at pH5.2 under conditions (a) and then 120 min at pH8.5 under conditions (b);
c - 30 μg fusion protein digested for 120 min at pH8.5 under conditions (c).

α-hANF. It has been demonstrated that in the early minutes of cleavage about 40% of the maximal yield of α-hANF were released. Extension of time of incubation to 25-35 min had not an appreciable effect on the cleavage yield. A change to conditions (b) and (c) of reaction mixture also did not lead to further increases in the accumulation of α-hANF (**Fig. 5**). Therefore, the conditions (a) were chosen for processing of the fusion protein and release of α-hANF.

Isolation of the recombinant α-hANF.

The reaction mixtures were analysed and recombinant α-hANF was isolated from other cleavage reaction peptides on a reverse phase HPLC UltroPack TSK-120T column using a GTi system (Pharmacia LKB Instruments). Following application of the cleavage products to the column a gradient of increasing acetonitrile gradient (0% to 60% v/v) in 0.1% TFA was applied to resolve the peptides (**Fig. 5**).
The recombinant α-hANF eluted from column was superimposed and quantitated with authentic synthetic α-hANF(Boehringer Mannheim).The peak with α-hANF was collected and lyophilized twice to remove any TFA and acetonitrile and then resolubilized in water. The resultant α-hANF peptide was about 90% free from other protein fragments and was useful for further biological tests.
Although increase of 10 - 20 times of amount of cleavable fusion protein and volume of reaction mixture offers no difficulties, the conditions (a) of cleavage, certainly, are not suitable for large scale preparation of α-hANF. It may be advantageous for industrial purposes to use recirculation of solution of fusion protein through the membrane or resin system containing immobilized endopeptidase Lys-C. After cleavage with immobilized Lys-C, α-hANF can be separated from other cleavage products by ultrafiltration and second cation exchange chromatography or preparative scale reverse phase high-performance liquid chromatography and lyophilized to remove organic solvents.

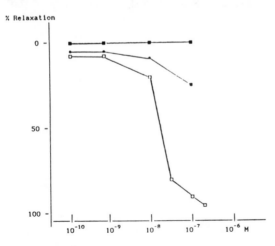

Figure 6. *Vasorelaxant activity of recombinant α-hANF purified on HPLC column. Concentration - response curves for relaxation of precontracted rabbit aorta rings induced by increasing concentrations of recombinant α-hANF.*
- recombinant α-hANF, n=8;
- fr CP cleaved with the Lys-C endopeptidase, n=6;
* - uncleaved phage fr coat-α-hANF fusion protein, n=5.

The biological activity of the recombinant α-hANF.

The activity of the recombinant α-hANF was assayed in a preliminary *in vitro* test. It showed vasorelaxant, antihypertensive action on rabbit aorta contracted by norephedrine similar to that of synthetic α-hANF (Boehringer Mannheim) **(Fig. 6)**. It should be noted that uncleaved fr coat-α-hANF fusion protein exhibits partial relaxant activity that can be explained by gradual digestion of fusion protein and liberation of functionally active α-hANF due to the action of aortal wall proteases.

SUMMARY

1. A method of purification of recombinant fusion protein containing human α-atrial natriuretic factor (α-hANF) joined to bacteriophage fr coat protein via Lys-C endopeptidase cleavable linker is described.
2. Optimised cleavage conditions resulting in release 40-50% of the maximal yield of α-hANF from the fusion are elaborated.
3. The homogenous and biologically active recombinant α-hANF from the Lys-C endopeptidase digests was purified by reverse phase high-performance liquid chromatography (HPLC).

REFERENCES

1. A. Rosenzweig and C.E. Seidman, Annu. Rev. Biochem., 1991, 60, 229.
2. Y. Saito, Y. Ishii, S. Koyama et al., J. Biochem., 1987, 102, 111.
3. C.W. Dykes, A.B. Bookless, B.A. Coomber et al., Eur. J. Biochem., 1988, 174, 411.
4. H. Sachse, G. Hagendorft, K.D. Prenss et al., Nucleosides and Nucleotides, 1988, 7, 61.
5. V. Berzins, I. Jansone, A. Skangals et al., J. Biotechnology, 1993, 30, 231.

Large-scale Production and Purification of L-Phenylalanine Dehydrogenase from *Nocardia SP.* 239

G. M. Brearley, C. P. Price,[a] R. S. Campbell,[a*] T. Atkinson, and P. M. Hammond

DIVISION OF BIOTECHNOLOGY, PHLS, CAMR, PORTON DOWN, SALISBURY, WILTSHIRE, UK

[a]DEPARTMENT OF CLINICAL BIOCHEMISTRY, THE LONDON HOSPITAL MEDICAL COLLEGE, TURNER STREET, LONDON, UK

[*]PRESENT ADDRESS – PORTON CAMBRIDGE LTD, KENNETT, NEWMARKET, CAMBRIDGESHIRE, UK

1 ABSTRACT

Alterations to the composition of the growth medium which included increasing the concentration of the primary carbon source, L-phenylalanine, and the addition of a vitamin solution significantly improved the growth levels of *Nocardia* sp. 239. Consequently, culture yields of the inducible, intracellular enzyme, L-phenylalanine dehydrogenase, were increased approximately 9-fold. The enzyme in a cell-free extract was bound to the biomimetic dye ligand, Procion Red HE-3B and was biospecifically eluted with the enzyme's cofactor, NADH. This simple, single step procedure gave homogeneous enzyme, as determined by a single protein band on a Coomassie blue-stained SDS-PAGE gel. The cultivation and purification techniques were both scaled-up and permitted the large-scale production of pure enzyme.

2 INTRODUCTION

L-phenylalanine dehydrogenase (L-phenylalanine: NAD^+ oxidoreductase, deaminating, EC 1.4.1.-) catalyses the reversible reaction:

$$L\text{-phenylalanine} + H_2O + NAD^+ \rightleftharpoons$$

$$\text{Phenylpyruvate} + NH_3 + H^+ + NADH \quad (1)$$

The enzyme has considerable potential in the field of biotechnology, both as a diagnostic reagent for the detection and monitoring of the inherited metabolic disorder, phenylketonuria and for the synthesis of L-phenylalanine[1] which is becoming increasingly significant due to its inclusion in the artificial sweetener, Aspartame.[2]

Phenylalanine dehydrogenase from *Nocardia* sp. 239 has been purified and partially characterised by de Boer *et al.*[3] We describe alterations to the growth conditions resulting in an approximate 9-fold increase in biomass (and total enzyme) and the scale-up from 8 litre fermentation to 400 litre. After cell disruption, the extract was clarified by centrifugation and the enzyme purified to homogeneity in a single column step by pseudo-affinity chromatography on a Procion Red HE-3B/Sepharose CL-4B matrix.[4]

3 MATERIALS AND METHODS

3.1 Materials

Chromatography matrices (Mono-Q, Sepharose CL-4B) and Phastgel media were purchased from Pharmacia-LKB. Procion Red HE-3B was a gift from ICI, Blackley. Mazu DF8005 antifoam was from PPG Speciality Chemicals, Manchester UK. All other chemicals were purchased from Sigma.

3.2 Microorganism and Cultivation

A sample of the original organism, *Nocardia* sp. 239[3], was generously donated by Dr. L. Dijkhuizen, University of Groningen, Haren, The Netherlands. A high yielding strain of the organism has been deposited by us at the NCIMB, number 40590 and was used throughout this work. All media were sterilised prior to inoculation by autoclaving at 121°C and 15 psi for 20 minutes and adjusted to pH 7.0 unless stated otherwise. The bacterium was grown aerobically in liquid cultures at 37°C, originally in the medium described by de Boer *et al.*[3] and referred to in this paper as **Medium O**, which contained the following (g per litre demineralised water): K_2HPO_4 1.0, $(NH_4)_2SO_4$ 1.5, $MgSO_4.7H_2O$ 0.2, L-phenylalanine 1.65 and 0.2ml/l of the trace element solution of Vishniac and Santer.[5]

An improved medium (**Medium I**) contained (g per litre of demineralised water): K_2HPO_4 4.0, KH_2PO_4 1.0, NH_4Cl 1.0, $CaCl_2.2H_2O$ 0.01, K_2SO_4 2.6, NaCl 1.0, L-phenylalanine 16.5 and Mazu DF8005 antifoam 0.5 ml/l. The following additions (filter sterile) were made post-autoclaving: 1.0 M $MgCl_2.6H_2O$ 1.0 ml/l, trace element solution 10 ml/l and vitamin solution 1.0 ml/l. The trace element solution contained (g per litre of demineralised water): $FeCl_3.6H_2O$ 3.9, $ZnSO_4.H_2O$ 0.58, $CoCl_2.6H_2O$ 1.0, $Na_2MoO_4.2H_2O$ 1.0, $CuSO_4.5H_2O$ 1.16, H_3BO_3 0.3, $MnSO_4.7H_2O$ 0.72, HCl (conc.) 7.2 ml/l. The vitamin solution contained (mg per litre of demineralised water): pyridoxin 0.125, biotin 0.125, thiamine 0.025, nicotinic acid 1.875 and riboflavin 0.125.

3.3 Enzyme Assay

Phenylalanine dehydrogenase activity (oxidative deamination) was assayed using a modification of the method described by Hummel *et al.*[6] The reaction mixture, incubated at 25°C, contained 200 mM glycine buffer pH 10.8, 3 mM NAD^+ and 5 mM L-phenylalanine and the reaction was initiated by the addition of 50 μl of enzyme solution. The formation of NADH was monitored spectrophotometrically at 340 nm and enzyme activity is expressed in units corresponding to the production of 1 μM of NADH per minute under the conditions outlined.

3.4 Enzyme Extraction

All extraction buffers contained 100 μM phenylmethylsulphonylfluoride (PMSF) and were chilled prior to use. Cells were resuspended (1 g wet weight per 5 ml of buffer) in 50 mM Tris/HCl pH 8.5 and disrupted using an APV Gaulin Lab 60 homogeniser operated at 8000 psi. Cell debris was removed by centrifugation (5000 x g, 4°C for 30 minutes) and protein concentrations were determined using the BCA assay system (Pierce Chemical Co., USA).

3.5 Enzyme Purification

Phenylalanine dehydrogenase from *Nocardia* sp. 239 had been purified previously on a small scale using a 1ml Mono-Q anion exchange column linked to an FPLC system.[3] Despite numerous attempts, the authors could not develop a satisfactory, scalable purification protocol based on anion exchange chromatography and could only achieve a partial clean-up of cell-free extract (maximum 25-fold) under the reported conditions using a Mono-Q anion exchange column. It should be noted that oxidative deamination activity was measured, not reductive amination activity as reported by de Boer *et al.*[3] Consequently, alternative chromatography matrices were investigated, particularly with a view to scale-up.

Pseudo-affinity chromatography was performed using Procion Red HE-3B/Sepharose CL-4B prepared according to the method of Atkinson *et al.*[7] The matrix was equilibrated in 50 mM Tris/HCl buffer pH 8.5 + 100 μM PMSF and loaded with cell-free extract. After washing with equilibration buffer to remove unbound substances, the enzyme was eluted with 1 mM NADH in equilibration buffer. All buffers contained 100 μM PMSF and all operations were carried out below 10°C unless stated. Enzyme purity was assessed by SDS-PAGE, performed using precast gradient 10-15% polyacrylamide gels and SDS buffer strips on a Phast System (Pharmacia) in accordance with the manufacturer's instructions.

4 RESULTS AND DISCUSSION

4.1 Fermentation

Figure 1 shows the data obtained from an 8 litre culture of *Nocardia* sp. 239 grown on **Medium O** and allowed to continue beyond the point where the available L-phenylalanine became depleted to observe the effect on phenylalanine dehydrogenase levels.

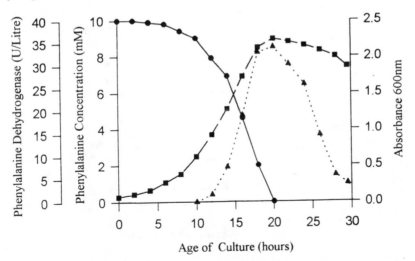

Figure 1. *Nocardia* sp. 239 fermentation data (8 litre) using Medium O (see text for details). **Key** - (●—●) phenylalanine concentration, (■—■) absorbance, (▲···▲) phenylalanine dehydrogenase.

Figures 2 and 3 show culture data obtained from 400 litre fermentations grown on **Medium O** and **Medium I** respectively. The growth and enzyme levels are significantly higher on **Medium I** and **Table 1** illustrates the improved biomass and enzyme yield achieved on **Medium I**.

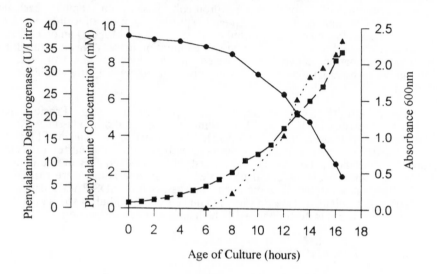

Figure 2. *Nocardia* sp. 239 fermentation data (400 litre) using Medium O (see text for details). **Key** - (●—●) phenylalanine concentration, (■—■) absorbance, (▲···▲) phenylalanine dehydrogenase.

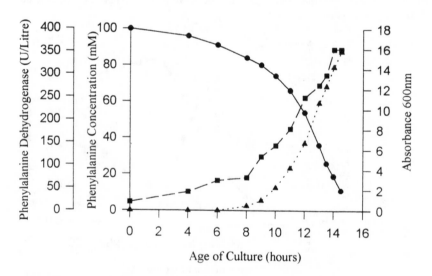

Figure 3. *Nocardia* sp. 239 fermentation data (400 litre) using Medium I (see text for details). **Key** - (●—●) phenylalanine concentration, (■—■) absorbance, (▲···▲) phenylalanine dehydrogenase.

Table 1. Comparison between fermentation data for cultures of *Nocardia* sp. 239 grown on **Medium O** and **Medium I** (see text for media composition)

Medium	Culture volume (litre)	Culture duration (hours)	Maximum OD_{600nm}	Biomass (wet wt.) (g)	Total enzyme (units)
Medium O	8	20.0*	2.4*	41*	272*
Medium I	8	18.0	21.0	406	2,880
Medium O	400	16.5	2.2	2,360	16,000
Medium I	400	14.5	16.0	14,000	140,000

*Values measured/calculated at the time when enzyme levels were highest.

The higher enzyme levels (typically about 9-fold) correspond closely to the increased biomass (measured as cell paste wet weight), indicating that the levels of enzyme expression are relatively unaffected by the changes to the growth medium. The improvement in biomass yield is almost certainly directly attributable to the 10-fold increase in concentration of the carbon and energy source, L-phenylalanine. Whilst none of the individual vitamins or salts were found to be essential, collectively they did help to sustain the increase in cell density (results not shown).

4.2 Purification of Phenylalanine Dehydrogenase

A 50 ml (35 mm i.d. x 52 mm high) Procion Red HE-3B column was equilibrated with 50 mM Tris/HCl pH 8.5 + 100 μM PMSF, loaded with 140 ml of crude enzyme extract, washed with the equilibration buffer and the phenylalanine dehydrogenase was bio-specifically eluted with 1mM NADH (**Figure 4**).

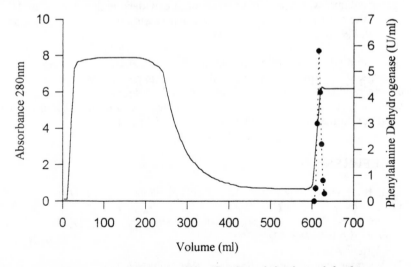

Figure 4. Affinity chromatography of phenylalanine dehydrogenase on Procion Red HE-3B/Sepharose CL-4B. **Key** - (——) protein, (●··●) enzyme. Note; the protein peak corresponding to enzyme elution is masked by the NADH in the elution buffer, which absorbs at 280 nm.

A summary of the purification data is given in **Table 2**, with data from the NADH-eluted column (column 1) compared to data from a similar column eluted with a combination of higher pH and ionic strength (50 mM glycine buffer pH 10.0 + 500 mM NaCl, column 2).

Table 2. Summary of phenylalanine dehydrogenase purification data.

Sample Volume (ml)	Total enzyme (units)	Total protein (mg)	Specific activity (U/mg)	Purification fold	Recovery (%)
Column 1					
Crude 1 140	92	990	0.09	1	100
Red He-3B 21	67	18	3.7	41	73
Column 2					
Crude 2 140	88	880	0.10	1	100
Red HE-3B 45	45	50	0.9	9	51

The material eluted with NADH appeared as a single band on a Coomassie stained SDS Phastgel (not shown) and was considered to be homogeneous, whereas the non-biospecifically eluted material had a lower specific activity and was not pure. The NADH-eluted material satisfied the purity criteria for a diagnostic reagent, being free from proteases and contaminating activity towards tyrosine. Although the use of NADH adds to the cost of running this column, the relatively small quantities required, allied to the savings in process time through use of a single-step   purification, off-sets this cost. The single step protocol using the dye-ligand column was selected because of the widespread use of Red HE-3B matrices in the purification of NAD/NADH dependent enzymes[8], and proved to be entirely scalable up to at least a 2.4 litre column.

The potential market for phenylalanine dehydrogenase as a diagnostic reagent is very large, with approximately 650,000 PKU tests performed in the UK and 10 million world wide in 1989[9]. A phenylalanine dehydrogenase based assay requires approximately one unit of enzyme per test, hence the production improvements described here offer significant advantages over existing processes.

5 REFERENCES

1. L. de Boer and L. Dijkhuizen, *Adv. Biochem. Biotech.*, 1990, *41*, 1.
2. G. A. Crosby, *C. R. C. Crit. Rev. Food Sci. Nutr.*, 1976, *7*, 297.
3. L. de Boer, M van Rijssel, G. J. Euvernink and L. Dijkhuizen, *Arch. Microbiol.*, 1989, *153*, 12.
4. UK Patent Application 9321764.4
5. W. Vishniac and M. Santer, *Bacteriol. Rev.*, 1957, *21*, 195.
6. W. Hummel, N. Weiss and M-R. Kula, *Arch. Microbiol.*, 1984, *137*, 47.
7. T. Atkinson, P. M. Hammond, R. D. Hartwell, P. Hughes, M. D. Scawen, R. F. Sherwood, D. A. P. Small, C. J. Bruton, M. J. Harvey and C. R. Lowe, *Biochem. Soc. Trans.*, 1981, *9*, 290.

8. M. D. Scawen and T. Atkinson, 'Reactive Dyes in Protein and Enzyme Technology', Y. D. Clonis, T. Atkinson, C. J. Bruton and C. R. Lowe (eds), The Macmillan Press Ltd., London, 1987, chapter 4, p. 51.

9. C.C. Mabry, *Ann. Clin. Lab. Sci.*, 1990, *20*, 392.

Protein Denaturation and Similar Effects Resulting from the Contact of Aqueous Protein Solutions with Compressed Carbon Dioxide

D. R. Butler, T. Lu, S. Meeson, M. B. King, T. R. Bott, and
A. Lyddiatt

NEAR CRITICAL EXTRACTION GROUP, SCHOOL OF CHEMICAL
ENGINEERING, UNIVERSITY OF BIRMINGHAM, EDGBASTON, BIRMINGHAM
B15 2TT, UK

1 INTRODUCTION

Compressed carbon dioxide both supercritical and marginally subcritical has been shown by several authors to be capable of extracting lipid material [1,2,3], whereas it has been found to be unable to extract proteins from aqueous solutions [4]. These facts suggest the possibility of separating lipid components from the aqueous protein solutions commonly found in fermentation broth.

However, there is some evidence that when compressed carbon dioxide is brought into contact with an aqueous protein solution the proteins undergo denaturation. (The term denaturation is here used in a broad sense as being "an alteration in the properties of a protein, e.g. decrease in solubility, loss of crystallisability or loss of specific activity if the protein is a hormone or enzyme"[5]. Elucidation of the structural changes leading to such effects is outside the scope of the present paper, but is an eventual objective of the work.) It is known that the denaturation of proteins can be produced by, for example, certain organic solvents, heat, extremes in pH, exposure to detergents or simply by vigorous shaking of a protein solution with an air phase until it foams.

Comparatively few have studied the effects of compressed CO_2 on protein solutions, though Balaban et al [6] reported that the enzyme responsible for the degradation of orange juice (pectinesterase) could be deactivated by contact with CO_2 when in solution both at ambient and at elevated pressures. Also the inactivation of polyphenoloxidase in fruit juice with CO_2 has been investigated [7]. Kamat et al [8] dispersed lipase in liquid hexane and also in supercritical (SC) CO_2 and other SC gases and studied the transesterification of methyl methacrylate. Observed reaction rates were lower in CO_2 than in the other solvents studied and they attributed this to specific interaction involving the CO_2 and the protein in the 'microaqueous environment associated with the protein.' On the other hand other workers [9-15] have found that $SCCO_2$ is a good solvent when used with immobilised enzymes.

In the present paper we report on the effect of compressed CO_2 on pepsin and also on several non-enzymic proteins (bovine serum albumin (BSA), casein, cytochrome c̲, ovalbumin and a commercial preparation ('Bipro')) both dry and in aqueous solution. The range of temperature and pressure considered was 40-50°C and 150-300 bar. In the case of pepsin, a decline in the enzymic activity which persisted when conditions were restored to their initial values provided a convenient measure of irreversible denaturation. The studies on the non enzymic proteins have so far consisted mainly in solubility tests (a result of irreversible denaturation which sometimes occurs is the formation of a permanent precipitate which persists when conditions are returned to their original conditions [14]), though functionality tests and UV absorbance scans are currently in progress.

Contributory effects which may lead to denaturation in protein solutions (but not in dry proteins) when contacted with compressed CO_2 include

1) the change in pH due to the hydrolysis

$$CO_2+H_2O \Leftrightarrow H_2CO_3 \Leftrightarrow HCO_3^- + H^+ \qquad (1)$$

2) the disruption of the gas/liquid interface which will almost inevitably occur to some degree when compressed CO_2 is contacted with a solution (the protein will tend to concentrate at the gas/liquid interfaces). The method of contact could, therefore, be important.

Control experiments carried out at ambient pressure to examine these factors were;

a) the protein solutions of interest were subjected to pH changes from 6 to 3.1 and then back to 6 in the absence of CO_2 and changes in properties were noted (3.1 was calculated to be the lowest pH to be anticipated over the range of temperatures and pressures considered).

b) the pepsin solutions were contacted in various ways with streams of both CO_2 and N_2 and the change in activity was noted as described in section 2.5 (below).

2 MATERIALS AND METHODS

2.1 High Pressure Contacting Equipment

Contact between the compressed CO_2 and the proteins took place in stainless steel cells with volumes of either 20ml or 500ml. The 20ml sample was used in conjunction with a modified Milton Roy Sample Preparation Accessory [4] and the 500ml cell on a similar but larger scale equipment.

Compressed CO_2 could either a) be passed upward through a bed of dry powder (surmounted by a de-entrainment pad) contained in the cell and then recirculated, or b) be 'bubbled' through an aqueous protein solution placed in the cell or c) passed continuously or stagnated over the solution. To minimise possible denaturation effects associated with rapid decompression [16] particular care was taken during this step to avoid sudden pressure changes.

Enzymic activity of the pepsin (Sigma Ltd, 1000 units/mg) was determined before and after contact and an "SDS-Page" gel test was carried out. In the case of the non-enzymic proteins, changes in solubility characteristics were noted.

2.2 Enzyme Assay [17]

400µl of acid-denatured haemoglobin (0.4mg/ml in citrate buffer, pH 3.2) was mixed in with 100µl (after appropriate dilution) of each of the test samples and incubated at $37^{\circ}C$ for 30 minutes. The reaction was then terminated by the addition of 50mg/ml of trichloroacetic acid solution. Precipitated protein was sedimented by centrifugation and the absorbance of the supernatant solution was compared with that of a corresponding 'blank' which had been prepared as above but subjected to zero incubation time. Enzyme activity is reported in units per ml of original sample where one unit is defined as sufficient to produce a rate of change of UV absorbance at 280nm of 0.001 absorption units/minute of the TCA soluble material in the above test.

2.3 SDS-Polyacrylamide gel electrophoresis test [18]

For the SDS-PAGE test, gradient (6 to 18%) gels were run, using the buffer system described by Laemmli [18].

2.4 pH tests at ambient pressure in the absence of CO_2

A MacIlvane buffer [19] was used to obtain the desired pH. 10ml samples of each of the protein solutions (10mg/ml BSA (Sigma Ltd. 98-99%) and 20mg/ml casein (BDH Ltd, soluble light white grade)) were dialysed (dialysis membrane: Medicell International Ltd, size 2) against 1 litre of appropriate buffer and the pH of the dialysed sample was measured (Corning pH monitor 240). The pH (initially 6) was brought to 3.1 for 4 hours and then returned to 6. The activities of the pepsin samples were measured initially and after being returned to a pH of 6. In the case of the non-enzymic proteins the UV absorbance of the (centrifuged) solution at 280nm was measured and the solutions examined for any other evidence of precipitation (a convenient and sensitive way of detecting whether a precipitate has been formed is to measure the UV absorbance of the centrifuged solution at 280nm (a characteristic protein absorption band) before and after contact with CO_2).

2.5 Pepsin denaturation tests at ambient pressure

Streams of both CO_2 and N_2 were passed downwards onto the surface of a pepsin solution (10mg/ml) in a glass vessel and then bubbled for 4 hours through a second sample of the solution and their activities were measured **(Figure 2)**. In other tests the activity of a stagnant solution of pepsin in water which had previously been saturated with CO_2 and which was stored at ambient pressure in an atmosphere of CO_2 was measured after a time interval of 4 hours.

3 RESULTS AND DISCUSSION

It was clear from the above tests that at least some proteins when in aqueous solution undergo permanent change when contacted with CO_2. In the case of pepsin this effect is very pronounced in terms of change in activity. However, we have found no positive evidence of any interaction between dry proteins and compressed CO_2. Five activity determinations were made on pepsin powder contacted with compressed CO_2 at 150 and

300 bar and 40 and 50 C. The contact time was 4 hours followed by decompression over 0.5 hours. Initial activity was 1000 units/mg and average activity after contact was not significantly different being 980 units/mg with a probable error [20] of 34 units/mg. Previous work [4] has shown that dry samples of BSA, cytochrome c, ovalbumin and 'Bipro' were apparently unaffected by contact with dry compressed CO_2 in that their solubility behaviour was unchanged ('Bipro' is a mixture of proteins marketed by Bio-isolates Ltd. It contains a serum albumin, lactoglobulin and lactalbumin).

In contrast aqueous solutions (10mg/ml) of both Bipro and ovalbumin solution produced a permanent precipitate which did not redissolve when conditions returned to their original values and we judged that in these cases irreversible denaturation had occurred. Casein showed an initial precipitate on exposure which did redissolve when conditions were returned to their original values. These results tie up quite well with the tests carried out at ambient pressure in which the pH of the casein solution was changed from 6 to 3.1 and then back to 6. UV absorbance measurements were made on the initial and final sample as well as on the one whose pH was 3.1. Again it was found that a precipitate formed at a pH of 3.1 and redissolved when the pH was returned to 6. The similarity in behaviour suggests that the effect of CO_2 on the casein solution was simply to change the pH and there was no evidence of a specific effect on the protein (in this context the isoelectric point of casein which corresponds to a minimum solubility is 4.1). In this case we judged that no permanent precipitate had occurred and that the solubility behaviour provided no positive evidence of irreversible denaturation.

High pressure tests on a 10mg/ml solution of BSA resulted in no visible precipitate and complimentary tests in which the pH was changed at ambient pressure from 6→3→6 showed no evidence of any precipitate at any stage. Although the solubility behaviour provided no evidence for irreversible denaturation in the BSA and casein solutions, more refined tests will be required before the possibility of this having occurred can be eliminated.

In the case of pepsin, pH tests at ambient pressure in the absence of CO_2 revealed little change in activity in the range 3-6 indicating that any permanent change brought about by contact with CO_2 cannot be attributed simply to change in pH.

When the pepsin solution was contacted with compressed CO_2 at 150bar and 40 C substantial decreases in activity were observed **(Figure 1)**. Contact was achieved both by passing the CO_2 vertically downwards onto the solution and by bubbling it up through the solution, the % fall in activity being lower in the former case. The fall in activity was associated with the formation of a permanent precipitate which did not redissolve when conditions were returned to their original values, indicating an irreversible change in the nature of the protein. In the case where CO_2 was bubbled through the solution for 12 hrs the denaturation was also associated with the virtual elimination of dissolved pepsin from solution as evidenced by SDS polyacrylamide gel electrophoresis. The same gel test revealed that serum albumin which was also present in trace amounts in the initial sample remained in solution.

Figure 1. Enzyme activity of aqueous pepsin vs time

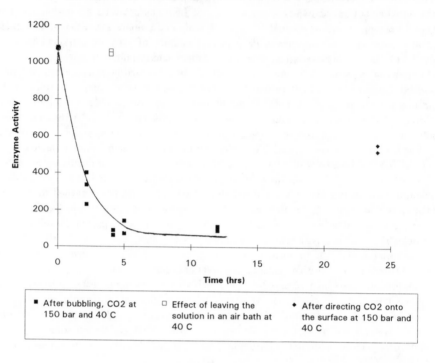

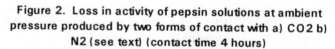

■ After bubbling, CO2 at 150 bar and 40 C	□ Effect of leaving the solution in an air bath at 40 C	◆ After directing CO2 onto the surface at 150 bar and 40 C

Figure 2. Loss in activity of pepsin solutions at ambient pressure produced by two forms of contact with a) CO2 b) N2 (see text) (contact time 4 hours)

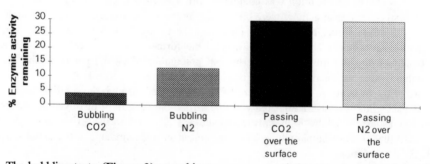

The bubbling tests **(Figure 2)** at ambient pressure revealed greater loss of activity with CO_2 than with nitrogen probably due to the greater foam height which was observed in the CO_2 case. The loss of activity with the CO_2 was similar to that found in the high pressure tests of the same duration. This suggests that foaming and the nature of the contact may be more important than the use of high pressure in producing the loss in activity. Formation of foam cannot be solely responsible since some loss of activity occurred both at high and ambient pressure when the CO_2 was passed vertically downwards onto the solution (rippling may have occurred). The loss of activity when using nitrogen at ambient pressure was then similar to that for CO_2.

In conclusion, initial work with dry proteins and solutions of BSA and casein has not so far given evidence of irreversible denaturation when these are contacted with compressed CO_2. Deactivation has been observed with aqueous solutions of lactalbumin, 'Bipro' and pepsin. Pepsin showed clear evidence of such an effect both at elevated and at normal pressures. This was probably, due to it's tendency to concentrate at surfaces and the inevitability of some surface disruption occurring when the solution is contacted with CO_2 and other gases.

Some aspects of the behaviour of the pepsin solutions appear to be qualitatively similar to those reported for pectinesterase [6].

4 ACKNOWLEDGEMENTS

The authors would like to thank Dr S Mohan for her support and advice, Dr R Santos, Mr M Burns and Dr B Al-Duri for their help during the experimental work.

5. REFERENCES

1. MB King and TR Bott , "Extraction of Natural Products using Near-Critical solvents", Blackie Academic & Professional, Glasgow, 1993, Chapter 6, p. 147
2. H Hammam , J Supercritical Fluids, 1992, 5, no 2, 101
3. M McHugh and V Krukonis , 'Supercritical Fluid Extraction: Principles and Practice', Butterworths, USA, 1986, chapter 10, p. 181
4. SR Meeson , MB King and A Lydiatt , ICheme Research Event, 1993, 1, p. 141
5. TC Collocott and AB Dobson eds, "Chambers Science and Technology Dictionary", 1984
6. M Balaban , S Pekyardimci , M Marshall , A Arreola , JS Chen and CI Wei , Proceedings of the 2nd-International Symposium on Supercritical Fluids, 1991, Boston, p. 114
7. GP Zemel , M.Sc Thesis, University of Florida, 1989
8. S Kamat , J Barrera , EJ Beckman and AJ Russell , Biotech & Bioeng, 1992, 40, p. 158
9. AMM van Eijs , JPL de Jong , HJ Doddema and DR Lindeboom , Proceedings of the International Symposium on Supercritical Fluids, Oct 1988, Nice, p. 933
10. A Marty , W Chulalaksananukul , JS Condoret , RM Willemot and G Durand , Biotech Letts,1990, 12, no 1, p. 11
11. C Vieville , Z Mouloungui and A Gaset , Ind Eng Chem Res, 1993, 32, p. 20 12. K Nakamura , Y Min Chi , Y Yamada and T Yano , Chem Eng Commun, 1986, 45, p207
13. TW Randolph , HW Blanch , JM Prausnitz and CR Wilke , Biotech Letts, 1985, 7, no 5, p. 325
14. E Cernia,C Palocci, F Gasparrine and D Misiti, Chem Biochem Eng Q, 1994,8, p. 1

15. DA Hammond , M Karel and AM Klibanov , <u>Appl Biochem Biotech</u>, 1985, <u>11</u>,
     p. 393
16. V Kasche, R Schlothauer and GBrunner, <u>Biotech Letts</u>, 1988, <u>10</u>, no 8, p.569-574
17.   B Kassell and PA Meitner , <u>Methods in Enzymology</u>, 1970, <u>19</u>, p. 337
18. UK Laemmli, <u>Nature</u>, 1970, <u>227</u>, p. 680
19. C Hewitt , PhD Thesis University of Birmingham, 1993
20. JB Scarborough, "Numerical Mathematical Analysis. Sixth Edition", 1966, p. 502

Determination of Liquid Membrane Resistance to Amino Acid Transport by Ion-pairing

M. M. Cardoso,[a] C. M. Mendes,[a] M. J. T. Carrondo,[a,c] K. H. Kroner,[b] W.-D. Deckwer,[b] and J. P. S. G. Crespo[a]

[a]DEPART. DE QUÍMICA, FACULDADE DE CIÊNCIAS E TECNOLOGIA, UNIVERSIDADE NOVA DE LISBOA, 2825 MONTE DA CAPARICA, PORTUGAL

[b]GBF-GESELLSCHAFT FUR BIOTECHNOLOGISCHE FORSCHUNG MBH, MASCHERODER WEG I, 3300 BRAUNSCHWEIG, GERMANY

[c]INSTITUTO DE TECNOLOGIA QUÍMICA BIOLÓGICA (ITQB)/INSTITUTO DE BIOLOGIA EXPERIMENTAL E TECNOLÓGICA (IBET), APARTADO 127, 2780 OEIRAS, PORTUGAL

1 ABSTRACT

This work reports studies of extraction of aspartic acid by ion pairing using a supported liquid membrane with trioctylmethylammonium chloride (Tomac) as carrier. Membrane interfacial potentials are measured and specific conductance of each form of aspartic acid is determined for different conditions of transport. The results show that this procedure permits a rapid selection of ion pair extraction conditions.

2 INTRODUCTION

The separation and purification of amino acids is commonly accomplished by evaporative crystallization or ion exchange. The presence of trace contaminants is the most important problem in the recovery of amino acids produced by fermentation. The first step of separation clarifies the fermentation broth but all the soluble media constituents - salts, proteins, residual sugars, secondary metabolites - still remain in the clarified broth. Consequently it is necessary to accomplish successive purification operations: crystallization-washing-recrystallization to get the final product with a high purity grade. The purification process might be greatly simplified and the production costs reduced if the initial separation steps were selective to the desired species excluding undesired media constituents [1]. Recovery of amino acids from fermentation broths by liquid-liquid extraction processes, using an extractant with a high specificity to the desired product, has been suggested [2].

Amino acids are always present as charged molecules and, in some pH ranges, they have a zwitterionic character which makes them insoluble or with a very low solubility in conventional organic solvents [3]. A promising recovery process is extraction using a water insoluble agent in the solvent phase able to carry the charged amino acid with higher selectivity than the other ionic species present in the fermentation media. To overcome the problem of emulsion formation and loss of solvent to the aqueous phase it may be necessary to introduce decanters in the process sequence [2]. A more efficient extraction/reextraction process is to couple extraction and reextraction in an integrated form. If the organic phase (containing the carrier) contacts on one side with the feed phase (fermentation broth) and on the other side with the stripping phase it is possible to have both processes in one operation. In this case, the organic phase can be considered as a liquid membrane. Microporous hydrophobic membranes can be used to immobilize the extractant phase [4].

Most work on reactive extraction of amino acids has been performed using quaternary ammonium salts as carriers. Due to the anion exchange properties of these carriers extraction can only take place if the amino acids are present as anions. This work discusses the transport of amino acids across liquid membranes using ion pairing.

3 TRANSPORT MECHANISM

The ion-pairing transport of an amino acid through a liquid membrane is represented in Figure 1.

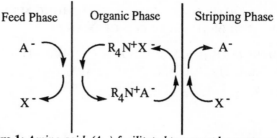

Figure 1: *Amino acid (A-) facilitated transport by quaternary ammonium salts.* $(R_4N^+X^-)$

The chemical equilibrium observed in both interfaces: feed/organic and organic/stripping phases can be described by (1):

$$R_4N^+X^-_{(org)} + A^-_{(aq)} \rightleftharpoons R_4N^+A^-_{(org)} + X^-_{(aq)} \qquad (1)$$

The forward reaction is more favourable at the feed phase / organic phase interface and the opposite reaction at the organic phase / stripping phase interface (e.g.: using a low pH or a high salt concentration in the stripping phase). This mechanism permits the recovery of the amino acids and its concentration if the right conditions are selected.

This type of mechanism involves electrostatic interactions and the transport of charged molecules through the membrane. Consequently an electrical potential gradient may be established inside the membrane. In that case the overall transport process is a result of two driving forces: chemical potential gradient and electrical potential gradient [5]. The measurement of membrane interfacial potentials during the extraction/reextraction process permits the determination of the membrane transport resistance for different ionic species. Using this procedure it is possible to predict the membrane behaviour concerning the transport of different species. The best operating conditions (e.g. pH) and the most selective extraction system (e.g. carrier/solvent/membrane) can be chosen using this technique.

The amino acid charge plays an important role on these electrostatic interactions. Equation 2, relating the net charge (Z) of aspartic acid with pH (see Figure 2) was developed from a material balance and using the equilibrium constant expressions.

$$Z = \frac{10^{(pKa1-pH)} - 10^{(pH-pKa2)} - 2 \times 10^{(2pH-pKa2-pKa3)}}{10^{(pKa1-pH)} + 1 + 10^{(pH-pKa2)} + 10^{(2pH-pKa2-pKa3)}} \qquad (2)$$

For aspartic acid : $pKa_1 = 2.1$; $pKa_2 = 3.9$; $pKa_3 = 9.8$.
For glutamic acid: $pKa_1 = 2.2$; $pKa_2 = 4.3$; $pKa_3 = 9.87$.

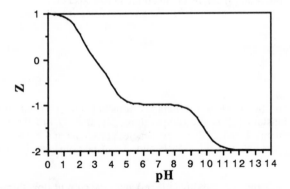

Figure 2: *Aspartic acid net charge against pH.*

Depending on pH different forms of amino acid, able to establish different interactions with the carrier, are present (Figure 3). From a material balance, an expression relating the molar fractions of different aspartic acid forms with pH was developed (Figure 4).

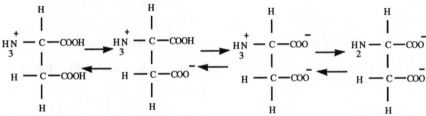

Figure 3: *Aspartic acid equilibrium forms*

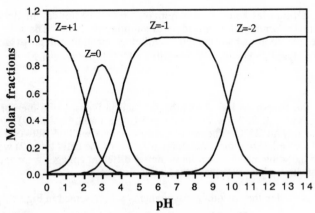

Figure 4: *Aspartic acid forms molar fractions against pH.*

Using the Nernst-Planck equation, that describes the flux (J) of ions in free solutions and considering a linear voltage (V) profile across the membrane, and a constant diffusivity (D), expression (3) can be obtained for steady state:

$$J = \frac{DBZFV}{m_t RT} \frac{(C_1(\frac{ZFV}{RT}) - C_2)}{(e^{(\frac{ZFV}{RT})} - 1)} \tag{3}$$

where B is the partition coefficient, F is the Faraday constant, m_t is the membrane thickness, R is the gas constant , T is the temperature, C_1 and C_2 are the solute concentrations in compartments 1 and 2. For short time experiments and equal solute concentration in both compartments of the cell, B and D can be considered constant during the experiment.

Defining Pm as the membrane permeability (BD/m_t) and I the current flow through a unit area of membrane, equation (4) is obtained when the concentration is the same in both compartments.

$$I = ZFJ = Pm\frac{(ZF)^2 VC}{RT} = GV \tag{4}$$

where G is the conductance.

The plot of I vs. V permits the determination of G using equation 4.

When more than one ionic species is present the total ionic membrane conductance is the sum of individual conductances.

$$G = \sum_i C_i G_i \tag{5}$$

4 MATERIALS AND METHODS

Aspartic acid, Glutamic acid and n-heptane were obtained in 99% from Merck.Tricaprylmethylammonium chloride was supplied by Fluka. The solid support membranes were microporous hydrophobic membranes made of polypropylene with a area of $5.73cm^2$: Celgard 2500 with 45% porosity and 0.075 µm pore size.

4.1 Methods

To prepare the supported liquid membrane each polypropylene membrane was immersed in a mixture of 35% (w/v) of carrier (Tomac) and 65% (w/v) diluent (n-heptane) for one day at 25°C. The membrane was sandwiched between the two compartments of the cell. After that, both compartments of the cell (1 and 2) were filled with a solution of the solute under study. Immediately different voltages were applied and the corresponding current intensities were measured.

4.1.1 Apparatus

The apparatus used in the potential measurements is represented in Figure 5.

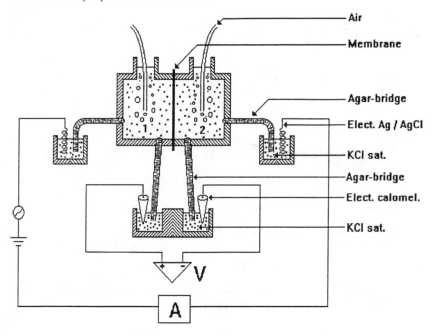

Figure 5: *Apparatus used in potential measurements*

5 RESULTS AND DISCUSSION

The measured values of V and I have a linear relation as is shown in Figure 6. The high value obtained for G for the net charge Z= - 2 is due to the contribution of the OH⁻ ion present at the working pH (Z=-2, pH= 12). However this contribution can be quantified using the specific conductance G_{spc} of the OH⁻ species determined by the same experimental procedure and solving equation 5. It was experimentally found that the total membrane conductance is the sum of the specific conductances of each ionic species present.

The relation between conductance and concentration for each species is linear as can be expected from equation 5. For net charge Z = -1 and Z = -2 this relation is presented in Figure 7. In Table 1 the values of specific conductance and membrane permeability are presented. The values presented for a net charge Z=-2 were calculated by subtracting the OH⁻ contribution. The specific conductances show a clear dependence with aspartic acid net charge. This fact shows that the type of interaction between this amino acid and Tomac is mainly electrostatic as expected due to its hydrophilic character. A similar behavior was found for glutamic acid which has a very similar structure with aspartic acid and close pI and pK_a values.

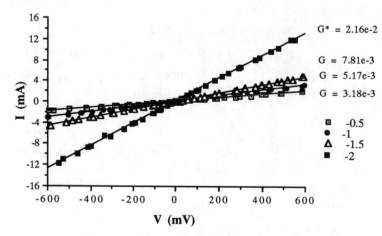

Figure 6: *I against V for aspartic acid for Tomac/n-heptane/Celgard 2500 system. (G* - conductance including OH⁻ contribution)*

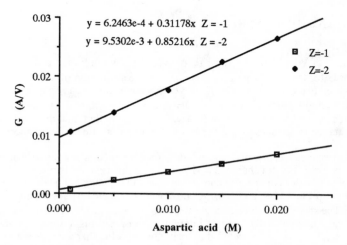

Figure 7: *Conductances against Aspartic acid concentration for Tomac / n-heptane/Celgard 2500 system.*

Table 1: *Specific conductances of aspartic acid, glutamic acid and OH$^-$*

Species	Z	-0.5	-1	-1.5	-2
Aspartic acid	Gspc $\Omega^{-1}M^{-1}$		0.345		0.687
	Pm $cm^{-1}s^{-1}$	4.01e-5	1.59e-5	1.06e-5	7.98e-6
Glutamic acid	Gspc $\Omega^{-1}M^{-1}$		0.341		0.697
	Pm $cm^{-1}s^{-1}$	4.01e-5	1.59e-5	1.08e-5	8.1e-6
OH$^-$	Gspc $\Omega^{-1}M^{-1}$	1.246			
	Pm $cm^{-1}s^{-1}$	5.7e-2			

6 CONCLUSIONS

These results show that the determination of liquid membrane resistances to the transport of different solutes can provide useful information for selection and optimization of the organic phase (type of carrier, carrier concentration and diluent), supporting membrane (type of membrane, porosity, pore diameter and thickness), feed phase (solute concentration and pH) and stripping phase (pH and counter ion concentration).

A large number of reagent / carrier / pH combination can be assessed in a short time using very small amounts of reagents to form the membrane.

7 REFERENCES

1 - E. L. Cussler, Ber. Busenges. Phys. Chem., 1989 93, 944
2 - M. P.Thien and T. A.Hatton, Sep. Sci. Technol., 1988, 23, 819
3 - K. E.Goklen and T. A.Hatton, Biotechnol. Progr., 1985, 1, 69
4 - J.J.Pellegrino and R.D.Noble, Trends Biotechnol., 1990, 8, 216
5 - E. N. Lightfoot , "Transport Phenomena and Living System", John Wiley and & Sons., New York 1974
6- H. G. Ferreira and M. Marshall , "The Biophysical Basis of Excitability" Chapter 4

ACKNOWLEDGEMENTS: The authors acknowledge the support from GBF-Gesellschaft für Biotechnologische Forschung mbH and JNICT- Junta Nacional de Investigação Científica e Tecnológica.

Characterization of a Family of Polymeric Resins with Average Pore Diameters of 150Å, 300Å, and 1000Å for the Preparative Reverse Phase Purification of Polypeptides

Peter G. Cartier, Karl C. Deissler, John J. Maikner, and Michael Kraus[a]

ROHM AND HAAS COMPANY RESEARCH LABORATORIES, SPRING HOUSE, PA 19477, USA

[a]TOSOHAAS GMBH, ZETTACHRING 6, STUTTGART 80, GERMANY

ABSTRACT

A family of polymeric resins, designed for medium pressure, preparative reversed phase purification of polypeptides and biomolecules is characterized with respect to column capacity and average pore diameter. Column breakthrough capacity to 1% leakage as a function of loading velocity is reported for BSA, insulin, vancomycin and cephalosporin C. Molecular size, pore size and flow rate are correlated with saturation and breakthrough capacity. These studies are used to identify the optimum operational conditions and the molecular size niche for each resin.

1 INTRODUCTION

An understanding of the key parameters – resolution, speed and capacity – affecting reversed phase (RPC) purification is essential in order to design an optimum purification process. Such understanding is most important to the preparative purifier, who might not be able to use high pressure high theoretical plate count media, and must rely on choosing the optimum medium pressure packing.

This study uses column loading runs with a series of molecules of different molecular weights to probe resin performance. Similar studies have been reported on silica based media for affinity[1] and anion exchange purification[2] of proteins. The loading runs directly provide data useful for choosing sample loading conditions and estimating resin capacity. They are also a guide to the selection of the optimum pore size resin for a given purification, since column performance as a function of loading velocity is related to the "effective" pore size of the resin.

2 EXPERIMENTAL

Chemicals

Amberchrom® CG-161, CG-300 and CG-1000 resins were obtained from TosoHaas (Montgomeryville, PA USA). The resins are highly cross-linked, macro-porous, styrenic, reversed phase type. Resin physical properties are listed in Table 1.

Table 1   Summary of Resin Physical and Performance Properties and Column Capacity.
The maximum loading flow rate listed in the Table is defined as the highest flow at which loading efficiency remains above 80%.

	CG-161md				CG-300md				CG-1000sd			
	Ceph C	Vanco-mycin	INS	BSA	Ceph C	Vanco-mycin	INS	BSA	Ceph C	Vanco-mycin	INS	BSA
Physical Properties												
Particle Size Range		50-100 µm				50-100 µm				20-50 µm		
Surface Area		900 m²/g				700 m²/g				250 m²/g		
Average Pore Diameter (APD)		150Å				300Å				1000Å		
Plate Count at 40 cm/hr (1)		N = 2500				N = 2200				N = 6000		
Capacity												
Dynamic Saturation Capacity (2)	102	87	91	43	72	50	97	74	22	25	38	34
Max. Loading Flow Rate (3)	769	153	38	38	769	769	153	38	769	769	769	769
Capacity to 1% Leakage at Max. Loading Flow	88	67	26	7	60	40	70	57	20	22	36	25

(1) Method described in Reference 1
(2) All capacities are in mg/cc of resin
(3) All flows are empty column linear velocity/cm/hr

Trifluoroacetic acid (TFA), cephalosporin C potassium salt, vancomycin, bovine insulin and bovine serum albumin (BSA) were obtained from Sigma Chemical Company (St. Louis, MO USA). HPLC grade solvents and ACS reagent grade buffers and acids were obtained from Fisher Scientific (Fairlawn, NJ USA).

Feed for column runs

The cephalosporin C potassium salt was dissolved in water containing 3.5 % sodium chloride at a concentration of 10mg/cc, and the solution was adjusted to pH 2.5 with concentrated sulfuric acid. Vancomycin and insulin were dissolved in water containing 0.1% TFA at 5 mg/cc. BSA was dissolved in 0.05M Tris buffer at pH=8.0 at a concentration of 2mg/cc.

Apparatus

10 x 1.0 cm ID glass columns obtained from Omnifit (Atlantic Beach, New York USA) were packed as described in Ref. 3. Bed length was 6.2 cm. A Rainin HP pump (Rainin Instrument Co., Inc. MA, USA) with a 0.010 to 9.99 cc/min head was used for the column breakthrough studies. Fractions were collected with a Gilson model 201 fraction collector (Gilson Inc., Middleton WI, USA). Fraction analysis was run on a Series 4 Liquid chromatograph equipped with an ISS-100 auto-injector with refrigerated sampling tray, a Kratos Model 783 UV detector (Perkin-Elmer, Norwalk, CT USA). Data was handled by a P.E. Nelson (P.E. Nelson Systems Inc., Cupertino, CA USA) Model 2600 system.

Analytical methods

The cephalosporin C column runs were analysed by collection of fractions. Analysis was done in 8% acetonitrile/0.1% aqueous TFA at 2.0 cc/min on a 15 x 0.46 cm ID 8 mm, PLRP-S 100Å column (Polymer Laboratories, Foster City, CA USA). Detection was at the lambda maximum of cephalosporin C, 259nm. The injection size was $10\mu l$. The other column runs were followed directly by UV at 290nm for vancomycin and insulin, and 280nm for BSA. Nitrogen adsorption tests for surface area and porosity were run on a PMI BET-206AEL autosorptometer (Porous Materials, Inc. Ithaca, NY). Surface area was determined by the BET method and the pore analysis by the "Pierce" method, which originates from the Kelvin equation[4].

Regeneration

Column loading was continued until the effluent concentration reached that of the influent. The saturated column was regenerated and the amount recovered quantified as described in the Analytical Methods. In all cases more than 95% of calculated amount loaded was recovered. Regeneration solutions were 12% 2-propanol for cephalosporin C, 35% acetonitrile/65% water with 0.1% TFA for vancomycin and insulin and 50% acetonitrile/50% water with 0.1% TFA for BSA.

3 RESULTS AND DISCUSSION

Figures 1 through 4 show how column loading performance varies with increasing flow rate. The plots show how capacity at 1% leakage, the amount adsorbed by the resin when 1% of the influent concentration is detected in the column effluent, varies as flow rate is increased. Flow rate is expressed as empty column linear velocity. Also measured was the dynamic saturation capacity, the amount adsorbed when the column effluent concentration was equal to the influent concentration and the amount recovered on regeneration. In all cases, the regenerated capacity was at least 95% of the dynamic capacity. The dynamic saturation capacities are listed in Table 1 and are independent of loading flow rate.

As shown in Figure 1, the capacity of cephalosporin C to 1% leakage is seen to be almost independent of loading flow rate and resin average pore diameter. Loading efficiency at the highest flow rate studied is high, ranging from 83-91%. Loading efficiency is defined as ratio of the 1% capacity to the dynamic saturation capacity. The smallest molecule studied, cephalosporin C (415D), can access the pore structure of all three resins in an equivalent manner since the dynamic saturation capacity expressed per square meter of surface area is similar for all resins studied. CG-161 resin adsorbs 0.11 mg/m², CG-300 0.10mg/m² and CG-1000 0.09 mg/m².

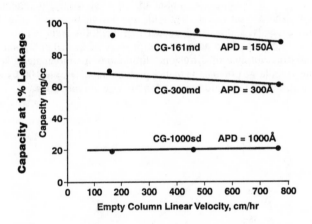

Figure 1 Effect of Loading Velocity on Cephalosporin C Capacity to 1% Leakage Capacity is determined by mass balance when 1% of the influent concentration is detected in the column effluent. Loading velocity is empty column linear velocity.

Figure 2 shows that for adsorption of the glycopeptide, antibiotic vancomycin (1449 D), the three resins do not perform equivalently. The capacity of the 150 Å average pore diameter (APD) resin drops as flow rate is increased. This drop is likely caused by pore diffusion limitations. In order to obtain 80% efficiency, flow must be kept below 153 cm/hr. The limitation defines the upper limit of application for the 150Å resin. The performance of the 300Å and 1000Å resins is unaffected over the flow rate range studied.

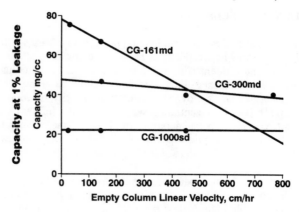

Figure 2 Effect of Loading Velocity on Vancomycin Capacity to 1% Leakage Capacity is determined by mass balance when 1% of the influent concentration is detected in the column effluent. Loading velocity is empty column linear velocity.

Figures 3 and 4 show that only the 1000Å resin maintains high adsorption efficiency for insulin (5700D) and BSA (66,000D) over the entire flow rate range studied. The pore diffusional limitations of the smaller pore diameter resins are evident in the sharp drop in capacity with increase in flow rate. The 150Å resin has very low capacity, and the high capacity of the 300Å resin is only available at reduced flow rates. The high adsorption efficiency of the 1000Å resin at all flow rates suggests the absence of diffusional limitations for molecules like BSA and insulin. Due to its large pores and smaller particle diameter, the 1000Å resin offers excellent potential for the rapid purification of large polypeptides.

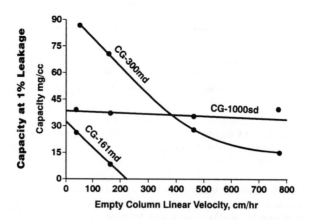

Figure 3 Effect of Loading Velocity on Bovine Insulin Capacity to 1% Leakage Capacity is determined by mass balance when 1% of the influent concentration is detected in the column effluent. Loading velocity is empty column linear velocity.

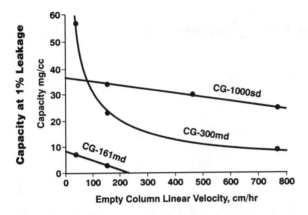

Figure 4 Effect of Loading Velocity on Bovine Albumin Capacity to 1% Leakage Capacity is determined by mass balance when 1% of the influent concentration is detected in the column effluent. Loading velocity is empty column linear velocity.

The 150Å resin also has potential for the purification of polypeptides. Taken together, Figures 1 through 4 show that for BSA and insulin sized molecules, the 150Å resin offers the potential for the removal of small molecular impurities by kinetic exclusion. Above a flow of 200 cm/hr, large molecules will pass through the column unbound while considerable capacity remains for low molecular weight impurities.

Table 1 lists the physical properties of the resins studied and a summary of the column capacity data. The maximum loading flow rate listed in Table 1 is defined as the highest flow at which loading efficiency remains above 80%.

4 CONCLUSION

The molecular size niche for each resin can be inferred from the way resin capacity responds to increasing flow rate. The 150Å resin shows its suitability for molecules like cephalosporin C, and reaches its limit with the decapeptide vancomycin. The 1000Å APD offers excellent performance for larger polypeptides like insulin and BSA. The 300Å resin fills the gap between the two, offering high capacity for vancomycin and insulin.

5 REFERENCES

1. S. R. Narayan, S. Knochs, Jr. and L.J. Crane, J. Chromatogr., 1990, 503, 93.
2. W. Kopacewicz, S. Fulton and S.Y. Lee, J. Chromatogr., 1987, 409, 111.
3. "Instructions for Hydrating, Conditioning, Packing and Cleaning Amberchrom® CG-71, CG-161, CG-300 and CG-1000 Resins" Publication 27A11, TosoHaas, Montgomeryville, PA USA May 1993.
4. S. Lowell, 'Introduction to Powder Surface Area', John Wiley & Sons, NY 1979.

Expanded Bed Adsorption for the Direct Extraction of Proteins

Y. K. Chang and H. A. Chase

DEPARTMENT OF CHEMICAL ENGINEERING, UNIVERSITY OF CAMBRIDGE, PEMBROKE STREET, CAMBRIDGE CB2 3RA, UK

Expanded bed adsorption (EBA), enables clarification, capture, and purification to be achieved in one step and will greatly simplify downstream processing flowsheets and reduce the costs of protein purification. The liquid hydrodynamics have been examined with bed expansion and residence time distribution tests. The equilibrium adsorption characteristics have been assessed by the measurement of adsorption isotherms. The rates of adsorption of protein onto STREAMLINE™ adsorbents have been studied in a batch stirred tank, a packed bed, and expanded bed system. It was demonstrated that there were no differences between the efficiencies of protein adsorption in packed or expanded bed modes. Glucose-6-phosphate dehydrogenase (G6PDH) has been successfully purified directly from a crude yeast homogenate using expanded bed adsorption with a purification factor of 16 and yield of 77 % in one step. The bed expansion and adsorption characteristics have been tested after a number of cycles of cleaning-in-place and compared with virgin adsorbents.

1. INTRODUCTION

In commercial downstream processing it is not uncommon to find the desired protein in a particulate containing feedstream where the presence of cells or cell debris totally prohibits the use of column chromatographic procedures. Such procedures (e.g. ion exchange and affinity chromatography etc.) are often relegated in sequence until removal of particulate material (e.g. centrifugation and microfiltration) has taken place. The use of these clarification techniques inevitably leads to loss of product yield, loss of valuable processing time and a general increase in downstream processing expenditure. It is obvious, therefore, that the use of a technique whereby clarification, concentration, and purification can occur in a single step would be both technically and economically valuable. Two novel technologies have been described for the direct extraction of proteins from fermentation broths and cell homogenates [1]. In this paper, we describe such a technique (EBA, Expanded Bed Adsorption) with particular reference to the purification of a commercially used diagnostic enzyme (G6PDH) from preparations of disrupted yeast cells. Early examples of expanded bed adsorption procedures [2-3] used conventional agarose based adsorbents designed for use in packed beds. However, as a result of their small particle size and low density, beds of these adsorbents expand to undesirable extents when operated with typical fermentation or cell homogenates even at low superficial velocities. Industrial applications of EBA would necessitate the use of adsorbents which would expand to 2-3 times their settled height when using high viscosity feedstocks at higher superficial flow rates (100-300 cm/h). Hence, larger beads of a greater density are required. To meet this need, Pharmacia Biotech has manufactured various specialised adsorbents (STREAMLINE™ adsorbents) which have higher densities and larger particle size. In this paper, the adsorption performance of expanded beds was characterised and the impact of this method for direct broth extraction was assessed.

2. MATERIALS AND METHODS

2.1 STREAMLINE™ adsorbents

The purpose designed adsorbent (STREAMLINE™ adsorbent, Pharmacia Biotech, Uppsala, Sweden) is based on the Sepharose matrix, but incorporates a quartz core, and is specifically modified to have well-defined particle size and density for use in expanded bed processes. STREAMLINE™ adsorbents, with a particle diameter range of 100-300 µm, are available with two different ion exchange ligand, SP (sulphopropyl) and DEAE (diethylaminoethyl).

2.2 Expanded bed adsorption apparatus

Our previous work has demonstrated that the design of expanded bed apparatus is critical in obtaining good adsorption performance [3-5].The purpose designed column, STREAMLINE 50, for use with STREAMLINE™ adsorbents in expanded bed adsorption provides an effective solution to the following criteria which are critical to the successful operation of expanded bed procedures:

• To create a stable expanded bed flow distribution, there is a specially designed liquid distributor and mesh cap (stainless steel) at the bottom of the column. The distributor design allows the unhindered passage of cells and cell debris.

• The position of the upper adapter of the column can be altered, to minimise the volume of liquid above the expanded bed. This adapter was also used to return the bed to a packed configuration for efficient elution.

2.3 Measurements of bed expansion characteristics

The characteristics of bed expansion were measured at 20°C in the STREAMLINE 50 column (Dc= 5.0 cm). The expansion of the bed of adsorbent with increasing superficial velocity of flow through the bed was measured as described previously [3].

2.4 Measurements of equilibrium adsorption isotherms

The adsorption isotherms of STREAMLINE™ adsorbents were determined by batch experiments in which a measured volume of settled adsorbent was equilibrated with different concentrations of protein. The equilibrium amount of protein bound per ml of adsorbent was calculated by mass balance by knowledge of the amount of soluble protein present at equilibrium.

2.5 Measurements of breakthrough curves using frontal analysis

Breakthrough curves were measured in a STREAMLINE 50 column in order to compare the efficiencies of the adsorption of protein onto the ion exchange adsorbents in expanded bed and packed bed operating modes. During the expanded bed runs, the top adapter in the column was lowered to just above the surface of the expanded bed to avoid the presence of a volume of liquid above the bed. For packed bed adsorption experiments, the top adapter was lowered onto the surface of the settled bed.

2.6 Preparation of disrupted bakers' yeast homogenate

Large scale disruption of *Saccharomyces cerevisiae* cells was carried out as described previously [6]. Protein concentration and G6PDH activity in disrupted cell proportions were carried out as previously described [7].

2.7 Direct extraction of G6PDH from unclarified yeast cell homogenate

The isolation of G6PDH from a bakers' yeast cell homogenate was carried out at 4-8°C in the STREAMLINE 50 column. STREAMLINE™ DEAE adsorbent was packed into the column to give a settled bed height of 20 cm. The top adapter of the column was positioned at about 2 times the settled bed height (45 cm), to minimise the volume of liquid above the expanded bed. Two Pharmacia syringe P-6000 pumps (pump A and pump B) were used in the EBA system and the liquid flow rate was controlled either manually, or automatically, using a Pharmacia FPLC Manger LCC 500 CI process controller. Pump A was used to pump buffer during equilibration, washing, and re-equilibration stages. Pump B was used to pump the eluent solutions. In addition, a peristaltic pump C was used to apply crude homogenate to the expanded bed.

Initially, the adsorbent bed was washed with 0.05 M phosphate buffer (pH 6.0), using pump A, until 5 bed volumes of buffer had passed through the bed and until the bed had expanded to the desired height of 40.0 cm at a superficial liquid velocity of 214 cm/h. Pump A was then switched off and simultaneously pump C started to pump the crude yeast cell homogenate from observation of the position of the float through the bed. Since the homogenate was more dense and viscous than the buffer, it was important that the bed remained stable when the crude suspension passed through it. Estimations of height of the expanded bed were made using a rotameter connected in line before the bed. To maintain the same extent of bed expansion and to prevent the bed from compacting against the top adapter in the column, the velocity (214 cm/h) was reduced to 30 cm/h during the application of the feedstock.

The optical absorbance of the outlet stream was recorded using a Pharmacia UV-1 monitor at 280 nm. As soon as the absorbance in the column outlet stream began to rise as a result of the cells breaking through the bed (540 ml), samples of the outlet stream were collected, and stored on ice at 4-8°C. Samples were taken at regular intervals (20 ml) during the runs. When 900 ml of the homogenate had passed through the bed, pump C was switched off and pump A was switched on, to deliver the wash solution to the column. A wash solution of 50% glycerol (v/v) in 0.05 M phosphate buffer (pH 6.0), was used to wash out the homogenate from the bed. The wash velocity was increased from 30 cm/h to 60 cm/h to maintain the same extent of bed expansion and application of the glycerol solution was continued until the homogenate was observed, on a qualitative basis, to have been washed out from the bed (2-3 bed volumes of 50% glycerol/buffer solution). 0.05 M phosphate buffer (pH 6.0) was then used to wash out 50% glycerol from the bed. The buffer flow rate was slowly increased in steps, until the 50% glycerol had been washed out from the bed, and the absorbance in the column outlet stream had decreased to zero after which the velocity was increased in discrete steps to 214 cm/h.

Once the absorbance had dropped to zero, the bed was returned to a packed bed mode. Elution was carried out in a downward direction at 50 cm/h firstly using 0.15 M NaCl/buffer to elute G6PDH. The process was continued until the optical density in the outlet stream had returned to zero. Secondly, 1M NaCl/buffer at 50 cm/h was used to elute non-G6PDH proteins until the absorbance returned to a steady minimum value. 8 ml samples of the eluate were collected and stored in ice water at 4-8°C in 10 ml fraction collector test tubes. Each sample collected during the experiment was analysed for its G6PDH activity and its total protein content .

Cleaning-in-place was carried out in a packed bed mode by using a sequence of solutions containing 1M NaCl, 1M NaOH, buffer, and 70% ethanol (3-5 bed volumes each) at 50 cm/h.

3. RESULTS AND DISCUSSION

3.1 Physical properties of STREAMLINE™ adsorbents

Table 1 demonstrates that the combination of increased particle diameter and density results in a two fold expansion of the bed being achieved at 8 times the liquid flow rate that produces an equivalent expansion of a bed of Sepharose FF. This feature enables feedstocks to be processed using STREAMLINE™ adsorbents at faster flow rates which would be more suited to industrial implementations of this technique.

Table 1. *Physical properties of STREAMLINE™ adsorbent*

Adsorbent	Average particle size (μm)	Particle density (kg/m³)	Flow rate (cm/h) to give 2xbed expansion in aqueous buffer
Sepharose FF	93	1130	32
STREAMLINE™	186	1180	240

3.2 Bed expansion characteristics

The variation of the expansion of the bed with flow rate is a function of the viscosity and density of the liquid flowing through it. All the measured bed expansion characteristics where found to follow the Richardson-Zaki correlation:

$$U = U_t \, \varepsilon^n$$

where U is the superficial velocity of liquid flow and ε is the voidage of the expanded bed. The value used for the Richardson-Zaki parameter (n) in the laminar flow regime is typically 4.8 [8], so that the experimentally determined values which range from 4.5-5.1 agree reasonably well. The bed expands far more in glycerol solutions of higher viscosity (Figure 1). The experimentally determined terminal velocities of particles agree well with calculated theoretical values from Stokes' equation.

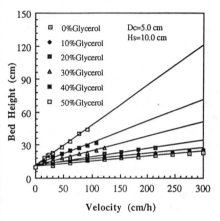

Figure 1. *Bed expansion characteristics for STREAMLINE™ DEAE*

3.3 Equilibrium adsorption isotherms

The ion exchange ligands, diethylaminoethyl (DEAE) and sulphopropyl (SP), are often chosen for protein purification as a result of their general applicability, robustness and stability to cleaning-in-place procedures. Langmuir isotherms characterised by maximum capacity (Q_m) and dissociation constants (K_d) were found for the adsorption of bovine serum albumin (BSA) in 0.02 M piperazine buffer, pH 6.0 and Egg-white lysozyme (Lys) in 0.1 M Na-acetate buffer, pH 5.0 at 25°C. Values of the Langmuir isotherm parameters were obtained from nonlinear squares regression analysis and shown in Table 2.

Table 2. *Adsorption equilibrium characteristics*

Adsorbent	Adsorbate	Qm (mg/ml)	Kd (mg/ml)
DEAE-Sepharose FF	BSA	68	0.025
STREAMLINE™ DEAE	BSA	60	0.20
S-Sepharose FF	BSA	113	0.13
STREAMLINE™ SP	BSA	44	1.37
S-Sepharose FF	Lysozyme	120	0.019
STREAMLINE™ SP	Lysozyme	70	0.002

Comparison of the values of Qm suggest that the maximum capacities of the STREAMLINE™ adsorbents are somewhat lower than their equivalents based on the Sepharose FF matrix. This may be due to the inclusion of the quartz core resulting in a reduction in the volume of adsorbent available for the protein adsorption. Alternatively the effect may be a result of the reduced surface area per unit volume of the adsorbent which has a larger particle diameter. Some differences were also noted in the values of the dissociation constants Kd, but these effects are not readily attributable to differences in the structure of the adsorbents.

3.4 Frontal analysis

Studies of the relative adsorption performance achieved with two different operating protocols were carried out by frontal analysis using the same amount of adsorbent either in a packed bed mode or an expanded bed mode which was loaded with protein at the same linear velocity in the STREAMLINE™ 50 column. A qualitative analysis of breakthrough curves for both packed and expanded bed adsorption of BSA onto STREAMLINE™ DEAE indicates that an ideal step breakthrough of BSA is almost achieved (Figure 2). These breakthrough curves for BSA are indistinguishable indicating that the adsorption performance achieved in the expanded bed is as good as that in a packed bed mode. Therefore, the expanded bed appears to behave like a packed bed except that its interstitial voidage is greater. It can be shown that the liquid dispersion in the expanded bed is small by measuring residence time distributions and the liquid appears to pass through the bed in close to plug flow. It also appears that there is little axial mixing of the adsorbent particles which favours efficient adsorption performance. These observations demonstrate that a stable expanded bed of adsorbent can be established without the need to stabilise the fluidized bed using a uniform magnetic field [9].

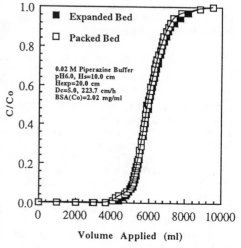

Figure 2. *Packed and expanded bed adsorption of BSA onto STREAMLINE™ DEAE*

Studies of the adsorption of lysozyme onto expanded beds of STREAMLINE™ SP from solutions of different viscosity were also carried out by frontal analysis at an appropriate flow rate that resulted in the same expanded bed height (2x height of settled bed). The results indicate that sharp breakthrough curves were obtained even when adsorption was carried out in high viscosity solutions or at high velocity, 300 cm/h. (Figure 3). Using frontal analysis, the dynamic capacity of the uptake of lysozyme onto STREAMLINE™ SP from 25% glycerol is greater than that from 0.1 M Na-acetate, pH 5.0. This is probably a result of the longer residence time of protein in the expanded bed when adsorption was carried out in the presence of glycerol.

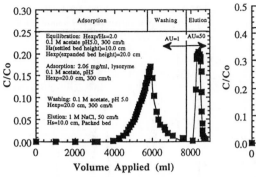 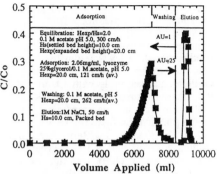

Figure 3a. *Expand bed adsorption from aqueous buffer @ 300 cm/h*

Figure 3b. *Expanded bed adsorption containing buffer from 25% glycerol @ 121 cm/h*

3.5 Direct extraction of G6PDH from yeast cell homogenate

Table 3. *Purification of G6PDH from bakers' yeast cell homogenate*

Purification Step	Volume (ml)	Total Activity (U)	Total Protein (mg)	Specific Activity (U/mg)	Purification Factor	Yield G6PDH (%)
Crude Homogenate	800	5648	16808	0.336	(1.0)	(100)
Flow through	800	308	2058	-	-	-
Washing	820	983	13040	-	-	-
Eluate (1)	492	4359	834	5.227	15.6	77.2
Eluate (2)	224	0	874	-	-	-

The potential for the use of EBA in direct extraction has been demonstrated in the purification of G6PDH from a crude bakers' yeast homogenate using STREAMLINE™ DEAE (Figure 4). Appropriate operating conditions had first been established from scouting experiments conducted with clarified homogenate and packed columns of the adsorbent. Purification of G6PDH from unclarified homogenate was carried out using an expanded bed of STREAMLINE™ DEAE (400 ml adsorbent, 2x height of settled bed). A steep rise in optical density at 280 nm was observed after 540 ml of homogenate had been applied to the column which indicated the front of the particulate material in the liquid leaving the bed. Total protein broke through the bed nearly at the same time. However, the G6PDH activity did not break through the bed until 660 ml of homogenate were applied to the bed. This is due to the somewhat higher selectivity of STREAMLINE™ DEAE for G6PDH under these conditions. At the end of adsorption stage, residual particulate material was completely washed out from expanded bed using 50% (v/v) glycerol (5 bed volumes) followed by 0.05 M Na-phosphate buffer, pH 6.0.

Following washing, the bed was returned to a packed bed mode and eluted firstly with 0.15 M NaCl/buffer, which was shown to elute all the adsorbed G6PDH, and secondly with 1.0 M NaCl to elute other remaining material. G6PDH was recovered in one step using EBA with a purification factor of 16 and yield of 77.2% (Table 3). Reproducibility studies confirm that the bed expansion and adsorption characteristics of the adsorbent are completely restored following the subsequent cleaning-in-place (CIP) procedures. Overall, the results indicated that the use of stable expanded bed adsorbents is an effective method for the purification of proteins from crude feedstocks without prior removal of particulate matter.

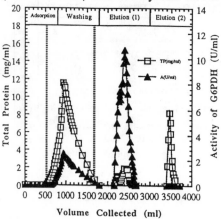

Figure 4. *Purification of G6PDH from bakers' yeast cell homogenate on an expanded bed of STREAMLINE™ adsorbent*

4. CONCLUSIONS

(1) STREAMLINE™ adsorbents have suitable properties for use in expanded bed procedures.
(2) STREAMLINE™ adsorbents used in a STREAMLINE™ EBA apparatus show that the adsorption performance in an expanded bed is as good as that obtained with the same matrix in a packed bed.
(3) Enzyme can be recovered directly from unclarified bakers' yeast homogenates without the need to remove cells or cell debris by centrifugation and filtration.
(4) The ability to be able to achieve clarification, capture, and purification in a single step will greater simplify downstream processing flowsheets and reduce the cost of protein purification.

References

1. Y.K. Chang, G.E. McCreath, N.M.Draeger, and H.A. Chase, *Trans IChemE*, 1993, **71** (Part A) 299.
2. N.M. Draeger and H.A. Chase, *Trans. Inst. Chem. Eng.*, 1991, **69** (Part C) 45.
3. H.A. Chase and N.M. Draeger, *J. Chromatogr.*, **597** (1992) 129.
4. Y.K. Chang and H.A. Chase, *Procedings from I. Chem. E. Res. Event.*, University College London, London, January 5-6, 1994, P.168-170.
5. H.A. Chase and Y.K. Chang, *Presented at 22 nd FEBS Congress*, Stockholm, Sweden, July, 1993.
6. G.E. McCreath, PhD Thesis, University of Cambridge, 1993.
7. G.E. McCreath, H.A. Chase, and C.R. Lowe, *J. Chromatogr.*, 1994, **659**, 275.
8. J.F. Richardson and W.N. Zaki, *Trans. Inst. Chem. Eng.*, 1954, **32**, 35.
9. A.S. Chetty and M.A. Burns, *Biotechnol. Bioeng.*, 1991, **38**, 963.

Acknowledgements
YKC and HAC would like to thank Pharmacia Biotech for the generous provision of chromatographic materials and Dr G.E. McCreath for helping with some important experiments. Also, YKC greatly acknowledges the Mingchi Institute of Technology (Taiwan, ROC) for financial support.

Solid–Liquid Separation of PHB by Froth Flotation

J. J. Cilliers, L. S. Johnson, and S. T. L. Harrison

DEPARTMENT OF CHEMICAL ENGINEERING, UNIVERSITY OF CAPE TOWN,
PRIVATE BAG, RONDEBOSCH 7700, SOUTH AFRICA

1 ABSTRACT

Poly-β-hydroxybutyrate (PHB) has emerging potential for use as a biodegradable thermoplastic. In its extraction from the bacterial source, the polymer remains in granular form. Hence the final stages of extraction require its rigorous washing and dewatering. Initial studies have shown that PHB is naturally floatable and frothing. In an unoptimised process, recoveries of 98 % and froth concentrations as high as 291 kg/m^3 have been achieved in separate trials. However, a compromise between recovery and concentration is required. Typical combined values of 75 % recovery and 200 kg/m^3 are attained.

In order to predict flotation performance, characterisation studies of the polymer granules and bulk suspension were undertaken. Electrophoretic mobility measurements indicate a negatively charged surface with an isoelectric point at pH 1.6. Particle agglomeration occurs with decreasing surface charge and a significant increase in the suspension viscosity occurs at conditions below pH 5. Flotation was studied as a function of pH. Little effect of pH on froth concentration was found; however recovery increased with decreasing pH and increased particle size. As a consequence of these observations, a hypothesis is put forward that flotation is enhanced by the increased size of particle agglomerates in the range 2 to 45 microns and that high suspension viscosities are detrimental to the process.

2 INTRODUCTION

The bacterial storage product poly-β-hydroxybutyrate (PHB) is of commercial interest as a biodegradable thermoplastic that is produced from renewable resources. Under nutrient limited conditions, the bacterium *Alcaligenes eutrophus* can accumulate PHB to levels of 60 to 90 % of its biomass as a granular intracellular product [7]. In the extraction and purification of the polymer, cell disruption is followed by the removal of insoluble contaminants. Thereafter, soluble contaminants are removed from the PHB suspension and it is dewatered prior to conventional processing. This study centres on the potential of froth flotation for the de-watering of the polymer suspensions with concomitant removal of soluble contaminants.

Whereas centrifugal separations are based on the size and density of the particulates, flotation exploits its surface properties. Particles that are not easily wetted (hydrophobic) and carry little or no surface charge favour collection at the gas-liquid interface to existence in the bulk aqueous phase. In a sparged system, these particles attach to the surface of the rising air bubbles and accumulate in the froth phase. The natural hydrophobicity and frothing of PHB suggest flotation as a potential separation process [1].

This paper illustrates the potential of separation by flotation. The effect of the suspension pH on the process is investigated owing to its influence on the relative charge of the bubble and particle. Emphasis is placed on the concomitant influence of these parameters on the properties of the particle and its suspension.

3 EXPERIMENTAL

Reagents

Analytical grade acetic acid and ammonium hydroxide were used for pH control. All experiments were conducted using de-ionised water at 20 to 22°C or as specified. Washed and spray dried PHB in powder form as well as PHB slurry resulting from the solubilisation process were obtained from Zeneca Ltd., Billingham, UK. The dried powder was sized to -150 μm by screening. Three size fractions were generated by heating in an autoclave at 216 °C for 10 minutes and then sized by sieving.

Flotation Experiments

Flotation experiments were carried out in a bottom driven Leeds type batch flotation cell having a nominal volume of 3 l. The liquid volume used was 1.75 l. The froth depth was 90 mm, chosen in order to promote the drainage of water from the froth. Compressed air was used as the flotation gas at a flow rate of 2 l/minute. The impeller speed was set at 560 rpm. The PHB particles were conditioned for five minutes before floating at a suspension concentration of 130 kg/m^3 (10.4 % v/v). Flotation was carried out over the range pH 2 to pH 10 for five minutes. The froth was collected at 1 minute intervals. The product wet mass and dry mass were obtained by weighing. Hence the volume of water recovered, the PHB recovery and froth concentration were calculated. The error analysis showed that the maximum error was less than five percent at the 95 percent level of confidence.

Particle and Suspension Characterisation

Electrophoretic mobility measurements were carried out using a Rank Brothers Electrophoresis apparatus. A capillary cell and Phillips video 40 camera linked to a monitor were used to measure the motion of the particles under an applied voltage of 10 V/cm. The zeta potential (ζ mV) was calculated using $\zeta = 12.85$ U where U is the mobility of the particle (μm/s per V/cm). Measurements were performed at 25 °C over the range pH 1 to 11. Suspension viscosity was measured by a Mettler Rheomat RM 180 concentric cylinder viscometer. An immersion tube rotor system was used and suspension of the PHB ensured by agitation with a magnetic stirrer. Shear stress was monitored over the range pH 3 to 11. The shear rate was varied over the range 50 s^{-1} to 1000 s^{-1}. Particle sizes were determined by Malvern Instruments 2600 Laser Diffraction particle size analyser fitted with a flow through cell over the range pH 1 to 11.

4 RESULTS

Flotation trials under standard operating conditions illustrated that PHB suspensions are naturally frothing without chemical addition. The polymer granules favoured collection in the froth phase without the addition of collector molecules to alter the surface chemistry. The natural floatability of the PHB confirmed that it is hydrophobic. Recoveries obtained (i.e mass ratio of PHB collected to that present) ranged from 20 % to 98 % (pH range pH 2 to 10). The froth concentration attained was in the range 120 to 168 kg/m^3 (9.6 to 13.4 % v/v). Analysis of the data showed that the mass recovery of PHB could be modelled as a first order process (R^2 = 0.997). Typical rate constants were of the order of 0.006 s^{-1}.

Flotation of PHB as a Function of pH.

The affinity of a particle for a gas-liquid interface is influenced by the relative charge of these surfaces. Approach toward the point of zero charge (PZC) is expected to favour flotation [4,5]. Hence the floatability of the PHB granules is expected to be a function of the zeta potential. The zeta potentials of the PHB samples before and after drying are the same and are shown as a function of pH at 25 °C in Figure 1. This shows that a large negative potential exists at neutral pH indicating a large negative surface charge. The point of zero charge is found at pH 1.6.

Flotation experiments were performed over the range pH 2 to 10. Figure 2 illustrates the recovery of PHB after two minutes of flotation as a function of pH. On decreasing the pH in the range pH 6 to 4, PHB recovery was increased. This followed the trend expected on the basis of surface charge. However the decrease in PHB recovery on further decreasing the pH from pH 4 to 2 suggested an opposing influence from another suspension property. The general trend in water recovered into the froth phase follows that of PHB. Froth concentration lay in the range 159 to 168 kg/m^3 (12.7 to 13.4 % v/v) over the range pH 3 to 10. No significant change in froth concentration was found in response to change in

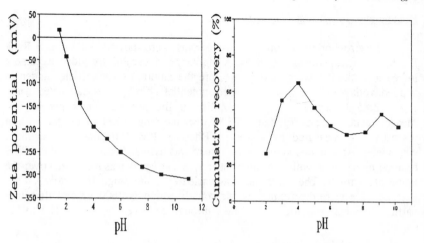

Figure 1 Influence of pH on zeta potential at 25 °C.

Figure 2 Influence of pH on the recovery of PHB after two minutes of flotation.

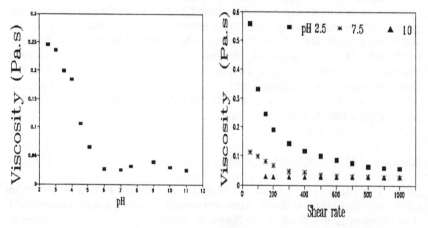

Figure 3 Viscosity of PHB as a function of pH. (C = 300 kg/m³ and 22 °C).

Figure 4 Viscosity of PHB suspension as a function of pH and shear rate showing the shear thinning nature of PHB. (C = 300 kg/m³ and 22 °C).

flotation pH in this range. A decrease in concentration to 122 kg/m³ was observed at pH 2.

The viscosity of the PHB suspension was measured as a function of pH. In Figure 3 it is shown that the viscosity increased dramatically as the pH decreased from pH 6 to 2.5. Little effect was seen in the range pH 6 to 11. Figure 4 shows the shear thinning nature of the PHB suspension on exposure to increased shear.

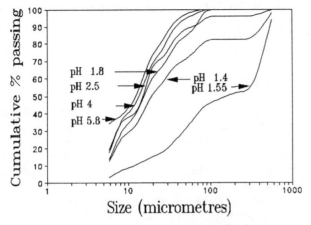

Figure 5 Influence of pH on the size distribution.

A study was conducted on the possible aggregation of PHB particles at low pH and therefore reduced surface charge. Little effect was noted in the pH range above pH 6. Figure 5 illustrates the agglomeration of PHB granules with decrease in pH below pH 6. Maximum size was seen at pH 1.55, corresponding to the point of zero charge. On further reduction of the pH, a reversal of the effects was observed, confirming the dependence of aggregation on the surface charge.

Flotation as a Function of Size.

It has been reported frequently that an optimum size range exists for the maximum rate of recovery. At either extreme of this size range, the rate of flotation decreases [2,4,5,6]. Flotation experiments were performed using 5 size fractions of PHB. The suspension concentration was 100 kg/m^3 (8 % v/v). The results in Table 1 show that the rate of recovery increases with larger particle size. However the froth concentration profile showed that a maximum

Table 1   Average flotation performance as a function of particle size

Nominal particle size (μm)	90 % passing smaller than (μm)	Recovery %	Froth concentration (kg/m^3)	First order rate constant (s^{-1})
38.0	51.8	97.8	153	.0167
28.4	45.7	83.6	164	.0052
18.5	41.6	74.4	168	.0050
18.4	28.4	70.2	143	.0027
5.4	7.02	48.1	128	.0024

concentration was reached between 18 to 30 microns nominal size. Extending these studies beyond 38 μm (90 % passing 52 μm) was hindered by unstable froth affecting the flotation results.

5. DISCUSSION

In optimising flotation conditions, attraction between the bubble and particle is controlled by manipulation of their relative charge and the particle hydrophobicity. The former can be achieved by use of pH to affect the charge at the Stern plane (zeta potential). The existence of a point of zero charge at pH 1.6 was anticipated to arise from the low level of phospholipid associated with the polymer rather than from the molecular structure of the PHB itself. As it is generally maintained that bubbles carry a small negative charge, improved flotation was seen to correspond with the decrease in absolute zeta potential in the pH range 4 to 6 (Figure 2). However the reversal of this effect before the PZC is reached suggests the opposing influence of another suspension property. To this end, particle size distribution and suspension viscosity were considered.

In the same way that hydrophobic particles in aqueous suspension have an affinity for the bubble surface, they tend to attract one another. In Figure 5, it was seen that this aggregation was at a maximum at the PZC owing to both the absence of electrostatic repulsion and the decreased energy of hydration of the reduced hydrophobic surface area. Generation of either a positive or a negative zeta potential reduced the degree of aggregation. For a specific particle size, a critical contact angle exists below which flotation does not occur [3]. Conversely, for a particular hydrophobicity, flotation is found for a particular size range only. In Table 1, data are shown to support the improved flotation found with increasing particle size in the range 2 to 45 microns. The results suggested that the formation of aggregates of the PHB granules in this size range augmented the flotation trends predicted by zeta potential studies. The reversal of flotation performance in the pH range 2 to 4 may have resulted from the presence of larger aggregates shown to reduce froth formation or from viscosity effects discussed below.

The attractive interparticle forces responsible for aggregation also contribute towards a resistance of the suspension to flow *i.e.* increased suspension viscosity. This is illustrated in Figure 3 in which a significant increase in viscosity was seen to coincide with the decreasing magnitude of zeta potential illustrated in Figure 1. Both the presence of fine particles [5] and increased suspension viscosity [4,5] have been shown to be detrimental to froth drainage. Hence this may be expected to decrease flotation, both in terms of recovery and concentration, on approach toward the PZC.

6 CONCLUSIONS

PHB suspensions have been shown to be naturally frothing. In addition, the PHB granules were collected at the gas liquid interface without the requirement of the addition of collector molecules to enhance hydrophobicity.

Maximum recovery and froth concentration of 98 % (pH 7) and 291 kg/m³ (pH 9) have been shown. A PZC at pH 1.6 was found for the extracted PHB granules at 25 °C. Granule aggregation and increased suspension viscosity resulted on approach toward the PZC. On investigating pH effects on PHB flotation, optimal recovery was found at pH 4 (94 % recovery and 138 kg/m³) and optimal froth concentration at pH 9 (205 kg/m³ and 64 % recovery). The improved flotation performance with decreasing pH in the range pH 4 to 7 could be predicted from the zeta potential and aggregation studies. A decrease in flotation performance at lower pH was contrary to the expected trends. It is hypothesised that this results from the increased suspension viscosity causing reduced froth drainage or from the generation of aggregates (dominant size 60 to 564 μm) which have a reduced ability to float [3] and to generate a sufficient froth phase (unpublished data). The increase in froth concentration at high pH results from the lower viscosity facilitating drainage. The decrease in recovery is a result of small particles which float slowly as substantiated by the flotation experiments on particle size (Table 1).

Acknowledgements

The authors wish to acknowledge the financial support and the supply of polymer samples provided by Zeneca Bioproducts, Billingham U.K.

REFERENCES

1.   S.R. Amor, T. Rayment and J.K.M. Sanders, Macromolecules, 1991, 24, 4583.

2.   N.N Arbiter and C.C Harris, 'Froth Flotation 50th Anniversary volume', D.W. Fuerstenau, ed., The American Institute of Mining, Metallurgical, and Petroleum Engineers, Inc., 1962, Chapter 8, p. 239.

3.   R. Crawford and J. Ralston, International Journal of Mineral Processing, 1988, 23, 1.

4.   E.G. Kelly, and D.J. Spottiswood, 'Introduction to Mineral Processing', John Wiley and Sons, New York, 1982.

5.   R.P. King, editor, 'Principles of Flotation', South African Institute of mining and metallurgy, Cape Town, 1982.

6.   W.J. Trahar, International Journal of Mineral Processing, 1980, 298, p. 289.

7.   A.C. Ward, B.I. Rowley and E.A. Dawes, J. Gen. Microbiol., 1977, 102, 61.

Effect of Solvent Toxicity on Lactic Acid Fermentation Kinetics

I. M. Coelhoso,[a] J. P. S. G. Crespo,[a] P. Silvestre,[a] C. V. Loureiro,[a] K. H. Kroner,[b] W.-D Deckwer,[b] and M. J. T. Carrondo,[a,c]

[a]DEPART. DE QUÍMICA, FACULDADE DE CIÊNCIAS E TECNOLOGIA DA UNIVERSIDADE NOVA DE LISBOA, 2825 MONTE DE CAPARICA, PORTUGAL

[b]GBF-GESELLSCHAFT FÜR BIOTECHNOLOGISCHE FORSCHUNG MBH, MASCHERODER WEG I, D-3300 BRAUNSCHWEIG, GERMANY

[c]INSTITUTO DE TECNOLOGIA QUÍMICA BIOLÓGICA (ITQB)/INSTITUTO DE BIOLOGIA EXPERIMENTAL E TECNOLÓGICA (IBET), APARTADO 127, 2780 OEIRAS, PORTUGAL

1 INTRODUCTION

A wide range of chemicals, namely organic acids, are produced by fermentation. Given the inhibitory aspects of all acidogenic fermentations, extractive fermentation seems very promising. In this case, an immiscible organic solvent is used to extract *in situ* the inhibitory product, thus improving the productivity of the process. However, solvent toxicity and incompatibility with the fermentation pH have so far limited the application of this alternative[1,2].

Extraction of organic acids with tertiary amines and phosphorus - based extractants is only possible when the pH is lower than the acids pKa value[3-5] and so, extractive fermentation is only feasible for microorganisms that are not significantly inhibited by a pH < pKa.

As the most appropriate pH range for lactic acid bacteria is between 5 and 7 and the pKa of lactic acid is 3.86, these extractants are not adequate for integrated extraction/fermentation process.

At neutral pH, only anion exchangers are able to extract lactate. The lactate extraction mechanism with a quaternary amine can be described by,

$$R_4 N^+ X^-_{(org)} + La^-_{(aq)} \rightleftharpoons R_4 N^+ La^-_{(org)} + X^-_{(aq)}$$

A quaternary ammonium salt (Aliquat 336) has already been reported for the recovery of amino acids, not coupled with the fermentation process[6,7], but its use was not attempted in extractive fermentation of organic acids, with the exception of preliminary studies abandoned due to toxicity problems[8].

Extraction using hydrophobic microporous membranes to immobilize the organic phase looks promising because back mixing effects and loss of carrier are minimized, reducing solvent toxicity[9-12].

The aim of this work is to evaluate the toxicity effect of the organic phase on the lactic acid fermentation kinetics. The effect of three different aqueous phase / organic phase contact conditions, namely, dispersion of the organic phase, bulk liquid membrane and supported liquid membrane, was also studied.

2   MATERIALS AND METHODS

2.1 Microorganism and media

2.1.1 *Microorganism*. The organism used was *Lactobacillus rhamnosus* NRRL B445, a homofermentative lactic acid producer.

2.1.2 *Media*. A medium with the following composition was used: 5g/L of yeast extract, 1 g/L of beef extract, 10 g/L of tryptone, 2 g/L of KH_2PO_4, 10 g/L of sodium acetate, 2 g/L of ammonium citrate, 0.2 g/L of $MgSO_4.7H_2O$, 0.05 g/L of $MnSO_4.H_2O$ and 1 mL/L of Tween 80. Glucose concentration was 60 g/L. The media pH was 6.3.

2.2 Solvent toxicity tests

Two sets of experiments were performed in order to:
(i) evaluate the effect of solvent concentration below saturation and (ii) test different conditions of aqueous phase / organic phase contact on lactic acid fermentation kinetics.

2.2.1 *Organic phase*. A quaternary ammonium salt Aliquat 336 (Fluka, A.G.) was used as a carrier and Shellsol A (Shell, Portugal) as diluent.

2.2.2 *Effect of solvent concentration*. In order to study the effect of solvent concentration on lactic acid fermentation three solvent concentrations were tested: 100%, 10% and 1% of broth saturation.

Prior to inoculation, the quantities of Aliquat 336 and Shellsol A refereed as the solubilities in water were added to the fermentation media in order to obtain a saturated broth. The 10% and 1% saturation were obtained by successive dilution with fresh media. Then, the bottles were inoculated.

2.2.3 *Effect of different conditions of contact*. Three different configurations were used : dispersion of the organic phase, bulk liquid membrane and supported liquid membrane. **Figure 1** shows the supported liquid membrane cell used. The liquid membrane is immobilized in a porous support polypropylene membrane,with a pore diameter of 0.1 μm (Gelman Sciences).

In the tests of dispersion of organic phase in the broth, equal volumes of broth and organic phase (30% (w/w) Aliquat 336 + 70% (w/w) Shellsol A) were mixed for 4 h. An emulsion was formed and after decantation the broth was transferred to the serum bottles and inoculated.

In the bulk and supported liquid membrane configurations, dispersion of the organic phase in the broth was avoided. In both cases, 10 mL of organic phase were brought in contact for 4 h, with 150 mL of broth. After this time the broth was transferred to the serum bottles and inoculated.

2.3 Fermentation procedure

All experiments were performed in duplicate and for each of the tested conditions a control fermentation was carried out for comparison with the solvent toxicity tests. The fermentations were performed in 100 mL serum bottles equipped with septa and were agitated at 120 rev min^{-1} at a constant temperature of 41°C. They were inoculated at 10% (v/v) with *Lactobacillus rhamnosus* in the exponential phase of growth.

2.4 Analytical Methods

2.4.1 *Lactic acid and glucose concentration*. Lactic acid and glucose concentrations were determined by HPLC. The column used was a Shodex SH 1011 (Showa Denko K.K., Japan) and the eluent was sulfuric acid 0.01 N. A refractive index detector (ERC, ERMA INC., Japan) was used.

122

Separations for Biotechnology 3

2.4.2 *Cell concentration*. Cell concentration was measured by optical density at 610 nm. Optical densities were related to dry cell concentrations by using a calibration curve.

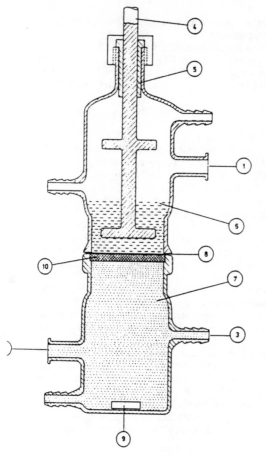

LEGEND:

1. Sampling point

2. Sampling point

3. Connection for continuous operation

4. Mechanical stirrer

5. Teflon seal

6. Organic phase

7. Fermentation media

8. Membrane

9. Magnetic stirrer

10. Porous glass support

Figure 1 Supported liquid membrane cell.

3   RESULTS AND DISCUSSION

3.1 Effect of Solvent Concentration

At 100% and 10% of broth saturation it was observed that the microbial culture did not grow. At 1% of broth saturation the fermentation was not significantly affected. Substrate consumption (S_0-S) and lactic acid produced (P) are similar to those obtained in the control fermentation (**Figure 2** and **Figure 3**). S_0 represents the initial substrate concentration and S the substrate concentration during the course of fermentation.

Cell concentration and growth are more affected. The maximum cell concentration (X_{max}) was obtained after 30 h from the beginning of the fermentation and remained constant after that time. For the 1% of broth saturation experiment $X_{max}=3.65g/L$ which is lower than the value obtained for the control experiment, $X_{max}=4.24g/L$, (**Figure 4**)

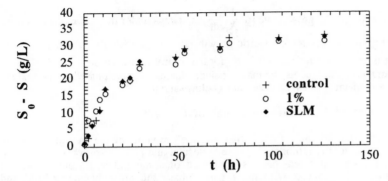

Figure 2 Comparison of glucose consumption (S_0-S) in the control, broth at 1% saturation and broth after contact in the supported liquid membrane cell

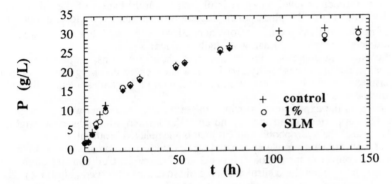

Figure 3 Comparison of lactic acid produced (P) in the control, broth at 1% saturation and broth after contact in the supported liquid membrane cell

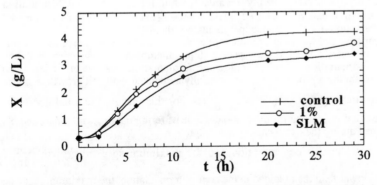

Figure 4 Comparison of cell dry weight concentration (X) in the control, broth at 1% saturation and broth after contact in the supported liquid membrane cell

and the curve of specific growth rate ($\mu = \frac{1}{X}\frac{dX}{dt}$) is always under the curve of the control

(**Figure 5**) giving a maximum specific growth rate (μ_{max}) of 0.49h^{-1}, which is about 10% lower than the value of μ_{max}=0.54 h^{-1} for the control fermentation.

From these experiments, we may conclude, that the organic phase is quite toxic to these microorganisms even at concentrations below saturation.

3.2 Effect of Different Conditions of Contact

As expected, in the case of dispersion of organic phase in the broth the fermentation was totally inhibited, because molecular and phase toxicity occurred. The broth was solvent saturated and also depleted of some nutrients due to protein denaturation. After contact with the organic phase the broth became turbid and precipitates could be noticed.

At this point, a distinction between two levels of solvent toxicity, molecular and phase toxicity, has to be stressed. In fact in a broth with higher solvent concentration than the saturation, (for example, due to droplets of emulsified solvent), the toxic effect would be not only at molecular level (enzyme inhibition or modification of the membrane permeability) but also at the phase level (depletion of nutrients due to extraction or blockage of nutrient diffusion due to a solvent coating) [13,14]. The latter type of toxicity may be eliminated if emulsion formation is totally avoided.

In the bulk and supported liquid membrane configurations, dispersion of the organic phase in the broth was totally avoided. It was observed that no growth occurred in the experiments in which the broth was in contact with the organic solvent in the bulk liquid membrane cell.

However, in the experiments in which the broth was in contact with the organic solvent in the supported liquid membrane cell, the fermentation was only slightly affected. The use of the hydrophobic support membrane delayed broth saturation.

In fact, both substrate consumption and lactic acid production are similar to those obtained in the control fermentation (**Figure 2** and **Figure 3**). Cell concentration and growth rate are lower than the control and a lag phase can be observed (**Figure 4**). The maximum cell concentration (X_{max}) obtained is 3.43 g/L.

Due to the lag phase, the curve of specific growth rate is affected and the value of μ_{max}=0.47h^{-1} is about 15% lower than the control (**Figure 5**). **Figure 6** shows that after a lag phase of approximately 1 h the ratio specific growth rate/control specific growth rate (μ/μ_0) increases steeply towards 1, while for the broth at 1% saturation a slow increase to 1 can be observed. Data from **Figures 5** and **6** was obtained by numerical derivation of cell growth experimental results.

A comparison of the yields, productivities and specific rates obtained in the control and in the broth after contact in the supported liquid membrane cell is shown in **Table 1**. Lactic acid fermentation, as other acidogenic fermentations follows a Gaden II type behaviour, exhibiting a product formation partially associated with cell growth, according to Luedking-Piret expression, $\frac{1}{X}\frac{dP}{dt}=\alpha\frac{1}{X}\frac{dX}{dt}+\beta$. For this reason, the yield coefficient, $Y_{P/S}$, is evaluated using data for the complete time of fermentation, while $Y_{P/X}$ and $Y_{X/S}$ are determined during the exponential phase of growth.

Although the cell concentration and the maximum specific growth rate, ($\frac{1}{X}\frac{dX}{dt}$)max, are lower than the obtained in the control fermentation, the maximum lactic acid specific product rate ($\frac{1}{X}\frac{dP}{dt}$)max, the maximum specific substrate rate ($-\frac{1}{X}\frac{dS}{dt}$)max and substrate conversion yield, $Y_{P/S}$, are similar. An increase on the culture specific activity

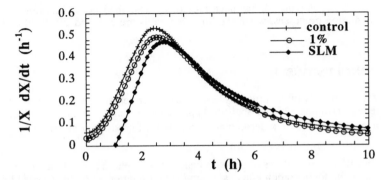

Figure 5 Comparison of specific growth rate in the control, broth at 1% saturation and broth after contact in the supported liquid membrane cell

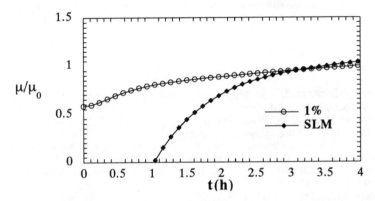

Figure 6 Comparison of the ratio specific growth rate (μ) / control specific growth rate (μ_0) in the broth at 1% saturation and broth after contact in the supported liquid membrane cell

Table 1 Substrate consumption, final lactic acid concentration, maximum cell concentration, yields, productivities and specific rates obtained in the control and broth after contact in the supported liquid membrane cell

	S_0-S (g/L)	P (g/L)	X_{max} (g/L)	$Y_{P/S}$ (g/g)	$Y_{P/X}$ (g/g)	$Y_{X/S}$ (g/g)
Control	33.3	31.3	4.24	0.852	3.38	0.265
SLM	32.8	28.7	3.43	0.825	3.91	0.232

	$(P/t)_{max}$ (g/Lh)	$(\frac{1}{X}\frac{dP}{dt})_{max}$ (g/gh)	$(-\frac{1}{X}\frac{dS}{dt})_{max}$ (g/gh)	$(\frac{1}{X}\frac{dX}{dt})_{max}$ (h^{-1})
Control	0.911	2.38	3.47	0.54
SLM	0.831	2.46	3.32	0.47

even if the growth rate is reduced, has been previously described in the literature for different fermentation processes, when the microbial cells are exposed to osmotic or chemical stress[15].

4   CONCLUSIONS

The effect of toxicity of the organic phase on lactic fermentation was studied. The organic phase used (Aliquat 336 and Shellsol A) proved to be quite toxic, in accordance with other authors work[8,14]. The fermentation was not affected, only at 1% of broth saturation.

However different conditions of aqueous phase/organic phase contact may lead to distinct effects on the fermentation kinetics. If emulsion formation is totally avoided by using supported liquid membrane contact, toxicity can be drastically reduced. The results indicate that the product yield and rate are not affected and a tolerable small decrease in the cell growth rate is obtained.

Establishment of the conditions to a sucessful *in-situ* extraction of lactate, using the organic phase immobilized in hydrophobic microporous hollow fiber modules, is under study.

REFERENCES

1.   T.B. Vickroy, Lactic acid, in M. Moo-Young (Ed.) Comprehensive Biotechnology, Pergamon Press, vol. 3, 761, 1985.
2.   V.M.Yabannavar and D.I.C.Wang, Biotechnol. Bioeng., 1991, 37, 1095.
3.   A.S. Kertes and C.J. King, Biotechnol. Bioeng., 1986, 28, 269.
4.   J.A. Tamada, A.S. Kertes and C.J. King, Ind. Eng. Chem. Res., 1990, 29,1319.
5.   J. Chaudhuri and D.L. Pyle, Liquid membrane extraction, in M. S. Verrall and M.J. Hudson (Eds.), Separations for Biotechnology, Ellis Horwood Publ., Chichester, U.K., 241, 1987.
6.   M.P.Thien, Ph.D. Thesis, MIT, U.S.A, 1988.
7.   S.Schlichting, W.Halwachs and K.Schügerl, Chem. Eng. Comm., 1987, 51, 193.
8.   S.R.Roffler, Ph.D. Thesis, University of California, Berkeley, U.S.A., 1986.
9.   P. Nuchnoi, T. Yano, N. Nishio and S. Nagai , J. Ferment. Technol., 1987, 65, 301.
10. J.J. Pellegrino and R.D. Noble, Trends Biotechnol., 1990, 8, 216.
11. A. Sengupta, R. Basu, R. Prasad and K.K. Sirkar, Sep. Sci. and Technol., 1988, 23, 1735.
12. R.Basu and K.K.Sirkar, AIChE Journal, 1991, 37, 383.
13. T. Cho and M.L. Shuler , Biotech. Prog., 1986, 2, 53.
14. R. Bar and J.L. Gainer, Biotech. Prog. ,1987, 3, 109.
15. S.J. Pirt, 'Principles of microbe and cell cultivation', Blackwell Scientific Pub., Oxford, U K, 1975 .

ACKNOWLEDGEMENTS

We acknowleged the financial support of Gesellschaft für Biotechnologische Forschung (GBF) and Junta Nacional de Investigação Científica e Tecnológica (JNICT).

Towards a Mathematical Model for the Flocculation of *Escherichia coli* with Cationic Polymers

R. H. Cumming, P. M. Robinson, and G. F. Martin

SCHOOL OF SCIENCE AND TECHNOLOGY, UNIVERSITY OF TEESSIDE,
MIDDLESBROUGH, CLEVELAND TSI 3BA, UK

1. INTRODUCTION

Flocculation by a polymer is thought to be brought about by the charge of the polymer (charge neutralisation) or its molecular weight (bridging flocculation). In the case of charge neutralisation the polymer is of opposite charge to the bacterial cells and its adhesion to the cells surface is thought to bring about an overall neutralisation of the surface charge or a *localised* reversal of charge (a patch) which allows effective particle collisions and floc growth. In both situations it is possible to "overdose" the cell surface when an excess of polymer is adsorbed and the overall charge of the cell is reversed, preventing effective particle collisions and floc growth. In the case of bridging flocculation, the optimum surface coverage has been estimated as 50%, and thus excess polymer adsorption can lead to a reduction in the extent of flocculation. Hence it is customary to perform a dose curve with a flocculant to determine the concentration which gives the maximum extent of flocculation (the optimum). Some earlier work on the method of polymer addition gives an insight into the dynamics of the flocculation process [1]. These workers found that the extent of flocculation of kaolin with a range of polymers increased if the polymer was added dropwise over the mixing period compared with when the polymer was added as a slug. We have confirmed this finding in work with flocculation of *E. coli* with a range of polymers [2].

The work presented here shows that overdosing with polymer is not a problem with the polymer addition technique used here, and that a general model can be formulated for the dose curve of bacteria flocculated with four different polymer types.

2. MATERIALS AND METHODS

Cell cultivation and flocculation procedure

Continuously grown cells of *E. coli B* were cultivated as described previously [2]. Flocculation of the cells was achieved by adding 2ml of the appropriate dilution of the polymer to give the desired final concentration to 50 ml of cell suspension in a 100 ml beaker. Mixing was achieved with a stirrer bar 24 mm in length and 10 mm in diameter. The polymer was added dropwise over the entire 10 minute mixing period while stirring was maintained at a constant speed of 500 rpm. The pH of the flocculation experiments was equal to that of the broth (pH 6.5). At the end of the mixing period, 4 ml samples were removed with a wide bore pipette and placed in plastic cuvettes. After a settling period of 1h, the optical density(OD) at 560 nm was measured. Controls were run at the same time with cells in the absence of the polymer to measure any background settling. The percentage reduction in OD was calculated from the final OD of the test and its control. The cell dry weight was 0.37 ± 0.01 g l^{-1}(expressed as ± 1 standard deviation).

Polymers:

Four cationic polymers labelled A,B,C and D were supplied by Allied Colloids Ltd, Bradford. The intrinsic viscosity and charge density was measured for each[3]. The charge density and molecular weight of the polymers used is given in table 1.
The polymers used in this study were dissolved in de-ionised water using acetone as a wetting agent (Allied Colloids), and made to a 1% ^{w}/$_v$ stock solution. With a knowledge of the charge density and mass added to the cells ,it was possible to calculate the dose in terms of milliequivalents of polymer/g cells.

Table 1 : Properties of the polymers used in the study . The molecular weight of the polymers is based on the Mark-Houwink correlation, with constants of k=3.73 10^{-4} cm^3 g^{-1}and α= 0.66 (supplied by the manufacturer).:

polymer	intrinsic viscosity cm^3 g^{-1}	Molecular mass	charge density meq g^{-1}
A	12.9	7.6×10^6	1.4
B	12.8	7.4×10^6	3.3
C	7.5	3.8×10^6	1.3
D	7.5	3.8×10^6	3.6

3. RESULTS AND DISCUSSION

The dose curves for each polymer type are given in fig 1. It can be seen that the dose curves follow the same trend over the dose range, but give different amounts of flocculation according to their type. It is evident that the polymers having low charge

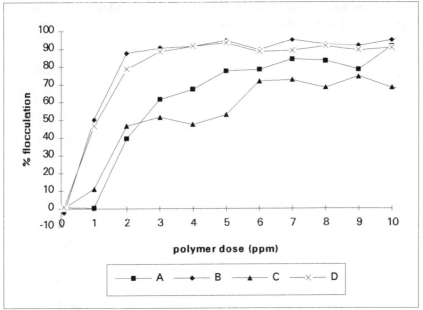

Fig 1. The polymer dose curve for the four polymers with the dose expressed in ppm for the 4 polymers A-D

density (A and C) give the poorest flocculation . Fig 2 shows the effect of expressing the polymer dose as charge added per gram of cells. It is apparent that the dose curves become more similar in shape, irrespective of their molecular weight.

The conclusion which can be made from the fact that the dose curves can be combined to produce one reasonable dose curve, providing the amount of polymer added is expressed as charge added per gram of cells, is that the molecular weight of the polymer is fairly unimportant in the flocculation of *E. coli*, under the conditions of the assay. It seems that the charge density of the polymer is its most important attribute, and that flocculation is thus brought about by charge neutralisation.

A model for the dose curve.

For a system containing cells of negative charge to which polymer of opposite charge is added, there should be equivalence during the neutralisation process. Thus the amount of microbial charge neutralised by the polymer is directly proportional to the amount of added charge, i.e.

$$Q_x = mQ_p \qquad\qquad ----- \quad 1$$

where Q_x is the total cell charge neutralised (meq.) and Q_p is the total charge added (meq.) through the polymer addition. Ideally $m=1$ for equivalence. The total charge present on the cells is the product of the mass of cells , X (g) and the specific charge density q_x (meq/g cells). Similarly, the total polymer charge added is the product of the mass of polymer added P (g) and its specific charge density q_p (meq./g). Thus:

$$Xq_x = mPq_p \qquad\qquad ----- \quad 2$$

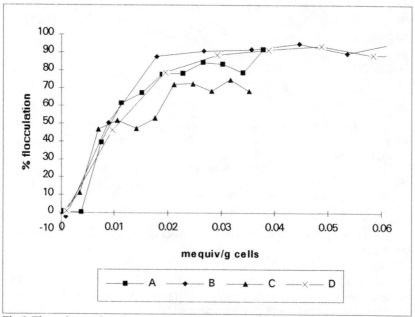

Fig 2 The polymer dose curve expressed in mequivalents/g cells for the 4 polymers A-D

For a finite mass of cells in the system, with an initial mass of X_0 (g), the fraction of cells neutralised is given by X/X_0. Hence the fraction of initial cells neutralised is given by:

$$\frac{X}{X_0} = \frac{mPq_p}{X_0 q_x} \qquad\qquad ----- \quad 3$$

If the fraction of cells neutralised is equivalent to the fraction of cells flocculated, then the percentage of cells flocculated, F (%), is

$$F = \frac{100X}{X_0} = \frac{100mPq_p}{X_0 q_x} \qquad\qquad ----- \quad 4$$

If we combine Pq_p/X_0 to give the amount of polymer added per gram of cells to give Q'_p (meq./g cells), the percentage flocculation is given by

$$F = \frac{100 m Q_p'}{q_x} \qquad \text{----- } 5$$

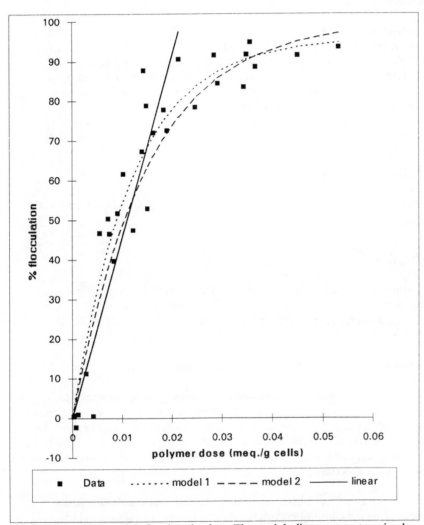

Fig 3 The various models fitted to the data. The straight line represents a simple linear model. The two curves represent the two logarithmic models discussed in the text. Model 1 is with a maximum flocculation set at 95%, model 2 is with a maximum flocculation set at 100%

Thus a plot of percentage flocculation versus Q'p should give a straight line, up to the point where 100% flocculation has occurred. The equation does not predict the extent of flocculation beyond this point. Increasing the dose beyond the dose which gives the maximum extent of flocculation does not change the extent of flocculation (i.e. there is no overdosing) Thus in Fig 3, the dose curve should be linear, with a slope equal to the inverse of the specific cell charge density(for $m =1$).

It can be seen that the dose curve does, to some extent, follow this predicted trend. However, the percentage flocculation never quite reaches 100% , levelling at a plateau of around 93-95%. From the slope of this line drawn through the lower doses, before a constant maximum extent of flocculation is reached, q_x can be calculated as 3.3×10^{-6} meq. /g cells. It is proposed that the specific charge density will be dependent on the environmental conditions such as pH and ionic strength.

More of the dose curve can be encompassed by fitting a logarithmic model to the data:

$$F =100(1 -e^{-kQ'_p})$$ ----- 6

This is also shown in Fig 2, together with a model which does not assume 100% flocculation:

$$F =F_{max}(1 -e^{-kQ'_p})$$ ----- 7

It can be seen that a good fit over a wide dose range is achieved. Adoption of a logarithmic model which predicts a lower maximum extent flocculation (F_{max}) instead of a maximum of 100% , could be explained by steric hindrance .This would be caused by the adsorbed polymer molecules impeding access to the cell surface of further polymer molecules.

The constant k has units of g cells/meq., and is the reciprocal of q_x .

It was noted that 100% flocculation was rarely achieved with the experimental technique used. It was possible to remove the remaining cells by centrifugation, which then gave a residual OD equivalent to 99.5% removal. However, it was noted that sometimes flocs adhered to the walls of the cuvette during the settling tests. These would lead to a higher residual OD than expected, and may be the reason why the OD never reached zero when total flocculation had occurred.

Other workers have noted a stoichiometry between charge added and flocculation performance. Leu and Ghosh (1988) studied a polymer-bentonite system and observed that there was a linear stoichiometric relationship between the bentonite concentration (mg/L) and the optimal dose expressed in units of charge added(coul/L). We are currently attempting to fit these model to data collected at different pH's and ionic strengths.

ACKNOWLEDGEMENTS

We gratefully acknowledge the helpful comments and support from Allied Colloids during this work, and to the SERC for the CASE studentship for PMR.

4. REFERENCES

1.   R.O Keys and R. Hogg, <u>AIChe Symposium series Water</u> No. 90,63, 1978
2.   P.M.Robinson, G.F. Martin,   and R.H.Cumming, ,1993, Accepted for <u>Bioseparations</u> ,1994
3.   Anon, Pollution control Division Training manual,, Allied Colloids, Bradford,UK,1982
4.   R-J. Leu and M.M. Ghosh, 1989, <u>Journal AWWA</u>, April, 159-167

Macro-Prep Ceramic Hydroxyapatite—New Life for an Old Chromatographic Technique

L. J. Cummings, T. Ogawa,[a] and P. Tunón

BIO-RAD LABORATORIES, 2000 ALFRED NOBEL DRIVE, HERCULES, CA 94547, USA

[a]PENTAX ASAHI OPTICAL CO. LTD., 2-36-9, MAENO-CHO, ITABASHI-KU, TOKYO, 174 JAPAN

1 INTRODUCTION

Two new types of chemically pure hydroxyapatite for chromatography of biomolecules have been developed, Macro-Prep® Ceramic Hydroxyapatite and Macro-Prep Ceramic Hydroxyapatite, type II. The physical properties of these new materials have been greatly improved to overcome the limitations of crystalline hydroxyapatite. The new materials are macroporous, spherical, and have been used at high flow rate and pressure. Columns have been scaled from bench top size to process size and used over a large number of runs. The Macro-Prep Ceramic Hydroxyapatite has high protein binding capacity ($\geq$ 20 mg BSA/gm and $\geq$ 25 mg lysozyme/gm) with elution characteristics similar to that of crystalline hydroxyapatite yet without the inherent problems of crystalline hydroxyapatite. The Macro-Prep Ceramic Hydroxyapatite, type II has a lower binding capacity for proteins relative to lysozyme, usually about half that of the former, and has chromatographic properties dissimilar to other types of hydroxyapatite.

Hydroxyapatite is a unique form of calcium phosphate, which has proven itself an advantageous tool for the fractionation and purification of a wide variety of biological molecules, such as subclasses of IgG, enzymes, antibody fragments, and nucleic acids. It provides a separation approach different from that of any other established technique such as ion exchange, hydrophobic interaction, affinity or size exclusion chromatography. The separation is strongly influenced by the isoelectric point of the sample molecules, with acidic proteins eluting first. Hydroxyapatite chromatography can be used as a polishing step to separate closely co-purifying molecules[1,2,3,4] or for the initial purification of crude samples[5], since it adsorbs virtually no low molecular weight substances such as nucleotides, salts or amino acids, yet binds larger molecules and separates them. Hydroxyapatite can be used in the presence of detergents and glycerol[6]. The material can be sanitized with sodium hydroxide, and autoclaved. In its crystalline form, manufactured using the original method[7], hydroxyapatite was commercialized as Bio-Gel® HT hydroxyapatite by Bio-Rad Laboratories in 1963. The dry powdered form, Bio-Gel HTP hydroxyapatite, was introduced in 1964. Crystalline hydroxyapatite is widely used in analytical and preparative biomolecule research, as well as in industrial production of biologically active substances. The fragility of the crystals limits flow rates especially during multiple uses of the packed columns. Porous spheroidal particles of hydroxyapatite were developed in the early 1970's[8] and commercialized by BDH, Ltd., Dorset, England. The BDH hydroxyapatite is larger than 100 microns.

High performance liquid chromatography applications using hydroxyapatite were somewhat elusive because of the fragility of the crystalline forms and the unavailability of smaller spheroidal particles. In 1986 the properties of a newly developed hydroxyapatite column containing completely spherical and porous beads were studied[9]. Greater than 90%

of the amount of each protein applied to the 0.8 cm X 10 cm column was recovered. The chromatographic behavior of small and large particle sizes (2 to 3 μm and 6 to 8 μm) of this new spherical and porous hydroxyapatite were compared in 0.8 cm X 10 cm columns[10]. Twenty-one proteins were used in the study and for the first time, the material was described as a porous-spherical ceramic hydroxyapatite. The synthesis and chromatographic properties of larger particle sizes (10 μm, 20 μm, and 40 μm) were described in 1989 in terms of resolution and sample load as a function of microstructure and particle size distribution[11].

The newly developed types of ceramic hydroxyapatite overcome the limitations of crystalline hydroxyapatite. The duration and temperature of the sintering[12] step determines whether the ceramic hydroxyapatite is Macro-Prep Ceramic Hydroxyapatite or Macro-Prep Ceramic Hydroxyapatite, type II. The particle size distribution of these two types of ceramic hydroxyapatite is determined during the particle forming step which is tuned to produce high percentages of either 20, 40 or 80 μm size particles. These different particle size ranges allow developmental and analytical applications using the two smaller size ranges and preparative applications using the larger size range. High resolution preparative applications may be achievable with the 40 μm particle size range.

2 MATERIALS AND METHODS

All buffer reagents were purchased from Spectrum Chemical Manufacturing Corporation, Gardena, California, U.S.A. Proteins and calf thymus deoxyribonucleic acid (DNA) were purchased from Sigma Chemical Company, St. Louis, Missouri, U.S.A. Ten millimolar phosphate buffer (pH 6.8) was sterile filtered then freshly boiled and cooled before use. Bio-Gel HT hydroxyapatite (HT), Bio-Gel HTP hydroxyapatite (HTP), Macro-Prep Ceramic hydroxyapatite (MP) and Macro-Prep Ceramic hydroxyapatite, Type II (MPII), were obtained from Bio-Rad Laboratories, Hercules, California, U.S.A, or Pentax, Asahi Optical Co., Ltd., Itabashi-ku, Tokyo, Japan. The Automated Econo System, a multitasking low pressure chromatography system, was purchased from Bio-Rad Laboratories. A Perkin-Elmer Lambda 3 UV/Visible Spectrophotometer, Norwalk, Connecticut, U.S.A., was used to determine protein and DNA binding capacities.

Protein mixtures for chromatography on MP were prepared by weighing approximately 500 mg ovalbumin, 220 mg myoglobin, 170 mg α-chymotrypsinogen A, and 160 mg cytochrome c. The proteins were dissolved in 100.00 ml of 10 mM buffer. Protein mixtures for chromatography on MPII were prepared by weighing approximately 500 mg ovalbumin, 220 mg myoglobin, 160 mg ribonuclease A, 170 mg α-chymotrypsinogen A, and 160 mg cytochrome c and dissolving them in 100.00 ml 10 mM buffer. The comparative chromatograms between HT and MP were obtained using a protein mixture containing 133 mg bovine α-lactalbumin, 220 mg myoglobin, 170 mg α-chymotrypsinogen A, and 160 mg cytochrome c dissolved in 100.00 ml 10 mM buffer. In each of the described protein mixtures, the proteins elute in the order listed.

A solution of DNA was prepared by adding 10 mg to 50.00 ml of 10 mM buffer and dissolving it overnight at 4 °C. A DNA mixture consisting of single stranded (ss-DNA) and doubled stranded (ds-DNA) was prepared by heating 5.00 ml of the dissolved DNA in a 15 ml polypropylene centrifuge tube to 85°C for 60 minutes. The solution was quenched by immersing the tube in an ice bath for 10 minutes. Finally, 5 ml of the boiled DNA was mixed with 5 ml of DNA in a 15 ml polypropylene centrifuge tube. Aliquots of this solution were used within 4 hours to test the ability of the hydroxyapatite samples to separate ss-DNA from ds-DNA.

Solutions containing 2 mg/ml lysozyme or 1 mg/ml bovine serum albumin were prepared in 10 mM buffer and used to determine protein binding capacity of the hydroxyapatite samples. The DNA binding capacity of the hydroxyapatite samples was determined using a solution containing 20 μg/ml DNA prepared by diluting 20 ml of the previously prepared DNA solution to 200 ml with 10 mM buffer.

The separation of protein mixtures and ss-DNA/ds-DNA was accomplished using linear gradients of sodium phosphate. Samples were loaded onto columns containing 2.8 ml of hydroxyapatite (1.0 cm I.D. X 3.5 cm) which had been equilibrated with the 10 mM sodium phosphate buffer (pH 6.8). After a brief wash with 10 mM buffer, the samples were eluted using the terminal sodium phosphate buffer (pH 6.8) and conditions described in the figures.

Capacities for BSA, lysozyme, and DNA were determined by a batch method using approximately 400 mg of dry hydroxyapatite in each of the tests. The amount of sample adsorbed was determined by the difference between the amount in the initial solution and the amount remaining after equilibrating the solution with the hydroxyapatite. This difference is divided by the amount of hydroxyapatite used for the test. The absorbancy was measured at 280 nm for BSA and lysozyme and at 260 nm for DNA.

Flow rates were determined using a 40 cm hydrostatic head applied to 1.5 cm I.D. columns packed to a height of approximately 14 cm with hydroxyapatite. Flow rates were recorded within a few minutes of achieving a stable bed height and determined by measuring the volume collected within the first 10 minutes. Packed density was determined by calculating the quotient of the dry weight of hydroxyapatite and the volume of the hydroxyapatite in the column. The density of the hydroxyapatite supports was also determined by centrifuging a suspension of the powder and 10 mM buffer for 15 minutes. The centrifuged density was determined by calculating the quotient of the dry weight of hydroxyapatite and the volume of the packed hydroxyapatite in the centrifuge tube.

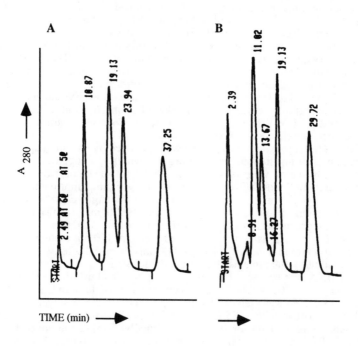

Figure 1 *Chromatograms obtained on MP (A) and MPII (B) using the following conditions: 200 µL sample size, 10 mM loading buffer, 2 minute wash with 10 mM buffer, linear gradient to 200 mM buffer in 50 minutes. The flow rate was constant at 2.0 ml/min or 152 cm/hr. Chromatographic peaks in order of elution: (A) ovalbumin, myoglobin, α-chymotrypsinogen A, and cytochrome c. (B) ovalbumin, myoglobin, ribonuclease A, α-chymotrypsinogen A, and cytochrome c.*

3 RESULTS AND DISCUSSION

One of the differences in chromatographic properties between MP and MPII is shown in **Figure 1**. The 20 μm particle size range of the base material, lot number CEC0819, was sintered separately to make the two types of ceramic hydroxyapatite, MP **Figure 1A**, and MPII **Figure 1B**. Retention times for proteins were longer with MP and are measured from the time of injection. Ribonuclease A is not included in the protein sample for MP because it generally co-elutes with myoblobin using the conditions described. Ovalbumin is retained on MP (10.87 minutes) while unretained on MPII (2.39 minutes). The experienced chromatographer would question the retention of 2.39 minutes as being disproportionate to the column volume. The longer than expected time can be explained by the volume of the tubing in the Econo System. The same retention behavior is seen with bovine serum albumin.

The Econo System was programmed to cycle columns through the gradient, a rinse with 500 mM buffer for 10 minutes and a rinse with 10 mM buffer for 20 minutes. The two columns were challenged with 50 cycles without an apparent loss in performance.

HT and HTP are essentially the same material according to the manufacturer. HT is simply a slurry of hydroxyapatite in 10 mM phosphate buffer, whereas, HTP is the dried product made by processing HT to dryness. Columns (1.0 cm I.D. X 3.5 cm) were packed with Bio-Gel HT, lot number 40439 and MP 40 μm, lot HJK0510D. The retention for each protein was determined relative to the gradient profile shown with each chromatogram in **Figure 2**. The two materials match surprisingly well.

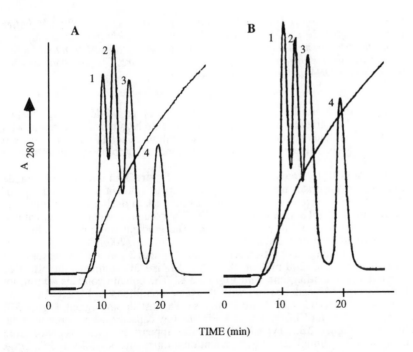

Figure 2   *Chromatograms obtained on MP (A) and HT (B) using the following conditions: 100 μL sample size, 10 mM loading buffer, 10 minute rinse with 10 mM buffer, linear gradient to 400 mM buffer in 26 minutes. The flow rate was constant at 0.8 ml/min or 60 cm/hr. Chromatographic peaks: (1) bovine α-lactalbumin, (2) myoglobin, (3) α-chymotrypsinogen A, and (4) cytochrome c.*

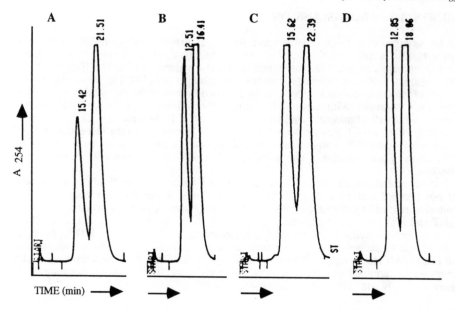

Figure 3 *Chromatograms of ss-DNA/ds-DNA on 20 μm MP (A) with 200 mM buffer terminal buffer, 20 μm MP (B) with 280 mM terminal buffer, 40 μm MP (C) with 200 mM terminal buffer, and 40 μm MPII with 280 mM terminal buffer. Flow rates were constant at 2.0 ml/min or 152 cm/hr. The first peak in each example is ss-DNA and the last is ds-DNA.*

The fragility of the crystalline material, HT, was observed while cycling the two columns for additional separations of the protein test mix. Both columns were rinsed with 10 mM buffer for 15 minutes before repeating the separation. The HT column collapsed after the fourth cycle. The MP column was cycled 35 times with no observable change in performance.

The separation of ss-DNA from ds-DNA is shown in **Figure 3** on three lots of ceramic hydroxyapatite and demonstrates the effects of particle size, gradient profile, and type of ceramic hydroxyapatite relative to retention time. **Figure 3A and 3B** are chromatograms obtained with 20 μm, lot number CEC0819 type MP. The column was equilibrated with 10 mM buffer, then injected with 200 μL of sample. The gradient was initiated after a 2 minute rinse with 10 mM buffer, reaching terminal concentration, 200 mM for 3A and 280 mM for 3B, within 20 minutes. The difference between **Figures 3A and 3B** is the selection of the terminal buffer, either 200 mM or 280 mM. **Figure 3C** compares the effect of particle size relative to retention time with **Figure 3A**. The column contains 40 μm, lot number CEC0818 type MP and was run with the same conditions as **Figure 3A**. The retention times for ds-DNA and ss-DNA are very close. The last chromatogram, **Figure 3D**, was achieved with 40 μm, lot CEC0818 type MPII, the low capacity material run with the same conditions as **Figure 3B**. Type MPII material appears to retain ds-DNA more strongly than type MP. Though not shown, retention times for ss-DNA and ds-DNA obtained with HT and HTP were similar to those obtained with MPII.

The chromatographic properties of MP appear to be similar to HT and HTP relative to protein separations. On the other hand, the MPII seems to behave like HT and HTP relative to DNA separations. **Table 1** lists the binding capacities of BSA, lysozyme and DNA and may help explain these observations.

The data in **Table 1** show that MP has very high binding for BSA and lysozyme. With exception of lot 45407, HT and HTP also have high binding capacities for BSA. On the other hand, MPII has a much lower binding capacity for BSA than MP, HT and HTP and about 75% the binding capacity for lysozyme as MP. The data show that only HT and HTP have high binding capacity for DNA. One could conclude that MP, HT and HTP behave similarly when separating proteins because binding capacities for BSA are similar. The results in **Figure 2** help support this conclusion.

The differences observed in DNA separations is difficult to explain based on the data in **Table 1**. Clearly, the DNA binding capacity of HT and HTP is much greater than either the MP or MPII. Yet, the retention behavior of MPII is similar to that of HT and HTP. The differences may be related to the porosity of these types of hydroxyapatite as a function of their sintering conditions[13].

The data in **Table 1** also show some interesting differences between the ceramic and the crystalline types of hydroxyapatite relative to packing densities. MP and MPII have higher packing densities than HT and HTP as measured by the two methods. The differences are noticeably greater with the column method. As mentioned previously, columns of HT and HTP collapsed after a few uses. These columns were used at 60 cm/hr, exceeding their minimum specified flow rate of 35 cm/hr and 25 cm/hr. When columns used for flow rate determination were cycled alternately with 50 ml each of 10 mM buffer and 400 mM buffer, the HT and HTP columns collapsed to about 88% of their initial volume and the flow rates declined to less than 25 cm/hr. Eventually, the HT and HTP columns stopped flowing. The flow rates of MP and MPII columns remained constant but collapsed to about 90% of their initial volume.

HT, lot number 46194, and HTP, lot number 47647, were removed from their columns and dispersed in 10 mM buffer. Two phases were noted for each, cloudy supernatants and densely settled layers. The cloudy supernatant consisted of numerous small irregular particles. Cloudy layers were decanted and the settled layers suspended with 10 mM buffer to further remove the fine particles. After 4 repetitions, the supernatants were clear. Columns were repacked and flow rates determined. The repacked HT flowed much slower than it did originally, 45 cm/hr. The HTP column flowed 30 cm/hr. When recycled with 10 mM and 500 mM buffer, the flow rates declined as previously described. The columns did not collapse as they did when first used.

The high linear flow rates used while separating proteins or ss-DNA from ds-DNA on MP and MPII columns suggest that high flow rates should be achievable with larger columns. The larger columns tested at 40 cm of hydrostatic head were connected to the Econo System and challenged with 10 mM and 500 mM buffer using the system's EP-1 Econo Pump. The 40 μm MP and MPII achieved linear flow rates up to 500 cm/hr at less than 15 psi. The 80 μm ceramic hydroxyapatites achieve linear flow rates up to 640 cm/hr, the maximum flow rate of the system.

The two new types of hydroxyapatite, Macro-Prep Ceramic Hydroxyapatite and Macro-Prep Ceramic Hydroxyapatite, Type II, retain many of the unique properties of crystalline hydroxyapatite allowing bench and process scale chromatographers to continue using well established buffer systems. Both types may be used with greater ease because they retain flow properties and do not degrade while cycling repeatedly with buffers having large differences in concentration. MP and MPII behave differently, however, and may have different applications. Bio-Gel HT and Bio-Gel HTP, though more difficult to use in bench and process scale columns, are useful for 3 to 4 cycles, and perhaps more cycles by the experienced user.

Table 1 *Density, Binding Capacity and Flow Rate*

Lot No.	Type	Particle Size μm	Density Centrifuge g/ml	Density Column g/ml	BSA binding mg/g	Lysozyme binding mg/g	DNA binding μg/g	Flow rate cm/hr
BEC2419-20-II	MPII	20	0.54	-	2.2	14.8	338	-
CEC0819-20-II	MPII	20	0.52	-	2.0	15.0	342	-
BEC2419-20	MP	20	0.55	-	25.7	21.0	428	-
CEC0819-20	MP	20	0.55	-	25.2	21.8	415	-
CEC0818-40-II	MPII	40	0.53	0.48	2.9	15.9	237	15.3
CEC2219-40	MP	40	0.59	0.49	23.9	22.8	197	17.3
CEC0819-40-II	MPII	40	0.53	0.48	5.7	13.9	196	17.3
BEC2419-40	MP	40	0.54	0.48	25.8	20.7	254	17.7
CEC2218-40	MP	40	0.57	0.50	22.5	22.8	184	21.4
CEC0818-40	MP	40	0.58	0.49	24.8	22.4	257	17.0
CEC0819-40	MP	40	0.56	0.49	24.3	21.1	240	17.3
CEC0813-80-II	MPII	80	0.56	0.50	7.2	14.8	144	56.2
CEC2213-80-II	MPII	80	0.58	0.51	4.5	15.1	96	45.9
CFC0804-80-II	MPII	80	0.56	0.51	6.7	13.9	126	51.3
CFC0805-80-II	MPII	80	0.60	0.50	6.4	14.0	117	58.1
CFC0806-80-II	MPII	80	0.56	0.51	6.3	13.7	116	50.3
CEC0813-80	MP	80	0.56	0.50	22.6	21.6	142	59.4
CEC2213-80	MP	80	0.59	0.52	21.2	21.1	117	44.5
45407	HT	-	0.45	0.29	13.0	-	780	146.0
46194	HT	-	0.42	0.31	16.0	-	700	138.0
46244	HT	-	0.40	0.39	20.0	-	960	114.0
40439	HT	-	0.43	0.37	17.0	-	840	135.0
46856	HTP	-	0.46	0.30	20.0	-	940	107.0
47647	HTP	-	0.45	0.30	21.0	-	687	45.0
48611	HTP	-	0.46	0.40	30.0	-	990	25.3

References

1. Yalchelini, P., et al, *J. Immunology,* 1990, **145**, 1382
2. Schott, K., et al, *J. Biol. Chem.*, 1990, **265**, 4204
3. Mahoney, C. W., et al, *J. Biol. Chem.*, 1990, **265**, 5424
4. Newton, A.C. and Doshland, D.E. Jr., *Biochemistry*, 1990, **29**, 6656
5. Ganong, B.R., *Biochemistry*, 1990, **29**, 6904
6. Nguyen, L.B., et al, *J. Biol. Chem.*, 1990, **265**, 4541
7. Tiselius, A., Hjertén, S., and Levin, O., *Arch. Biochem. Biophys.*, 1956, **65**, 132
8. Tompson, A., and Miles, B. J., *Methodological Developments in Biochemistry*, 1973, **2**, 95.
9. Kadoya, T., et al, *J. Liq. Chromatogr.*, 1986, **9**, 3543
10. Kadoya, T., et al, *J. Liq. Chromatogr.*, 1988, **11**, 2951
11. Ogawa, T., et al, Poster Presentation, Proceedings of the 10th Conference on Liquid Chromtography, June 25-30, 1989, Stockholm, Sweden
12. ibid
13. S. Inoue and N. Ohtaki, *J. Chrom.*, 1990, **515**, 193.

Scale-up and Intensification of the Isolation of Abnormal PrP Present in Bovine Brain Infected with Spongiform Encephalopathy

C. J. Dale, S. G. Walker, and A. Lyddiatt

BIOCHEMICAL RECOVERY GROUP, UNIVERSITY OF BIRMINGHAM,
EDGBASTON, BIRMINGHAM B15 2TT, UK

1 INTRODUCTION

Transmissible spongiform encephalopathies (TSE) are a group of fatal neurodegenerative diseases affecting mammals (sheep, cow, deer, mink and man). The causative agent of these diseases has yet to be conclusively established however most studies (6,7) exclude a viral or bacterial origin of infection. Transmissible spongiform encephalopathy infectivity has been found to co-purify with aggregates of an abnormal isoform of a host cellular protein (PrP) (5). Indeed a large body of convergent evidence suggests that this protein is the sole component of the infectious agent or prion. If this hypothesis is correct then PrP is a unique pathogen.

The normal cellular form of PrP (PrP^c) is a membrane bound glycoprotein of relative molecular mass Mr 35 000. The protein comprises a single polypeptide chain with a C-terminal glycophosphatidyl inositol moiety (GPI membrane anchor). The function of this protein within the cell is unknown. Studies with mice lacking the PrP gene suggest that the protein is either non-essential for normal growth and development or that its function in PrP ablated animals is compensated by other cellular components (2). Animals affected by spongiform encephalopathy are characterised by intracellular accumulation of PrP. This isoform of PrP (PrP^{sc}) appears to be of identical amino acid sequence to normal cellular PrP^c but exhibits characteristic properties of infectivity and partial resistance to the action of proteinase K. Attempts to attribute these changes in the behaviour of PrP^{sc} to post-translational covalent modification of PrP^c have so far failed. One possible explanation may be that the transition of PrP^c to PrP^{sc} is mediated by a change in protein conformation (1).

The incidence of natural transmissible spongiform encephalopathy in mammals is probably rare. However, the potential for widespread dissemination of transmissible spongiform encephalopathies via the food chain was demonstrated by the outbreak of bovine spongiform encephalopathy (BSE) in

dairy cattle within the United Kingdom (>100 000 recorded cases). The severity
of this outbreak has prompted government legislation to control the spread of
the disease by eradication of affected animals and the prohibited use of potential
high risk tissue products (offal) in food. At present clinical diagnosis of
transmissible spongiform encephalopathy can only be conducted by post mortem
histopathological examination of brain tissue. A rapid diagnostic methodology
capable of assessment of extant animals and tissue products would clearly be
beneficial to the agricultural and food industries. The most suitable candidate for
the development of a diagnostic tool would be the detection of PrPsc due to the
unambiguous presence of PrPsc in the diseased state and its potential role as the
infectious agent of transmissible spongiform encephalopathy. The generation of
antisera for immunodetection of PrPsc requires the preparation of relatively pure
antigen. Additionally, large amounts of antigen would be required for
standardisation of quantitative assays. We have investigated established
methodologies for the isolation and purification of PrPsc from bovine brain
tissue in the context of process intensification and scale-up and the development
of potential alternative purification regimes.

2 RESULTS AND DISCUSSION

The conventional methodology for the isolation and purification of PrP
(Figure 1) exploits the property of aggregation of PrP into fibril structures
termed scrapie associated fibrils (SAF). Clarified tissue homogenate is subjected
to an initial high speed ultracentrifugation stage and the resulting pellet
(intermediate pellet) contains a complex mixture of protein components (Figure
2). This material is resuspended and subjected to a second high speed
ultracentrifugation stage to yield a final SAF containing pellet. Alternatively the
intermediate stage pellet may be treated with proteinase K. This procedure
results in the formation of aggregates of an N-terminally truncated form of
PrPsc (termed PrP$^{27-30}$).

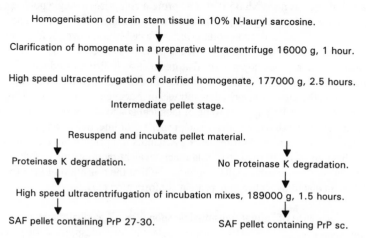

Figure 1. Preparation of SAF from bovine brain tissue affected with Bovine
Spongiform Encephalopathy.

The source material for the isolation and purification of PrP^{sc} and PrP$^{27-30}$ was bovine brain stem tissue affected with bovine spongiform encephalopathy. The supply and positive diagnosis of this material was conducted by the Central Veterinary Laboratory in conjunction with the Veterinary Investigation Centres. All manipulative operations with BSE positive tissue were conducted under stringent conditions to minimise the formation of aerosols and the risk of operator auto inoculation.

SAF Preparations Contain Relatively High Levels of Contaminating non PrP proteins.

The presence of PrP^{sc} or PrP$^{27-30}$ in SAF pellets was demonstrated by SDS PAGE fractionation followed by immunodetection on western blots with a polyclonal antiserum raised against a peptide corresponding to part of the human PrP protein sequence. However, analysis of SAF final pellet material on Coomassie Blue stained SDS PAGE gels or Western Blots suggested that PrP was a minor component of a complex mixture of proteins (Figure 2). A comparison of SAF preparations derived from TSE affected bovine and murine brain tissue suggested that the levels of PrP in BSE affected brain tissue were considerably lower than in scrapie (another TSE disease) affected laboratory animals. Proteolytic treatment of intermediate pellet material did reduce the number of contaminant protein components in the final SAF pellet however a number of contaminants exhibited the property of proteinase K resistance. The principal protein contaminants of both proteinase K and non -proteinase K treated SAF fractions (Mr 22 000-23 000 and Mr 24 000-26 000) were identified by amino acid sequence analysis, amino acid analysis and immunorecognition as comprising H and L subunits of the iron storage protein ferritin (4).

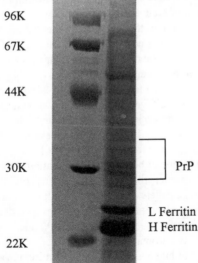

Figure 2. SDS PAGE analysis of a non-proteinase K treated SAF final pellet.

Both ferritin subunits survived prolonged exposure to proteinase K (up to 24 hours) without significant levels of degradation (4). This suggested that purification of PrP by proteolysis of non-PrP protein contaminants was of limited scope. Furthermore, both H and L ferritin subunits also showed positive cross-reactivity with the anti-PrP antiserum. This severely compromises quantitation of PrP and has the potential for false positive detection of PrP in brain tissue extracts.

A Mechanistic Explanation of the Conventional PrP Purification Methodology.

The conventional PrP purification methodology is dependent on the aggregation of PrP into fibril structures and their sedimentation under high speed ultracentrifugation. It is unclear whether these structures are assembled *in vivo* (9) or are artefacts resulting from tissue homogenisation in the presence of a detergent (N- lauryl sarcosine) (8). The sedimentation of protein molecular aggregates probably accounts for the co-purification of ferritin in SAF since ferritin may readily assemble into multimeric globular or ring structures (3). However, we have also characterised non-PrP proteins in SAF which do not form large molecular aggregates *in vivo* (creatine kinase, glyceraldehyde-3-phosphate dehydrogenase). The presence of these and other uncharacterised protein contaminants in SAF preparations remains unexplained. One possible explanation may be that hydrophobic aggregation of such proteins is promoted under the conditions employed during the ultracentrifugation stages. Preliminary experiments have demonstrated that it is possible to sediment bovine serum albumin (BSA) under these conditions.

Scale-up of the Conventional SAF Preparation Procedure.

Homogenisation of brain tissue presented few processing problems due to the relatively soft texture of the tissue. The potential problems of generation and release of aerosols were countered by containment of all operations in a class II lamina flow cabinet. The rate and batch limiting steps in the conventional methodology were the centrifugation stages (see Figure 1). Clarification of the homogenate was limited to batch sizes of 2-3 l using fixed angle rotors in a preparative ultracentrifuge. However, the two high speed ultracentrifugation stages severely compromised throughput due to the constraints of rotor capacity. This effectively limited batch processing to 320 ml of clarified homogenate originating from approximately 160 g of brain stem tissue. Clearly circumvention of one or both ultracentrifugation steps would increase productivity. However, the complexity of the protein composition of intermediate stage pellets and the relatively low abundance of PrP necessitate that alternative purification strategies possess high degrees of selectivity. The obvious candidate for the purification of PrP under these circumstances would be immunoaffinity chromatography. However, the current lack of relatively large quantities of highly specific anti-PrP antisera is a major obstacle to the development of this technique. Further purification of SAF by conventional separation methodologies was antagonised by the insolubility of PrP aggregates in aqueous buffers (Table 1). The yield of PrP from solubilised SAF under these conditions is currently too low to operate conventional molecular charge based methodologies. Attempts to increase solubilisation by physical disruption and shear of SAF (glass beads, ultrasonication) resulted in a marginal increase the yield of solubilised PrP.

Effective PrP solubilisation necessitated treatment of SAF preparations with 2% w/v SDS at elevated temperatures. Clearly this results in total denaturation of PrP tertiary structure. However, such preparations may be subjected to SDS PAGE fractionation of PrP from other contaminant proteins. Preparative SDS PAGE and electroelution of PrP from gel fractions therefore represents the only current viable methodology for the purification of PrP^{sc} and PrP^{c} from SAF.

100 mM Sodium phosphate, pH 7
50 mM Tris-HCl
TFA (100%, 50% and 30% v/v in water)
50 mM Tris + 1% w/v CHAPS
50 mM Tris + 2% v/v Triton X-100
50 mM Tris + 2% v/v Phosphatidyl inositol choline
50 mM Tris + 4 M Urea
8 M Urea
4 M Urea
0.2% w/v N-lauryl sarcosine
2% w/v N-lauryl sarcosine
1% v/v Nonidet NP40

Table 1. Solubilisation regimes applied to SAF preparations. None of these regimes resulted in significant solubilisation of PrP as estimated by analytical SDS PAGE and immunoblotting.

3 CONCLUSIONS

1. PrP is a low abundance protein in the extracts of bovine brain tissue affected with spongiform encephalopathy.
2. SAF preparations from bovine brain tissue sources are heavily contaminated with non-PrP proteins (ferritin, creatine kinase, glyceraldehyde-3-phosphate dehydrogenase).
3. The poor solubility characteristics of aggregated abnormal PrP (SAF) severely constrain the use of conventional chromatographic media based methodologies for protein purification.
4. Scale-up of the conventional methodology for the isolation and purification of PrP in the form of SAF is limited by the available batch capacity of the ultracentrifugation stages. However, alternative methodologies for the fractionation of PrP from clarified tissue homogenates or intermediate stage pellets necessitate methodologies with high degrees of selectivity. One possible solution would be immunoaffinity purification of PrP. However, the lack of sufficient quantities of anti-PrP antisera and the cross-reactivity of existing antisera currently hamper the development of this technique.

REFERENCES

1. Baldwin, M.A. and Prusiner, S.B. (1993), Structural studies on prion proteins. Proceedings for the Royal Society meeting on Molecular Biology of Prion Diseases, in press

2. Bueler, H., Fischer, M., Lang, Y., Bleuthmann, H., Lipp, H.P., DeArmond, S.J., Prusiner, S.B., Aguet, M. and Weissmann, C. (1992), Normal development and behaviour of mice lacking the neuronal cell-surface PrP protein. Nature, 356, 577 - 582.

3. Cho, H.J., Grieg, A.S., Corp, C.R., Kimberlin, R.H., Chandler, R.L., Millson, G.C., (1977), Virus-like particles from both control and scrapie-affected mouse brain. Nature, 267, 459 - 460.

4. Dale, C.J, Walker, S.G., Percy, C. and Lyddiatt, A. (1994), Characterisation of non-PrP proteins in Scrapie Associated Fibril preparations from bovine brain tissue affected with spongiform encephalopathy.(manuscript in preparation)

5. Diringer, H., Hilmert, H., Simon, D., Werner, E. and Ehlers, B. (1983), Towards purification of the scrapie agent. European Journal of Biochemistry, 134, 555 - 560.

6. Gabizon, R. and Prusiner, S.B. (1990), Prion liposomes. Biochemical Journal, 266, 1 - 14.

7. McKinley, M.P., Bolton, D.C. and Prusiner, S.B. (1983), A protease resistant protein is a structural component of the scrapie prion. Cell, 35, 57 - 62.

8. Prusiner, S.B., McKinley, M.P., Bowman, K.A., Bolton, D.C., Bendheim, P.E., Groth, D.F. and Glenner, G.G. (1983), Scrapie prions aggregate to form amyloid like birefringent rods, Cell, 35, 349 - 358.

9. Somerville, R.A., Ritchie, L.A. and Gibson, P.H. (1989), Structural and biochemical evidence that scrapie associated fibrils assemble *in vivo*. Journal of General Virology, 70, 25 - 35.

Use of Non-specific Interactions of Gelatin and Pectin for Concentration and Separation of Proteins and Formulation of New Nutritive Products

P. G. Dalev and L. S. Simeonova

DEPARTMENT OF BIOLOGY, UNIVERSITY OF SOFIA, 8 DRAGAN TSANKOV BLVD., 1421 SOFIA, BULGARIA

1. INTRODUCTION

Proteins and polysaccharides are the main types of food macromolecules and their interactions are of a great importance for the composition, structure and physico-chemical properties of natural and artificial food systems. For more than twenty five years many scientists have investigated the nonspecific interactions between the main classes of food polymers[1-4]. These studies are aimed mainly at revealing the functional properties of protein-polysaccharide mixtures which are determined by two phenomena: thermodynamic incompatibility and complex formation between these macromolecules.

Out of all proteins used in artificial food systems gelatin is the most common food hydrocolloid and the most intensively studied food biopolymer. Pectin is a valuable health care product, also used in dietetics. That is why its incorporation into nutritive products is of major importance. Pectin acts as an inhibitor of the saprophyte microflora in the intestines. It also absorbs exo- and endogenic toxic substances and heavy metals, prevents the absorption of cholesterol and reduces the sugar content in the blood of diabetics[5]. This paper outlines the nonspecific interactions of gelatin and pectin used to develop a method for concentration and separation of these biomacromolecules and the employment of the resulting products in the production of new foodstuffs.

2. NONSPECIFIC INTERACTIONS BETWEEN GELATIN AND PECTIN

In a solution of gelatin and pectin in aqueous medium two types of macromolecule interactions may occur: attraction or repulsion which are responsible for complex formation or for the incompatibility of the polymers respectively. At certain conditions they both may result in phase separation of the system. The direction of the polymers interaction depends on pH, ionic strength, the charge density and the concentration of the macromolecules.

Despite the fact that gelatin - pectin interactions are well studied the understanding of their physico-chemical relationships is still rather unsatisfactory. The knowledge of these interactions is poorly employed for practical purposes, namely for concentration and separation of biopolymers, as well as for the formation of new nutritive products.

2.1 Thermodynamic incompatibility of gelatin and pectin

At pH values above the gelatin isoelectric point both types of macromolecules are negatively charged and due to electrostatic repulsion between them each molecule tends to be surrounded by molecules of its own type. When the bulk polymer concentration is low the system is homogeneous but when it exceeds a critical value phase separation takes place and each of the coexisting phases is enriched in one of the polymers. The polymer which shows lower hydrophilicity (gelatin) can be concentrated considerably. The process of concentration resembles osmosis as it is due to movement of water between the two phases through the interphase surface playing the role of a semipermeable membrane.

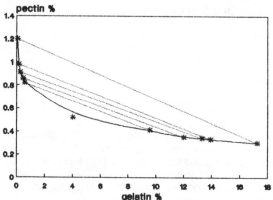

Figure 1 *Phase diagram of the system gelatin-pectin (pH 6.0, 0.5M NaCl, 40ºC)*

The phase diagram of the system gelatin-pectin depicts very well the state of their incompatibility (Figure 1). The binodal curve separates regions of one-phase systems (below) and of two-phase systems (above it). The minimum bulk concentration at which the system breaks down into two phases, i.e. the phase separation threshold for the given conditions (pH 6, 0.5M NaCl, 40ºC) is 3.4%. This means that gelatin exhibits relatively low compatibility with pectin, compared to globular proteins.

2.1.1 *Application of thermodynamic incompatibility for concentration of proteins.* Antonov et al.[6] first employed the thermodynamic incompatibility for concentration of proteins from skimmed milk on a technological scale. Using high-methoxyl pectin (degree of esterification 75% and molecular mass 45 kD) they succeeded in obtaining a liquid protein rich phase with 20-30% concentration with protein yield in this phase 80-85%. At lower temperatures the yield increased and at 10ºC it reached 91%. The protein phase contained the major amount of milk whey proteins while the pectin phase contained only α-lactalbumin and a small amount of ß-lactoglobulin. It is possible to control the ratio between the phases by changing the parameters of the system.

Other authors[7] have reported a successful application of this method for concentration of various solutions, including legumes and oil seed storage globulins, protein of green leaves, etc. Of interest is the purification of baker's yeast protein, lipids and nucleic acids by a stepwise addition of a polysaccharide solution to the yeast extract.

2.1.2 *Results of our experience.* Our experience in the application of the method is related to the purification of collagen from leather industry wastes to produce gelatin of high quality. The waste was the lowest layer of the dermis together with some of the underlying fat. It remains after splitting of the leather before tanning and contains a significant amount of inorganic salts. The collagen was extracted according to our method described recently[8]. The diluted (about 2.5%) collagen extracts were brought to pH 8.0-8.2 and 1% solution of pectin (technical grade of purity) was added at a temperature of 40-45°C. The mixture was stirred and was left at 5-10°C for several hours. Then the upper liquid layer was poured off. The remaining lower layer represented a jelly containing about 20% collagen. The alternative of this method is vacuum evaporation which is an expensive process. The concentration of collagen by means of pectin requires less energy and simple procedures and equipment.

2.2 Complexing of Gelatin and Pectin

The complexing of gelatin and pectin occurs at pH values below the isoelectric point of gelatin, where its polypeptide chain is positively charged and behaves like a polycation, while the pectin molecule carries negative charge and acts like a polyanion. These polyions interact and form an intermolecular electrostatic complex which is electrically neutral and can be regarded as a new type of biopolymer with properties substantially differing from those of the initial polymers[9,10]. This process requires low ionic strength of the medium as the inorganic ions change the charge density around the polyions and hinder it.

2.2.1 *Formation of insoluble complexes.* The formation of electrically neutral insoluble complexes between gelatin and pectin was studied in a pH range below the isoelectric point of gelatin which was found to be 5.1-5.2 by viscometry. The complex is characterized by its composition (gelatin/pectin ratio) and content of biopolymers (% of their initial quantity in the system).

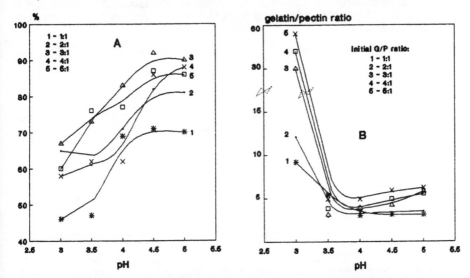

Figure 2 *Complexing of gelatin and pectin at differet pH, A-yield of the complex, B - gelatin/pectin ratio in the complex*

The maximum yield of the insoluble complex should be reached when the mass ratios of gelatin and pectin are equal to the ratio of their charges. The maximum yield depending on pH at different initial concentrations of gelatin and pectin is shown in Figure 2A and is expressed for different gelatin/pectin ratios. It is seen that the maximum yield for all ratios is slightly under the isoelectric point of gelatin - 4.5-5.0 and lowering pH the yield decreases dramatically. It is evident that the yield depends on the charge of the two polymers.

The composition of the complex (gelatin/pectin ratio) depends also strongly on pH of the system, while it is not influenced by the initial ratio of the polymers within a certain pH range (Figure 2B). But at pH 3 the relative content of gelatin in the complex rises several fold. At this pH the complex composition depends on the initial gelatin/pectin ratio. Probably at low pH values the complex loses its electrically neutral character.

Two stages may be distinguished in the process of complexing of proteins and polysaccharides: complex formation and phase separation. On molecular level the first stage can be considered as gradual attachment of the gelatin macroions (ligands) to the anions of the polysaccharide (nucleus of the complex). It is stimulated by the decrease of the free electrostatic energy of the system because of the neutralization of the oppositely charged groups, as well as by the decrease of entropy because of dehydration.   That is why this stage is fast and it strongly depends on pH of the system.

During the second stage - phase separation the complex particles form floccules and globules and finally thick precipitates. This is the stage of complex coacervation which results in phase separation of the system and concentration of the complex in one of the phases. Insoluble complexes can be formed at low polymers concentrations (0.01%) and in this way proteins and polysaccharides can be precipitated from very diluted solutions. The complex coacervation is a slow process and this gives us the opportunity to enclose some disperse systems in the forming coacervats thus fixing them included in a protein-polysaccharide cover.   Encapsulation can be expected for inert materials, oils and microorganisms[11-15].

2.2.2 *Application of gelatin-pectin complex formation.* Our experience is related to the microencapsulation of bacterial bodies and in the formation of stable oil emulsions.

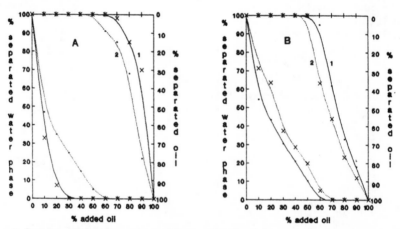

Figure 3 *Stability diagrams of emulsions stabilized by gelatin-pectin complexes (1) and gelatin (2), A - 1% gelatin and 0.33% pectin; B - 0.1% gelatin and 0.033% pectin*

The bacterial bodies were Lactobacterium bulgaricum derived from the market product "LB-Lactobact" of LB-Engineering (Sofia, Bulgaria). It is designed for a domestic preparation of yoghurt. The product was washed for the removal of milk proteins and of the low molecular substances present in it and after centrifugation at 12000 rpm for 20 minutes the bacterial bodies were suspended in 2% water solution of pectin. Then 6% solution of gelatin was added, pH was adjusted to 4.2 and the mixture was stirred for 30 minutes. The mixture was centrifuged at 3000 rpm for 10 minutes and the sediment containing the encapsulated bacterial bodies was suspended in 2% water solution of sucrose. The suspension was freeze-dried. Biological examinations show at least 3-fold higher vitality compared with the "free" bacterial bodies (vitality was checked in curdling of the milk).

The formation of stable emulsions of gelatin and pectin with sunflower oil was performed by mixing the components in a Waring blender for 2 minutes at 3000 rpm. The diagrams of emulsion stability of gelatin-pectin emulsions are compared to those of emulsions stabilized only by gelatin (Figure 3). The gelatin-pectin ratio and pH of the system (4.5) were chosen in terms of the best conditions for complex formation. It is evident that pectin raises the emulsifying activity of gelatin and it moves the point of phase inversion of the emulsion with 10% more oil. The stability diagrams make possible the choice of the composition of stable emulsions within a wide range of oil content, i.e. to compose foodstuffs as creams, mayonnaises, etc. of various oil, gelatin and pectin content.

Thus the application of the nonspecific interactions of gelatin and pectin - thermodynamic incompatibility and complex formation enables different separation tasks to be carried out as well as the formation of new foodstuffs and products with useful properties.

References

1. P.A.Albertson, 'Partition of Cell Particles and Macromolecules', Almqvist & Wiksell, Stockholm, 1971.
2. V.B.Tolstoguzov, 'Food Structure - its Creation and Evaluation', Blaushard and Mitchell, Butterworths, London, 1980.
3. D.A.Ledward, 'Polysaccharides in Food', Blaushard and Mitchell, Butterworths, London, 1979, p.205.
4. A.F.Imeson, P.K.Watson, J.R.Mitchel and D.A.Ledward, *J.Food Technol.*, 1978, **13**, 329.
5. W.Munn Rankin and E.M.Hildreth, 'Food and Nutrition', Mills & Boon, London, 1974.
6. Yu.Antonov, V.Grinberg, N.Zhuravskaya and V.Tolstoguzov, *Carbohydr.Polymers*, 1982, **2**, 81.
7. T.Bogracheva and V.Tolstoguzov, *Carbohydr.Polymers*, 1982, **2**, 163.
8. P.Dalev, *Biotechnol.Letters*, 1992, **14**, 6, 531.
9. H.L.Booly and H.G. de Yong, 'Biocolloids and their interaction', Springer, Wien, 1956.
10. H.G.B. de Yong, 'Colloid Science', H.R.Kruyt, Elsevier, New York, 1949, Vol.2.
11. A.P.Jmeson, D.A.Ledward and J.R.Mitchel, *J.Sci.Fd.Agric.*, 1977, **23**, 661.
12. G.Stainsby, *Food Chem.*, 1980, **6**, 3.
13. J.A.Rees and E.S.Welsh, *Angew.Chem.Int.Ed.*, 1977, **16**, 214.
14. O.Smidsrod, H.Grasdalen, *Carbohydr.Polymers*, 1982, **2**, 270.
15. G.R.Chilvers, A.P.Grunning and V.J.Morris, 1988, *Carbohydr.Polymers*, **8**, 55.

Magnetic Solid Phase Supports for Affinity Purification of Nucleic Acids

Martin J. Davies, Ian J. Bruce, and Diane E. Smethurst

DEPARTMENT OF BIOLOGICAL AND CHEMICAL SCIENCES, UNIVERSITY OF
GREENWICH, WELLINGTON STREET, LONDON SE18 6PF, UK

ABSTRACT

A simple, small-scale, non-hazardous procedure for the production of a magnetizable solid phase support (MSPS) has been developed based on the extrusion of molten agarose/iron oxide mixtures, which enables manufacture of a range of differently-sized spherical agarose/iron oxide beads. MSPS was cross-linked for thermal stability then derivatized with a range of ligands which possess affinity for nucleic acids. MSPS less than 150µm diameter performed best in affinity chromatography applications, and tertiary amine-containing supports showed the highest affinities for nucleic acids. MSPS derivatized with DEAE groups were used to adsorb mixtures of different types of nucleic acids from solution, which could then be eluted individually from the MSPS by application of different elution solutions. A mixture of total RNA and digested λ phage DNA adsorbed to the DEAE support could be speciated by eluting the RNA in 0.3M NaCl / 50mM arginine (free base), then eluting the DNA in 1.0M NaCl / 50mM arginine (free base). A protocol has also been developed for the purification of plasmid DNA direct from cell lysates. Plasmid pUC 18 has been isolated after boiling lysis using DEAE-MSPS, the DNA being eluted using 1M NaCl / 50mM arginine (free base) providing plasmid DNA of a purity comparable with other methods of plasmid isolation.

INTRODUCTION

In recent years, the emergence of <u>magnetic</u> solid phase supports (MSPS) for the affinity purification of biological molecules has lead to the development of a new generation of affinity chromatography media.[1] MSPS, when derivatized with a suitable affinity group, can be used for the routine separation and purification of biomolecules using equipment as simple as a magnet (**Figure 1**). Magnetic separation technologies have proved to be advantageous to many biotechnological problems, including cell-sorting, enzyme immobilization and the purification of antibodies.[2] The isolation of nucleic acids using magnetic separation is also becoming increasingly common. Most procedures are designed for specific isolations of nucleic acid species, such as messenger RNA, biotinylated DNA fragments from PCR amplifications (using streptavidin bound to magnetic beads) or plasmid DNA by specific triple helix affinity capture.[3-6] However, few methods are currently available for the *general* separation and purification of nucleic acids using a single support.

We have developed a simple method for the manufacture of beaded MSPS based on the extrusion of molten agarose/iron oxide mixtures into an immiscible organic phase, and have succeeded in producing highly uniform spherical magnetic agarose particles of a size suitable for affinity purifications.[7,8] In particular we have addressed the general isolation and purification of nucleic acids and oligonucleotides using our agarose MSPS.

1 . Biological molecule to be isolated (〰)
   in solution with unwanted species (▱)

2 . Addition of MSPS for which
   the biomolecule to be isolated
   has an affinity.

3. The molecules bind to
    the affinity ligands which
    are attached to the
    surface of the MSPS.

4 . Application of magnet
   immobilizes the MSPS leaving
   unwanted species in solution.

5. Supernatant containing
    the unwanted species
    is removed.

6 . Addition of elution solution
   frees the purified molecules
   from surface of the MSPS;
   They can now be removed
   and used in further applications.

Figure 1 *Purification of biological molecules using magnetic solid phase supports*

EXPERIMENTAL PROCEDURES

MSPS Preparation.

A solution of molten agarose (10ml; 2% w/v; type XII, low viscosity for beading, Sigma) containing paramagnetic iron oxide, Fe_3O_4 (4% w/v; Aldrich), and sodium azide (0.02% w/v) was extruded from a hole drilled at the end of a sealed 10ml syringe into vegetable oil (100ml), rapidly stirred by an overhead paddle stirrer. Stirring was continued for 1min after extrusion was completed then 100ml of deionized water was added and the two-phase mixture left to stand on a slab magnet for 16h. The majority of the cleared oil phase was poured off and the aqueous phase containing the beaded MSPS re-washed with water. The resultant suspension of MSPS was initially sized by sieving through a series of Endecott sieves (mesh size 500μm, 250μm, 200μm, 180μm and 150μm) using a Fritsch sieve shaker. All fractions collected were then washed successively with 30:70 v/v acetone:water, 70:30 v/v acetone:water, 100% anhydrous acetone. The amount of each sized fraction produced was measured in terms of its "moist weight", i.e. MSPS was filtered under water pump suction on a sintered glass funnel until the surface of the MSPS began to crack and no more filtrate was collected.All MSPS derivatives were stored in 20% aqueous methanol to act as an anti-bacterial agent.

General procedure for adsorption and elution of nucleic acids using MSPS.

25mg (250µl of 100mg ml^{-1} suspension) of the agarose-MSPS derivative was placed in an Eppendorf tube. The MSPS was magnetically immobilized using a magnetic stand/concentrator and the supernatant removed. Sterile distilled water (1ml) was added, the suspension shaken, then the water removed. The nucleic acid solution (1ml) was added, and the suspension shaken gently at 23°C for 30min. The MSPS was magnetically immobilized, the supernatant removed and its A_{260} and A_{280} values recorded. These were compared with those of the stock solution of nucleic acid, and the amount of nucleic acid adsorbed by the MSPS calculated and expressed as a percentage of the theoretical maximum. To elute adsorbed nucleic acids, the relevant elution solution (1ml) was added to the MSPS and the suspension incubated at 65°C for 30min with occasional shaking. The MSPS was immobilized, the supernatant removed and its A_{260} and A_{280} values recorded to calculate the amount of adsorbed nucleic acid eluted (expressed as percentage of the theoretical maximum).

Speciation of RNA and DNA using DEAE-MSPS.

A sample of DEAE-MSPS (2-3mg) was immobilized in the well of a microtitre plate using a BeadPrep® magnetic microtitre plate concentrator/shaker (Techne (Cambridge) Ltd., Duxford, UK). The MSPS was washed with sterile distilled water, then incubated with a mixture of λ phage DNA/Hind III digest (0.1µg µl^{-1}; 10µl) and total RNA (type IV from Torula yeast; 5µg µl^{-1}; 3µl) for 30min at 23°C with shaking. The MSPS was then immobilized and the supernatant removed. The MSPS was washed with sterile distilled water, then treated with 50µl of 0.3 M NaCl/50mM arginine (free base) for 30min at 23°C with shaking, followed by 50µl of 1.0M NaC/50mM arginine (free base) at 65°C for 30min with shaking The second elution step was performed in an oven at 65°C for 30min. The microtitre plate was removed periodically and replaced in the concentrator/shaker for short pulses in order to resuspend the particles before returning to the oven. The MSPS was immobilized and the supernatant removed. The supernatants from each step were analysed directly by agarose gel electrophoresis.

Isolation of plasmid DNA from cell lysates.

1.5ml of *E. coli* strain JM 109 cell culture, containing plasmid pUC 18 was spun in a microcentrifuge for 20s in an Eppendorf tube. The supernatant was removed and the cell pellet resuspended in 350µl STET buffer (0.1M NaCl, 10mM Tris-HCl pH 8,1mM EDTA, 5% triton X-100) containing 2µl of RNAse A solution (10mg ml^{-1}). 25µl of a freshly prepared solution of lysozyme (10mg ml^{-1}) was added in 10mM tris-HCl, pH 8. The solution was placed in a boiling water bath for 40s, then spun in a centrifuge at 13,000rpm for 10min. The pelleted cell debris was removed using a disposable pipette tip, and the lysate solution added to 25mg of DEAE-MSPS (250µl of 100mg ml^{-1} suspension) in an Eppendorf tube. The suspension was shaken gently for 30min at 23°C, the MSPS magnetically immobilized and the supernatant removed. The DEAE-MSPS was washed twice with sterile distilled water, then washed twice with 0.1M NaCl / 50mM arginine (free base) at 23°C for 5min to remove any oligoribonucleotides present in the lysis mixture which may become adsorbed to the MSPS. 1ml of elution solution (1.0M NaC/50mM arginine (free base)) was added and the suspension incubated at 65°C for 15min with occasional shaking. The MSPS was magnetically immobilized and the supernatant containing purified plasmid DNA removed. The DNA can be precipitated, analysed directly by agarose gel electrophoresis, or manipulated in other ways.

RESULTS AND DISCUSSION

The beaded agarose MSPS produced using the extrusion methodology was examined using a light microscope set up for Köhler illumination (**Figure 2**). The beads are highly

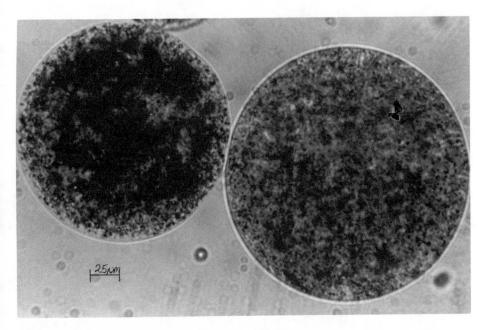

Figure 2 *Light micrograph of beaded agarose MSPS.*

spherical and the iron oxide is incorporated regularly throughout the bead matrix, and a range of sizes (>500µm - <150µm) were produced (**Table 1**). MSPS of a diameter less than 150µm were selected for affinity chromatography applications as their greater surfacearea/volume ratio enhances their ligand loading capacity and subsequent ability to take up and release ligates during chromatography. Agarose is particularly suited to affinity chromatography due to its porosity and biological inertness, and its abundance of hydroxyl groups means that ligands can easily be conjugated to it. To engender thermal and chemical stability in the matrix, agarose-MSPS was cross-linked (epichlorohydrin/1M NaOH/60°C/2h) and once cross-linked could be autoclaved (121°C, 15psi, 15min) without damage or alteration to the bead stucture. This chemical stability allows the derivatization of agarose-MSPS under a range of conditions, and the thermal stability provided by cross-linking allows affinity chromatography applications to be performed at temperatures above that at which agarose would normally melt.

TABLE 1 *Amounts of differently sized agarose MSPS formed by extrusion of a molten agarose/iron oxide mixture into oil.*

Diameter of MSPS/µm	Moist weight/g
>500	2.78
250-500	1.87
180-250	1.01
150-180	0.75
<150	1.99

Amount of molten agarose/iron oxide used is as described in experimental section.

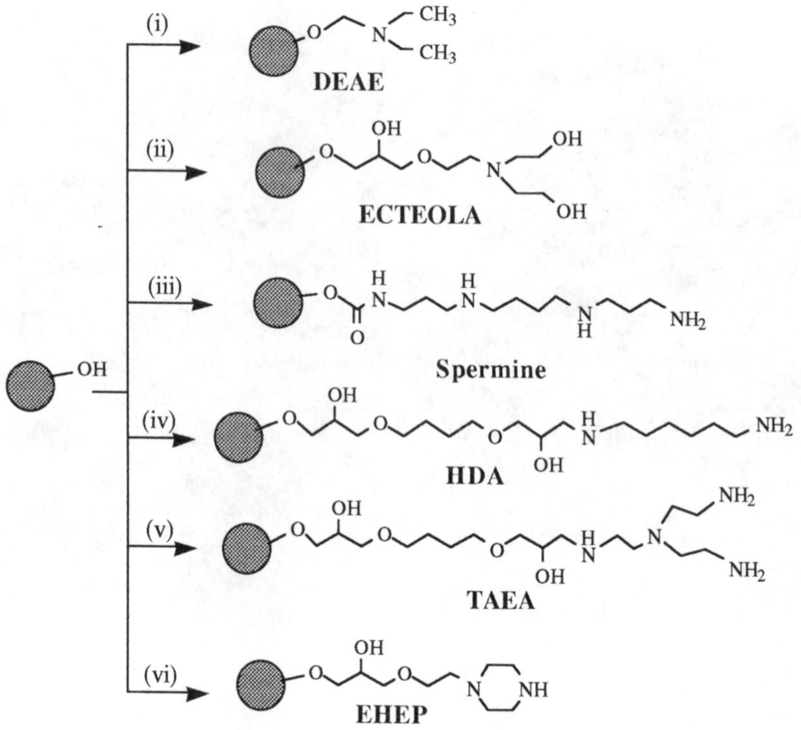

Figure 3 *Derivatives of agarose-MSPS for nucleic acid purification.*
Reagents: (i) 2-(diethylamino)ethyl chloride hydrochloride/NaOH; (ii) epichlorohydrin/
triethanolamine/NaOH; (iii) 1,1'-carbonyldiimidazole/spermine; (iv) butane-1,4-diglycidyl
ether/NaOH then tris(2-aminoethyl)amine; (v) butane-1,4-diglycidyl ether/NaOH then
hexane-1,6-diamine; (vi) epichlorohydrin/2-hydroxyethylpiperazine/NaOH.

The MSPS was functionalized with a range of nucleic acid binding ligands including diethylaminoethyl groups(DEAE), epichlorohydrin/triethanolamine (ECTEOLA) and spermine, as well as other ligands based on hexane-1,6-diamine (HDA), tris(aminoethyl)amine (TAEA) and epichlorohydrin/2-hydroxyethylpiperazine (EHEP) (**Figure 3**).[9] The DEAE, ECTEOLA, EHEP, HDA, TAEA and spermine derivatives are all amino group-containing supports, and have a positively charged surface in solution due to protonation of the nitrogen atoms; all will take up nucleic acids from solution (*via* the negatively charged phosphate backbone) to varying degrees. The affinity for nucleic acids is highest with tertiary amine containing ligands (such as DEAE, ECTEOLA and TAEA), and is less with secondary and primary amino ligands. This is understandable as tertiary amines are more basic than secondary or primary amines, and solid phase supports derivatized with amine-containing ligands will have a higher positive charge density and therefore a greater affinity for nucleic acids. Thus, the DEAE-, ECTEOLA- and TAEA-MSPS performed best in tests of uptake of salmon sperm DNA from solution (**Table 2**), whilst the EHEP and HDA supports showed decreasing affinity for nucleic acids, not because of low degree of surface derivatization, but because of the lower affinity of the secondary and primary amine ligands on the surface for DNA. However, the spermine-MSPS displayed good affinity for DNA, possibly because its long chain polyamine nature enables it to bind DNA molecules by

TABLE 2 *Efficiency of adsorption of salmon sperm DNA by differently derivatized MSPS*

MSPS	Adsorption / %
DEAE	95
ECTEOLA	89
TAEA	95
Spermine	86
EHEP	67
HDA	40

25mg of MSPS incubated with 1ml of salmon sperm DNA (50µg ml^{-1}), 30min, 23°C. Values calculated from O.D. 260nm readings. (Average values of at least three experiments).

wrapping around them, and that the ionic charge attraction of the multi-protonated polyamines is greater.[10]

These MSPS derivatives are <u>non-specific</u> in their uptake of nucleic acids and oligonucleotides, but can be used <u>specifically</u> for the separation of different forms of nucleic acids. Using DEAE-MSPS, mixtures of DNA and RNA can be separated into their component parts upon application of different concentrations of an elution solution. DEAE-MSPS was incubated with a mixture of Hind III-digested λ phage DNA and total RNA then treated with an elution solution of 0.3M NaCl/50mM arginine (free base) at 23°C to elute the RNA. Further treatment with an elution solution of 1M NaCl/50mM arginine (free base) at 65°C cleanly eluted the digested λ DNA, as analysed by agarose gel electrophoresis. DNA integrity is unaffected by exposure to MSPS: λ phage DNA has been successfully digested using Hind III restriction endonuclease after adsorption to and elution from DEAE-MSPS, or whilst still adsorbed to the solid phase support.

DEAE-MSPS can also be used for the purification of plasmid DNA. Plasmid pUC 18, prepared by standard boiling lysis incorporting an RNAse treatment step, has been purified directly from the lysis solution by adsorption to and elution from DEAE-MSPS. By incorporating a 0.1M NaCl wash step in the protocol it can be eluted free of any RNA contaminants with a purity comparable to that afforded by other methods for plasmid purification. A_{260}/A_{280} ratios for solutions of the plasmid DNA were 1.8-2.0, values comparable with those obtained for solutions of plasmid DNA purified by caesium chloride gradient ultracentrifugation or other methods of plasmid preparation such as alkaline lysis. The plasmid DNA isolated using this method has also been digested with a range of restriction endonucleases. Furthermore, full sequence data has been obtained for the pUC 18 DNA purified from DEAE-MSPS.

These procedures for nucleic acid purification can be scaled up or down according to the amount of material to be isolated. As described above for the speciation of RNA and DNA, applications can be performed in a microtitre plate which reduces greatly the amounts of MSPS, reagents and elution solutions required. Alternatively, large-scale isolation of nucleic acids can be carried out by increasing the amount of MSPS used to provide milligram quantities of nucleic acid, which would be of particular use for large scale isolation of plasmid DNA from cell lysates. Therefore, MSPS can be used either for analytical or preparative purposes. Future applications of our MSPS to nucleic acid separations will include the purification of messenger RNA, and isolation of specific gene sequences by capture with an MSPS-bound oligonucleotide probe. The versatility of agarose MSPS is such that the purification of many other types of important biological molecule will also be possible.

As well as developing new protocols for bio-molecule purification using MSPS, we are investigating new reagents for the modification of agarose matrices which, as well as lending thermal and chemical stability to the MSPS, possess other desirable characteristics such as colour and/or bio-affinity. Highly coloured organic dye molecules such as Reactive Blue 4

(1) have been used as cross-linkers, giving a coloured support which also has a non-specific affinity for proteins. Studies on the incorporation of silicon-containing molecules, which confer thermal stability on the matrix, are also currently in progress. For example, bis(chloromethyl)dimethyl silane (2) can cross-link the agarose matrix, and also impart it with physical characteristics which enable improved ease of handling.The degree of cross-linking and thermal stability is similar to that afforded by epichlorohydrin (3), the cross-linking group most commonly used; the silicon cross-linked matrix can be autoclaved without damage. Importantly, this cross-linking method does not introduce any charged species into the agarose matrix which may interfere with its performance due to non-specific adsorption effects. Studies on these new cross-linked matrices are continuing, in conjunction with the use of other polymeric magnetic matrix materials such as perfluorocarbons, instead of agarose.

Ultimately, we hope to design and synthesize cross-linking reagents which possess all the desired properties of colour, thermal stability and bio-affinity in one molecule, and incorporate them in a new class of solid phase supports.

REFERENCES

1. E. T. Menz, J. Havelick, E. V. Groman and L. Josephson, Am. Biotech. Lab., 1986, 46.
2. G. M. Whitesides, R. J. Kazlauskas and L. Josephson, Trends in Biotechnology, 1983, 1, 144.
3. K. S. Jakobsen, E. Breivold and E. Hornes, Nucl. Acids Res., 1990, 18, 3669.
4. T. Hawkins, DNA Seq., 1992, 3, 65.
5. J. Wahlberg, A. Holmberg, S. Bergh, T. Hultman and M. Uhlen, Electrophoresis, 192, 13, 547.
6. H. Ji and L. M. Smith, Anal. Chem., 1993, 65, 1323.
7. Patent applied for.
8. M. P. Ennis and G. B. Wisdom, Applied Biochem. Biotech., 1991, 30, 155.
9. Many common procedures for derivatization of solid phase supports can be found in P. D. G. Dean, W. S. Johnson and F. A. Middle, "Affinity Chromatography - a practical approach", IRL Press, Oxford, 1990.

An Integrated Approach for the Cloning, Overexpression, and Use of Lytic Glucanases for the Selective Protein Recovery from Yeast Cells

P. Ferrer,[a,b] N. Mir, [a,b] C. Shene,[a] T. Halkier,[c] I. Diers,[c] D. Savva,[b] and J. A. Asenjo[a]

[a]BIOTECHNOLOGY AND BIOCHEMICAL ENGINEERING GROUP, DEPARTMENT OF FOOD SCIENCE AND TECHNOLOGY;

[b]DEPARTMENT OF BIOCHEMISTRY AND PHYSIOLOGY, UNIVERSITY OF READING, READING RG6 2AP, UK

[c]NOVO-NORDISK, BAGSVAERD, DENMARK

This paper reviews the potential use of yeast "lytic" glucanases as cell wall permeabilizing agents in large scale protein separation processes for the selective release of proteins from yeast cells. It also describes the approach currently investigated in order to produce an inexpensive lytic β-1,3-glucanase by carrying out the molecular cloning and high-level expression of the lytic β-1,3-glucanase gene from Oerskovia xanthineolytica LLG109 in Bacillus subtilis.

1 INTRODUCTION

An increasing number of heterologous proteins are being expressed in yeast[1,2,3,4]. However, the secretion of such proteins is in many cases limited, representing a bottleneck in the overall bioprocess. The unit operation of cell disruption appears as an essential first step for intracellular product separation and downstream processing of valuable intracellular products. Large scale cell disruption is generally achieved by mechanical methods, leaving the target protein with a very complex mixture of contaminants. In this context, extensive industrial implementation of alternative approaches to conventional microbial cell disruption techniques is becoming of increasing relevance[5,6].

Selective Cell Permeabilization (SCP) and Selective Protein Recovery (SPR) as a means of bioprocess integration to increase bioprocess productivity, economy and product quality by simplifying the downstream processing of intracellular products have proved to be very attractive in terms of their delicacy and specificity[5,7]. SCP and SPR involve the use of pure preparations of cell-wall-degrading ß-glucanases to increase yeast cell wall porosity (with very limited cell lysis) and facilitate the release of intracellular proteins. In this way, SCP gives a primary separation of the target product from some of its major contaminants. A major limitation to this approach is the relatively low levels of expression of yeast lytic enzymes presently obtained in the bacteria used for the production of these enzymes (e.g. *Oerskovia xanthineolytica*[8]).

The cell wall of *Saccharomyces cerevisiae* (and most yeasts) consists of glucans, mannoproteins and a small amount of chitin[9]. Glucans are predominantly ß-1,3-linked with some branching via ß-1,6-linkages[10]. Mannoproteins overlay the glucans and cover the cell surface. The glucans are essential structural components of the cell walls, conferring rigidity and shape to the cell wall and the mannoproteins limit its porosity[11,12].

The glucanase-extractable fraction of the mannoproteins are likely to be covalently linked to glucan[13].

Many microorganisms produce extracellular lytic complexes able to lyse viable yeast cells[14]. These lytic enzyme systems usually contain endo-β-1,3-glucanases, β-1,6-glucanases, proteases, mannanases and chitinases, which are thought to function synergistically to degrade the yeast cell walls[15]. The lytic enzyme system of *O. xanthineolytica* LLG109 has been isolated and purified and the glucanase and protease components have been characterised[16]. Although *O. xanthineolytica* LLG109 seems to secrete a single molecular species of lytic β-1,3-glucanase, other *Oerskovia* strains seem to have multiple β-1,3-glucanase systems[17]. However, this multitude of enzyme species produced by *Oerskovia* may be partially due to proteolytic processing resulting from one or more isoenzymes. The genetic relationships between these enzymes is still unclear, as the number of yeast lytic enzymes so far cloned is very limited. As a result, there is still limited knowledge about the gene structure and protein function relationships[18, 19].

In this paper, we review the potential use of yeast lytic glucanases as cell permeabilizing agents in large scale protein separation processes. Also, we describe the approach currently investigated in order to produce an inexpensive lytic β-1,3-glucanase. The large scale availability of a pure lytic β-1,3-glucanase will find use in the development and improvement of large scale processes for selective recovery of intracellular proteins. In addition, it will provide an excellent tool to further characterise the yeast cell wall properties related to cell wall porosity and protein release.

2 USE OF A PURE LYTIC GLUCANASE TO RELEASE RECOMBINANT PROTEINS FROM *Saccharomyces cerevisiae* CELLS

Studies on the resistance of yeast cells to Zymolyase (an enzyme preparation from *O. xanthineolytica* containing ß-1,3-glucanase activity)[12] have suggested a major role for glucanase-soluble proteins in determining cell wall porosity. Porosity is an important property of the cell wall, because it limits the secretion of homologous (e.g. periplasmatic proteins) and heterologous proteins[12].

Recently, Asenjo *et al.*[16] studied the release of recombinant proteins from *S. cerevisiae* using a pure lytic β-1,3-glucanase component of the lytic enzyme complex from *O. xanthineolytica* LLG109 as an improvement on the use of crude lytic enzyme preparations[20,21] and conventional mechanical methods. The model system used to assess the pure lytic glucanase (containing negligible protease activities) in the release of intracellular recombinant protein particles from yeast was a hybrid yeast retrotransposon Ty virus-like particles (Ty-VLP)-producing *S. cerevisiae* strain BJ2168/pMA5620. In these studies, it was shown that yeast cell cultures were able to selectively release recombinant 60 nm protein particles (Ty-VLPs) following permeabilization of the cell wall with β-1,3-glucanase in the absence of 2-mercaptoethanol. Moreover, the pure β-1,3-glucanase did not show any degradative action on the target product, as opposed to the protease components of the enzyme lytic system of *O. xanthineolytica* LLG109 or the use of mechanical disruption methods (Figure 1). In addition, the relative cell lysis due to β-1,3-glucanase action was very limited (ca. 17%) (Figure 2) when compared to the total (crude) lytic enzyme preparation[16]. The fact that complex structures, such as the 60 nm particles of VLP, can be released by *S. cerevisiae* into the medium with only a fraction of contaminant proteins, represents an improvement over presently used mechanical or enzymatic cell disruption processes. Also, it brings up the potential of

SCP and SPR as a primary separation of a target product produced intracellularly in yeast.

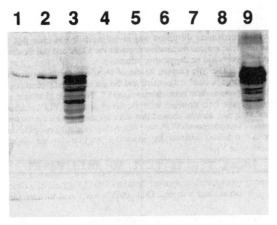

Figure 1 *Western blot assays showing the release of the recombinant protein particles (VLPs) after incubation with purified glucanases, unpurified enzymes and unpurified enzymes with PMSF (phenylmethylsulphonyl fluoride). Cells were pretreated with mercaptoethanol. Final glucanase concentrations: Oerskovia; purified 0.83 U/ml, crude 1.2 U/ml, Cytophaga; purified 0.083 U/ml, crude 0.11 U/ml. VLP release from spheroplasts in osmotic support buffer: Lane 1: Control, no enzyme treatment. Lane 2: Pure Oerskovia glucanase. Lane 3: Pure Cytophaga glucanase. Lane 4: Crude Cytophaga enzyme. Lane 5: Crude Oerskovia enzyme. Lane 6: Crude Cytophaga enzyme + PMSF. Lane 8: Control, no enzyme. Lane 9: Mechanical disruption. (From Asenjo et al, 1993).*

3 MOLECULAR CLONING AND OVEREXPRESSION IN *B. subtilis* OF THE LYTIC ß-1,3-GLUCANASE GENE FROM *O. Xanthineolytica* LLG109 AS A MEANS OF MAXIMIZATION OF ENZYME PRODUCTION.

In order to have practical amounts of the lytic ß-1,3-glucanase available to investigate further practical applications of selective and controlled yeast cell permeabilization, it is essential to exploit modern methods for overproduction of lytic ß-1,3-glucanase (Figure 3). Our work is concerned with the cloning and high-level expression of the genes for the yeast lytic enzymes as a means of maximizing enzyme production. Also, the heterologous production of lytic ß-1,3-glucanase circumvents problems associated with purifying individual ß-1,3-glucanases from a complex mixture of similar activities.

3.1 Molecular Cloning of the Lytic ß-1,3-Glucanase Gene

A lytic ß-1,3-glucanase has been previously isolated from *O. xanthineolytica* LLG109 culture broth[16]. The purified protein, with a molecular mass of about 30 KDa when estimated by Sodium dodecyl sulfate - Polyacrylamide gel electrophoresis (SDS-PAGE), was subjected to automated amino acid sequence analysis to obtain the first 12 N-Terminal residues of the native product. Also, the N terminus of several internal peptides generated from the native product were sequenced.

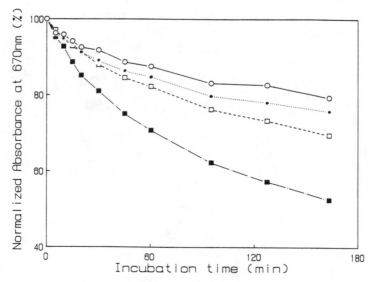

Figure 2 *Yeast lytic activity of β-1,3-glucanase and protease. Yeast suspension (A_{670}= 1.3) was incubated with glucanase (0.2 U), protease IIa (100 U azocaseinase, 0.018 U esterase), or protease IIb (75 U azocaseinase, 0.12 U esterase) to a total volume of 3 ml in the following combinations: control, no additions (○), β-1,3-glucanase (●), β-1,3-glucanase + protease IIa, (□), β-1,3-glucanase + protease IIb (■). (From Ventom & Asenjo, 1991).*

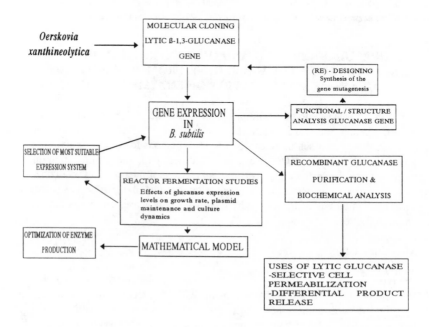

Figure 3 *Flow diagram of the integrated approach for the cloning, overexpression and use of lytic glucanases for the selective protein recovery from yeast cells.*

On the basis of the partial amino acid sequences of the purified enzyme, two degenerate oligonucleotides were synthesized and used as PCR (Polymerase Chain Reaction) primers to amplify a 0.2 Kb fragment of corresponding *O. xanthineolytica* LLG109 DNA. The PCR product was used as a radiolabelled probe to screen a genomic *O. xanthineolytica* LLG109 DNA library. This library consisted of *Bam*HI-cleaved fragments ligated into the *Bam*HI site of the vector pUC18. *E. coli* JM109 cells were transformed with these resulting plasmids and grown on Nutrient Broth plates containing Ampicillin, isopropyl-β-D-thiogalactopyranoside (IPTG) and 5-bromo-4-chloro-3-indolyl-β-galactoside (X-gal). The library was screened by probing duplicate colony replicas on nitrocellulose filters with the radiolabelled oligonucleotides. A number of positive genomic clones have been isolated from the plates and characterised by restriction mapping and Southern blot analysis, followed by sequence analysis. This will allow us to subclone the ß-1,3-glucanase gene into *B. subtilis* expression systems to achieve high-level expression levels of lytic ß-1,3-glucanase.

3.2 Characterization of an Industrial *Bacillus subtilis* Expression System for its Use in the Production of Lytic ß-1,3-Glucanase

In order to characterize an industrial *B. subtilis* expression system where large scale production of the lytic ß-1,3-glucanase will be carried out, continuous culture fermentation studies using a *B. subtilis* 168 protease-weak strain DN1885[22] expressing the heterologous endo-ß-1,4-glucanase gene *cel*A from *Bacillus lautus* PL236 are being performed. The parameters presently under study are plasmid stability as function of dilution rate and heterologous expression levels. For the plasmid pCH7 at a dilution rate of 0.2 h[-1], no plasmid loss was observed over 25 generations, whereas at a dilution rate of 0.4 h[-1], plasmid loss was observed over 25 generations.

A mathematical model of synthesis of a recombinant enzyme subject to catabolite repression by glucose in *Bacillus* is presently being developed to simulate and optimize recombinant enzyme synthesis.

4 CONCLUSIONS

The cloning of the lytic ß-1,3-glucanase from *O. xanthineolytica* LLG109 expands the number of yeast lytic ß-glucanases so far cloned[17,18,23]. The availability of the nucleotide sequences of such a family of genes will allow further understanding on the role and mode of action of these enzymes on the yeast cell wall degradation. In addition, a more extensive study on the structure and functional relationships of these enzymes will allow us to engineer new "taylor-made" lytic ß-1,3-glucanases for use in new and improved large scale selective cell permeabilization (SCP) and selective protein recovery (SPR) from yeast cells, not only from *S. cerevisiae* but also from alternative yeast expression systems such as *Hansenula polymorpha*[7], *Pichia pastoris*[24] and others[4] which are becoming of an increasing importance in Biotechnology.

5 MATERIALS AND METHODS

Bacterial strains and plasmids: *Oerskovia xanthineolytica* LLG109, described previously[25], was obtained from Rutgers University, USA. *Escherichia coli* strain

JM109[26] (*recA1 endA1 gyrA96 thi hsdR17 supE44 relA1 Δ(lac-proAB) F'[traD36 proAB⁺ lacI^q lacZΔM15*) and *Bacillus subtilis* 168 strain DN1885[22] *amyE*, a derivative of *B. subtilis* 168 RUB200, were used as host organisms for DNA manipulation and expression, respectively. The plasmid vector pUC18[26] has been utilized for cloning, subcloning and determination of DNA sequences. The plasmid pCH7[22] is composed of the promoter of the maltogenic α-amylase-encoding gene from *B. stearothermophilus*, the chloramphenicol resistance gene from pC194 and the *celA* gene from *B. lautus* PL236, which encodes an endo-β-1,4-glucanase. *B. subtilis* DN1885 and plasmid pCH7 were a gift from I. Diers (Novo-Nordisk, Denmark).

Enzymes and chemicals: Restriction endonucleases, DNA modifying enzymes and RNAseA were obtained from Pharmacia LKB Biotechnology, Boehringer Mannheim, Promega and USB, and used in accordance with the manufacturers instructions.

Preparation of genomic DNA: Chromosomal DNA from *O. xanthineolytica* was prepared as described previously[27] with minor modifications: Total DNA was isolated from a 10-ml 24h culture grown in Nutrient Broth (Oxoid) at 30°C.

Construction and screening of the genomic library for *O. xanthineolytica* LLG109: Construction of the genomic library for *O. xanthineolytica* LLG109 and the screening of colony replicas from library clones were carried out following standard procedures[28].

Continuous cultivation of *B. subtilis*: Continuous cultivations were made with 750 ml Bioflo C30 fermenter (New Brunswick Scientific) using a working volume of 350 ml. The bioreactor was equipped with pH and dissolved oxygen sensor. The defined medium used for bioreactor studies was described previously[29].

ACKNOWLEDGEMENTS

We gratefully acknowledge the financial and material support of Novo-Nordisk, EBEN Network of the EC and the CIRIT (Generalitat de Catalunya, Catalonia) for support of P. Ferrer. The support of the SERC (N.M.) and the University of Chile (C.S.) is also gratefully acknowledged.

REFERENCES

1. P. Valenzuela, A. Medina, W.J. Rutter, W.J. Ammerer and B.D. Hall, Nature, 1982, 298, 347.
2. P. Valenzuela, D. Coit, M.A. Medina-Selby, C.J. Kuo, G. Van Nest, R.L. Burke, P. Bull, M.S. Urdea and P.V. Graves, Bio/technol. 1985, 3, 323.
3. S.M. Kingsman, A.J. Kingsman and J. Mellor, Tibtech., 1987, 5, 53.
4. R.G. Buckholz and M.A.G. Gleeson, Bio/technol., 1991, 9, 1067.
5. J.A. Asenjo, A.M. Ventom, R.-B. Huang and B.A. Andrews, Bio/technol., 1993, 11,214.
6. J. de la Fuente, A. Vázquez, M.Mar González, M. Sánchez, M. Molina and C. Nombela, Appl. Microbiol. Biotechnol., 1993, 38, 763.
7. S.-H. Shen, L. Bastien, T. Nguyen, M. Fung and S.N. Slilaty, Gene, 1989, 84, 303.
8. B.A. Andrews and J.A. Asenjo, Biotechnol. Bioeng., 1987, 30, 628.
9. C.E. Ballou, In: "Molecular Biology of the yeast *Saccharomyces* Vol. 2", J. Strathern, E.W. Jones and J.R. Broach, eds., Cold Spring Harbor, 1991, 335.

10. D.J. Manners, A.S. Masson and J.C. Patterson, Biochem J., 1973, 135, 19.
11. J.G. de Nobel, F.M. Klis, T. Munnik, J. Priem and H. Van den Ende, Yeast, 1990, 6, 483.
12. J.G. de Nobel, F.M. Klis, J. Priem,T. Munnik, and H. Van den Ende, Yeast, 1990, 6, 491.
13. M.P. Schreuder, S. Brekelmans, H. Van den Ende and M. Klis, Yeast, 1993, 9, 399.
14. S. Bielecki and E. Galas, Cr. Rev. Biotechnol., 1991, 10, 275.
15. T. Obata, K. Fujioka, S. Hara and Y. Namba, Agric. Biol. Chem., 1977, 41, 671.
16. A.M. Ventom and J.A. Asenjo, Enzyme Microb. Technol., 1991, 13, 71.
17. K. Doi and A. Doi, J. Bacteriol., 1986, 168, 1272.
18. S.-H. Shen, P. Chrétien, L. Bastien and S.N. Slilaty, J. Biol. Chem., 1991, 266, 1058.
19. H. Shimoi and M. Tadenuma, J. Biochem., 1991, 110, 608.
20. J.A. Asenjo, B.A. Andrews and J.M. Pitts, Annals New York Acad. Sci., 1988, 542, 140.
21. R.-B. Huang, B.A. Andrews and J.A. Asenjo, Biotechnol. Bioeng., 1991, 38, 977.
22. C.K. Hansen, B. Diderichsen and P.L. Jørgensen, J. Bacteriol., 1992, 174, 3522.
23. M.J. Fiske, K.L. Tobey-Fincher and R.L. Fuchs, J. Gen. Microbiol., 1990, 136., 2377.
24. J.M. Cregg, J.F. Tschopp, C. Stillman, R. Siegel, M. Akong, W.S. Craig, R.G. Buckholz, K.R. Madden, P.A. Kellaris, G.R. Davis, B.L. Smiley, J. Cruze, R. Torregrossa, G. Veliçelebi and G.P. Thill, Bio/technol., 1987, 5, 479.
25. M.P. Lechevalier, J. System. Bacteriol., 1972, 22(4), 260.
26. C.J. Yanisch-Perron, J. Vieira, J. Messing, Gene, 1985, 33, 103.
27. H.M. Meade, S.R. Long, G.B. Ruvkum, S.E. Brown, F.M. Ausubel, J. Bacteriol, 1982, 149, 114
28. J. Sambrook, E.F. Frisch, T. Maniatis, "Molecular cloning. A laboratory manual", Cold Spring Harbor Laboratory Press. Cold Spring Harbor, New York, 1989.
29. J. Lee, S.J. Parulekar, Biotech. Bioeng., 1993, 42, 1142.

Development of an Automated Displacement Chromatography Purification Process for Biotechnologically Produced Proteins

R. Freitag and J. Breier

INSTITUT FÜR TECHNISCHE CHEMIE, UNIVERSITÄT HANNOVER, CALLINSTR. 3, 30167 HANNOVER, GERMANY

1. INTRODUCTION

The importance of chromatographic separations for the isolation and purification of the more valuable products of modern biotechnology has been demonstrated before.[1-3] Ion exchange and affinity chromatography as well as gel filtration are part of most of these downstream processes. At present, non-linear elution chromatography dominates the field, even though it is known that neither the stationary nor the mobile phase are used economically in such processes.[4,5] The stationary phase capacity is exploited more efficiently in displacement chromatography. In this case the mixture to be purified is loaded onto the column until saturation. Afterwards the displacer, a substance with an exceptionally high affinity for the stationary phase, is introduced at high concentration. Due to the competition for the adsorption sites ensuing between the various substances and the displacer, the so-called displacement train is formed. Consecutive zones of the pure substances move through the column at the speed of the displacer front. The concentration found in the various zones depends on the displacer concentration and the substance's adsorption isotherm in relation to the displacer isotherm. When suitable operating conditions are chosen, the final substance concentration will be considerably higher than those found originally in the feed.[6-10]

While the superiority of displacement chromatography over elution chromatography in terms of throughput, recovery, final product concentration, and waste production has been recognised for some time now, this technique is seldom used in downstream processing. [6,7,11,12] In the case of biotechnological products, the lack of reliable, readily available protein displacers contributes to this.[6,13-18] Concomitantly, other than in elution chromatography, process automation may be a problem in displacement chromatography. As the purified proteins appear consecutively in highly concentrated zones at the column outlet, monitoring of the composition of the column effluent by other means than a simple UV-detector becomes necessary. Usually this is done by laboriously collecting and analysing fractions.

2. MATERIAL AND METHODS

All chemicals used were of the highest available purity. Proteins and fine chemicals were from Sigma, bulk chemicals for buffer and eluent preparation were from Fluka. Human

Antithrombin III (AT III, KyberninTM) was donated by the Behring-Werke, FRG. Polymeric displacers were from Polysciences, USA. The Hydroxyapatite (HA) stationary phases were donated by PENTAX, Asahi Optical Co. Ltd., Japan. Non-porous 2 μm Micropell C-18 beads from Glycotech Inc., USA were used as stationary phase in the analytical chromatographic separations.

The preparative chromatographic system was assembled from a Gynkotek 300 CS pump (Gynkotek, FRG), a Shodex pulse damper (Showa Denko, Japan), and a Valco 10-port-valve (Valco, USA). A 1 ml sample loop was used for sample injection, while the displacer was injected from a preparative sample loop (Knauer, FRG) that could hold up to 11 ml. 50 μl fractions were collected and analysed by reversed phase HPLC. Columns were packed at 150 bar from 2 μm and 10 μm HA beads. The average pore size was 1000 A in both cases. Unless indicated otherwise column dimensions of 4.6 x 100 mm and flow rates of 100 μl/min were used in the Displacement experiments. For column regeneration a 400 mmol/l phosphate buffer (pH 6.8) was used. In the case of the protein displacements by EGTA (Ethyleneglycol-bis(β-aminoethylether)-N,N,N'N'-tetraacetic acid) a 100 mM $CaCl_2$ solution was used for column regeneration instead.

The isotherms were determined as described in Reference 19. Column dimensions were 2 x 50 mm. A mobile phase flow rate of 100 μl/min and a detection wavelength of 214 nm were used.

The high pressure gradient system for the analytical HPLC was assembled from two IRICA S 871 pumps controlled by an Autochrom gradient controller and a dynamic mixing chamber (250 μl, Knauer, FRG). The SF 757 UV-detector (ABI, USA) equipped with a 0.5 μl / 1 mm micro flow cell (0.1 sec filter time) with the detection wavelength set to 214 nm was employed. For sample injection (5 μl) either an electrically driven EC6W valve (Valco, USA) or a pressure driven 7010 valve (Rheodyne, USA) was used. The electrically driven valve caused a high pressure pulse during injection. Buffer A consisted of deionized water with 0.1 % (v/v) TFA added, Buffer B of acetonitrile with 0.08 % (v/v) TFA added. The gradient was run from 30 % B to 70 % B. By using a flow rate of 4 ml/min and an operating temperature of 60°C, an analysis was completed within 20 seconds. Column dimensions were 4.6 x 30 mm. Data collection and interpretation was carried out with a SP4270 Spectra Physics integrator or by PC using the CAFCA (Computer Assisted Flow Control and Analysis) software developed by B. Hitzmann et al. at the Institut für Technische Chemie, Universität Hannover, FRG.

3 RESULTS AND DISCUSSION

For some time now the aim of our group has been the development of chromatographic separations in the displacement mode for the isolation of pharmaceutically applicable substances (e.g. recombinant h-AT III, monoclonal antibodies) from mammalian cell culture supernatants. A fully automated system was developed, where the preparative chromatographic unit is monitored by an extremely fast analytical HPLC-system, Figure 1. A number of macroporous (perfusion chromatography) and nonporous reversed phase stationary phases as well as macroporous ion exchange resins have been investigated in regard to their suitability for providing such fast analytical columns. Best results were achieved with columns packed from nonporous 2 μm Micropell C-18 beads from Glycotech; Inc., USA. By using the conditions described above, the composition of effluent of the preparative column could be analysed within 20 seconds, Figure 2.

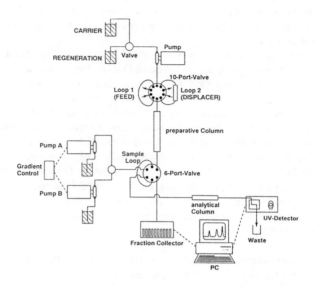

Figure 1   *Schematic drawing of the chromatographic system*

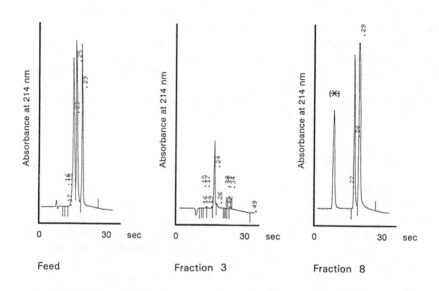

Figure 2   *Analysis by reversed phase chromatography of the feed and two fractions collected during a separation of b-transferrin (retention time 0.24 min.), BSA (bovine serum albumin, retention time 0.26 min.) and AT III (retention time 0.29 min) by polyglutaminic acid (*)*

A flow rate of 100 µl/min was used in the preparative unit. The analytical frequency thus allows the quasi on line monitoring and documentation of the purification and makes accurate control of the fractionation possible. As only 5 µl of sample are required, little product is lost to the analysis. In addition to the product quantification by analytical chromatography, the biological activity of the products was tested using a Heparin co-factor activity assay (Boehringer Mannheim, FRG) in the case of AT III and an immuno-assay in the case of monoclonal antibodies. Displacement chromatography was found to be a fast and gentle means for protein purification.

Many of the high valued products of modern biotechnology are produced by mamma-lian cells. The nutritional and environmental requirements of such cells are complex. Most culture media contain, e.g., significant amounts of foetal calf serum (FCS). The product is thus contaminated by a large number of proteins, peptides and other substances, many of which have never been isolated and characterised. BSA (bovine serum albumin, IEP: 4.9) b-IgG (bovine immunoglobulin G, IEP: 7.3), and bovine transferrin (IEP: 5.5) must be expected as major product contaminants. On hydroxyapatite (HA) stationary phases only well defined biological macromolecules (proteins, DNA, ...) are retained at all. As HA becomes unstable below a pH of 6, a mobile phase pH of at least 6.5 is usually maintained. The complex interaction behaviour of proteins with HA has been described as mixed mode ion exchange.[20-24] Proteins with an isoelectric point (IEP) below the mobile phase pH are assumed to interact with the surface calcium ions, while the more basic proteins are re-tained by electrostatic interaction with the negatively charged phosphate groups. The dis-placement of basic and acidic molecules thus requires different displacers.

In our case basic proteins such as monoclonal antibodies have been successfully dis-placed by basic proteins such as cytochrome C and polycations such as PEI (polyethylen-imide). When a 10 mM phosphate buffer (pH 6.8) was used, basic protein such as myo-globin (IEP: 7.4), lysozyme (IEP: 10.5) and monoclonal mouse/mouse IgG (IEP: 7.2) were well separated when displaced by cytochrome C. Presumably, however, a conditioning of

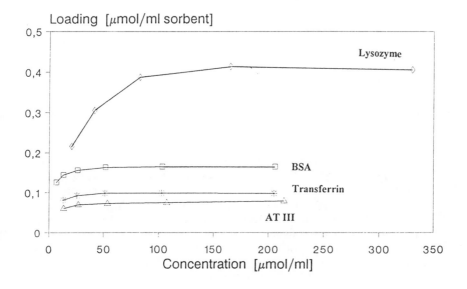

Figure 3   *Selected protein adsorption isotherms on HA*

the HA-surface by the phosphate ions of the buffer is necessary, since a displacement carried out in a TRIS buffer of similar pH-value and concentration yielded less satisfactory results. When 1 mM of $CaCl_2$ was added to the TRIS buffer, lysozyme was not retained anymore, even though the myoglobin, which in the original displacement appeared in a zone preceding the lysozyme, i.e. was less strongly adsorbed, is still retained on the column and displaced well by the cytochrome C.

Most proteins and consequently most biotechnological products, including e.g. the anticoagulant antithrombin III (AT III, IEP: 4.6-5.1), have isoelectric points below a value of 7. In order to retain such acidic proteins on HA, conditioning with a 10 mM $CaCl_2$ should be carried out and TRIS rather than phosphate buffers should be used. While the retention of one of the major product contaminants, namely the more basic b-IgG, can be prevented under the conditions suitable for AT III adsorption, e.g. by adding $CaCl_2$ to the mobile phase, b-transferrin and BSA must be removed by the displacement step. In Figure 3 the adsorption isotherms of BSA, AT III, and b-transferrin on HA are shown together with that of lysozyme.

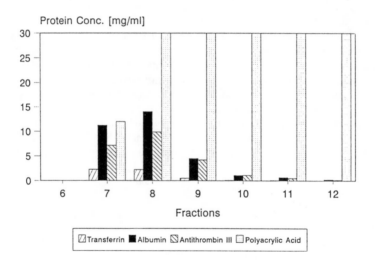

Figure 4   *Displacement of b-transferrin, BSA, and AT III using polyacrylic acid as displacer*

Acidic proteins should be displaced from HA by ionic polymers such as polyacrylic acid (Figure 4) and polyglutaminic acid or by proteins carrying carboxyl clusters such as β-casein (Figure 5). Moreover, any substance that acts as a complexing agent towards Ca^{2+} should also displace acidic proteins. Here EGTA (Ethyleneglycol-bis(β-Aminoethylether)-N,N,N',N'-tetraacetic acid) was investigated as displacer (Figure 6). Neither polyacrylic acid nor polyglutaminic acid was found to be a successful protein displacer. Some degree of separation could be achieved with β-casein as displacer, while clearly the best results were obtained for in the case of protein displacement with EGTA.

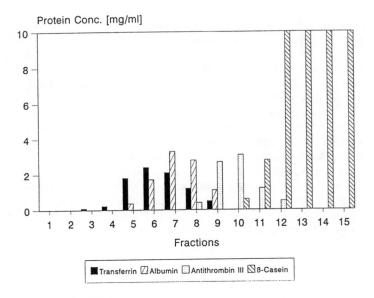

Figure 5   *Displacement of b-transferrin, BSA, and AT III using β-casein as displacer*

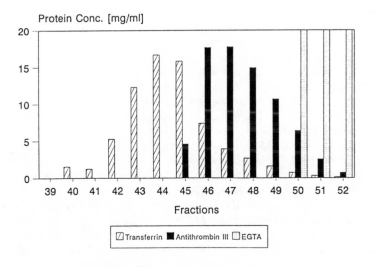

Figure 6   *Displacement of b-transferrin, BSA, and AT III using EGTA as displacer*

CONCLUSIONS

Displacement chromatography was found to be an efficient means for the isolation of biotechnological products. Hydroxyapatite was used as stationary phase, as all proteins interact with this material, unless denatured, but differ in their adsorption mechanism depending on whether they carry a positive or negative net-charge during the separation. Thus acidic contaminants such as BSA do not interfere with the isolation of more basic products such as immunoglobulins and *vice versa*. Quite important is the modulation of the adsorption through the composition of the mobile phase. The most important aspect in successful protein displacement is the displacer itself. Here basic proteins such as cytochrome C were used for the displacement of basic proteins, while certain polymers, proteins, and Ca^{2+} complexing agents were used to displace acidic proteins. Polymers were unsuited to this task, while proteins such as β-casein had some success. The best results were, however, seen with the complexing agents. To avoid the task of collecting and analysing fractions to distinguish between the successive zones of the displacement train, we used a fast analytical RPC-system for quasi on line monitoring of the composition of the eluate of the preparative column.

REFERENCES

1.   J. Bonnerjea, S. Oh, M. Hoare and P. Dunnill, *Bio/Technology*, 1986, **4**, 954
2.   M.A. Taipa, M.R. Aires-Barros and J.M.S. Cabral, *J. Biotech*, 1992, **26**, 111
3.   T. Burnoff, *Bioseparation*, 1991, **1**, 383
4.   S.M. Cramer and G. Subramanian, *Sep. Purif. Methods*, 1990, **19**, 31
5.   J.L. Dwyer, *AIChE Symp. Series*, 1984, **80**, 120
6.   J. Frenz and C. Horvath, in: C. Horvath (Ed.) 'High Performance Liquid Chromatography - Advances and Perspectives', Vol 8. Accademic Press, 1988
7.   A.R. Torres and E.A. Peterson, *J. Chromatogr.*, 1990, **499**, 47
8.   G. Subramanian, M.W. Phillips, S.M. Cramer, *J. Chromatogr.*, 1988, **439**, 341
9.   F. Antia and C. Horvath, *Ber. Bunsen-Ges. Phys. Chem.*, 1989, **93**, 962
10.   A.M. Katti, E.V. Dose and G. Guiochon, *J. Chromatogr.*, 1991, **540**, 1
11.   B.E. Dunn, S.E. Edberg and A.R. Torres, *Anal. Biochem.*, 1988, **168**, 25
12.   A.R. Torres, G.G. Krnger and E.A. Peterson, *Anal. Biochem.*, 1985, **144**, 469
13.   S. Ghose and B. Mattiasson, *J. Chromatogr.*, 1991, **547**, 145
14.   A.R. Torres and E.A. Peterson, *J. Biochem. Biophys. Methods*, 1979, **1**, 349
15.   A.L. Lee, A.W. Liao and C. Horvath, *J. Chromatogr.*, 1988, **443**, 31
16.   G. Subramanian, M.W. Phillips, G. Jayaraman and S.M. Cramer, *J. Chromatogr.*, 1989, **484**, 225
17.   J.A. Gerstner and S, M. Cramer, *BioPharm*, 1992, **5**, 42
18.   S.-C.D. Jen and N.G. Pinto, *J. Chromatogr.*, 1990, **519**, 87
19.   J.X. Huang and C. Horvath, *J. Chromatogr.*, 1987, **406**, 275
20.   G. Bernadi, M.G. Giro and C. Gaillard, *Biochim. Biophys. Acta*, 1972, **278**, 409
21.   M.J. Gorbunoff, *Anal. Biochem.*, 1984, **136**, 425
22.   M.J. Gorbunoff, *Anal. Biochem.*, 1984, **136**, 433
23.   M.J. Gorbunoff and S.N. Timasheff, *Anal. Biochem.*, 1984, **136**, 440
24.   T. Sato, T. Okuyama and M. Ebihara, *Bunseki Kagaku*, 1989, **38**, 34

Multi-dimensional Affinity Membrane Chromatography in Down-stream Processing of Antithrombin III

R. Freitag and O.-W. Reif

INSTITUT FÜR TECHNISCHE CHEMIE, UNIVERSITÄT HANNOVER, CALLINSTR. 3, 30167 HANNOVER, GERMANY

1 INTRODUCTION

Antithrombin III (AT III) is an important blood factor whose main function is to counter-act blood coagulation factors such as Thrombin. The AT III activity is enhanced conside-rably in the presence of Heparin. A number of congenital and acquired diseases have been linked to abnormalities in the blood level or the activity of AT III. AT III is also given du-ring or after surgery to reduce the risk of unwanted blood clogging.[1] As a consequence AT III has become a highly valued pharmaceutical, which is routinely isolated by introdu-cing a chromatographic step, usually a Heparin affinity column, into the blood fractiona-tion scheme according to Cohn[1] or Kistler and Nitschmann.[3,4] Concomitantly, the high value of the blood factor together with the infection risk inherent to blood derived substan-ces has made the production of human AT III by recombinant mammalian cells (rh-AT III) an attractive option for biotechnologists.

The requirements of mammalian cells in regard to their culture media are complex. Often the product has to be separated from a multitude of other components (proteins, peptides, by-products, product variants) of similar physico-chemical character and perhaps even biological affinity. Liquid chromatography is most suited to such a task. However, since the purity requirements for a recombinant human therapeutic are much more strin-gent than those for a similar substance derived from human plasma, a single chromatogra-phic step, e.g. a Heparin affinity separation, will likely not suffice in the case of rh-AT III. In addition, the rh-AT III levels currently produced in cultures of recombinant baby ham-ster kidney (BHK) or chinese hamster ovary (CHO) cells are much lower than those found in normal human blood.[5,6]

For some time now membrane adsorbers (MA), i.e. filter membranes that have been functionalized by the attachment of interactive groups (ion exchanger, chelating agents, affinity ligands) have been discussed as stationary phases in liquid chromatography.[7-10] Favourable mass transfer properties, compatibility to high mobile phase flow rates, easy handling and scale up have been mentioned as advantages of such systems. Nevertheless, MA have been used mainly in single step separations, often called affinity filtration, rarely in true multi stage chromatographic separation processes.[11] The advantages and disadvan-tages of several increasingly complex MA based chromatographic systems are discussed in this contribution, taking the isolation of rh-AT III produced by BHK cells as an example

2 MATERIAL AND METHODS

All chemicals used were of the highest purity available. Supernatants of BHK-21 (clone 13)[12] cell cultures performed in 500 ml spinner flasks were kindly donated by R. Weidemann (Cell Culture Technology Group, Institut für Technische Chemie, University of Hannover, Germany). Unless indicated otherwise cell culture media contained 10% foetal calf serum (FCS). For the chromatographic separations either an FPLC (Pharmacia) or a self assembled HPLC system was used. Unless indicated otherwise a flow rate of 2 ml/min was adjusted. All gel electrophoresis experiments were performed on the respective Pharmacia equipment following the manufacturers instructions. Trace levels of impurities were quantified by colloid gold staining of gels from a single radial immunodiffusion.[13] Concentrations of rh-AT III were established by ELISA,[14] the Heparin co-factor activities of the rh-AT III according to Reference 15.

The MA used (Sartorius, Göttingen, Germany) consisted of a synthetic co-polymer with an average thickness of 180 µm and an average pore size of 0.45 µm. Filtration areas were either 3.4 cm² or 20 cm². Strong ion exchanger (sulfonic acid groups, quaternary ammonium groups) and affinity MA (Heparin and Cibacron Blue ligands) were used. Stacks of up to ten MA were encased in suitable filter holders fitted with Luer lock connectors and thus integrated into the chromatographic system. If necessary ion exchanger MA were regenerated by washing with 1 N NaOH followed by 1 N HCl or with 0.5 N NaOH at 50°C for 30 minutes. Affinity MA were rinsed with 0.1 N NaOH for 10 to 30 minutes instead.

3 RESULTS AND DISCUSSION

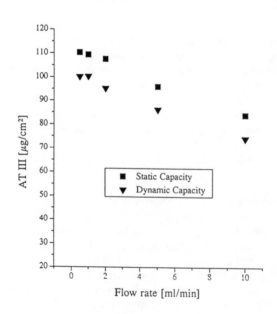

In Figure 1 the static (i.e. the maximun amount of AT III, which could be loaded onto the MA at that flow rate) and the dynamic (i.e. the amount of AT III retained at the moment of break through) capacities of the Heparin MA are given as a function of the flow rate. A "column" consisting of 5 layers of 3.4 cm² MA, representing 17 cm² of adsorptive area, was used. If regeneration was carried out regularly, no decrease in capacity was observed during the experiments (well over 30 runs).

Figure 1   *Flow rate dependency of the AT III capacity of the Heparin Affinity MA*

In Procedure 1, the Heparin MA (stack of ten 3.4 cm^2 MA) were used to recapture the rh-AT III from 250 ml of a cell culture supernatant (10 % FCS) after clarification by centrifugation and diafiltration (cut off: 2000 Da). The results are given in Table 1. 81.07 % of the original rh-AT III concentration were recovered. The overall concentration factor was 81, the concentration factor of the Heparin MA was 101. The AT III activity increased from 773.3 $U_{AT\ III}/mg_{AT\ III}$ to 857.4 $U_{AT\ III}/mg_{AT\ III}$ during the process. A final purity of 64 % was established by SDS-PAGE and immuno-diffusion. The major remaining impurity was bovine serum Albumin (BSA). Bovine transferrin and bovine immunoglobulin G (b-IgG) were also found in significant amounts. The entire process required 90 minutes, 60 of which were necessary for the chromatographic separation.

Table 1   *Purification of rh-AT III by Heparin Affinity MA*

	Cell Culture Supernatant	Diafiltration	Heparin-MA
AT III [µg/ml]	7.5	6	608
AT III [µg]	1875	n.d.	1520
Activity [$U_{AT\ III}/mg_{AT\ III}$]	773.3	766.6	857.4
Total Protein Concentration [mg/ml]	6.1	5.64	0.95
BSA [mg/ml]	3.64	3.36	0.167
b-Transferrin [mg/ml]	0.21	0.187	0.043
b-IgG [mg/ml]	0.77	0.68	0.007

In order to improve the final purity of the rh-AT III isolate, a two-step chromatographic procedure was designed, which incorporated a cation or an anion exchanger MA in addition to the Heparin affinity MA, Figure 2. The results are compiled in Table 2. Both types of ion exchanger are found to remove contaminating proteins. Due to the resulting more favourable concentration ratio of rh-AT III to the total protein content, the final

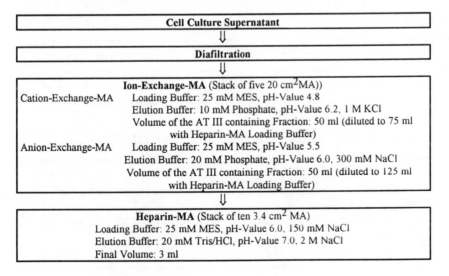

Figure 2   *Purification scheme for rh-AT III*

Heparin affinity step is more effective and final purity levels of 85 % (anion exchanger) and 87 % (cation exchanger) are achieved. The additional chromatographic step increases the total process time to 225 and 210 minutes in case of the anion and the cation exchanger respectively.

Table 2   *Purification of rh-AT III by a two step chromatographic process*

	Cell Culture Supernatant	Diafiltration	Cation-Exchanger-MA	Heparin-MA
AT III [µg/ml]	14	8.7	29.04	416
AT III [µg]	1750	1740	1452	1248
Activity [$U_{AT\ III}/mg_{AT\ III}$]	657.14	643.68	588.85	507.45
Total Protein Concentration [mg/ml]	5.6	3.46	9.69	0.478
BSA [mg/ml]	3.28	2	7.34	0.032
b-Transferrin [mg/ml]	0.18	0.11	1.22	0.032
b-IgG [mg/ml]	0.88	0.55	0	0

	Cell Culture Supernatant	Diafiltration	Anion-Exchanger-MA	Heparin-MA
AT III [µg/ml]	9	8.9	38.3	574.3
AT III [µg]	2250	2225	1915	1723
Activity [$U_{AT\ III}/mg_{AT\ III}$]	811.11	775.3	725.8	641.13
Total Protein Concentration [mg/ml]	5.4	5.24	14.4	0.676
BSA [mg/ml]	3.12	3.01	11.2	0.087
b-Transferrin [mg/ml]	0.15	0.11	2.21	0.006
b-IgG [mg/ml]	0.82	0.81	0.002	0

In the above purification attempts, BSA proved to be most difficult to remove. This is to be expected as the major physiological function of BSA as a transport protein requires adherence to the various blood factors including AT III. To relieve this problem a cibacron blue (CB) affinity MA was integrated into a three step chromatographic procedure,

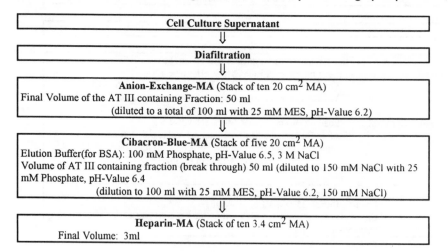

Figure 3   *Purification scheme for rh-AT III*

Table 3   *Purification of rh-AT III by a three step chromatographic process*

	Cell Culture Supernatant	Diafiltration	Anion-Exchanger-MA	CB-MA	Heparin-MA
AT III [µg/ml]	12	9.8	51.8	49.8	738.7
AT III [µg]	3000	2940	2590	2490	2216
Activity [$U_{AT\ III}/mg_{AT\ III}$]	716.7	704.1	660.2	636.5	568.7
Total Protein Concentration [mg/ml]	6.1	4.98	17.93	0.072	0.88
BSA [mg/ml]	3.64	2.95	11.16	0.016	0.127
b-Transferrin [mg/ml]	0.21	0.147	0.095	0.006	0.001
b-IgG [mg/ml]	0.77	0.56	0.008	0.001	0

which included also an anion exchanger MA and a Heparin affinity MA. The affinity of BSA towards cibacron blue has been well established. When the CB-MA was used between an anion exchanger MA and the Heparin affinity MA (Figure 3, Table 3), the AT III's final purity was only 84 % while the total process time increased to 400 minutes.

Better results were obtained if the CB-MA were used immediately after the diafiltration, i.e. if the BSA concentration was lowered at the earliest possible moment (Figure 4, Table 4). Now a final purity of 94 % is found together with an overall yield of 71.7 %. A whole purification cycle took 300 minutes in this case.

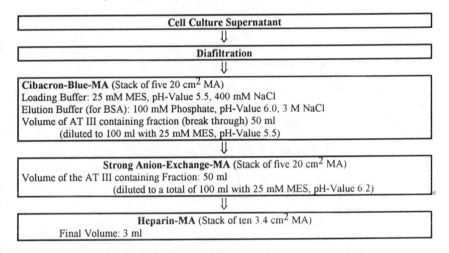

Figure 4   *Purification scheme for rh-AT III*

The purification cycle time can be reduced considerably by increasing the volumetric flow rate in the system. If the flow rate is increased step wise from the 0.12 l/h used in the experiment described above to a maximum of 4.8 l/h, the rh-AT III throughput is concomitantly increased from 1.2 mg/h to 13.7 mg/h. The recovery is reduced from 69.3 % to 52.4 %. Final purities between 94 % and 96 % are found. No correlation was observed between the final purity and the flow rate.

Table 4   *Purification of rh-AT III by a three step chromatographic process*

	Cell Culture Supernatant	Diafiltration	CB-MA	Anion-Exchanger-MA	Heparin-MA
AT III [μg/ml]	11	10.9	48.1	43.8	657
AT III [μg] Activity	2750	2725	2405	2190	1971
$[U_{AT III}/mg_{AT III}]$ Total Protein	790.9	798.2	684	630.2	577.8
Concentration [mg/ml]	5.7	5.5	8.1	0.36	0.699
BSA [mg/ml]	3.4	3.3	0.22	0.19	0.015
b-Transferrin [mg/ml]	0.16	0.16	0.76	0.021	0.001
b-IgG [mg/ml]	0.79	0.8	3.6	0.003	0

The serum content of the culture medium is also of significance. In the above mentioned experiments a rather high FCS concentration of 10 % was used. If culture supernatants containing 5 % or 3 % FCS were used as raw material, final purities of 97 % respectively 99.9 % could be obtained by the final three step chromatographic procedure. However, the specific activity was much reduced in such cases. Whereas a specific activity of 577.8 U/mg was found for the rh-AT III produced in the presence of 10 % FCS, this value was 512.4 U/mg in the case of 5 % FCS and only 319.9 U/mg if 3 % FCS were used. The adverse effect of low serum concentration on product quality was seen repeatedly.

REFERENCES

1.   R.L. Bick, 'Seminars in Thrombosis and Hemostasis', Vol. 8, No 4, Thieme Stratton Inc., N.Y. (USA), 1982.
2.   E.J. Cohn, L.E. Strong, W.L. Hughes, D.J. Mulford, J.N. Ashworth, M. Melin and H.L. Taylor, *J. Am. Chem. Soc.*, 1950, **72**, 465.
3.   L. Kistler and H. Nitschman, *Vox Sang.*, 1962, **7**, 414.
4.   T. Burnouf, *Bioseparation*, 1991, **1**, 383.
5.   M. Wirth, S.Y. Li, L. Schuhmacher, J. Lehmann, G. Zettelmeissl, H. Hauser, in: R.E. Spier, J.B. Griffith, J. Stephane and P.J. Crooy: 'Advances in Animal Cell Culture'. Butterworth & Co. Publ., 1989.
6.   G. Zettelmeissl, H. Ragg, H.E. Karges, *Bio/Technology*, 1987, **5**, 720
7.   T.B. Tennikova and F. Svec, *J. Chromtogr.*, 1993, **646**, 279.
8.   D. Josic, J. Reusch, K. Löster, O. Baum and W. Reutter, *J. Chromatogr.*, 1992, **590**, 59.
9.   J. Gerstner, R. Hamilton, S.M. Cramer, *J. Chromatogr.*, 1992, **596**, 173
10.   O.-W. Reif and R. Freitag, *J. Chromatogr. A*, 1993, **654**, 29.
11.   B. Champluvier and M.-R. Kula, *Bioseparation*, 1992, **2**, 343.
12.   M. Wirth, W. Reiser, G. Zettelmeissl, in: R.E. Spier and J.B. Griffith, 'Modern Approaches to Animal Cell Technology'. Butterworth & Co. Publ., 1987.
13.   G.S. Bailey, in: J.M. Walker, 'Methods in Molecular Biology', Vol. 1 (Proteins), Humana Press, 1984.
14.   R. Weidemann, Diploma Thesis, 1992, University of Hannover, Germany
15.   U. Abildgard, M. Lie, O.R. Odegard, *Thromb. Res.*, 1977, **11**, 549.

Polymer-shielded Dye-affinity Chromatography

I. Yu. Galaev and B. Mattiasson

DEPARTMENT OF BIOTECHNOLOGY, CHEMICAL CENTER, LUND
UNIVERSITY, PO BOX 124, S-221 00 LUND, SWEDEN

1 INTRODUCTION

Affinity chromatography has customarily been used in the last stages of a purification process due to the high costs of matrices and ligands as well as the instability of ligands to degradative enzymes present in very crude extracts[1]. Recent efforts have been directed toward the application of this high-resolution technique to the earlier stages of a purification scheme so as to reduce working volumes and processing time, thereby reducing the overall costs. One approach has been to replace expensive protein ligands with cheaper triazine dyes which serve as group-specific ligands[2]. Purification of nucleotide dependent enzymes, such as dehydrogenases and kinases, as well as non-nucleotide dependent proteins, such as BSA, on dye-linked supports is well documented[3-9]. The commonly used triazine dyes provide the possibility for a range of different interactions between the target protein and the dye. Since a well-defined biological affinity site is not always present in the protein a number of nonspecific binding interactions may occur, which can lead to the irreversible binding of the protein thereby obstructing the selectivity of the process[10].

The selectivity of affinity-chromatographic procedures is governed in general by the ratio:

$$\frac{\text{efficiency of specific interaction of the target protein with the ligand}}{\text{efficiency of nonspecific interaction of foreign proteins with the ligand.}}$$

The use of more specific ligands increases selectivity due to the increase of the numerator. Another way to improve the selectivity is to reduce the denominator, in other words to reduce nonspecific interactions. This approach is promising because it can also improve the recovery of the target molecule. In fact, target molecules interact with the ligand-containing matrix via specific binding sites and nonspecifically due to various comparatively weak hydrophobic and electrostatic interactions. Reduction of these nonspecific interactions would probably improve recovery, especially during the specific elution of the target protein. One of the possible ways to reduce nonspecific interactions is to block the ligand with an inert agent. This agent must bind to the ligand tightly enough to prevent nonspecific interactions but without binding so tightly that the specific interactions are disturbed. The multipoint attachment of such an agent to several ligands allows it to remain constantly on the column and protects it from displacement by the target protein. This agent must not interact with the proteins in the sample. Only water-soluble nonionic polymers satisfy these characteristics required in blocking agent.

2 CIBACRON BLUE INTERACTION WITH POLYMERS

We have studied the interaction of Cibacron Blue with different water soluble polymers[12]. The technique of choice was difference spectroscopy. The spectra of Cibacron Blue in the visible region are sensitive to the environment of the dye chromophore. Changes in the spectra are easily revealed in the difference spectra[11].

The relative efficiency of Cibacron Blue interaction with water-soluble polymers is presented in Table 1. We were interested in a polymer with a relatively strong, but not too strong multipoint interaction with the support containing dye-ligand as an appropriate blocking agent. Preferably, this polymer is a nonionic one to reduce nonspecific interactions of the polymer itself with proteins. Based on the data of Table 1 and the requirements mentioned above we restricted our choice to poly(vinyl alcohol) (PVA) and poly(vinyl pyrrolidone) (PVP).

Difference spectrum of PVA interaction with Cibacron Blue is characterised by a positive maximum at 675 nm, an isobestic point at 610 nm, and a minimum at 580 nm with a shoulder at 540 nm[12]. The calculated value of the binding constant was 16.7 μM and one Cibacron Blue molecule complexed with a molecular mass of PVA of about 10,000 . PVA with a molecular mass of 13,000 interacted with the dye nearly stoichiometrically. This polymer could not provide multipoint attachment when interacting with a Cibacron Blue-containing matrix, and for this reason it could not be used for polymer shielding in dye-affinity chromatography.

Cibacron Blue interaction with PVA reduces access to the ligand in dye-PVA conjugates. This explains previously reported data[13] that lactate dehydrogenase from bovine heart was inhibited ten times less efficiently by a Cibacron Blue-PVA conjugate than by a Cibacron Blue-dextran conjugate in which no interaction was detected between the dye and the polymer. The absence of interaction of Cibacron Blue with dextran, resulting in good access to the dye-ligand in such conjugates, is in agreement with the fact that the reported values of the inhibition constants of lactate dehydrogenase with free and dextran coupled triazine dyes are practically the same[14]. The interaction of BSA with free and PEG-coupled Cibacron Blue are characterised by practically the same half-saturation dye concentrations, while for PVA-Cibacron Blue this value is about 10 times greater[15].The absence of Cibacron Blue interaction with PEG is the reason for the extensive use of such conjugates in affinity partitioning[16]. The restricted access of the dye-ligand in Cibacron

Table 1 *Relative strength of interaction of Cibacron Blue with water soluble polymers*

Polymer	*Interaction*
Dextran	NO INTERACTION
Hydroxyethylcellulose	"
Poly(ethyleneglycol)	"
Poly(acrylamide)	"
Poly(methylmethacrylate-methacrylic acid)	"
Methylcellulose	WEAK
Poly(vinyl alcohol), PVA	MODERATE
Chitosan	"
Poly(vinyl pyrrolidone), PVP	STRONG
Poly(ethyleneimine)	VERY STRONG (PRECIPITATION)

Blue-PVA conjugates might be the reason for unsuccessful affinity precipitation of lactate dehydrogenase using these conjugates[13,17].

PVP interaction with Cibacron Blue resulted in difference spectra with a peak at 680 nm, a minimum at 580 nm with a shoulder at 550 nm and an isobestic point at 630 nm, indicating complex formation. This type of difference spectra is typical of dye interaction with proteins, and according to Subramanian[11] can be classified as "electrostatic interaction spectra". The difference spectra were significantly changed in the presence of 1.5 M KCl (traditionally used as a nonspecific eluent in dye-affinity chromatography) and characterised by a positive peak at 650 nm and a small negative contribution below 550 nm. This spectrum type is similar to "hydrophobic interaction spectra"[11]. PVP is well known to complex both with negatively charged and with hydrophobic substances[18]. The contribution of electrostatic interaction in the difference spectra is more pronounced at a low ionic strength. The increase in ionic strength suppressed the electrostatic component and made the contribution of the hydrophobic interaction more pronounced.

The binding constant for PVP complexing with Cibacron Blue in solution was calculated as 2.1 μM (Fig. 1)[12]. A molecule of the polymer with a molecular mass of 40,000 contained 30 sites capable of binding Cibacron Blue ligands. For PVP-10 with a molecular mass of 10,000 the calculated value of the binding constant was 6.1 μM and the number of binding sites per polymer molecule was 8. It should be noted that the calculated dependence of complex concentration on polymer concentration was very sensitive to the number of binding sites per polymer molecule. Even minimal changes in the number of binding sites resulted in significant changes of the curve (Fig. 1, dashed curves).

The dye ligand was bound by a polymer segment with a molecular mass of 1,000 - 1,300 regardless of the size of the polymer molecule. This fact supported the assumption of independent binding sites and is in agreement with the direct measurement of I_3 and 1-anilinonaphthalene-8-sulphonate complexation with vinyl pyrrolidone oligomers of different molecular masses. Complexation takes place only when the oligomer molecular mass is greater than 1,500.[19]

Though the type of the difference spectra was changed with the increase of ionic strength, the binding of PVP-40 with Cibacron Blue was relatively insensitive to 1.5 M KCl: the binding constant was 5.7 μM, and the number of binding sites was practically the same, 32.

Among the polymers studied only PVP could bind efficiently to Cibacron Blue ligands of Blue Sepharose via multipoint attachment and serve as an inert blocking agent of nonspecific protein-dye interactions. The binding constants of one PVP segment with the dye ligand were in the micromolar range. The binding constants of Cibacron Blue to the nucleotide-binding sites of enzymes are in the same range or even lower (Table 2). Thus, enzymes can successfully compete with PVP segments for the ligand. The nonspecific interaction of dye ligands with proteins is much weaker and characterised by higher binding constants. Thus, the PVP segments, when bound to the dye ligand, would protect it from relatively weak nonspecific interactions with proteins. In other words, PVP molecules bound to the Blue Sepharose could serve as a "lid", making the ligand available for strong specific interactions and protecting it from weaker nonspecific interactions. This forms the concept of polymer-shielded dye-affinity chromatography.[12,20,21]

3 POLYMER-SHIELDED DYE AFFINITY CHROMATOGRAPHY

Application of porcine muscle extract on a fresh Blue Sepharose column until break-through resulted in the binding of lactate dehydrogenase (LDH), along with a significant

Table 2   *Cibacron Blue interaction with various proteins and PVP*

Protein	Binding Constant, μM	Procedure	Reference
SPECIFIC BINDING			
Dihydrofolate reductase from chicken liver	0.8	Difference spectroscopy	22
Dihydrofolate reductase from *Lactobacillus casei*	0.13	Difference spectroscopy	23
Phospholipase A$_2$	2	Difference spectroscopy	24
	3.5	Inhibition	
Cytochrome b$_5$ reductase	1	Difference spectroscopy	25
Nucleotide phosphodiesterase	0.3	Inhibition	14
Lactate dehydrogenase from rabbit muscle	0.5	Difference spectroscopy	26
	0.1	Inhibition	
	0.7	Equilibrium dialysis	
	0.2	Analytical affinity chromatography	
Poly(vinyl pyrrolidone)	2-6	Difference spectroscopy	present work
NONSPECIFIC BINDING			
Bovine serum albumin	35-85	Difference spectroscopy	27
Cytochrome b$_5$ reductase	85	Difference spectroscopy	2

amount of foreign proteins which could not be eluted with the buffer. Nonspecific elution with 1.5 M KCl resulted first in the elution of foreign proteins followed by LDH elution with a recovery of 76%, as judged by activity measurement. LDH recovery increased gradually to nearly 100% only after subsequent purification cycles with applications of porcine muscle extract on the same column. Thus, pretreatment of Blue Sepharose with the homogenate resulted in the masking of sites capable of nonspecific irreversible binding of LDH. The same effect occurred during the specific elution of LDH with 0.1 mM NADH + 10 mM oxamate, but in contrast to nonspecific elution, foreign proteins adsorbed on the column were not eluted during specific elution. After specific elution the column needed to be regenerated; foreign proteins could be eluted with 1.5 M KCl.

The PVP shielding of the column resulted in a significant reduction of binding of foreign proteins and in an improvement of the effectiveness of elution of LDH both specifically and nonspecifically (Fig. 2). The reduced binding of foreign proteins to the PVP shielded column eliminated the need for a regeneration step after the specific elution. The column could be used repeatedly after reequilibration with buffer. The LDH recovery was nearly 100%, even during the first run on a fresh Sepharose Blue column shielded with PVP. Thus, PVP blocked the sites to which LDH irreversibly bound. Some proteins from the homogenate played the same role during the first application of porcine muscle extract on a Blue Sepharose column. It is preferable to use PVP to block binding sites on the matrix rather than an unidentified mixture of proteins from the homogenate. PVP is a cheap, stable, nontoxic, and highly biocompatible polymer. No polymer was detected in the eluate from the PVP-shielded column during chromatography of LDH (the sensitivity of the method of the PVP assay was 0.1 mg/ml).

PVP shielding resulted in a significant improvement of the effectiveness of elution.The HETP for untreated column was 1.3 cm (nonspecific elution) and 0.47 cm (specific elution). PVP shielding decreased the HETP to 0.13 cm for both types of elution.

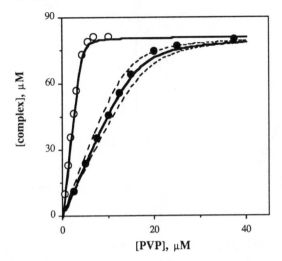

Figure 1 *PVP-40 (O) and PVP-10 (●) complexation with Cibacron Blue (81 μM Cibacron Blue, 50 mM Tris pH 8.0). The solid curves were calculated using the binding constants 2.1 and 6.1 μM and numbers of binding sites 30 and 8 for PVP-40 and PVP-10, respectively. The dashed curves were calculated using binding constant 6.1 μM and numbers of binding sites 7 (lower curve) and 9 (upper curve) for PVP-10*

LDH was eluted from a PVP shielded column as a symmetrical peak and about 95 % eluted enzyme could be collected in a volume of 8 ml. LDH was eluted from an untreated column of the same size with significant tailing, and only 95 % of eluted LDH could be collected in a volume of about 100 ml (Fig. 2). Volume reduction is preferable for further down-stream processing because the cost of the purification step is usually proportional to the feed concentration.

Application of an extract of *Thermoanaerobium brockii* on a Scarlet Sepharose column resulted in adsorption of secondary alcohol dehydrogenase (SADH) along with significant amounts of foreign proteins. This behaviour resembled that of LDH binding to Blue Sepharose, but in contrast to the case with Blue Sepharose, foreign proteins were eluted by 1.5 M KCl simultaneously with SADH, resulting in only a 5-fold purification. Specific elution with 0.5 mM NADP resulted in more pure enzyme preparation (14-foldpurification) with 70 % recovery; the remaining 30 % of SADH activity could be eluted by 1.5 M KCl as a rather crude preparation together with other adsorbed proteins. The specific elution at lower loadings gave lower recoveries and in zonal mode, SADH was not eluted by 0.5 mM NADP at all.

The PVP shielding of Scarlet Sepharose resulted in an improvement of effectiveness, with a 93 % recovery. Again, as in the case of Blue Sepharose, PVP shielding of Scarlet Sepharose prevented adsorption of foreign proteins and nonspecific interaction of SADH with the matrix (Fig. 3).

4 CONCLUSION

PVP shielding of dye affinity matrices significantly decreased the adsorption of foreign proteins as well as nonspecific binding of the target molecules, while not seriously impairing the specific binding. This chromatographic behaviour is in complete agreement with that expected on the basis of study of PVP interaction with Cibacron Blue.

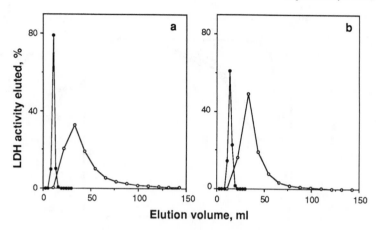

Fiure 2 *Elution profile of LDH activity with 1.5 M KCl (a) and with 0.1 mM NADH + 10 mM oxamate (b) from unmodified (O) and PVP treated (●) Blue Sepharose. Experimental conditions: 9.8x0.9 cm I.D. column was treated with 1% PVP-40 000 solution followed by washing with 1.5 M KCl, pH 3.4 until no PVP was detected in the eluate, and reequilibration with 20 mM Tris HCl buffer, pH 7.3. The porcine muscle extract was applied on the column until break-through of LDH occurred.The column was washed with buffer until no more protein (monitored as absorbance at 280 nm) eluted. LDH was eluted at a flow rate of 0.55 ml/min with 1.5 M KCl or with 10 mM oxamate + 0.1 mM NADH. Fractions were collected every 20 min when eluted from untreated column and 5 min when eluted from PVP protected column*

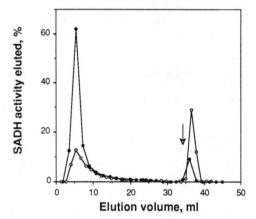

Figure 3 *Elution profile of SADH activity with 0.5 mM NADP from unmodified (O) and PVP treated (●) Scarlet Sepharose followed by elution with 1.5 M KCl. The arrow indicates when elution with 1.5 M KCl was begun. Experimental conditions: 2.8x0.9 cm I.D. column was treated with 1% PVP-40,000 solution followed by washing with 1.5 M KCl, pH 3.4 until no PVP was detected in the eluate, and reequilibration with 20 mM morpholinopropane sulphonate buffer pH 6.5 containing 30 mM NaCl and 2 mM MgCl₂. The Thermoanaerobium brockii extract was applied on the column until break-through of SADH occurred.The column was washed with buffer until no more protein (monitored as absorbance at 280 nm) eluted. SADH was eluted at a flow rate of 0.09 ml/min with 0.5 mM NADP. Fractions were collected every 15 min. The total amount of SADH eluted from the column was taken as 100 % in both cases for the sake of comparison*

Acknowledgements

The support of The Swedish Royal Academy of Sciences (KVA), the National Swedish Board for Technical and Industrial Development (NUTEK), the Swedish Agency for Cooperation with Developing Countries (SAREC), and The Swedish Research Council for Engineering Sciences (TFR) are gratefully acknowledged. The authors thank Professor Robert K. Scopes for generously providing Scarlet Sepharose, Nora Perotti for the synthesis of Blue Sepharose, Dr. Olle Holst and Åsa Rosenlund for cultivation of cells, Eva Linné-Larsson for the practical advice in electrophoresis, and Dr. Scott Bloomer for linguistic advice.

References

1. E. Stellwagen in M.P. Deutscher (Editor), 'Methods in Enzymology', Academic Press, San Diego, 1990, Vol.182, Chapter 28, p. 343.
2. R.K. Scopes, *Anal. Biochem.*, 1987, **165**, 235.
3. F.Qadri , *Trends Biotechnol.*,1985, 3, 7.
4. C.V.Stead, *Bioseparation*, 1991, 2 ,129.
5. M.A.Vijayalakshmi, *Trends Biotechnol.* , 1989, **7** , 71.
6. T.Makriyannis and Y.D.Clonis, *J. Chromatogr.*, 1993, **28** ,179.
7. Y.D.Clonis, T.Atkinson C.J.Bruton and C.R.Lowe (Editors) 'Reactive dyes in protein and enzyme technology', Macmillan Press, Basingstoke, UK, 1987.
8. Y.D.Clonis in M.T.W. Hearn (Editor) 'HPLC of proteins and polynucleotides',VCH Publishers, New York, 1991, Chapter 13, p.453.
9. M. Allary, J. Saint-Blancard, E. Boschetti and P. Girot, *Bioseparation*, 1991, **2**, 167.
10. R.K. Scopes, *J. Chromatogr.*, 1986, **376**, 131.
11. S.Subramanian, *Arch. Biochem. Biophys.*, 1982, **216**, 116.
12. I.Yu. Galaev and B. Mattiasson, submitted.
13. J.E.Morris and R.R.Fisher, *Biotech. Bioeng.*, 1990, **36**, 737.
14. A.R.Ashton and G.M.Polya, *Biochem. J.*, 1978, **175**, 501.
15. G.Johansson and M.Joelsson, *J.Chromatogr.*, 1991, **537**, 219.
16. G.Johansson, in H.Walter, D.E.Brooks and D.Fisher (Editors), Partitioning in aqueous two-phase systems: Theory, methods, uses and applications to biotechnology, Academic Press, New York, 1985, p. 161.
17. R.R.Fisher, B.Machiels, K.C.Kyriacou and J.E.Morris, in M.A.Vijayalakshmi and O.Bertrand (Editors), Protein-dye interactions: Developments and applications, Elsevier, London, 1989, p.190.
18. Yu.E.Kirsh, *Progr. Polym. Sci.*, 1985, **11**, 283.
19. Yu.E.Kirsh, T.A.Soos and T.M.Karaputadze, *Eur. Polym. J.*, 1983, **19**, 639.
20. I.Yu. Galaev and B. Mattiasson, *J.Chromatogr.*, 1993, **648**, 367.
21. I.Yu. Galaev and B. Mattiasson, *J.Chromatogr.*, 1994, **662**, 27.
22. S.Subramanian and B.Kaufman, *J. Biol. Chem.*, 1980, **255**, 10587.
23. B.B.Chambers and R.B.Dunlap, *J. Biol. Chem.*, 1979, **254** , 6515.
24. R.E.Barden, P.L.Darke, R.A.Deems and E.A.Dennis, *Biochemistry*, 1980, **19**, 1621.
25. D.Pompon, B.Guiard and F.Lederer, *Eur. J. Biochem.*, 1980, **110**, 565.
26. S.T.Thompson and E.Stellwagen, *Proc. Nat. Acad. Sci. USA*, 1976, **73**, 361.
27. G.Johansson and M.Joelsson, *J.Chromatogr.*, 1991, **537**, 219.

Solid Phases for Protein Adsorption in Liquid Fluidized Beds: Comparison of Commercial and Custom-assembled Particles

Gordon R. Gilchrist, Michael T. Burns, and Andrew Lyddiatt

BIOCHEMICAL RECOVERY GROUP, BBSRC CENTRE FOR BIOCHEMICAL
ENGINEERING, SCHOOL OF CHEMICAL ENGINEERING, UNIVERSITY OF
BIRMINGHAM, BIRMINGHAM B15 2TT, UK

1. INTRODUCTION

Adsorptive recovery in fluidised or expanded beds potentially offers the process engineer new options for the compression of operational sequences necessary to achieve concentration and fractionation of target products (commonly sourced in dilute feedstocks). In particular, the method facilitates solute adsorption from particulate feedstocks such as whole fermentation broths, cell disruptates or biological extracts. Direct and rapid product capture eliminates the necessity for costly solid-liquid separation (with or without the conditioning additives), and enhances the yield and molecular integrity of labile protein products (1,2). The technical philosophy of suspension processing is not new. Bartel *et al* (3) in 1958, and Belter *et al* (4) in 1973 directly recovered antibiotics from whole broths. However, developments in the field have been limited by the paucity of commercial solid phases with the physical and biochemical qualities necessary for effective performance in expanded and fluidised beds. The design of adsorbent matrices in general has focused upon production of spherical, mono-dispersed, small diameter (10-200µm) macroporous particles with high rigidity-requirements dictated by classical fixed bed chromatography (5,6,7). For protein recovery the expanded or fluidised particle should include significant density (1.2-2.0 g cm⁻³), suitable surface area and internal geometry (*eg.* pore diameter 50-100nm), flexible surface chemistry (for ease of derivatisation) and adequate sorption capacity (e.g. 20-100mg BSA equivalents/ml settled adsorbent).

To meet commercial shortfalls, work since 1989 in Birmingham has addressed the assembly of simple polysaccharide-ceramic composites and their application in single-stage, recirculating fluidised beds (SSRFB). Gibson (8) has shown that cellulose, widely used and effective in the field of chromatographic solid phase production (9, 10, 11), is a suitable material for the fabrication of such composites. Composite particles are formed by firstly solubilising the cellulose followed by the addition of a dense ceramic "filler" (TiO₂, density 4.17 g cm⁻³) (12). The cellulose matrix is then regenerated to encapsulate the "filler", and then milled and washed to produce a particle of enhanced density that can be derivatised for specific application.

Subsequent to 1989, a small number of suitable commercial solid phases have been identified, although their hydrodynamic behaviour, biochemical performance and generic applicability has yet to be widely publicised. Here, we report a comparison of the performance of simple cellulose composites with commercial solid phases (Streamline from Pharmacia, Spherodex and Spherosil from Sepracor). Physical behaviour in fluidised beds and the biochemical performance in respect of dynamic protein capacities and productivities are reported. The advantages of customised assembly of adsorbent particles for specific applications characterised by varied biomass and broth rheologies are discussed.

2. MATERIALS AND METHODS

Fibrous cellulose powder (CF11) was supplied by Whatman International Ltd., Maidstone, Kent. All other chemicals were analar grade.

Fabrication of Cellulose Composite Ion-exchangers.

A method reported by Kuga (13) and adapted by Gibson (8) was adopted for composite fabrication purposes. The particles were prepared by the dissolution of cellulose (CF11 powder) in a 5% (w/v) aqueous solution of calcium chloride and potassium thiocyanate at 120°C. Titanium(IV) oxide (TiO$_2$, 2.5% (w/v), 1-5μm) was dispersed thoroughly in the mixture and the cellulose matrix regenerated by precipitation in cold methanol. The resulting solids were washed in deionised water to remove fines and residual methanol. Particle formation was achieved by milling and sieving to produce a selected size range. This material has been designated Ti-Cell-1 and was cross-linked and derivatised by the author utilising confidential commercial methods adopted by Whatman International Ltd. (14). The derivatisation introduced anion exchange ligands into the composite by chemically bonding diethylaminoethyl (DEAE) groups to yield DEAE Ti-Cell-1.

Physical and Biochemical Characterisations.

Compositional data was determined by dehydration at 120°C to a constant mass for moisture content whilst non-combustible content was estimated gravimetrically following heating at 900°C **(Table 1.)**. "Wet" composite density was determined by water displacement and particle size range by sieving. A comparison was made with commercial adsorbents DEAE Spherodex LS, DEAE Spherosil LS and DEAE Streamline **(Table 2.)**. Liquid fluidisation performance was characterised by a study of bed expansion in a single stage, recirculating fluidised bed (SSRFB; 2.2cm id x 20cm) fitted with a 106μm distributor mesh **(Figure 1.)**. Protein binding capacities were determined by Whatman International Ltd. Q.C. protocols (14). Adsorption of bovine serum albumin (BSA) was achieved by contacting the pre-equilibrated adsorbent in 10mM sodium phosphate buffer, pH 8.5 (4mg/ml BSA, 0.4dry g equivalents of adsorbent) for 90 minutes. Desorption was achieved with 30mM sodium phosphate buffer, pH 8.5 and the protein concentrations were estimated spectrophotometrically at 280 nm **(Table 2.)**.

Dynamic Protein Binding Capacities in a SSRFB

Anion exchange materials were tested in SSRFB for the adsorption of protein (BSA). A 500ml homogeneous liquid phase (10 mM Tris/HCl, pH 7.5, 3mg/ml BSA) was recirculated through a contactor (glass column, 2.2cm id x 20cm) at volumetric flowrates that fluidised the bed (10ml settled volume of pre-equilibrated adsorbent) to 100% expansion. Reservoir protein concentrations were monitored spectrophotometrically at 280 nm **(Figure 2.)**. Fluidised washing of loaded adsorbents was undertaken in single-pass mode at an expansion of 200%. Dynamic capacities are reported in **Table 3.**. Materials were also tested in the presence of suspended solids, for the adsorptive recovery of a simulated fermentation broth. The liquid feed (20dry g/L *Saccharomyces cerevisiae*, bakers yeast supplied by British Fermentation Products Ltd., suspended in 500ml, 10mM Tris/HCl, pH 7.5, dosed with 3mg/ml BSA) was recirculated through a contactor (see above) at volumetric flowrates that facilitated 100% bed expansion (initial bed volume as above). Reservoir protein concentrations of clarified samples were monitored by the method of Bradford (15) whilst bed height was recorded over a fixed time. The pH of the suspension was controlled by the addition of 0.1M HCl and 0.1M NaOH. Volumetric flowrates were confirmed during fluidisation by timed volumetric capture of the recirculating liquid feed. Cell counts and cell viabilities during the experimental run were determined (data not shown). Washing of the adsorbent was carried out in a single-pass, fluidised mode (200% expansion) until washes were free of solids. The adsorbent was then transferred to a fixed bed (1.0cm id x 10cm) and bound solutes desorbed by gradient elution (0-1M NaCl in 10mM Tris/HCl, pH 7.5). The collected fractions were assayed for protein content (15) and dynamic capacities and recoveries are reported in **Table 3.**.

3. RESULTS AND DISCUSSION

Physical Characteristics of Experimental and Commercial Adsorbents

The data in **Table 1.** confirms that ceramic (TiO_2) material has been successfully entrapped within the cellulose matrix to impart increased densities to particles, comparable with commercial materials **(Table 2.)**. Bed expansion in a SSRFB **(Figure 1.)** indicates that the hydrodynamic behaviour of the Ti-Cell-1 is acceptable within a range of flow velocities ($0-30 \times 10^{-4} ms^{-1}$). We conclude that this test yields a good indication of material fluidisation characteristics when compared to density measurement and determination of terminal settling velocity U_t (16). This is confirmed by the performance of Streamline (developed as an expanded bed particle) where **Figure 1.** clearly shows that it fluidises to a much greater degree than Ti-Cell-1 but possesses equivalent density **(Table 2.)**. Factors such as size and shape of material must be considered when testing fluidisation performance.

Protein Binding Capacities of Experimental and Commercial Adsorbents

Ti-Cell-1 was chemically cross-linked and derivatised with diethylaminoethyl (DEAE) anion exchange groups to show small-ion capacities that equal or better the commercial materials **(Table 2.)**. Ion exchange performance was carried out in protein binding, dynamic fluidised capacity (homogeneous system) and productivity (mg BSA recovered per ml adsorbent in the simulated broth system) experiments (See Materials and Methods). **Table 2.** confirms that Ti-Cell-1 compares well with the commercial materials and gives the highest ratio of protein binding per available anion exchange milliequivalent. This may be due to Ti-Cell-1's porous structure and the greater degree of polysaccharide material present **(combustible material, Table 1.).**

TABLE 1. : Compositional analysis of experimental and commercial adsorbents.

ADSORBENT	MOISTURE CONTENT (% w/w)	COMBUSTIBLE CONTENT (% w/w)	NON-COMBUSTIBLE CONTENT (% w/w)
DEAE Ti-Cell-1 (✥)	73.00	19.25	7.75
DEAE Spherodex (✦)	2.01	13.17	84.82
DEAE Spherosil (✦)	0.97	13.22	85.81
DEAE Streamline (✥)	73.06	11.68	15.26

Compositional data was determined by means of dehydration and combustion. Data for ✦ was based on dry packaged formulation. Data for ✥ was based on vacuum dried material.

TABLE 2. : Physical characterisation, small-ion capacity and fixed bed protein binding of experimental and commercial adsorbents.

ADSORBENT	DENSITY (g cm⁻³)	SIZE RANGE (µm)	SMALL-ION CAPACITY	MAX. BINDING CAPACITY (BSA)
DEAE Ti-Cell-1 (✦)	1.20	125-600	0.63 meq/dry g	650 mg/dry g
DEAE Spherodex (✥)(✪)	1.35	100-300	1.16 meq/dry g	1046 mg/dry g
DEA Spherosil (✪)	1.33	100-300	0.25 meq/dry g	115 mg/dry g
DEAE Streamline (✦)	1.20	100-300	0.87 meq/dry g	335 mg/dry g

Adsorbent densities were determined by means of water displacement. Ti-Cell-1 size range was determined by "wet" sieving, all other sizes as published (see17,18,19). Small-ion capacities for ✦ were determined by Whatman International Ltd. Q.C. protocols, (14) all other capacities as published (18,19). Small-ion capacity for ✥ was converted to (meq/dry g) by dry-wt. analysis. Maximum protein binding capacities for ✦ were determined by Whatman International Ltd. Q.C. protocols, (14) all other capacities as published (18,19). Binding capacities for ✪ were converted to (mg/dry g) by dry-wt. analysis.

TABLE 3. : Dynamic binding capacities and productive efficiency for experimental and commercial adsorbents in a SSRFB.

ADSORBENT	DYNAMIC CAPACITY FOR BSA IN CELL-FREE SYSTEM (Driving Force 3mg/ml) (mg/ml)	DYNAMIC CAPACITY FOR BSA IN SIMULATED BROTH (Driving Force 3mg/ml) (mg/ml)	PERCENTAGE BOUND BSA RECOVERED FROM THE SIMULATED BROTH (%)	PRODUCTIVITY FROM THE SIMULATED BROTH (mgBSA/ml adsorbent)
DEAE Ti-Cell-1	60.10	48.24	88.52	42.70
DEAE Spherodex	120.50	76.99	96.20	74.07
DEAE Spherosil	31.50	4.49	25.12	1.13
DEAE Streamline	78.20	51.10	84.50	43.18

The above data was recovered as described in Materials and Methods.

Figure 1. : Variation of bed expansion with linear velocities for experimental and commercial adsorbents in a SSRFB.

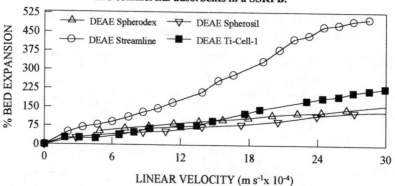

The dynamic capacities for BSA in the homogeneous/cell-free system **(Figure 2.,Table 3.)** indicate that Spherodex performs better than the other materials but that Ti-Cell-1 is again comparable with Streamline. However, in the presence of suspended-solids, it is observed that dynamic capacities are reduced **(Table 3.)**. **Figure 3.** shows that the binding performance of Spherodex is closer to that of Streamline and Ti-Cell-1 which together clearly exceed Spherosil. It should also be noted that the presence of suspended-solids has the least effect on the binding capacity of Ti-Cell-1 **(Table 3.)**. In terms of protein recovery Ti-Cell-1 exhibits comparable results with Streamline, but is out-performed by Spherodex. Spherosil is shown to be an unacceptable material for use in this suspended solids system **(Table 3.)**. It should be recorded that the presence of cells had no gross effect on bed height, system pH, cell counts and viabilities throughout experimental runs (data not shown). Data accumulated in Birmingham by Wells (20) and Morton (16) utilising fermentations of up to 100dry g/L (as compared to the 20dry g/L herein) showed similar results.

Figure 2. : Dynamic binding profile of BSA in a homogeneous system within a SSRFB.

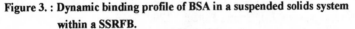

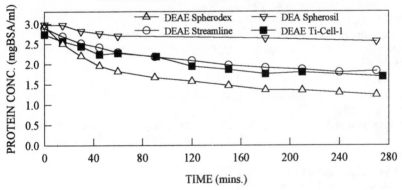

Figure 3. : Dynamic binding profile of BSA in a suspended solids system within a SSRFB.

4. CONCLUSIONS

The fabrication of an enhanced density cellulose composite has yielded macroporous particles that possess physico-chemical attributes and hydrodynamic characteristics required for use in a SSRFB. Chemical derivatisations have imparted mechanical strength through cross-linking and specific anion exchange properties by the introduction of diethylaminoethyl (DEAE) ligands. The composite has comparable BSA protein binding capacities with a range of commercially available adsorbents (Spherodex, Spherosil and Streamline). DEAE Ti-Cell-1 performed well when operated in a SSRFB for dynamic binding with a homogeneous feed <u>and</u> a suspended solids feed yielding capacities and recoveries comparable with the best commercial materials. These results encourage the further development of custom-built matrices designed for specific recovery applications exhibiting problems of varied biomass and broth rheologies.

5. ACKNOWLEDGEMENTS

The authors thank Whatman International Ltd. for the financial support of GRG in a SERC CASE studentship.

6. REFERENCES

1. C.P. Bowden, Proc. Biotech., 1984, 2, 39.
2. P.H. Morton and A. Lyddiatt, "Ion Exchange Advances", (Ed. M.J. Slater), Elsevier Applied Science, 1992, p. 237.
3. C.R. Bartel *et al*, Chem. Eng. Prog., 1958, 54, 49.
4. P.A. Belter *et al*, Biotech. Bioeng., 1973, 15, 533.
5. B.J. Miles and A.R. Thomson, Proc. Biochem., 1974, 11, 11.
6. S.R. Navayanan and L.J. Crane, Tibtech., 1990, 8, 12.
7. Y.D. Clonis, Bio/technology, 1987, 5, 1290.
8. N.B. Gibson, PhD Thesis, University of Birmingham, 1992.
9. P.R. Levison, in "Cellulosics : materials for selective separations and other technologies", (Eds. J.F. Kennedy, G.O. Philips, P.A. Williams), Ellis Horwood, 1993, p. 25.
10. D.G. Kilburn *et al*, Bio/technology, 1993, 11, 1570.
11. E.A. Peterson, "Cellulosic Ion Exchangers", Elsevier/North Holland Biomedical Press, Amsterdam, 1980.
12. N.B. Gibson and A. Lyddiatt, in "Cellulosics : materials for selective separations and other technologies", (Eds. J.F. Kennedy, G.O. Philips, P.A. Williams), Ellis Horwood, 1993, p. 55.
13. S. Kuga, J. Chromatogr., 1980, 195, 221.
14. P. Levison, Personal Communication, 1993.
15. M.M. Bradford, Anal. Biochem., 1976, 72, 249.
16. P.H. Morton, PhD Thesis, University of Birmingham, 1993.
17. Pharmacia Product Data File, 1993.
18. IBF Product Information, No. 205101 IBF, 1993.
19. Sepracor Product Bulletin, 1993.
20. C.M. Wells, PhD Thesis, University of Birmingham, 1990.

An Integral Approximation for the Optimization of Size-exclusion Chromatographic Column Performance

Massimiliano Giona and Daniela Spera[a]

DIPARTIMENTO DI INGEGNERIA CHIMICA, UNIVERSITÀ DI CAGLIARI, PIAZZA D'ARMI, 09123 CAGLIARI, ITALY

[a]CONSORZIO DI RICERCHE APPLICATE ALLA BIOTECNOLOGIA, STRADA PROVINCIALE 22, 67051, AVEZZANO (AQ), ITALY

1 INTRODUCTION

Chromatography is an important tool in the separation of biomolecules. A broad literature has been published focusing on the different aspects of chromatographic processes, but little work has been performed on process optimization and control in the application of chromatographic columns as separation units[1]. This paper concerns the application of short-cut methods for the characterization, optimization and control of Size-Exclusion Chromatographic (SEC) units. SEC separation is characterized by the diffusional partition of a solute between the mobile and the stationary phase[2] so that transport properties inside the porous packing are controlled exclusively by the geometric factors (ratio between solute and average pore-radius) affecting solute dynamics in the porous structure. A detailed analysis of size-effects, dispersion properties and dynamics can be found elsewhere[3-7].

2 EXPERIMENTAL SECTION

The biomolecules purchased from Sigma, Poole, U.K., and Bio-Rad Richmond Ca, USA, and used for chromatographic analysis were: bovine Thyroglobulin (669 kDa), horse spleen Apoferritin (443 kDa), bovine γ-globulin (160 kDa), mouse IgG (150 kDa), chicken egg Ovalbumin (44 kDa), horse Myoglobin (17 kDa) and Cyanocobalamin (1.35 kDa). A Bio-Sil TSK 250 column and a Bio-Sil TSK 125 column (length L=60 cm, inner section area S=0.44 cm^2, from Toyo-Soda) were used. The Bio-Sil 250 column consisted of G 3000 SW packing (average particle size 10 μm, average pore size 250 Å) and the Bio-Sil 125 column of G 2000 SW packing (average particle size 10 μm, average pore size 125 Å). The TSK 250 column was eluted with 0.1 M Na$_2$SO$_4$, 0.1 M NaH$_2$PO$_4$ and 0.02 % Sodium Azide (w/v) ph 6.8, and the TSK 125 column was eluted with 0.05 M Na$_2$SO$_4$, 0.02 M NaH$_2$PO$_4$ and 0.02 % Sodium Azide (w/v) ph 6.8. The mobile phase was prepared with deionized water and was filtered through a 0.45 μm filter by means of an all-glass Millipore filtration unit (Millipore, U.K.). Samples containing typically 1-10 mg/ml of biomolecule were injected via Rheodyne 7125 (20 μl) loop injector after centrifugation. The mobile phase flow rate was varied from 0.1 to 1.0 ml/min using a twin-headed reciprocating pump model 510 (Millipore U.K.). All the experiments were performed at 25° C. Detection was performed by U.V. absorbance at 280nm (481 Lambda Max,Millipore U.K.). The data from the U.V. monitor were collected by using a Baseline 810 Chromatographic Workstation (Dynamic Solution, Dividion of Millipore).

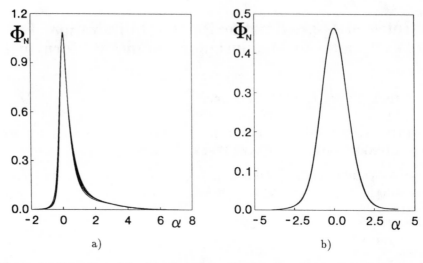

Figure 1: *Behaviour of the invariant Green function Φ_N. In the figures, Φ_N has been calculated from the experimental data through eq. (2) for different values of the elution flow rate in the range 0.1-1.0 ml/min. a) Thyroglobulin on TSK 250; b) Cyanocobalamin on TSK 125. In the latter case all the invariant chromatograms Φ_N at different flow rates are almost identical and are indistinguishable.*

3 INVARIANT FORMULATION

The dynamics of solute transport in SEC columns can be described and greatly simplified by means of an experimentally found invariant principle[4]. Let $c(t, t_r)$ be the outlet chromatogram (associated with an pulse injection) of a biomolecule having peak retention time t_r and variance $\sigma^2(t_r)$ (the variance depends on the elution flow rate and consequently on the retention time). It can be shown experimentally that in SEC columns for all the biomolecules considered $c(t, t_r)$ can be expressed as

$$c(t, t_r) = \frac{1}{\sigma(t_r)} \Phi_N \left(\frac{t - t_r}{\sigma(t_r)} \right) . \tag{1}$$

Equation (1) can be rewritten as

$$\Phi_N(\alpha) = \sigma(t_r) c(t_r + \sigma(t_r)\alpha, t_r) , \tag{2}$$

which means that the function $\Phi_N(\alpha)$, $\alpha = (t - t_r)/\sigma(t_r)$, is independent of the flow conditions and characterizes the dynamics of a given biomolecule in the column. As an example, figure 1 a)-b) shows the experimental validation of the invariant principle (eq.(2)) in the case of thyroglobulin (TSK 250) and cyanocobalamin (TSK 125) for elution flow rates in the range 0.1-1.0 ml/min. The function $\Phi_N(\alpha)$ can be called the *Invariant Green Function* associated with a given column and given biomolecule. The details and the mathematical implications of eq. (1)-(2) can be found in Giona et al.[4]. This paper is mainly concerned with developing the practical consequences and the optimization strategy related to this result.

Since $\Phi_N(\alpha)$ is independent of the flow condition (i.e. of the elution flow rate), this function fully characterizes solute dynamics in the limit of linear response (linearity of response was always satisfied in the considered range of concentrations and flow rates).

It should be observed that eqs. (1)-(2) are valid for high Peclet numbers (a condition practically always satisfied in SEC columns) and for high Sherwood numbers. The simplest way of representing $\Phi_N(\alpha)$ and $c(t, t_r)$ for optimization and control purposes is to introduce the integral quantities

$$q_N^{(+)}(\alpha) = \int_0^\alpha \Phi_N(\tau)d\tau \ , \quad q_N^{(-)}(\alpha) = \int_{-\alpha}^0 \Phi_N(\tau)d\tau$$

$$q^{(+)}(t, t_r) = \int_{t_r}^{t_r+t} c(\tau, t_r)d\tau \ , \quad q^{(-)}(t, t_r) = \int_{t_r-t}^{t_r} c(\tau, t_r)d\tau \ . \tag{3}$$

The functions $q_N^{(+)}$, $q_N^{(-)}$ represent the total amount of solute (normalized to unity) at a distance α from the peak of the invariant Green function respectively to the right and to the left of the peak. Similarly for $q^{(+)}(t, t_r)$ and $q^{(-)}(t, t_r)$ with respect to $c(t, t_r)$. The advantage in adopting the integral representation (3) lies in the fact that the functions q can be easily represented by means of simple expressions and in particular by means of Toth-isotherm-functions[6] (these functions arise in the analysis of adsorption phenomena)

$$q_N^{(+)}(\alpha) = \frac{Q_{N+}}{(\gamma_{N+}^{m+} + \alpha^{m+})^{1/m+}} \ , \quad q_N^{(-)}(\alpha) = \frac{Q_{N-}}{(\gamma_{N-}^{m-} + \alpha^{m-})^{1/m-}} \ , \tag{4}$$

with the closure conditions

$$Q_{N+} + Q_{N-} = 1 \ , \quad \gamma_{N+} = Q_{N+}/\Phi_N(0) \ , \quad \gamma_{N-} = Q_{N-}/\Phi_N(0) \ . \tag{5}$$

Given the invariant Green function $\Phi_N(\alpha)$, the four parameters $Q_{N+}, Q_{N-}, \gamma_{N+}, \gamma_{N-}$ can be obtained directly from the experimental data and only the two exponents m_+, m_- need to be determined by means of optimization techniques. Figure 2 a)-b) shows the reconstruction with Toth-isotherm-function approximation of the invariant Green function and of its integral representation in the case of Apoferritin (TSK 125) and of Thyroglobulin (TSK 125). It is important to observe that the integral representation (4-5) is particularly suitable for an accurate representation of the overall quantity of biomolecule flowing through the outlet up to a given time instant.

4 RECONSTRUCTION AND FILTERING OF OUTLET CHROMATOGRAMS

An important aspect in the analytical application of chromatography is the estimation of solute quantities in a mixture, i.e. in those cases in which the outlet chromatogram consists of the overalapping of two or more peaks. It is convenient to define the quantitative identification of single peaks from overlapped chromatograms as the *filtering problem of chromatographic data*. This problem can be tackled in the framework of the approach presented above in terms of Toth-isotherm-function approximation by means of an iterative optimization method.

Let us consider the case of a two-peak outlet chromatogram, figure 3 a). Initially we define a cut instant t_c, usually equal to the time instant at which the experimental chromatogram shows a local minimum. Iteratively, we reconstruct the integral quantities $q_1^{(+)}$, $q_1^{(-)}$, $q_2^{(+)}$, $q_2^{(-)}$ associated with the chromatograms $c_1(t, t_{r1})$, $c_2(t, t_{r2})$ of the single biomolecules (the subscript stands for the biomolecule) by taking into account the influence exerted on these quantities by the presence of the other biomolecule. It is worth noting that from the closure condition on Toth-isotherm-function approximation, for each function q it is necessary to determine by means of an optimization method only parameter m (exponent appearing in eq. (4)). Given all the q's, the fractions $\beta_1, \beta_2 = 1-\beta_1 = \beta$ of each biomolecule

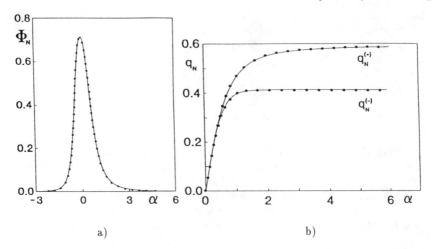

Figure 2: *Reconstruction of invariant chromatograms* Φ_N *and invariant integral functions* $q_N^{(+)}$, $q_N^{(-)}$ *by means of Toth-isotherm-functions. The points are the experimental results and the curves the Toth-isotherm-function reconstruction a)* Φ_N *of apoferritin on TSK 125; b)* $q_N^{(+)}$, $q_N^{(-)}$ *of apoferritin on TSK 125.*

can be estimated by minimizing the cost function $\Psi(\beta, \nu)$

$$
\begin{aligned}
\Psi(\beta, \nu) = \ & \nu \int_{t_{min}}^{t_{max}} [c_{exp} - (1 - \beta)c_1^{(n)}(t, t_{r1}) - \beta c_2^{(n)}(t, t_{r2})]^2 dt \\
& + (1 - \nu)[c_{exp}(t_{r1}) - (1 - \beta)c_1(t_{r1}, t_{r1}) - \beta c_2(t_{r1}, t_{r2})]^2 \\
& + (1 - \nu)[c_{exp}(t_{r2}) - (1 - \beta)c_1(t_{r2}, t_{r1}) - \beta c_2(t_{r2}, t_{r2})]^2
\end{aligned}
\tag{6}
$$

with respect to β. $c_i^{(n)}$ are the reconstructed chromatograms of each biomolecule evaluated from the Toth-isotherm-function approximation at the n-th iteration of the reconstruction process and c_{exp} is the experimental chromatogram. The parameter ν plays the role of a weight. We found that the optimal choice of ν is $\nu = 0$ (optimization on the peaks). Figure 3 b) shows the reconstruction of the chromatogram of IgG from impurities of larger molecular weight (therefore having a shorter retention time). The above approach can be extended to more than two peaks and represents a useful integration of standard software packages in the application of SEC columns for analytical purposes.

5 OPTIMIZATION AND CONTROL OF SEC PERFORMANCES

From the above discussion and from previous results[3-5] it follows that a complete characterization of solute dynamics in SEC columns can be achieved from the knowledge

- of the accessibility factor G associated with the hindered solute motion in the packing[5,7] and related to the behaviour of the retention time t_r;

- of the behaviour of $\sigma^2(t_r)$ as a function of the retention time[3,8-9] (or equivalently of the elution flow rate);

- of the parameter characterizing the Toth-isotherm-function approximation of the invariant Green function[4].

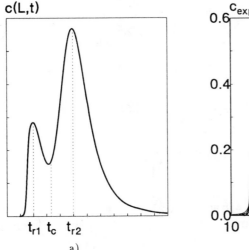

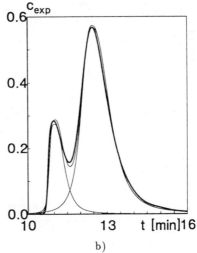

a)   b)

Figure 3: *a) Schematic representation of the filtering problem in the presence of two biomolecules; b) Reconstruction of the chromatogram of IgG for an elution flow rate of 1.0 ml/min on TSK 250. The heavy line is the experimental result.*

Given all these parameters, the purity and the optimal performances of SEC columns can be estimated.

The classical expression for the resolution factor in the presence of two solutes widely adopted in chromatographic literature is given by

$$R_s = \frac{t_{r2} - t_{r1}}{2(\sigma_1 + \sigma_2)} . \tag{7}$$

The validity of this expression increases the more the outlet concentration profile tends to assume a Gaussian shape. If the outlet profile exhibits a pronounced asymmetry, definition (7) proves to be inadequate to define column performances. For this reason it is convenient to define column performances starting from a more realistic definition of the degree of purity and of separation . Let us consider the test case of figure 3 a) and let us suppose that we want to separate two biomolecules in order to obtain two cuts: the first (up to t_c) enriched in the first biomolecule; the latter enriched in the second.

The purity ζ_1 of the first solute in the first cut, and conversely the purity of the second solute in the second cut ζ_2, can be defined as the solute fraction in the cut and are therefore given by

$$
\begin{aligned}
\zeta_1 &= \frac{Q_{c1}}{Q_{c1} + \chi[Q_{N-,2} - q_2^{(-)}(t_{r2} - t_c, t_{r2})]} \\
\zeta_2 &= \frac{Q_{c2}}{Q_{c2} + (1/\chi)[Q_{N+,1} - q_1^{(+)}(t_c - t_{r1}, t_{r1})]}
\end{aligned}
\tag{8}
$$

where $\chi = K_2/K_1$ is the ratio between the quantities of the second and first biomolecule and Q_{c1}, Q_{c2} are given by

$$
\begin{aligned}
Q_{c1} &= Q_{N-,1} + q_1^{(+)}(t_c - t_{r1}.t_{r1}) \\
Q_{c2} &= Q_{N+,2} + q_2^{(-)}(t_{r2} - t_c, t_{r2})
\end{aligned}
\tag{9}
$$

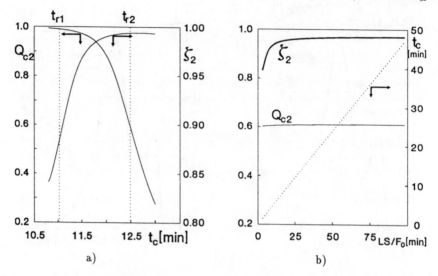

Figure 4: *a) Behaviour of the purity ζ_2 and of the efficiency Q_{c2} as a function of the cut time t_c of the case of figure 3 b); b) ζ_2 and Q_{c2} vs LS/F_o in the separation of IgG from impurities. The figure also shows the value of t_c chosen (dashed line).*

and represent the fraction of solute in the cut with respect to the overall solute quantity (they can be called for the sake of brevity *efficiency factors*).

Referring to the above quantities, it is easy to optimize the performances of SEC units. As an example, let us consider the case of figure 3 b), i.e. the problem of the purification of IgG. By applying the filtering technique discussed in the previous section it is possible to obtain from the overlapped output (see fig. 3b)) the invariant parameters of Φ_{Ni} ($i = 1, 2$) and the physical parameters (accessibility and variance) characterizing solute transport. From eqs. (7-8) it is possible to determine for every flow condition the optimal cut instant t_c depending on the prescribed values of the purity and of the efficiency factor. Figure 4 a) shows the behaviour of the purity and of the efficiency in the separation of IgG from impurities. Of course the choice of t_c is always a compromise between high purities and acceptable product recovery (as expressed by the efficiency Q_{c2}). Conversely, figure 4 b) shows the behaviour of ζ_2 and Q_{c2} as a function of the parameter LS/F_o (F_o is the eluent flow rate [ml/min] and $S = 0.44$ cm^2 the inner section area). In this figure the value of the cut time instant t_c is shown on the second y-axis. As can be seen from this figure the purity and the efficiency reach a saturation value for $LS/F_o \simeq 60$. This means that if $F_o = 1.0$ ml/min (which is the highest elution flow rate allowed for the specific columns considered throughout this article), $L \simeq 136$ cm, which corresponds to twice the length of the column. This situation can be achieved by a series-coupling of two SEC units. Of course a similar analysis can also be performed for other separation problems of industrial interest (e.g. the separation of IgG from bovine serum albumin (BSA) and impurities in the outstream of a bioreactor for monoclonal antibody production).

6 CONCLUDING REMARKS

We have proposed a practical and efficient method for treating analytical and optimization problems of SEC units. This method improves upon previous published approaches to the optimization problems of chromatographic units[1], since it is strongly related to a detailed (although simple) description of solute transport in SEC columns (invariant principle). The approach proposed is grounded on the formulation of the invariant principle (eqs. (1-2)). It should be observed that all the experimental results were obtained on preparative columns. Consequently, the next step towards the practical application of this approach is the validation of the proposed method in large-diameter industrial columns for protein separation and purification.

Acknowledgments

The authors thank Paola Verna for her help in performing experiments.

References

1. E. Suwondo, A.M. Wilhelm, L. Pibouleau and S. Domenech, European Symposium on Process Engineering-2 supplement to *Comp. Chem. Engng.*, 1992, **16**, 135.

2. W.W. Yau, J.J. Kirkland and D.D. Bly, 'Modern Size-Exclusion Chromatography', J. Wiley, New York, 1979.

3. A. Adrover, D. Barba, M. Giona and D. Spera, *Stochastic Analysis of Dispersion in Size-Exclusion Chromatographic Columns*, presented at the Third Symposium on the Characterization of porous Solids, Elsevier, Amsterdam, 1993 (in press).

4. M. Giona, A. Adrover, D. Barba and D. Spera, *Chem. Engng. Sci.*, 1994, **49**, 541.

5. A. Adrover and D. Barba, *Analysis of SEC Properties of Globular Biomolecules* presented at this Conference, 1994.

6. J. Toth, *Acta Chim. Acad. Sci. Hung.*, 1971, *69*, 311.

7. W.M. Deen, *AIChE J.*, 1987, **33**, 1409.

8. B. Anspach, H.U. Gierlich and K.K. Unger, *J. Chromat.*, 1988, **443**, 45.

9. K. Ostergren and C. Tragard, in 'Separations for Biotechnology 2', D.L. Pyle (Ed.), SCI Elsevier, London, 1990.

Rivanol–Silica: A Dye-ligand Solid Phase for Affinity Purifying Proteins

Miguel A. J. Godfrey, Piotr Kwasowski, Roland Clift,[a] and
Vincent Marks

SCHOOL OF BIOLOGICAL SCIENCES; [a]CENTRE FOR ENVIRONMENTAL
STRATEGY, UNIVERSITY OF SURREY, GUILDFORD, SURREY GU2 5XH, UK

1   INTRODUCTION

To date Rivanol $(C_{15}H_{15}N_3O.C_3H_6O_3$; 6,9-diamino-2-ethoxyacridine lactate, Merck 11,3668) has been exploited in protein fractionation as a semi-preparative precipitating agent, used in conjunction with others in order to increase the target protein content of a serum protein isolate[1]. In this study, Rivanol (Riv) has been immobilised on porous silica (PS), and evaluated as a potential dye-ligand solid phase for isolating proteins (particularly immunoglobulins) from serum. In doing so, we attempted to exploit the advantages offered by rigid silica-based matrices, and provide a tool for the rapid isolation of serum proteins applicable to process-scale applications.

2   EXPERIMENTAL

Preparation of Rivanol-Silica (Riv-PS) solid phases

The Preparation of Riv-PS. PS (Matrex™, particle size, 90 - 130 μm; pore size 1000 Å, Amicon Ltd UK) was diol-activated (PS-$(OH)_2$) and the degree of activation assessed[2]. 10 g of PS-$(OH)_2$ was then further aldehyde activated to PS-CHO and Riv immobilised by stabilised Schiff's base bonding[3]. Excess liquid was then aspirated and the Riv-PS solid phase washed with 3 x 100 ml of lactic acid (0.1 M), water, methanol and acetone successively. Finally, the Riv-PS solid phase was dried for 16 h at 80°C.

Determination of the dye content of Riv-PS. 50 mg of solid phase(s) were weighed into new 5 ml glass vessels and 2 ml of 2 M NaOH added. Dye standards were prepared by adding 2 ml of Riv (0 - 1 mg/ml) in 2 M NaOH to 50 mg samples of PS-CHO. Standard and sample solid phases were then incubated in a water bath at 90°C for 1 h and hydrolysates assessed for dye content by optical density at 450 nm (Riv $A1\%$ = 35).

Titration of Riv-PS solid Phases. Riv-PS samples (100 mg) were suspended in 100 ml of molar KCl and stirred with a magnetic stirrer at room temperature. To the slurry, 100 μl quantities of molar HCl (equivalent to 1 meq/g solid phase) were added at minute intervals, and the pH recorded after 30 s.

Protein binding of Riv-PS

The effect of pH on the serum protein binding of Riv-PS. 100 mg of Riv-PS were added to new 5 ml glass vessels and washed by roller-mixing and aspirating the washings with 5 ml of water, then twice with 5 ml of cit/phos pH buffer (0.1 M citric acid and 0.2 M Na$_2$HPO$_4$ mixed (v/v) to give pHs between 1.7 and 8.6). 5 ml of HSA or hIgG in cit/phos buffer solutions was then added to Riv-PS solid phases, giving pHs from 1.7 to 8.6 and serum concentrations of 40 and 17 mg/ml respectively. The solid phases were then washed (2 x 5 ml of cit/phos pH buffer) and adsorbed proteins eluted by roller-mixing with 5 ml of 2% NaCl (w/v) in 0.1 M lactic acid (eluting buffer). This batch purification process was repeated three times and the protein content of the eluates was determined by optical density (OD$_{280nm}$) or by Coomassie dye binding (CDB)[4].

The effect of ionic strength on protein binding of Riv-PS. Riv-PS columns were evaluated in order to determine the effect of ionic strength on the binding of serum proteins as above for pH, in solutions of Na$_2$HPO$_4$ (0 to 0.2 M).

The effect of contact time on Riv-PS protein binding. Riv-PS columns were evaluated in order to determine the effect of contact time on the binding of HSA as above for pH, in 0.1 M Na$_2$HPO$_4$ for between 0.25 and 60 min.

Determination of the protein capacity of Riv-PS by batch isothermal analysis. Riv-PS (25, 50, 100, 200, and 400 mg) was weighed into new screw-cap 5 ml glass vessels, in duplicate to form two batches. To one batch, 5 ml of HSA (2 mg/ml) was added, to the other 5 ml of hIgG (2.1 mg/ml), in 0.1 M Na$_2$HPO$_4$ and the vessels roller-mixed for 3 h at room temperature. The level of protein remaining in solution was assessed (OD$_{280nm}$ and CDB). The adsorption isotherms were plotted and the solid phase capacity (Q or q_{max}), for HSA and hIgG calculated. The K_d and q_m values for the interactions of HSA and hIgG on Riv-PS were determined from straight-line plots of C^*/q^* against C^*. From these plots, the intercept on the C^* axis was taken as being at $-K_d$, and the gradient of the line as $1/q_m$[5].

The effect of ionic strength on Riv-PS separation of serum proteins. This was carried out as above for the effect of pH evaluations, except that the pH solutions were substituted by Na$_2$HPO$_4$ (10 mM) pH 7.6, containing NaCl (0 to 512 mM).

Hydrophobic Interaction Chromatography (HIC) on Riv-PS, HIC[6] was attempted on Riv-PS as follows: a 1 ml Riv-PS column was conditioned by purging under gravity with 10 vols of 0.1 M Na$_2$HPO$_4$ (buffer B), followed by 10 vols of 0.1 M Na$_2$HPO$_4$ containing 1.5 M (NH$_4$)$_2$SO$_4$ (buffer A). 5 vols of (NH$_4$)$_2$SO$_4$ precipitated monoclonal antibody (MAb, 13 mg/ml in buffer A) were then added, and the column washed with 5 vols of buffer A. Adsorbed MAb was then eluted by the stepped gradient addition of 3 vols of buffers of B in A, increasing by 10% (v/v) increments to 100% B. Eluted fractions were checked for MAb content by OD$_{280nm}$, blanking against the appropriate eluting buffers.

Applications of Riv-PS

High performance liquid affinity chromatography (HPLAC) of proteins. Serum, protein, and MAb solutions, were injected into a Riv-PS HPLAC system and evaluated for solid phase binding and protein resolution under varying buffer conditions (pH and ionic strength). Protein separations were then performed in order to assess the effects of column length, solvent flow rate, and gradient development.

Purification of MAbs on Riv-PS. A 1 ml Riv-PS column was primed (under gravity flow) with 5 vols of eluting buffer followed by 5 vols of cit/phos pH 7.8 buffer. 5 vols of MAb (2 mg/ml) in cit/phos pH buffer (pH 7.8) was then applied and allowed to drain through under gravity. The column was then washed with 10 vols of cit/phos buffer and adsorbed proteins eluted with 5 vols of eluting buffer. This process was carried out a further 4 times, using cit/phos buffering solutions at pHs 6.7, 4.9, 3.3, and 1.8. Eluted fractions and 5 vols of the unprocessed MAb solution were then dialysed against 5 L of PBS (15 mM, pH 7.2) and assessed by SDS-PAGE (PhastSystem, Pharmacia Ltd.).

3   RESULTS AND DISCUSSION

The binding of Riv to PS was achieved by aldehyde activation of the PS matrix followed by Schiff's base formation and finally reduction of these bonds to a stable covalent state. This approach was chosen as the binding of ligands to matrices via amine or amide bonds has previously been reported by the authors as the most stable means of their immobilisation [3, 7]. The degree of dye substitution achieved when immobilising Riv on PS-CHO was at a maximum when the dye applied was at molar ratios of greater than 3 to 1 with respect to the initial surface diol density of the organosilane coated silica. Riv immobilisation was best achieved at pHs less than 5, where excess H^+ ions promote the formation of Schiff's bases. By increasing the Riv concentration it was possible to achieve a greater level of matrix Riv substitution resulting in higher levels of serum protein binding. However, it has previously been reported that dye-ligand solid phases with higher immobilised dye densities are sometimes less "efficient" at protein binding than those with a lower level[8].

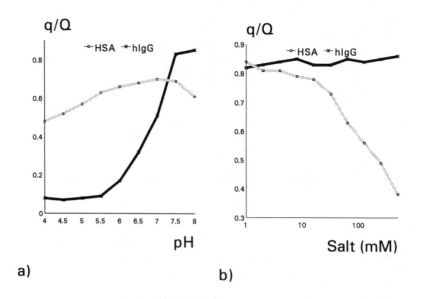

a)   b)

Figure 1.   The effect of pH (a) cit/phos and salt concentration (b) NaCl, pH 7 on the serum protein capacity of Riv-PS.

In addition to dye density, the binding of serum proteins to Riv-PS was also influenced by buffer pH and ionic strength, due to the ionic/hydrophobic nature of the Riv molecule. Riv-PS plasma protein binding was greatest in alkaline conditions, where HSA represents a polyvalent anion, and immobilised Riv represents a monovalent cation. Equilibrium studies on fixed-bed volumes of Riv-PS suggested its hIgG capacity was largely pH dependent, but remained relatively unaffected by variations in ionic strength. However, the HSA capacity of Riv-PS appeared to be dependent on both adsorption pH, and ionic strength (Fig. 1). Furthermore, HSA binding to Riv-PS was dependent on the time allowed for adsorption.

Increasing the ionic strength of buffers containing NaCl had the effect of lowering the protein binding of Riv-PS (Fig. 1b), though the opposite was found for buffers containing Na_2HPO_4 concentrations from 0 to 200 mM. This observation was possibly due to an increase in the adsorption solution pH at higher phosphate levels. Furthermore, high NaCl concentrations have a greater chaotropic effect than high concentrations of Na_2HPO_4, NaCl being much further up the Hofmeister series than Na_2HPO_4.

Titrational analysis of a Riv-PS solid phase indicated little variation in the net solid phase surface charge on addition of acid or alkali (Fig. 2). Protein separations on Riv-PS carried out in an HIC manner had only limited success: proteins emerging from Riv-PS columns as a single broad peak within 6 col vols. However, the ability to elute protein in a HIC manner would suggest a hydrophobic component to the adsorption mechanism. Hence, the binding of serum proteins to Riv-PS columns may be considered to occur as a result of the net surface charge of protein's at a given pH, which in turn is pI dependent.

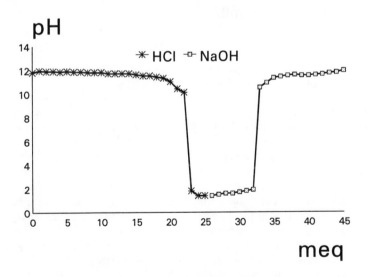

Figure 2. Titration of Riv-PS (1 g) in 1 M KCl.

The binding capacity of Riv-PS was lower for HSA than for hIgG when assessed by adsorption isothermal analysis (Fig. 3a). Batch equilibrium analysis of HSA and hIgG binding on Riv-PS gave capacities (Q or q_{max}) of 91 and 140 mg/g respectively, with dissociation constants (K_d) of 5.8 x 10^{-6} and 2.7 x 10^{-6} M respectively. When expressed in terms of an equilibrium isotherm, this further demonstrated the greater efficiency of HSA binding with respect to that of hIgG (Fig. 3b). The hIgG and HSA capacities of Riv-PS were differentially affected by variations in salt and pH. When at neutral pH, the affinity of Riv-PS binding to HSA was greater than that for hIgG; this difference was increased by more acidic conditions. This property may be exploited in the resolution of HSA and IgG from solutions containing both. However, due to the time dependence of Riv-PS binding, a stirred-tank mode of operation may prove the most appropriate for large sample volumes.

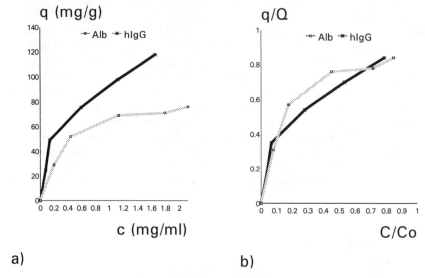

a)   b)

Figure 3.   Determination of the protein binding capacity of Riv-PS by a) batch isothermal analysis and b) Equilibrium isothermal analysis.

On injection of sheep serum into an HPLAC system containing a Riv-PS column (1 ml), most of the proteins present were adsorbed. Proteins eventually emerged as a single broad peak on applying a gradient of increasing NaCl. When the quantity of protein applied was below Q, the separation profile of breakthrough and elution peaks remained unchanged with respect to size/area ratio and distance, irrespective of the sample size, column, or rate of gradient development. On applying a gradient of eluting buffer, the single elution peak was further resolved into a number of peaks which appeared at the same points in the development of the salt/pH gradient, irrespective of system flow rate or the rate of gradient development. Attempts to further resolve the protein elution peaks by adjusting the HPLAC separation procedure (altering: pH, salt, amino acids, PEG, flow rate, sample loop size) proved unsuccessful.

The most appropriate operating conditions for the Riv-PS HPLAC system were to use a 0.1 M Na_2HPO_4 adsorption/washing buffer and to elute samples with a gradient of increasing salt and decreasing pH. In these conditions, proteins with a high pI did not bind to Riv-PS when injected into an HPLAC system. Human serum proteins emerged from the Riv-PS HPLAC system as a series of discrete peaks (Fig. 4a). When examined by IEF electrophoresis, consecutive elution peaks had a sequential increase in pI cut off at which point proteins were released or bound. Proteins with pI's above 7 made up the majority of the breakthrough. During the process of eluting serum proteins from Riv-PS by applying salt/lactic acid gradients, the protein fraction for those proteins with a lower pI was found to have increased with each peak (Fig. 4b).

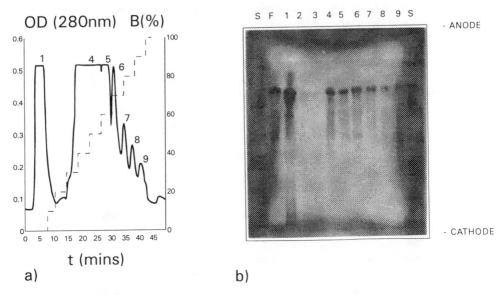

Figure 4. HPLAC of human serum on Riv-PS: buffer A was Na_2HPO_4 a (0.1 M), and B was NaCl (1 M), lactic acid (0.1 M); a) elution profile, and b) IEF electrophoretic analysis of peaks (S = Std.; F = Feed).

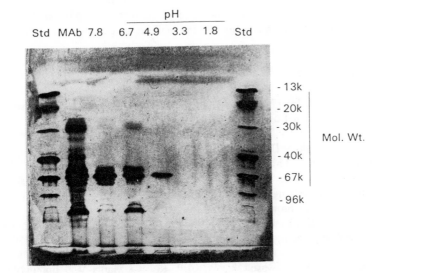

Figure 5. The effect of adsorption pH on the purity of MAbs eluted from a Riv-PS column.

On applying a MAb (mIgG, in 10% new born calf serum, 2 mg/ml) in cit/phos buffer, pH 4.9, to a Riv-PS column (1 ml), the eluted fraction appeared to consist exclusively of BSA on electrophoretic analysis. When the adsorption pH was increased to 6.7, IgG and HSA were detected on silver staining electrophoretic gels, but only HSA on Coomassie staining. This would indicate IgG to be present at a concentration of about 0.4 to 25 µg/ml (for 1 µl band vols). On increasing the adsorption pH still further to 7.8 (slightly above the mean PI of IgG), only the IgG heavy chain protein band was visible on silver staining; the mass of light chain present was too low to be visibly stained, suggesting an IgG presence of between 0.9-1 and 70 µg/ml (Fig. 5).

No antibody binding and reduced albumin binding ($< 0.3 \times Q$) was found on applying a murine MAb supernatant to a Riv-PS column at pH 4 under gravity flow. The use of Riv-PS to extract albumin from MAbs may also be expected to be applicable to all the IgG isotypes as none are precipitated by free Riv[9].

From our findings, it may be concluded that the binding of proteins to Riv-PS occurs largely as a result of electrostatic attraction. When in close proximity to the dye stationary phase, secondary adhesive dye-protein and lateral cohesive inter-protein hydrophobic interactions may further stabilise protein binding. On applying a gradient of increasing salt and decreasing pH to proteins adsorbed on Riv-PS, free Cl⁻ competes with proteins for available Riv binding sites and the force of electrostatic attraction shifts to repulsion as the pH falls to below the protein pI. The main protein binding force then becomes the relatively weak hydrophobic bonding. Consequently, proteins are eluted from a Riv-PS column in broad peaks about their pI.

In conclusion, porous microparticulate silica-immobilised Riv was found to be a general purpose protein binding adsorbent. By virtue of its inherent robustness, this solid phase shows great promise as a versatile HPLAC adsorbent. Its low cost, solvent resistance, mechanical stability, reusability, and its ability to be autoclaved are of particular value. Its greatest value may possibly be its selective use in the stirred-tank processing of biologicals for the semi-preparative clean up of isolated proteins. This method is of particular use in the process-scale production of proteins for therapeutic applications as neither the ligand nor the immobilisation chemistry are toxic.

REFERENCES

1   Steinbuch, M. In: J.M. Curling (Ed.),"Methods of plasma protein fractionation". Academic Press, Lon., 1980.
2   Walters, R.R. In: P.D.G. Dean, W.S. Johnson, and F.A. Middle (Eds), "Affinity Chromatography", IRL Press Limited, Oxford, 1985.
3   Godfrey M.A.J., PhD Thesis, University of Surrey (UK), 1993.
4   Scopes, R. In: C.R. Cantor (Ed),"Springer Advanced texts in chemistry, Protein Purification", Springer-Verlag, Berlin, 1982.
5   Chase, H.A. J Chromatogr, 1984, 297, 179-202.
6   Roe, S. In: E.L.V. Harris and S. Angal (Eds.) "Protein purification methods a practical approach". IRL Press at Oxford University Press, Oxon, UK, 175244, 1989.
7   Godfrey, M.A.J., Kwasowski, P., Clift, R., and Marks V. J Immunol Methods, 1993, 160, 97-105.
8   Clonis, Y.D. CRC Crit Rev Biotech, 1988, 7, 263-279.
9   Franek, F. Methods Enzymol, 1986, 121, 631-638.

Production and Purification of L-Asparaginase from *Vibrio (Wolinella) succinogenes*

R. J. Hinton, R. F. Sherwood, and J. R. Ramsay

DIVISION OF BIOTECHNOLOGY, PHLS, CENTRE FOR APPLIED
MICROBIOLOGY AND RESEARCH, PORTON DOWN, SALISBURY SP4 0JG, UK

1 ABSTRACT

L-Asparaginase with anti-lymphoma activity, from the anaerobe *Vibrio (Wolinella) succinogenes*, was purified to homogeneity by a simple two step process suitable for process-scale purification. An overall yield of 86% was achieved providing low-pyrogen cGMP material for clinical trials. The native and sub-unit molecular weight were determined at 140 kD and 35 kD respectively. The isoelectric point was 8.6. The pure enzyme was freeze-dried and stored at 4^OC with no loss in activity over a period of 24 months.

2 INTRODUCTION

L-Asparaginase (L-Asparaginase amidohydrolase EC 3.5.1.1) catalyses the reaction:

$$\text{L-Asparagine} + H_2O \text{ ---------> } \text{L-aspartate} + NH_3$$

L-Asparaginase in Guinea-pig serum has been shown to have anti-lymphoma activity.[1] Subsequently, the investigation of L-asparaginases from more readily available microbial sources, including *E. coli*, *Erwinia chrysanthemi*, *Serratia marcescens*, *Citrobacter* sp. and yeast, showed that the enzymes from *E. coli*(EC) and *Erwinia chrysanthemi*(ERW) had the necessary biochemical properties to be valuable in lymphoma chemotherapy.[2-4] Both the EC and ERW enzymes are today used in tandem, as they are antigenically distinct and do not show significant immunological cross-reactivity.[5]

The mechanism by which asparaginase works is simple; lymphoblastic leukaemic cells, which lack asparagine synthetase, require a supply in the serum of the essential amino-acid L-asparagine for their proliferation. Destruction of this supply with intravenous asparaginase is therefore cytotoxic for the cancer cells. This treatment in conjunction with other cytotoxic drugs, has resulted in overall remission rates of 93%.[6,7]

Of all the asparaginases available for chemotherapy, only two from Guinea pig serum and *Vibrio (Wolinella) succinogenes* have been shown to lack hepatotoxicity[8] which can result in liver dysfunction and pancreatitis. In addition both asparaginases lack significant glutaminase activity; the toxicity of asparaginases may be linked to their glutaminase activity.[9] This is probably due to the fact that (i) all animals require an exogenous supply of glutamine to sustain growth, (ii) glutamine is a constituent of many neurotransmitters and (iii) the cerebrospinal fluid contains a high concentration of glutamine.

Clearly therefore asparaginase from *Vibrio succinogenes* potentially has significant therapeutic advantages. It was the purpose of this study to develop a method to purify the enzyme to homogeneity from anaerobically grown cells of the organism. The method would have to be performed under cGMP conditions and suitable for process scale purification.

3 MATERIALS AND METHODS

3.1 Materials

Chromatography matrices, columns, gel electrophoresis materials and protein standards were obtained from Pharmacia Biotech Ltd., (St. Albans, UK). Chemicals and biochemicals were obtained from Sigma (Poole, UK), Boehringer Mannheim (Lewes, UK), or BDH (Lutterworth, UK). Protein diafiltration and concentration was performed using spiral-wound 10,000MW cartridges obtained from Amicon Ltd (Stonehouse, UK). Microbial growth media form Unipath Ltd.(Basingstoke, UK) and Aldrich (Poole, UK).

3.2 Methods

All extraction and purification procedures were performed at 4°C and used sterile depyrogenated distilled water.

3.2.1 Culture Conditions. Vibrio (Wolinella) succinogenes DSM 1740 was grown anaerobically at 37°C in a 25 L fermentation vessel containing (g/L) yeast extract, (4.0); fumaric acid, (13.9); formic acid [sodium salt], (6.8); ammonium chloride, (2.0). The pH was maintained at 7.0 with the automatic additions of 20% (w/v) H_3PO_4 and 3.2 M NaOH. The culture was maintained under anaerobic conditions with a nitrogen gas flow of 4.0 L/min, stirrer speed of 100 r.p.m and a back pressure of 3×10^4 Pa. The vessel was seeded with a 1 L inoculum and harvested between 24 - 30 hours with typical cell yields of approximately 1 g/L.

3.2.2 Enzyme Assay. L-asparaginase activity was determined using a method for the Technicon Autoanalyser.[10] incorporating the Berthelot colour reaction.[11] One unit of L-asparaginase was defined as one micromole of ammonia produced per minute at 37°C.

3.2.3 Protein Assay. Protein concentration was determined by the Folin method of Lowry,[12] with bovine serum albumin as the standard. Column eluates were monitored by the absorbance at 280 nm.

3.2.4 Enzyme Extraction. Frozen cell-paste (200 g) was resuspended in 50 mM succinic acid/NaOH pH 6.0 buffer (600 mL) and disrupted by liquid shear using a Lab 60 Manton Gaulin homogeniser (APV Baker Ltd, Derby, UK) at 550 Kg/cm²; breakage was confirmed by microscopy. Cell debris was removed by centrifugation for 30 min at 13000 x g. The cleared cell supernatant was sterile filtered using 0.22 μm low protein binding filters (Pall Process Filtration Ltd., Portsmouth, UK).

3.2.5 Enzyme Purification. Purification procedures were performed in a suite maintained to BS5295 Category 2 regulations. All columns were packed and sanitised with 500 mM NaOH and stored in 50 mM NaOH prior to use. Column dimensions are expressed as internal diameter x bed height.

Cation-exchange chromatography: A 11.3 x 24 cm column was packed with approximately 2.5 L CM-Sepharose and equilibrated with 250 mM succinic acid/NaOH + 1 M NaOH pH 6.0 and finally with 5 mM succinic acid/NaOH pH 6.0. The cleared cell extract was diluted with water to below a conductivity of 2 mS and loaded on to the column at a linear flow rate of 24 cm/h. Unbound material was washed from the column with equilibration buffer and enzyme eluted with a linear gradient of 0 - 300 mM NaCl in a total of 16 L equilibration buffer. Fraction sizes of 150 mL were collected.

Anion-exchange chromatography: A 5.0 x 10.2 cm column was packed with 200 mL Q-Sepharose FF and equilibrated with 1 M Tris-HCl pH 9.5 and finally with 50 mM Tris-HCl pH 9.5. The pooled fractions from CM-Sepharose were diafiltered against 50 mM Tris-HCl pH 9.5 and loaded on to the column at a linear flow rate of 15 cm/h. Non-binding material was washed from the column with equilibration buffer and the enzyme eluted with a linear gradient of 0 - 200 mM NaCl in a total of 2 L equilibration buffer. Fraction sizes of 25 mL were collected.

3.2.6 Freeze-drying. Pooled fractions from Q-Sepharose were diafiltered against 5 mM Tris-HCl pH 9.5 and concentrated to approximately 500 U/mL. Lactose was added to a final concentration of 1%(w/v). Freeze-drying was performed using glass vials of 37 mm height x 15 mm external diameter in an Edwards Minifast 3400 two-shelf freeze-drier (Edwards High Vacuum, Crawley, UK) with a 1 mL fill volume.

3.2.7 Estimation of Native Molecular Weight. The estimation of asparaginase native molecular weight was performed by gel-permeation chromatography using the Pharmacia FPLC system fitted with a HR 10/30 Superose-12 column. The column was equilibrated in 50 mM potassium phosphate buffer containing 100 mM NaCl pH 7.0 and loaded with 200 μL sample at a linear flow rate of 14 cm/h. The molecular mass of the homogeneous enzyme was estimated from the plot of M_r vs Ve/Vo of gel-filtration molecular mass standards.

3.2.8 Polyacrylamide Gel Electrophoresis. SDS-PAGE was performed at 15°C (60 Vh) using the Pharmacia Phast system, based on the method of Laemmli,[13] with samples solubilised under reducing conditions. Isoelectric focusing was performed at 15°C (500 Vh) with homogeneous gels containing carrier ampholytes for the generation of a 3.0 - 9.0 pH gradient.

Proteins were visualised by staining with Coomassie Brilliant Blue R-250 and scanned with a Chromoscan-3 optical densitometer (Joyce-Loebl, Gateshead, UK) for comparison with protein standards to estimate the apparent sub-unit M_r and pI respectively.

4 RESULTS & DISCUSSION

4.1 Enzyme Purification

4.1.1 CM-Sepharose. L-asparaginase eluted at approximately 160 mM NaCl, elution volumes between 7.8 - 8.8 L were pooled (**Figure 1**) and diafiltered against 50 mM Tris-HCl pH 9.5.

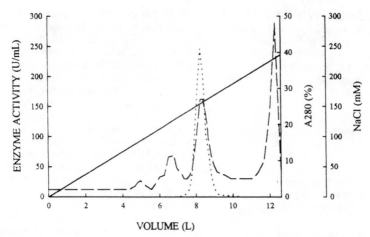

Figure 1 *Cation-exchange of L-asparaginase on CM-Sepharose.*
(_ _ _ _ Protein, ... Enzyme)

4.1.2 Q-Sepharose. L-asparaginase eluted from the anion-exchanger at approximately 40 mM NaCl. Elution volumes between 320 - 700 mL were pooled **(Figure 2).**

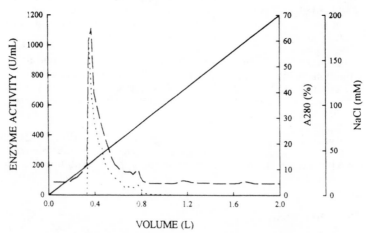

Figure 2 *Anion-exchange of L-asparaginase on Q-Sepharose.*
(_ _ _ _ Protein, ... Enzyme)

The pooled enzyme was homogeneous as indicated by a single band Coomassie blue stained on a SDS-PAGE gel. **Table 1** summarises the results.

Table 1 *Summary of purification results*

Sample	Total enzyme (units)	Total protein (mg)	Specific activity (U/mg)	Purification fold	Recovery (%)
Cell extract	184,000	12,962	14	1	100
CM-Sepharose	167,000	432	387	28	91
Q-Sepharose	160,000	247	647	47	87
Concentration	158,000	244	647	47	86

4.2 L-Asparaginase Properties

4.2.1 Native Molecular Weight. The native molecular weight of L-asparaginase was determined by gel-filtration as 140 kD **(Figure 3)**.

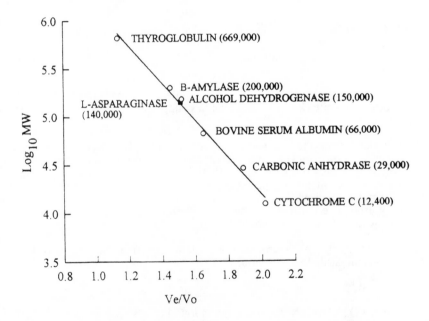

Figure 3 *Determination of L-asparaginase native molecular weight*

4.2.2 Sub-unit Molecular Weight. The sub-unit molecular weight was determined by SDS-PAGE as 35 kD showing that the native enzyme exists as a tetramer **(Figure 4)**.

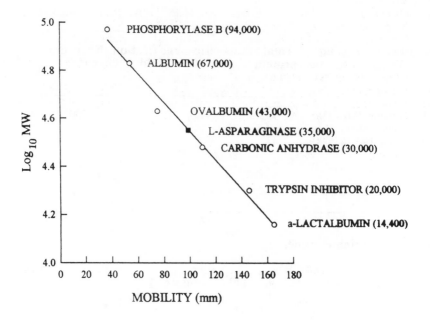

Figure 4 *Determination of L-asparaginase sub-unit molecular weight*

4.2.3 Isoelectric Point Estimation. The isoelectric point of L-asparaginase was estimated as 7.6.

4.2.4 Freeze-drying. No loss in activity was detected after lyophilisation. The material formed a white cake with no signs of collapse and excellent reconstitution characteristics.

The cell yields from the anaerobic fermentation of *Vibrio succinogenes* were very low, however sufficient enzyme was obtained for preliminary clinical trials to commence. It was important therefore that the purification protocol developed gave high recoveries of homogeneous enzyme. This was achieved, including final product formulation providing stable, sterile low-pyrogen (LAL < 1.1 EU/mg protein) material. The method suitable for scale-up should also provide a regime for the purification of the recombinant enzyme.

Acknowledgements- We thank Gerald Adams for freeze-drying and Ian McEntee for small-scale fermentation.

References

1. Broome, J.D., *Trans. N. Y. Acad. Sci.*, 1968, **118**, 99.
2. Howard, J. B. & Carpenter, F. H., *J. Biol. Chem.*, 1972, **247**, 1020.
3. Ho, P.P., Milikin, E.B., Bobbitt, J.L.,Grinnan, E.L., Burck, P.J., Boeck, L.D. & Squires, R.W., *J. Biol. Chem.*, 1970, **245**, 3708.
4. Wade, H.E., Elsworth, R., Herbert, D., Keppe, J. & Sargeant, K., *Lancet*, 1968, **2**, 776.
5. Cammack, K.A., Marlborough, D.I. & Miller, D.S., *Biochem. J.*, 1972, **126**,

361.
6. Ortega, J.A., Nesbit, M.E., Donaldson, M.H., Hittle, R.E., Weiner, J., Karon. M. & Hammond, D., *Cancer Res.*, 1977, **37**, 535.
7. Caprizzi, R.L., Griffin, F., Cheng, Y.C., Bailey, K. & Rudnick, S.A., *Proc. Am. Assoc. Cancer Res.*, 1980, **21**, 148.
8. Distasio, J.A., Niederman, R.A., Kafkewitz, D. & Goodman, D., *J. Biol. Chem.*, 1976, **251**, 6929.
9. Durden, D.L., Salazar, A.M. & Distasio, J.A., *Cancer Res.*, 1983, **43**, 1602.
10. Wade, H.E. & Phillips, B.W., *Anal. Biochem.*, 1971, **44**, 189.
11. Gordon, S.A., Fleck, A., & Bell, J., *Annals Clin. Biochem.*, 1978, **15**, 270.
12. Lowry, O.H., Rosebrough, N.J., Farr, A.L. & Randall, R.J., *J. Biol. Chem.*, 1951, **193**, 265.
13. Laemmli, U.K., *Nature*, 1970, **227**, 680.

The Recovery and Dewatering of Microbial Suspensions Using Hydrocyclones

S. T. L. Harrison, G. M. Davies, N. J. Scholtz, and J. J. Cilliers

DEPARTMENT OF CHEMICAL ENGINEERING, UNIVERSITY OF CAPE TOWN, PRIVATE BAG, RONDEBOSCH 7700, SOUTH AFRICA

ABSTRACT

Small diameter hydrocyclones have had increasing use in performing difficult separations between phases, due to the large centrifugal forces generated in them. The potential use of hydrocyclones in the dewatering of microbial suspensions is attractive as they are continuous, high capacity devices requiring low maintenance while having the additional benefit that they can be readily sterilised.

Results are reported on the de-watering of Bakers' yeast in a 10 mm diameter hydrocyclone to quantify the separation process and to determine the effect of the feed concentration and the cyclone outlet dimensions.

Relationships for predicting cyclone performance are presented. These relationships, in turn, allow the design of dewatering circuits yielding acceptable recovery levels and concentrating effects at specified feed concentrations by using multiple treatment stages and recycles. A circuit is presented to upgrade a yeast suspension from 50 g/l to 81.3 g/l at 80% recovery. Higher concentrations (up to 150 g/l) can be achieved with a concomitant loss in recovery.

INTRODUCTION

The separation of microbial cells from the culture medium is required in most microbial processes, regardless of whether the desired product is the cell itself, an intracellular or an extracellular compound. The challenge of these separations is the small particle size (typically 1 to 10 μm in diameter), the low density difference with respect to the suspending medium, the heat labile nature of many products and the non-Newtonian, concentration-dependent rheology of concentrated cell suspensions.

Most commonly used separation processes include filtration or high speed centrifugation. Centrifuges and hydrocyclones function on the same principle of a magnified centrifugal field to effect a separation between solids and liquids, solids of different sizes or different liquids. Traditionally hydrocyclones function at lower centrifugal fields than centrifuges. However, by decreasing the diameter of the hydrocyclone, the centrifugal fields produced can be increased. For example, fields of up to 50 000g can be produced in a 10 mm diameter hydrocyclone at a capacity of approximately 0.15 m³/hr [1]. In this study, the use of small diameter hydrocyclones is investigated owing to their potential as continuous, high capacity devices that require low maintenance and are readily sterilised.

The design of the hydrocyclone is that of an upper cylindrical section with a tangential inlet, and a lower conical section with an opening at the base, called the spigot. A tube, called the vortex finder, extends into the hydrocyclone from the closed top of the cylindrical section to below the inlet opening. Figure 1 shows the essential features and flow pattern within a hydrocyclone.

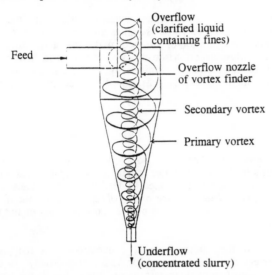

Figure 1 *Diagram of a hydrocyclone*

The feed slurry enters the hydrocyclone tangentially, resulting in a vortex flow moving down the cylindrical section. Due to the reduction in volume in the conical section, the flow is constrained and a secondary vortex of smaller diameter, moving upward, is formed. Solids are, due to the large centrifugal forces, forced into the outer vortex and exit the cyclone at the spigot. This product is called the underflow. The inner vortex, consisting primarily of the fluid, exits the cyclone through the vortex finder, and is referred to as the overflow. While the diameter of the cylindrical section of the cyclone is the major variable determining the size of particle that can be separated, the diameters of the vortex finder and spigot can be varied to alter the separation. Hydrocyclone diameters range from 10 mm to 2.5 m, with separating sizes of 1.5 μm to 300 μm [2].

EXPERIMENTAL METHODS

Microbial Suspension

In this study, *Saccharomyces cerevisiae* (Bakers' yeast) was used as a model suspension. Suspensions were obtained from Anchor Yeast (Epping, Cape Town) as yeast cream. The cells showed a narrow size distribution. The dominant diameter was measured as 5.5 μm by laser light scattering and 4.5 μm from the settling velocity. Reported densities [3] vary from 1072.5 to 1095.2 kg/m³. The isoelectric point for most microbial cells occurs at a pH of 2 [4].

The Hydrocyclone

A 10 mm Mozley hydrocyclone connected to a 1.1 kW Monopump was used. The hydrocyclone was fitted with a spigot of diameter 1.0 mm or 1.5 mm and a vortex finder of diameter 2.0 mm, as specified. The suspension was maintained at a constant temperature of 21°C in a constant temperature waterbath.

Analytical Methods

Timed samples of 30 seconds were taken simultaneously from the over- and underflow streams of the hydrocyclone. These samples and a sample of the feed were weighed, dried and the yeast concentration of the feed, underflow and overflow streams determined gravimetrically. Suspension viscosity was determined using a U-tube capillary viscometer. Cell breakage was determined by the release of soluble protein (measured by the Lowry method) as well as by direct counting using a light microscope.

RESULTS

The separation in a hydrocyclone between the solids and liquids is not complete. A fraction of the solids reports to the vortex finder rather than the spigot and a fraction of the water in the feed reports to the concentrated underflow stream. Thus quantifying the performance requires two values, the *recovery* of solids from the feed to the underflow and the *concentration ratio*, the ratio of the concentrations of the underflow to the feed.

Figures 2 and 3 show the effect of yeast concentration in the feed to the hydrocyclone on the recovery and the concentration ratio, respectively. In each case, both the 1.0 mm and 1.5 mm spigot diameters were tested. For these tests, a feed pressure of 700 kPa was used. It can be noted that as the feed concentration is increased, both the recovery of solids to the underflow and the concentration ratios decreased. The effect of feed concentration was reduced significantly beyond feed concentrations of approximately 40 g/l.

The effect of the spigot diameter is of great significance. From Figures 2 and 3 it can be seen that as the spigot diameter was increased, the recovery of yeast to the underflow increased, while the concentration ratio decreased. This was as expected, a larger outlet opening allowing a larger fraction of both yeast and water to exit through the spigot. On average, the hydrocyclone fitted with the 1 mm spigot had a capacity of 0.122 m³/hr, while using the 1.5 mm spigot increased the capacity to 0.162 m³/hr, at 700 kPa. The volumetric flow split is given by the product of the recovery and concentration ratio.

It is of interest to note that no cell breakage was observed in the samples, even after extended recycling. This is in agreement with the findings of Rickwood et al. [5].

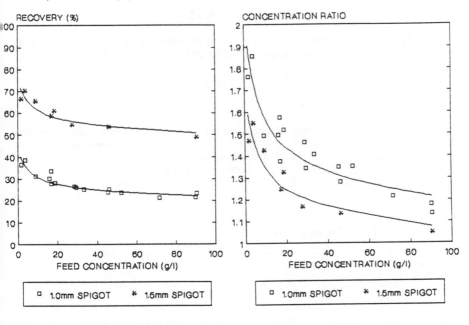

Figure 2   *The effect of feed concentration on the recovery obtained (700 kPa, 21°C)*

Figure 3   *The effect of feed concentration on the concentration ratio (700 kPa, 21°C)*

DISCUSSION OF RESULTS

The results indicate that at a low feed concentration (say 5 g/l), the dewatering of the hydrocyclone is quite effective. A concentration ratio of 1.67 can be achieved in a single pass when using a 1.0 mm spigot, yielding a product of 8.4 g/l. Unfortunately, this is achieved at the expense of a relatively low recovery, 33.6%. Increasing the spigot diameter to 1.5 mm increases the recovery to 64.7%, but decreases the concentration ratio to 1.42, yielding a product of 6.1 g/l. At a higher feed concentration of 50 g/l, using a 1mm spigot, a recovery of 24.1% and a concentration ratio of 1.30 can be achieved, yielding a product of 64.8 g/l, while when using a 1.5 mm spigot, a recovery of 53.3% and a concentration ratio of 1.14 (57.0 g/l) is possible.

While these figures may appear too low to allow commercial exploitation of hydrocyclones for this duty, it must be borne in mind that this performance is for a single pass and that, due to their relatively low operating costs, multi-stage hydrocyclone circuits can be used to improve the performance. An example of such a circuit will be given in the following section.

MULTI-STAGE HYDROCYCLONE SEPARATION

In order to design a suitable multi-stage circuit for dewatering of yeast, the results shown in Figures 2 and 3 were mathematically fitted to functions of the form

$$R = A.x^b \dots\dots\dots\dots\dots\dots\dots\dots\dots 1$$

and
$$C = D.x^f \dots\dots\dots\dots\dots\dots\dots\dots\dots 2$$

where R and C are the recovery and concentration ratio, respectively, and x the feed concentration in g/l. This form of equation is often used to describe hydrocyclone performance empirically [6]. Table 1 summarises the values of the coefficients in the equations for the tests using the 1.0 mm diameter and the 1.5 mm diameter spigots. The goodness of fit of these equations is illustrated by the solid lines in Figures 2 and 3.

Table 1 *Coefficients of equations 1 and 2*

Spigot diameter	Recovery		Concentrating ratio	
	A	b	D	f
1.0 mm	42.28	-0.1436	1.989	-0.1096
1.5 mm	74.12	-0.0842	1.652	-0.0950

Using these equations, it is possible to simulate the performance of circuits consisting of various combinations of hydrocyclones and to optimise the recovery and the concentration ratio.

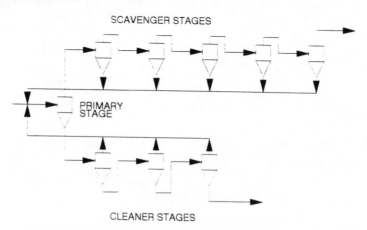

Figure 4 *Example of a hydrocyclone circuit design containing 5 scavenger stages and 3 cleaner stages.*

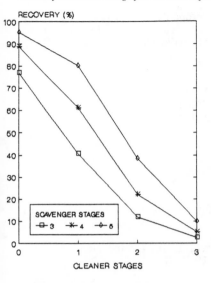

Figure 5 *The effect of the circuit design on recovery obtained*

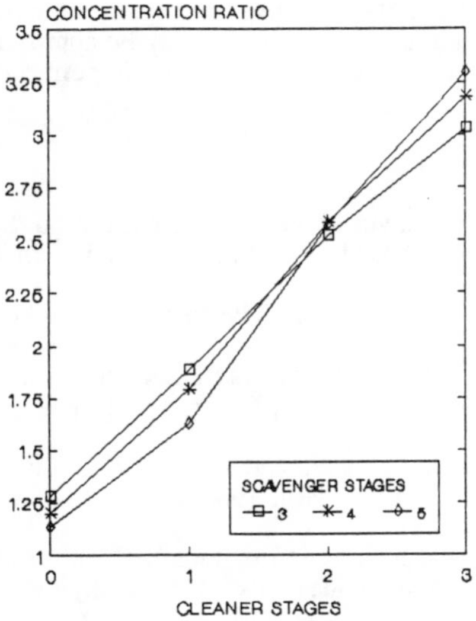

Figure 6 *The effect of the circuit design on concentration ratio*

An example of such a circuit is given in Figure 4. The circuit consists of a primary stage, with five scavenger stages. The underflow streams from all scavenger stages are recycled to the primary hydrocyclone. The role of the scavenger stages is to increase the overall recovery, and hence the larger spigot diameter would be used. The underflow from the primary hydrocyclone can be treated in a similar fashion by cleaner stages to increase the concentration ratio; hence the smaller spigot diameter is used. In this case the underflow is re-treated and the overflow recycled.

Results obtained by varying the number and combination of scavenger and cleaner hydrocyclones in the circuit are shown in Figures 5 and 6. By increasing the number of cleaner stages, an increasing concentration effect was found with a concomitant decrease in the recovery. Increasing the number of scavengers enhanced recovery. The effect of number of scavenger stages on the concentration ratio was dependent on the number of cleaner stages in the design circuit as shown in Figure 6.

Using the circuit shown in Figure 4 with one cleaner stage and treating a feed concentration of 50 g/l, a concentration ratio of 1.63 (82 g/l) and a recovery of 80.2% could be achieved. Should a higher recovery be required, the cleaner stage could be removed to yield a recovery of 95.2% and a concentration ratio of 1.13, yielding a product of 57 g/l.

CONCLUSIONS

The feasibility of attaining separation of microbial cells from the culture suspension has been illustrated without cell disruption occurring. Both the recovery of micro-organisms and the concentrating effect attained are a function of the feed concentration and the cyclone geometry. In a single stage, simultaneous achievement of acceptable recovery levels and concentration ratios can only be achieved when treating low feed concentrations. However, by the design of an appropriate hydrocyclone circuit, the required operating range may be approached for more concentrated feed slurries. In the circuits proposed, a concentrating effect of 1.63 was obtained at a recovery of 80%.

REFERENCES

1. Trawinski H., In "Solid-Liquid Separation Equipment Scale-up", Purchas D. (ed.), Uplands Press, Croydon, Ch 7, pp241-287, 1977.

2. Svarosky L., "Hydrocyclones", Technomic, 1984.

3. Reuss M., Josic D., Popovic M. and Bronn W.K., European J. Appl. Microbiol. Biotechnol., 1979, 8, 167-175.

4. Bailey J.E. and Ollis D.F., "Biochemical Engineering Fundamentals", 2nd ed, McGraw-Hill, 1986.

5. Rickwood D., Onions J., Bendixen B. and Smyth I., In "Hydrocyclones: Analysis and Applications", L Svarosky (ed.), pp109-119, 1992.

6. Plitt L.R., CIM Bulletin, 1976, December, 114-122.

Factors Affecting the Selectivity of Liquid–Liquid Ion Exchange of Anions

Nazar Ibrahim, Angela McGillivary, Bryan Reuben, and Lawrence Dunne

SCHOOL OF APPLIED SCIENCE, UNIVERSITY OF SOUTH BANK, LONDON
SE1 0AA, UK

Aqueous iodide, bromide, nitrate and penicillin V ions have been extracted into organic solvents by reactive ion exchange with tetraalkylammonium salts. Dichloromethane is better than toluene is better than octanol. Thermodynamic equilibrium is reached readily. Simultaneous extraction of two ions enables extraction selectivities and free energies of ion exchange to be estimated. For the inorganic ions, these were consistent with the simple theoretical model of Abraham et al[8] but penicillin V showed structure-specific effects. Extraction levels are not dependent on the lipophilicity of the extractant, suggesting that the ions in the organic phase are dissociated.

1. INTRODUCTION

Solvent extraction by ion pair extraction or liquid-liquid ion exchange is a well-established technique for separation of metals. Metals are extracted as cations with cationic reagents such as 2-ethylhexylphosphoric acid, and as complex chloride anions with anionic or neutral reagents such as quaternary ammonium salts. High selectivities can be achieved with chelating extractants.

Most anions of commercial interest (e.g. β-lactam antibiotics) are less easily extracted. High selectivities are difficult to achieve and the factors affecting them are not fully understood. There has nonetheless been recent work involving extraction of β-lactams[1], small peptides[2] and carboxylic acids[3]. The aim of this project is to investigate factors affecting selectivity in extraction of simple anions and more complex β-lactams.

2. THEORY

Consider an aqueous solution of V_{aq} dm^3 of a mixture of N singly charged anions, A_1^-, A_2^- ...A_n^-...A_N^- present initially at concentrations of a_1, a_2...a_n...a_N mol dm^{-3}. Suppose this solution is brought into equilibrium with V_{org} dm^3 of an immiscible organic solvent containing a_o mol dm^{-3} of an anionic extractant such as a quaternary ammonium salt $Q^+A_o^-$. This may be present as an ion pair or ionised. If the quaternary ion is highly lipophilic, it will not enter the aqueous phase in appreciable quantities and a series of N phase transfer reactions will take place of the type

$$Q^+A_o^- \text{ (org)} + A_n^-\text{(aq)} \rightarrow Q^+A_n^- \text{ (org)} + A_o^-\text{(aq)} \qquad (1)$$

Each is accompanied by a free energy change ΔG°_n kJ mol^{-1}. So long as the quaternary compound is dissociated and the extraction is purely an ion exchange, this is the difference between the free energies of phase transfer of the ions A_n^- and A_o^-.

Univalent ions were used throughout the work and solutions were dilute. Under these circumstances it was thought that equation (1) represented the stoichiometry, that clusters would not form and that activities could be equated to concentrations. The mass action equation for each equilibrium then is

$$K_n = \frac{[QA_n]_{org}[A_o]_{aq}}{[QA_o]_{org}[A_n]_{aq}} \tag{2}$$

If x_1, x_2...x_n...x_N are the fractions of QA_o that go to QA_1, QA_2...QA_n...QA_N then the equilibrium concentration of the typical ion A_n in the organic layer will be $a_o x_n$ mol dm^{-3} and in the aqueous layer $a_n - a_o x_n V_{org}/V_{aq}$ mol dm^{-3}. The concentrations of A_o however are augmented by contributions from all the other ions being transferred. Hence equation (2) becomes N equations of the form

$$K_n = \frac{a_o x_n \dfrac{V_{org}}{V_{aq}} \displaystyle\sum_{n=1}^{n=N} x_n}{\left(1 - \displaystyle\sum_{n=1}^{n=N} x_n\right)\left(a_n - a_o x_n \dfrac{V_{org}}{V_{aq}}\right)} \tag{3}$$

Values of K for each ion exchange may be calculated from the standard relationship $\Delta G^\circ_n = -RT \ln K$. The composition of both phases for any initial mixture of ions may be derived by solution of the simultaneous equations. Rather than Lagrangian multipliers, we have used the Levenberg-Marquardt algorithm incorporated in the PC program Mathcad 4.0[4].

For any two ions A_1 and A_2 initially in the aqueous layer, the difference in free energy of phase transfer between them is given by the ratio of the equilibrium constants, that is

$$\Delta G^\circ_{A_1 - A_2} = -RT \ln \frac{K_1}{K_2} = \frac{x_1\left(a_2 - a_0 x_2 \dfrac{Vorg}{Vaq}\right)}{x_2\left(a_1 - a_0 x_1 \dfrac{Vorg}{Vaq}\right)} \tag{4}$$

3. EXPERIMENTAL

The majority of solvent extraction experiments were carried out at room temperature by shaking together the aqueous solution to be extracted with an equal volume ($V_{aq} = V_{org}$) of a solution of the extractant in an organic solvent. The resulting mixtures were centrifuged and the aqueous layers separated and analysed. Chemicals were used as supplied by manufacturers except for tetraoctylammonium chloride (TOACl), which was prepared by shaking a standard solution of tetraoctylammonium bromide (TOABr) in an organic solvent with five successive samples of 2M potassium chloride. Aqueous layers were analysed on a Dionex 4000i ion chromatograph fitted with an Omni Pac PAX-100 column.

4. RESULTS AND DISCUSSION

4.1 Extraction of a single ion

Figure 1 shows the extraction of iodide ion from 10 mM potassium iodide with different concentrations of TOABr in toluene. The difference in free energy of phase transfer between iodide and bromide ions[5] is about 20 kJ mol^{-1} ($\Delta G°_{I\leftrightarrow Br}$ = -20 kJ mol^{-1}). As the theoretical lines indicate, the data are consistent with any value above about 15 kJ mol^{-1}.

Figure 2 shows the extraction of 5 mM bromide ion with TOACl in toluene, octanol and dichloromethane. While the points are scattered, it is evident that dichloromethane is more effective than toluene, which in turn is more effective than octanol. **Figure 2A** shows the extraction as measured by aqueous bromide concentrations and **Figure 2B** as measured by aqueous chloride concentrations. There is a slight difference between these curves because of experimental error. Least squares fitting of the joint curves gives $\Delta G°_{Br\leftrightarrow Cl}$ = -15, -8 and -6 kJ mol^{-1} for dichloromethane, toluene and octanol respectively, and the theoretical lines for these free energies are shown.

4.2 Extraction of two ions

Figure 3 shows the extraction of two ions of widely different lipophilicities, notably nitrate and iodide, with TOABr in toluene. Bromide is close to nitrate in lipophilicity, so that a clean separation of iodide might be predicted. The theoretical curves are plotted on the basis of $\Delta G°_{I\leftrightarrow Br}$ = -20 kJ mol^{-1} and $\Delta G°_{NO3\leftrightarrow Br}$ = -1.5 kJ mol^{-1}. Virtually all the iodide is extracted before any of the less lipophilic nitrate ions are drawn into the organic phase.

Figure 4 shows the reverse situation where the similar ions, bromide and nitrate, are extracted with TOACl in toluene. The extraction is much less clean. Under conditions where about 65% of nitrate has been extracted, about 35% of bromide has accompanied it. This level of separation is consistent with $\Delta G°_{NO3\leftrightarrow Br}$ = 3.8 kJ mol^{-1}.

The above experiments can be used to give free energies of ion exchange but the values are subject to large errors. For example, the difference between the extraction curves for free energy changes of 15 and 20 kJ mol^{-1} as shown in **Figure 1** are small compared with experimental errors. In **Figure 4**, "best fit" theoretical lines are shown but the experimental points still lie perceptibly below them. The two values for $\Delta G°_{NO3\leftrightarrow Br}$ are 2.3 kJ mol^{-1} apart. In general, internal checks indicated that mass balances and absolute concentrations better than 10% were difficult to achieve especially with commercial extractants such as Aliquat 336. Relative concentrations were ±2%.

We concluded that the systems that we were examining were coming very close to thermodynamic equilibrium and that approximate values of free energies of phase transfer could be obtained. The most reproducible and consistent values of free energies, however, were obtained by simultaneous extraction of the two relevant ions with a much more hydrophilic exchange ion. Thus we think the most reliable value above is the value of $\Delta G°_{NO3\leftrightarrow Br}$ = 3.8 kJ mol^{-1} obtained by simultaneous extraction of NO$_3^-$ and Br$^-$ with TOACl and the work below was based on this conclusion.

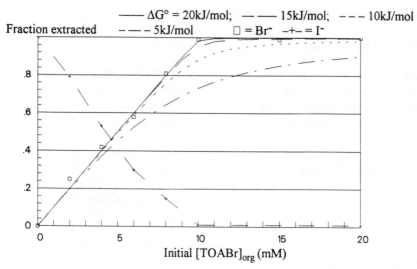

Figure 1 *Extraction of iodide with TOABr in toluene. Theoretical extraction curves for various free energies of phase transfer.*

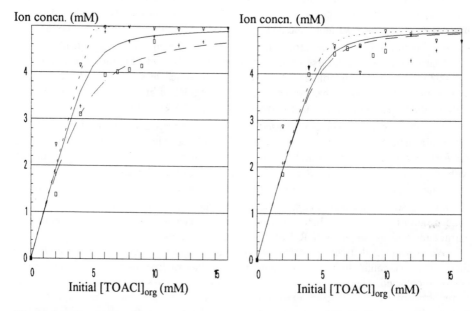

Figure 2A *Extraction of 5 mM bromide with TOACl in toluene (based on bromide concentrations).*

Figure 2B *Extraction of 5 mM bromide with TOACl in toluene (based on chloride concentrations).*

—+— *Toluene; --□-- Octanol; ···∇···Dichloromethane*

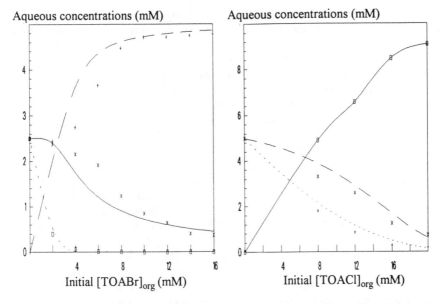

Figure 3 *Extraction of iodide and nitrate with TOABr in toluene.*
☐ = Br⁻; X = I⁻; + = NO₃⁻

Figure 4 *Extraction of bromide and nitrate with TOACl in toluene.*
☐ = Cl⁻; X = Br⁻; + = NO₃⁻

4.3 Effect of extractant structure

The selectivities of various quaternaries and solvents were measured by extraction of nitrate/bromide and nitrate/penicillin V mixtures with a series of tetraalkylammonium chlorides and trihexylamine hydrochloride. The molarities of the components of the feed mixture and the extractant solution were kept equal. We defined selectivity, S, for example, as $[NO_3]_{org}[Br]_{aq}/[NO_3]_{aq}[Br]_{org}$ where nitrate and bromide were being simultaneously extracted. Variation of nitrate/bromide selectivity with extracting ion lipophilicity is shown in **Table 1**. Free energies of ion exchange were obtained from equation (4).

There is virtually no difference in selectivity between tetraalkylammonium chlorides of different lipophilicities. For this series at least, the nature of the extracting ion is

Table 1: *Nitrate/bromide selectivities for various solvents and quaternary chlorides.*

Quaternary	Dichloro-methane		Octan-1-ol		Decan-1-ol†		Toluene*		Tributyl phosphate	
	Sel.	-ΔG°	Sel.	-ΔG°	Sel.	-ΔG°	Sel.	-ΔG°	Sel.	-ΔG°
tetrapentyl	4.50	3730								
tetrahexyl	3.68	3221	1.33	710	1.26	570	5.92	4410	6.09	4480
tetraheptyl	3.79	3301	1.27	590	1.36	760	5.58	4260	5.88	4390
tetraoctyl	3.35	3000	1.25	550	1.18	410	5.86	4380	5.21	4090
average	**3.82**	**3310**	**1.28**	**620**	**1.27**	**580**	**5.79**	**4350**	**5.73**	**4320**
trihexylamine	2.55	2320								

* Average of two runs. † Contained 10% toluene to aid solubility.

unimportant. One implication is that the tetraalkylammonium salts, whether extractants or extracted, are probably dissociated in the organic phase. This has been discussed for a range of tetraalkylammonium salts by Taft *et al.*[6] and for tetrabutylammonium thiocyanate by Gans *et al.*[7] The tetraalkylammonium ions are very large so that, even if ions pairs are formed, they might be expected to be only loosely bonded.

Whereas dichloromethane was the most powerful extracting solvent, both tributyl phosphate and toluene have higher and similar selectivities. Indeed the performance of toluene in this work was only slightly inferior to that of dichloromethane, which is not entirely predictable for a solvent with no dipole moment or hydrogen bonding capability. Octanol and decanol both have low selectivities. Hydrogen bonding is a dominant factor favouring solvent extraction yet apparently it is not effective here. Furthermore, the nitrate ion would be expected to hydrogen bond more strongly than the bromide ion, hence selectivity should be higher into a hydrogen bonding solvent. It seemed possible that the low selectivity of octanol and decanol was due to the high solubility of water in them (respectively 20.7 and 26.7 mole % at 298K). The theoretical case of co-extraction of water was considered by Abraham *et al* (see below).

Since the motivation behind this project was the extraction of β-lactams, it was desirable to see if such complex ions would interact differently with solvents and extractants. **Table 2** is the equivalent of **Table 1** for penicillin V/nitrate mixtures.

Table 2 differs from **Table 1** in that octanol has moved from least to most selective solvent, while decanol has improved slightly. Toluene and tributyl phosphate have moved from best to worst. The quaternary ion lipophilicity still seems to have little effect; the scatter in the dichloromethane and toluene points is probably not significant.

Abraham and Liszi[8] have described two methods for prediction of partition coefficients of ions. One assumes solvation by the dry organic solvent (unhydrated) and the other assumes a wet solvent in which the first ion solvation layer, even in the organic phase, is water (hydrated). **Table 3** compares the free energies of ion exchange in this work compared with the values we have calculated from their theories. Excellent agreement is found in cases 1,2,5 and 8. Agreement is worse in cases 3,4 and 6 but even these are reasonable in the sense that they represent the differences between large numbers. More advanced versions of the theories involve solvatochromic parameters[6] and introduce interactions due to dipolarity and hydrogen bonding. These will be applied to the discrepancies. Data were not available to predict penicillin V free energies but, if the free energy of phase transfer of the nitrate ion to dichloromethane is taken as 33 kJ mol[-1], we can claim a free energy of phase transfer of the penicillin V ion of 27 kJ mol[-1].

Table 2: *Penicillin V/nitrate selectivities for various solvents and quaternary chlorides*

Quaternary	Dichloro-methane*		Octan-1-ol		Decan-1-ol[†]		Toluene		Tributyl phosphate	
	Sel.	$-\Delta G°$	Sel.	$-\Delta G°$	Sel.	$-\Delta G°$	Sel.	$-\Delta G°$	Sel.	$-\Delta G°$
tetrapentyl	15.3	6760								
tetrahexyl	10.6	5850	23.2	7790	12.0	6160	3.49	3100	7.79	5090
tetraheptyl	9.95	5690	20.9	7530	11.1	5960	5.96	4420	6.34	4580
tetraoctyl	9.75	5640	20.3	7460	10.3	5780	1.68	1290	6.48	4630
average	**11.4**	**5990**	**21.5**	**7950**	**11.1**	**5970**	**3.71**	**2934**	**6.87**	**4770**

$\Delta G°$ in J mol[-1]. * Average of two runs. † Contained 10% toluene to aid solubility.

Table 3: *Predicted and experimental free energies of ion exchange*

	Ions (most lipophilic first)	Solvent	$-\Delta G°$ (this work) kJ mol^{-1}	$-\Delta G°$ (unhydrated) kJ mol^{-1}	$-\Delta G°$ (hydrated) kJ mol^{-1}
1	I$^-\leftrightarrow$Br$^-$	toluene	~20	17.4	16.9
2	Br$^-\leftrightarrow$Cl$^-$	dichloromethane	~15	14.9	13.5
3	Br$^-\leftrightarrow$Cl$^-$	toluene	8	15.1	14.7
4	Br$^-\leftrightarrow$Cl$^-$	1-octanol	6	15.6	13.5
5	NO$_3\leftrightarrow$Br$^-$	dichloromethane	3.3	4.3	4.3
6	NO$_3\leftrightarrow$Br$^-$	1-octanol	0.62	4.3	4.3
7	NO$_3\leftrightarrow$Br$^-$	1-decanol	0.58	n.a.	n.a.
8	NO$_3\leftrightarrow$Br$^-$	toluene	4.4	4.3	4.3
9	NO$_3\leftrightarrow$Br$^-$	tributyl phosphate	4.3	n.a.	n.a.

5. CONCLUSIONS

1. Liquid-liquid ion exchange of simple inorganic anions into organic solvents with the aid of quaternary ammonium extractants seems to approach thermodynamic equilibrium readily. Selectivities and levels of extraction are consistent with known free energies of ion exchange.

2. For the systems studied, extraction levels and selectivity are not dependent on the lipophilicity of the extractant, suggesting that the ions in the organic phase were dissociated.

3. Extraction levels and selectivity were dependent on solvent. The order of effectiveness of solvents for simple ions generally followed prediction but, with penicillin V, there appeared to be a number of structure specific effects.

6. ACKNOWLEDGEMENTS

We thank SERC for partial support for this work and for a studentship for A.McG.

7. REFERENCES

1 T.A.J. Harris, S.A. Khan, B.G. Reuben, T. Shokoya and M.S. Verrall, *in* 'Separations for Biotechnology 2,' Ed. D.L. Pyle, Elsevier, London, 1990, 172-180.

2 T. Hano, M. Matsumoto, T. Ohtake, and T. Kawazu, *Proc. ISEC '93*, 1993, **2**, 1018, Eds D.H. Logsdail and M.J. Slater, Elsevier.

3 S-T Yang, S.A. White and S-T Hsu, *Ind.Eng.Chem.Res.*, 1991, **30**, 1335.

4 Mathcad 4.0, Cambridge, Mass:Mathsoft, 1993

5 Based on M.H. Abraham, *JCS Perkin II*, 1972, 1343.

6 R.W. Taft, M.H. Abraham, R.M. Doherty and M.J. Kamlet, *J. Am. Chem. Soc.* 1985, **107**, 3105.

7 M. Bachelin, P. Gans, and B.J. Gill, *J. Chem. Soc. Faraday Trans*, 1992, **88**, 3327.

8 M.H. Abraham and J. Liszi, *J. Inorg. Nucl. Chem.* 1981, **43**, 143.

Effective Purification of α-Amylases from Fermentation Broth by Immunoaffinity Chromatography

Shigeo Katoh and Masaaki Terashima

DEPARTMENT OF SYNTHETIC CHEMISTRY AND BIOLOGICAL CHEMISTRY, KYOTO UNIVERSITY, KYOTO 606, JAPAN

1 INTRODUCTION

In recent years many high-value bioproducts have been produced by use of genetically engineered microorganisms. However, the concentrations of these products, especially secreted ones, in fermentation broth are usually very low (μg to mg per liter). For the purification in these cases, immunoaffinity chromatography is very effective because of the high affinity and specificity between antigens and antibodies.

For large-scale purification of recombinant proteins by immunoaffinity chromatography, there are two main problems encountered. One is to obtain antibody ligands suitable for purification without use of target proteins (antigens), because, in many cases, sufficient amount of antigens of suitable purity for immunization are not available. The other is to treat effectively a large volume of broth containing the target protein of a low concentration in a short time.[1]

In the first case, selection of a monoclonal antibody showing suitable binding characteristics as a ligand is one approach, but it might be a hard task. As alternatives, we propose utilization of cross-reactive antibody and anti-peptide antibody as affinity ligands in immunoaffinity chromatography. Proteins showing similar functions from various species often have high homology in their amino acid sequences, and antibodies against these proteins are expected to show cross-reactivity with each other. Therefore, an antibody which cross-reacts with a target protein should be obtained by immunizing with more readily available proteins, which show homology with the target protein. Recently it was shown that antibodies against short peptides, which consist of parts of sequences of proteins, can bind to the native proteins.[2] By use of a synthesized peptide, an anti-peptide antibody which reacts with the target protein may be obtained without the purified target proteins.

In affinity chromatography, which comprises highly specific adsorption and desorption steps, the overall performance is affected by several factors. A high feed rate decreases the operation time needed in one cycle and increase the productivity. A steep breakthrough curve, which results from a high mass transfer rate, will give an effective utilization of the adsorption capacity of fixed-beds at the breakpoint, at which the ratio of the concentration of adsorbed component in effluent to that of feed reaches 0.05 or 0.1. These considerations lead to utilization of support particles with high mass transfer rate and mechanical strength, *i.e.* HPLC type or perfusion type supports.[3]

In this work rice α-amylases produced by secretion from recombinant yeast cells were purified by use of cross-reactive and anti-peptide antibodies as ligands coupled on perfusion type supports. The ease of obtaining antibody ligands and their effective utilization will extend the applicability of immunoaffinity chromatography to downstream processes, especially at an early stage for concentrating and purifying proteins of very low concentrations in fermentation broth.

2 MATERIALS AND METHODS

2.1 Fermentation of Rice α-Amylases-Secreting Yeast

The cloned genes of two rice α-amylase isozymes (*RAmy1A* and *RAmy3D*) were subcloned into the yeast expression vector pMAC101 to give the yeast expression plasmids, pEno/RAmy1A and pEno/RAmy3D (These genes were kindly provided by Dr. R. L. Rodriguez). [4] A plasmid expressing a chimeric enzyme (RAmy1A/3D), which consists of 190aa of RAmy1A and 251aa of RAmy3D, was also produced. These α-amylases were expressed in *Saccharomyces cerevisiae* LL20 under the yeast enolase promoter and secreted to fermentation broth with their respective signal peptides. Mature RAmy1A and RAmy3D consist of 403 and 410 amino acids, respectively, and the former has one N-glycosylation site.[5]

The yeast was grown in 300 ml of YNBDH medium (0.67 % yeast nitrogen base without amino acids, 2 % glucose, 20 µg/ml histidine, 5 mM $CaCl_2$) and inoculated to 2 L of YEPD media (1 % yeast extract, 2 % bacto peptone, 8 % glucose, 5 mM $CaCl_2$). The inoculated culture was grown in a jar fermenter (Mitsuwa KMJ-5C) for 24 h at 30 °C under an agitation of 300 rpm and aeration of 2 L/min. The values of pH and DO were not controlled. After cells were separated by centrifugation (6500 rpm x 10 min), the supernatant containing α-amylase was concentrated to various concentrations by ultrafiltration with a hollow fiber module (Asahi Kasei ACP-0013, 13 kD molecular cut-off). For the measurement of the adsorption equilibrium, the supernatant was, in some cases, diluted by a buffer solution (50 mM Tris-HCl + 5 mM $CaCl_2$, pH 7.6).

2.2 Preparation of Cross-Reactive and Anti-Peptide Antibody-Ligands

Two types of antibody ligands were prepared for purification of rice α-amylases. One was anti-barley α-amylase antibody, which was expected to show cross-reactivity with rice α-amylase because of high homology between barley and rice α-amylases. The other was anti-peptide antibody against the eight amino acids of the C-terminal region (RVPAGRHL) of RAmy3D. Crude barley α-amylase (from barley malt, Sigma Chemical Co.) was partially purified by salting-out with ammonium sulfate and gel chromatography (Toyopearl HW55F, 20 mm inside diameter, 850 mm height, flow rate of mobile phase : 1.1 ml/min). The peptide (RVPAGRHL, PC-Am3D) was synthesized by the solid-phase method and coupled to keyhole limpet hemocyanin.[6] A mixture of the antigen and Freund's complete adjuvant (1 ml each) was immunized to rabbits. Booster injections were repeated twice in a similar way at 10 days interval. Specific antibodies were purified from pooled sera by affinity chromatography using barley α-amylase- or peptide-coupled Sepharose 4B.

The immunoadsorbents were prepared by coupling these antibodies to CNBr-activated Sepharose 4B (Pharmacia LKB Biotech) or POROS® (PerSeptive Biosystems). The immunoadsorbent was packed in an adsorption column , and the column was equilibrated with an equilibration buffer (50 mM Tris-HCl + 5 mM $CaCl_2$, pH 7.6) at 23 ± 2 °C. The fermentation broth (adjusted to pH 7.6) was applied to the column. After washing with the equilibration buffer, adsorbed α-amylase was eluted by eluents (2.5 M NaSCN containing 5 mM $CaCl_2$, pH 5.0 or 50 mM sodium acetate containing 5 mM $CaCl_2$, pH 3.6). The absorbance of the effluent solution at 280 nm was continuously measured by a spectrophotometer, and the activity of α-amylase in the collected samples was determined as stated below. The eluted fractions were dialysed against a sodium acetate buffer (50 mM, pH 5.0, 5 mM $CaCl_2$, 100 mM NaCl).

2.3 Measurement of Enzyme Activity and Protein Concentration

The activities of recombinant α-amylases were determined from the rate of increase in reducing ends from soluble potato starch.[7] The reaction was initiated by adding 10 μl of the enzyme solution to 90 μl of the substrate solution (pH 5.0, 30 °C) and terminated at various intervals by adding 100 μl of a 3,5-dinitro salicylic acid solution. After dipping in boiling water for 5 min, the solution was diluted to 1 ml, and the absorbance at 540 nm was measured. Maltose was used as a standard. One enzyme unit was defined as the activity to liberate 1 μmol of maltose per minute.

The protein concentration was determined by the dye method (Bio-Rad protein assay kit, Bio-Rad Lab). BSA was used as a standard. The purity of α-amylases was examined by SDS-PAGE stained with silver (Silver stain kit. Bio-Rad Lab).

3 RESULTS AND DISCUSSION

3.1 Fermentation of Recombinant Yeast

Alpha-amylases (RAmy1A, RAmy3D, RAmy1A/3D) were secreted in fermentation broth with increase in the cell concentration. After 24 hours, the activities of α-amylase in fermentation broth ranged from 1.6 to 3.7 U/ml. The supernatant was concentrated by ultrafiltration or diluted by Tris buffer and was purified by immunoaffinity chromatography in one step, as stated in Materials and Methods.

3.2 Adsorption Equilibrium

Figure 1 (a) shows an adsorption isotherm of rice α-amylase (RAmy1A) to anti-barley α-amylase antibody. The adsorption of RAmy1A by cross-reaction showed an adsorption isotherm of the Freundlich type. The low value of the exponent (0.19) indicates a relatively high affinity between RAmy1A and anti-barley α-amylase, as expected by high homology (about 50 %) between rice and barley α-amylases.[5] As expected, RAmy3D and RAmy1A/3D were also adsorbed by this antibody. In the case of the anti-peptide antibody (anti-PC-Am3D), RAmy1A/3D was adsorbed and showed a similar adsorption isotherm to anti-barley α-amylase-RAmy 1A, as shown in Figure 1 (b). RAmy 1A, which has a different C-terminal sequence from that of RAmy1A/3D, was not adsorbed by this anti-peptide antibody. These results show that with the antibody against 8 amino acids of the C-terminal region, the α-amylase molecules consisting of more than 400 amino acids were selectively adsorbed.

3.3 Immunoaffinity Purification of Rice α-Amylases

Figure 2 shows the activity of α-amylase (RAmy1A) in the effluent solution from the anti-barley α-amylase column. At first the activity of α-amylase was not detected in the effluent solution and then gradually increased (breakthrough). Then, the column was washed, and α-amylase was eluted as a sharp peak by 2.5 M

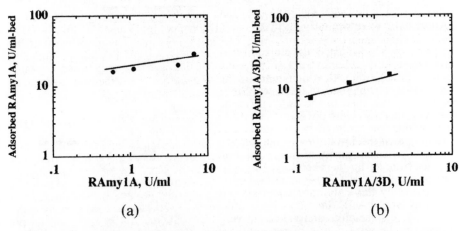

Figure 1 *Adsorption equilibria of rice α-amylases*
(a) adsorption of RAmy1A to anti-barley α-amylase antibody
(b) adsorption of RAmy1A/3D to anti-PC-Am3D antibody

NaSCN containing 5 mM $CaCl_2$, pH 5.0. In 10 ml of the eluate 75 % of adsorbed α-amylase was recovered. The specific activity of eluted RAmy1A after dialysis by the sodium acetate buffer (50 mM, pH 5.0, 5 mM $CaCl_2$, 100 mM NaCl) was 220 U/mg on average, which corresponded to 2000-fold purification from the fermentation broth of 0.11 U/mg-solid. Although feed loading was stopped at a suitable break point in this example, supply until saturation of the adsorbent gave simi-

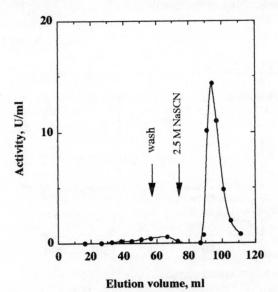

Figure 2 *Adsorption and elution of RAmy1A   anti-barley*
α-amylase-Sepharose 4B column, initial concn. : 2.66 U/ml
liquid flow rate : 2.0 ml/min

lar results (data not shown). RAmy3D and RAmy1A/3D were similarly purified.

As shown in **Figure 3**, the eluted rice α-amylase (RAmy1A) showed a single band by SDS-PAGE (lane A), which indicates the high purity of obtained α-amylase. Since impurities contained in the antigen (crude barley α-amylase) were different from those in fermentation broth of yeast, antibodies raised against the impurities in the antigen hardly affected the purity of rice α-amylase purified from the broth. Therefore, barley α-amylase similarly purified from crude barley α-amylase by this column showed several bands by SDS-PAGE (lane B). Thus, utilization of cross-reactive antibodies are very useful for the purification of recombinant proteins foreign to host cells.

Figure 4 shows breakthrough and elution curves for purification of RAmy1A/3D by the column of anti-PC-Am3D. RAmy1A/3D was adsorbed to the adsorbent of anti-peptide antibody against the C-terminal region. RAmy1A/3D was also eluted by 2.5 M NaSCN. The purity of the RAmy1A/3D was again very high. The slight leakage of activity from the column may suggest possibility of some processing at the C-terminal region of RAmy1A/3D.

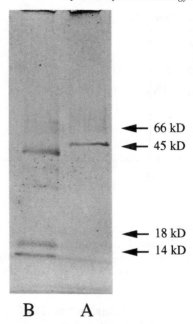

Figure 3 *SDS-PAGE of rice α-amylase and barley α-amylase purified by anti-barley α-amylase antibody (lane A : rice α-amylase, lane B : barley α-amylase)*

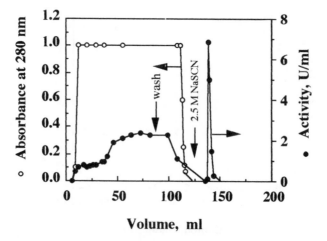

Figure 4 *Adsorption and elution of RAmy1A/3D anti-PC-Am3D-Sepharose 4B column, initial concn. : 2.37 U/ml, liquid flow rate : 0.97 ml/min*

3.5 High Speed Purification

By use of supports with high mass transfer rate and mechanical rigidity, it is possible to operate at a high flow rate with an effective utilization of the adsorption capacity of fixed-beds at the break point. The higher operating velocity reduces time needed not only for the adsorption step but also for the washing, elution and reequilibration steps. Further, in both analytical and preparative applications, speedy treatment improves the quality of unstable bioproducts purified.

In the case of soft gel, compaction of packed supports limits the maximum operating velocity. Though the rigidity of the synthetic polymer or silica supports may enable operation at higher flow rates, low mass transfer rates limit the velocity because of the low proportion of the adsorption capacity utilized at the break point. Therefore, the operating velocities with conventional supports are at most 500 cm/h.

By use of a perfusion type support, the shapes of breakthrough curves in the range from 700 to 3800 cm/h were similar, as shown in **Figure 5**. Thus operation with velocities several times higher than those of conventional supports was possible.[8] This makes it possible to use immunoaffinity chromatography at an early stage for concentrating and purifying secreted proteins of very low concentrations in fermentation broth.

4 CONCLUSION

By use of cross-reactive and peptide antibodies and perfusion type supports, α-amylases were effectively purified to a high purity from the fermentation broth in one step. These results show that cross-reactive and peptide antibodies have wide applicability for the purification of recombinant proteins secreted from microorganisms.

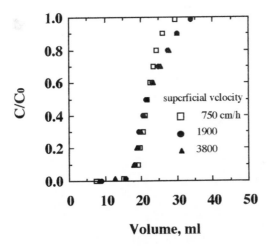

Figure 5   *Breakthrough curves from anti-barley α-amylase antibody coupled POROS® support, column size : 4.5 mm x 100 mm RAmy1A : 1 U/ml*

References

1. S. Katoh, <u>Trends Biotech.</u>, 1987, <u>5</u>, 328.
2. R. A. Larner, <u>Adv. Immunol.</u>, 1984, <u>36</u>, 1.
3. Y. Kamiya, S. Katoh *et al.*, <u>J. Ferment. Bioeng.</u>, 1990, <u>69</u>, 298.
4. M. H.Kumagai,   M. Terashima *et al.* , <u>Gene</u>, 1990, <u>94</u>, 209.
5. S. D. O'Neil, M. H. Kumagai *et al.*, <u>Mol. Gen. Genet.</u> , 1990, <u>221</u>, 235.
6. A. Kondo, S. Katoh *et al.*, <u>Biotech. Bioeng.</u>, 1990, <u>35</u>, 146.
7. B. Bernfeld,' Methods Enzymol. ' Academic Press, Florida, 1955.
8. N. B. Afeyan, S. P. Fulton *et al.*, <u>Bio/Technol.</u>, 1990, <u>8</u>, 203.

Partition of Soluble Proteins from *E. coli* in Polyethylene Glycol-salt Two Phase Systems

A. Kaul and J. A. Asenjo

BIOTECHNOLOGY AND BIOCHEMICAL ENGINEERING GROUP, DEPARTMENT OF FOOD SCIENCE AND TECHNOLOGY, THE UNIVERSITY OF READING, PO BOX 226, READING RG6 2AP, UK

The partition behaviour of soluble proteins from E.coli in polyethylene glycol-salt two-phase systems has been examined in order to investigate the effect of system variables on the partition coefficient, K. This was carried out to determine under which conditions the soluble proteins from E.coli partition preferentially to one phase. Systems which show this behaviour can be used as extraction systems for the separation of recombinant proteins provided, that under the same conditions, the target protein partitions to the opposite phase. Factors such as PEG MW, pH, concentration of NaCl and the stability ratio were all found to influence K; the protein concentration had no effect upto an overall system concentration of 1 g l^{-1} after which the bottom phase started to show saturation behaviour. For PEG-phosphate systems values of K were below 1 using stability ratios below 0.18 at 4% w/w and 8.8% w/w NaCl, a low pH and high molecular weight of PEG. For PEG-sulphate systems all the stability ratios studied gave values of K well below 1.

1 INTRODUCTION

The purification of enzymes and other biologically active proteins using aqueous two-phase technology is now available as an industrial unit operation and is used in industry for the production of pharmaceutical proteins, analytical enzymes and biocatalysts[1-3]. *Escherichia coli* is commonly one of the hosts for the production of recombinant proteins. For the recovery of a foreign protein from *E.coli* the two-phase extraction method can be used for cell debris separation and initial enrichment (where a significant proportion of the soluble contaminants are removed). If greater purity is required extraction can be coupled to chromatographic steps. Separation will depend on the interaction of the physico-chemical properties of both the target protein and the system employed. Aqueous two-phase systems such as those formed by PEG and salts offer greater selectivity in partitioning in comparison to polymer-polymer systems, however, a need to study the effect of system variables in order to develop suitable extraction conditions is required. Table 1 shows the typical composition of various macromolecules in microbial cells[4]. To permit the selection of an extraction system for the separation of a protein product, the behaviour of *E.coli* soluble contaminants in the system needs to be characterised. Phase system parameters include molecular weight of the polymer, pH of the system, type of phase forming salt, ionic strength of added salt (e.g. NaCl) and distance of the system from the binodial. Factors which influence or which can be utilised to partition the protein product to a

Table 1 *Content of various macromolecules in microbial cells*

Macromolecule	Amount (g/100g cells)
Protein	52.4
RNA	15.7
Lipid	9.4
Polysaccharide	6.6
DNA	3.2
Other compounds e.g K & Mg	1.7

different phase from that of the associated contaminants include hydrophobicity, charge, size and affinity. The exploitation of aqueous two-phase systems for protein concentration lies in their inherent advantages which include minimal product degradation and mixing time due to the low interfacial tension, a high capacity, linearity of scale up and the ability to handle large volumes in a continuous mode. For industrial application polymer-salt systems are preferential to polymer-polymer systems due to the lower viscosity, phase separation time and chemical cost.

In this study the potential of aqueous two-phase systems as a tool for extraction of a recombinant protein from *E.coli* was investigated. Various factors were examined to see if a favourable partition of *E.coli* K12 total proteins in PEG phosphate and sulphate systems could be obtained and therefore if these systems could be used for the purification of a recombinant protein expressed in *E.coli*. The effect of the stability ratio and NaCl concentration was studied for both these systems. Effects of concentration of protein, pH and M.W. of PEG were also studied for PEG-phosphate systems. These variables are commonly used to manipulate the partition coefficient of a target protein in order to achieve a desirable partition and it is therefore important to see their effect on the overall partition of total proteins from *E.coli*.

2 MATERIALS AND METHODS

2.1 Materials

Polyethylene glycol (PEG) with MW 4000 Da. was obtained from Fluka Chemicals LTD., Switzerland. All other chemicals were analytical grade. A cell paste of the organism *E.coli* K12 was provided by SmithKline Beecham and was resuspended in 50mM Tris HCl buffer at pH 7.5 (1 mg/ml lysozyme). The suspension was sonicated using an XL 2000 series ultrasonic liquid processor with a high grain Q horn. After centrifugation the supernatant (soluble fraction) and cell debris (insoluble fraction) were separated and stored at -20°C.

2.2 Selection and Preparation of Phase Systems

Systems chosen were at set distances from the binodial. This distance is defined as the stability ratio illustrated in Figure 1. At a stability ratio of 0.18 and 0% NaCl the volume ratio is approximately 1. When changing pH or adding NaCl the stability ratios were recalculated due to a shift in the binodial.

Systems, prepared in duplicate, had a final weight of 1.8g and were prepared from stock solutions of PEG 4000 (50%w/w), phosphate (40%w/w) (mixture of K_2HPO_4 & NaH_2PO_4 to obtain a pH of 7), magnesium sulphate (23% w/w) and NaCl (25%w/w). pH was set at 7 and stocks were stored at 4°C and prior to use temperature equilibrated by

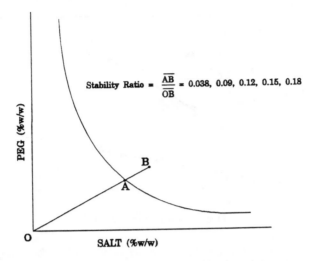

Figure 1 *Illustration of the stability ratio concept. B: system composition, A: point on the binodial, O: origin*

standing at room temperature (20°C). *E.coli* soluble fraction was added to give a final concentration of 0.5mg/ml protein. Systems were mixed thoroughly using a vortex and centrifuged (3,000 rev min⁻¹,3 min) to assist phase separation. Samples of 0.1ml were removed from the top and bottom phases, diluted with deionised water and assayed for protein concentration.

2.3 Protein Determination

Protein concentration was determined by the method of Sedmack and Grossberg[5], using Coomassie Brilliant Blue G250 (0.06% w/v diluted in 2.2% v/v HCl) which was modified to give a total volume of 3.5ml (0.1ml sample, 2.4ml water and 1ml dye). A standard curve of BSA was constructed which was linear within the range of protein concentration measured. Protein free systems were used for reference where the protein sample was replaced with 50mM tris HCl at pH7. Assays were carried out in triplicate.

3 RESULTS AND DISCUSSION

3.1 Effect of Protein Concentration on K_{app}

Figure 2 shows the protein concentration in the top and bottom phases of a system, with a stability ratio of 0.20, as the overall protein concentration is increased. Above 0.6 g/l in the bottom phase increase in protein concentration is no longer linear with an increase in the overall system concentration; this non-linear behaviour will eventually result in saturation of the phase. Log K_{app} (K_{app} represents an average K for soluble proteins from *E.coli*) decreases slightly as the overall protein concentration is increased. This is most likely due to precipitation of some of the proteins. For an extraction system it is important to determine the maximum load for protein as this will eventually influence the size of the system to be used.

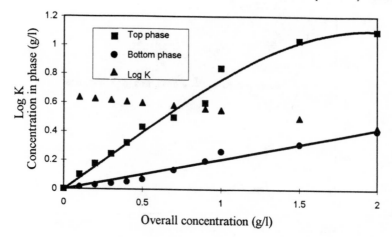

Figure 2 *Protein concentration in the top and bottom phases in a PEG 4000 phosphate system with a stability ratio of 0.20*

3.2 Effect of NaCl at different Stability Ratios.

For a given extraction system the chosen point of operation relative to the binodial is of major importance. If a chosen point is too far from the binodial, where concentrations of polymer and salt are high, the protein might precipitate out of solution. If the point is too near the binodial a small dilution of the system might cause a shift of the system composition to the left of the binodial and therefore the formation of a single phase where separation can not be achieved. Consequently for the application and scale-up of aqueous two phase systems, extraction systems must be 'robust' so that variations in process streams will not destroy the system by causing the formation of one phase[6]. System robustness is largely reflected in the length of the tie-line but, even simpler, in the value of the stability ratio.

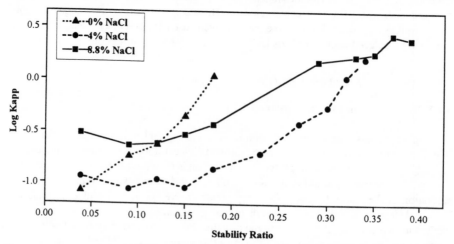

Figure 3 *Effect of the stability ratio on the partition of soluble proteins from E.coli as a function of NaCl concentration in PEG 4000-phosphate systems at pH7*

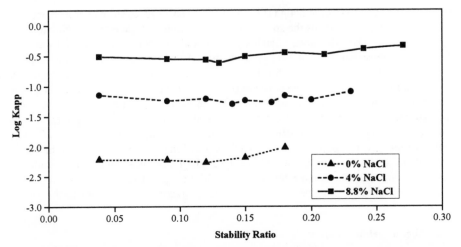

Figure 4 *Effect of the stability ratio on the partition of soluble proteins from E.coli as a function of NaCl concentration in PEG 4000- sulphate systems.*

This ratio is much easier to determine than the tie-line length as it only requires a titration with water and does not require the construction of a binodial.

3.2.1 PEG-Phosphate Systems. Figure 3 shows the partition of *E.coli* soluble protein contaminants at different stability ratios as a function of NaCl concentration in PEG-phosphate systems. Log K_{app} at 0% NaCl increases as the stability ratio increases. At 4 and 8.8%w/w NaCl, Log K_{app} remains relatively unchanged up to a stability ratio of 0.18, where the change in PEG and salt concentration have no apparent effect on K. However, above a stability ratio of 0.18 Log K_{app} increases as the stability ratio increases.

3.2.2. PEG-Sulphate Systems. Figure 4 shows the partition of *E.coli* soluble protein contaminants at different stability ratios as a function of NaCl concentration in PEG-sulphate systems. Log K_{app} at 0, 4 and 8.8% w/w NaCl remains relatively constant as the

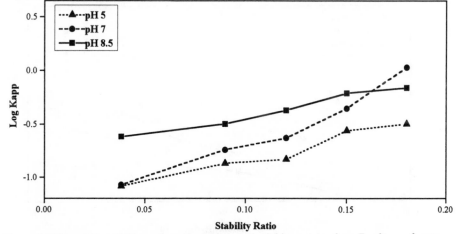

Figure 5 *Effect of the stability ratio on the partition of soluble proteins from E.coli as a function of pH in PEG 4000-phosphate systems*

stability ratio is increased, although the presence of NaCl causes an overall increase in the values for K.In these systems the stability ratio seems to have little effect on K_{app} for *E.coli* soluble proteins. Over the range of stability ratios studied, values for K are still well below 1. For both phosphate and sulphate systems, at 0, 4 and 8.8 %w/w NaCl, the respective systems have the same stability ratio. Log K_{app}, however, increases as a function of NaCl concentration suggesting that the presence of NaCl is changing the character of the phases and thus causing an increase in the value of K. An inert salt such as NaCl can be used to increase the partition coefficient for a protein (usually hydrophobic in character) while that for the contaminants remains unaffected[7,8]. Phosphate systems at 4 and 8.8% NaCl with a stability ratio of below 0.18 can all be used for extraction of a target protein as the majority of contaminants are in the bottom phase. Over the range of stability ratios studied for sulphate systems, at 0, 4, and 8.8% w/w NaCl, the values of K are well below 1. These systems would also be suitable for partitioning target recombinant proteins to the top PEG phase.

3.3 Effect of pH at different Stability Ratios.

Figure 5 shows the partition of *E.coli* soluble protein contaminants at different stability ratios as a function of pH. At pH 5, 7 and 8.5 Log K_{app} increases as the stability ratio increases, however, as the pH is decreased the values for K also decrease. As a general rule if the pH is increased above the pI of a protein the value of K also increases and vice versa.This suggests that a majority of the *E.coli* total proteins have pI's below 7. Thus for an extraction system a low pH is favourable to partition the contaminants to the bottom phase and vice verse.

3.4 Effect PEG Molecular Weight at different Stability Ratios.

Figure 6 shows the partition of *E.coli* soluble protein contaminants at different stability ratios as a function of PEG molecular weight. As the stability ratio increases so does the value for Log K_{app}. The value of K also increases as a function of PEG molecular weight. PEG 1500-phosphate systems may not be suitable for removal of protein contaminants as K

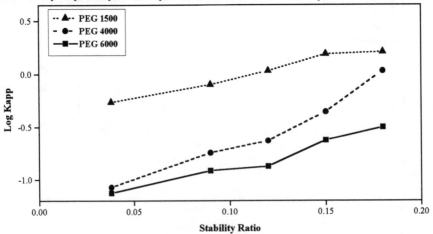

Figure 6 *Effect of the stability ratio on the partition of soluble proteins from E.coli as a function of PEG molecular weight in PEG-phosphate systems at pH 7*

values are around 1; however, if required, this can be overcome by employing a suitable volume ratio. On the other hand PEG 4000 and 6000-phosphate systems exhibit more favourable partitioning to the bottom phase for the protein contaminants over a range of stability ratios. For an extraction system, a high PEG molecular weight is favourable to partition the contaminants to the bottom phase and a low molecular weight PEG to partition the contaminants to the top phase. In Figures 3,4,5 and 6 as the stability ratio decreases towards 0, Log K_{app} does not in turn tend towards 0. It is generally accepted that at the critical point K= 1. For the systems employed in this study, point A (Fig 1) is close to the critical point, but Log K_{app} does not seem to tend to 0. It is evident that to the left of the binodial K is always 0 as there is only one phase. At the binodial there is a significant discontinuity and the values of K change quite dramatically from 0.

4 CONCLUSIONS

The partition behaviour of total proteins from *E.coli* as a function of the stability ratio was studied for PEG-phosphate and PEG-sulphate systems. Protein and NaCl concentration, pH and PEG molecular weight effects were also studied for PEG-phosphate systems.

As a function of protein concentration, K remained relatively constant up to an overall system concentration of 1 g/l after which saturation of the bottom phase caused an increase in the value of K.

As a function of the stability ratio, values for K generally increase except for stability ratios between 0.04 and 0.18 at 4 and 8.8% NaCl. However for sulphate systems, values for K remain constant.

As a function of NaCl concentration at 4 and 8.8%, a decrease in the value of K up to a stability ratio of 0.18 was observed after which it increases. For sulphate systems there is an overall increase in K but values still remain below Log K = 0 (K=1). As the systems get closer to the binodial values of Log K do not tend to 0 for both phosphate and sulphate systems.

As a function of pH, as the pH is increased K also increases.

As a function of PEG MW, K decreases as the molecular weight is decreased from 6000 Da. to 1500 Da..

Acknowledgements

This work was supported by SERC and SmithKline Beecham to whom thanks are due.

5 REFERENCES

1. H. Hustedt, K.H. Kroner, U. Menge, & M.-R. Kula, *Trends Biotechnol.*, 1985, **3**, 139.
2. M.-R. Kula, K.H. Kroner, H. Hustedt, S.H. Granda, & W. Stach, German Patent 26.39.129: US Patent 4 144 130.
3. K. Hayenga, M. Murphy, R. Arnold,J. Lorch, & H. Heinsohn, 7th Int. Conf. on partitioning in Aqueous two-phase systems, New Orleans, USA, #16
4. A.H. Stouthamer, *Antonie van Leeuwenhoek*, 1973, 545.
5. J.J Sedmak, S.E. Grossberg, *Anal.Biochem.*, 1977, **79**, 544.
6. O.Cascone, B.A. Andrews, J.A.Asenjo, *Enz. Microb. Technol.*,.1991, **13**, 629.
7. A. S. Schmidt, A. M. Ventom, J. A. Asenjo, *Enz.Microb. Technol.*, 1994, **16**,.131.
8. S. L. Mistry, J. A. Asenjo, C.A. Zaror. *Biosep.* , 1993, **3**, 343

Affinity Isolation of Proteins Based on Precipitation of Eudragit-ligand–Protein Complexes

R. Kaul, D. Guoqiang, A. Lali, M. A. N. Benhura, and B. Mattiasson

DEPARTMENT OF BIOTECHNOLOGY, CHEMICAL CENTER, LUND
UNIVERSITY, BOX 124, S-221 00 LUND, SWEDEN

1 INTRODUCTION

Biospecific affinity sorption represents the most powerful means of resolution of proteins. So far, the common mode of using affinity sorption has been packed bed chromatography, a technique which suffers from low throughputs and is unsuitable for use with unclarified feed. Moreover, in order to avoid the fouling of expensive matrix, affinity chromatography is commonly used as a final polishing step in the production process when it is required to separate proteins resembling each other in most physico-chemical properties. However, if affinity interactions could be used for the capture of protein directly from crude feedstock, the procedure is likely to have an impact on simplifying further downstream processing operations. For this to be realistic, the two major prerequisites are a separation technology capable of handling large throughputs of unclarified feedstream without causing denaturation of the protein, and stability of the ligand molecule. Efforts are being made to integrate affinity interactions with other separation technologies like ultrafiltration, precipitation, extraction and also fluidized bed adsorption, [1,2] which are well established techniques for processing large crude volumes. In view of the sensitive nature of most biological ligand molecules, the choice of ligands for affinity isolation of proteins from crude feed may become limited to the non-biological molecules like dyes, metal ions, hydrophobic groups etc., which have the required biological and chemical stability. Most of these ligands are group-specific, and it is likely that they would interact with some other proteins besides the target protein. However, further purification to homogeneity is possible in subsequent high resolution step(s), even affinity chromatography, at a smaller scale.

2 AFFINITY PRECIPITATION

Precipitation is ordinarily a non-selective precipitation technique often used for clarification and concentration of the feed and sometimes as a prefractionation step in downstream processing of proteins. The more selective form of the technique, affinity precipitation has been performed in two modes. One involves the selective precipitation of proteins using bifunctional ligands e.g. bis-dyes.[3] The requirement here is the presence of more than one specific binding site on the protein. The other mode of affinity precipitation, which forms the theme of this paper, employs the ligand coupled to a polymer having a property of reversible solubility and precipitation with respect to variation in a particular environmental parameter such as pH, temperature, ionic strength, specific metal ions etc.[1,4] The soluble form of the ligand is used for affinity interaction with proteins, and precipitation is induced for effecting separation. A number of reversibly soluble-insoluble polymers, both of natural and synthetic origin, and even polyelectrolyte complexes, have been used as carriers of affinity ligands and/or enzymes.[1,4] The efficiency of precipitation and solubilization of the synthetic polymers has been found to be higher than that of the natural polymers; the latter are also more polydisperse. Enteric coating polymers comprise a group of non-toxic, relatively inexpensive polymers with a sharp soluble-insoluble transition profile and high precipitation efficiency, and also possessing reactive groups for ligand attachment.

2.1 Eudragit as a Ligand Carrier in Affinity Precipitation

Eudragit (Röhm Pharma, Germany), an enteric coating polymer, is a copolymer of methacrylic acid and methylmethacrylate, having a molecular weight of 135 000. The polymer is soluble in neutral pH range and insoluble under acidic conditions. Eudragit has been used earlier as a polymer matrix for the immobilization of enzymes acting on macromolecular substrates.[4] This laboratory has been studying the use of Eudragit as a carrier of ligand molecules for isolation of proteins by affinity precipitation. Eudragit S 100, having a carboxyl to ester group ratio of 1: 2 has been employed for all the studies reported here. The polymer itself has been seen to bind a number of proteins by ionic and/or hydrophobic interactions. These interactions seem to be predominant as the pH is lowered resulting in the coprecipitation of the proteins with the polymer at pH 4.3. This has been exploited for the purification of D-lactate dehydrogenase from *Lactobacillus bulgaricus*,[5] and xylanase from *Trichoderma viride*.[6]

During our studies, it was noted that precipitation of Eudragit could be induced even at neutral pH by the addition of high concentration of Ca^{2+} ions, which was most likely due to the complexing of carboxyl groups of the polymer with the metal ions. This precipitation was facilitated at elevated temperatures. In the present report, we describe some of our observations on affinity precipitation of enzymes using Cibacron blue, a triazine dye coupled to Eudragit S 100. The triazine dyes have been extensively used as ligands in protein purifications. The binding of the dye ligand to the proteins involves a wide range of interactions including hydrophobic, ionic, charge transfer effects, and for several dehydrogenases and kinases they are presumed to behave as analogues of nucleotides or coenzymes for binding the enzymes.

2.1.1 Eudragit-Cibacron blue. Cibacron blue 3GA (Sigma, USA) was bound to the carboxyl groups of the polymer by the carbodiimide coupling procedure. Prior to coupling, the dye was modified with an amino group using 1,6-diaminohexane according to the procedure described earlier.[7] A 2% (w/v) Eudragit solution was prepared in water with pH adjusted to 7.0. The activation of the polymer with water-soluble carbodiimide (14 mg/ml) was done at pH 4.5 for 30 min at room temperature, followed by incubation with the modified dye (8 mg/ml) at pH 10 for 20 h. Eudragit with the coupled dye was separated as described earlier.[8] A yield of 195 mg Cibacron blue bound per gram Eudragit preparation was obtained.[9]

The precipitation behaviour of the derivatized Eudragit (0.1 % w/v) with respect to change in pH, and temperature (5 - 10 min incubation) at pH 7.5 in the presence of 50 mM $CaCl_2$ respectively, was compared with that of the native polymer (**Fig. 1**). The precipitation of the native polymer was measured as turbidity at 470 nm, and that of Eudragit-Cibacron blue by subtracting the dye content remaining in the supernatant after centrifugation of the precipitate, from the total dye content determined by absorbance at 625 nm.[9] The native Eudragit was completely precipitated at pH 4.7. However, after immobilization of Cibacron blue, the soluble-insoluble transition region of the polymer was shifted upward on the pH scale, and total precipitation was observed at pH 5.0 (**Fig. 1a**).

With respect to precipitation in the presence of Ca^{2+} ions, the native Eudragit remained soluble at the metal ion concentration of 50 mM upto 50 °C. Further increase in temperature up to 65 °C resulted in complete precipitation. In contrast, Eudragit-Cibacron blue started to precipitate at low temperatures, with total precipitation at 40 °C (**Fig. 1b**). The increased efficiency in precipitation of the modified polymer could arise from the increased metal ion complexing capacity provided by the sulfonic acid residues of the dye molecules. The precipitate formed at 40 °C was stable for several hours at room temperature in the presence of Ca^{2+}.

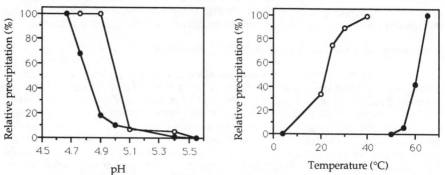

Figure 1 *Precipitation profiles of Eudragit (•) and Eudragit-Cibacron blue (o) with respect to (a) pH variation and (b) temperature change in presence of 50 mM Ca^{2+} ions.*

2.2 Examples of Affinity Precipitation of Enzymes with Eudragit-Cibacron blue

Eudragit-Cibacron blue was utilized for the affinity isolation of enzymes, lactate dehydrogenase (LDH, EC 1.1.1.27) and pyruvate kinase (PK, EC 2.7.1.40) from porcine muscle,[9] and alcohol dehydrogenase (ADH, EC 1.1.1.1) from *Saccharomyces cerevisiae*.[10] The combination of Ca^{2+} ions and incubation at 40 °C was used as the precipitation mode because of the sensitivity of the enzymes to low pH. All these enzymes were found to be stable during thermoprecipitation, and no unspecific binding to the polymer chain was observed.

The activity of the enzymes was determined according to established procedures.[11] Protein measurements were done by bicinchoninic acid method,[12] and the purity of the samples was checked by SDS-polyacrylamide gel electrophoresis (SDS-PAGE).[13]

2.2.1 Affinity thermoprecipitation of lactate dehydrogenase and pyruvate kinase from porcine muscle extract. Both LDH and PK are known to bind to Cibacron blue, the former having a stronger affinity to the dye.[14] The homogenate of porcine muscle in twice the amount of 25 mM Tris-HCl buffer, pH 7.5 was first centrifuged at 10 400 x g for 10 min, and the supernatant was then filtered through glass wool. The muscle extract thus prepared possessed per ml total protein concentration of 12.5 mg, and LDH and PK activity of 361 and 84 units respectively. The extract was incubated with 0.1% (w/v) Eudragit-Cibacron blue in 25 mM Tris-HCl buffer, pH 7.5 (total volume of 2 ml) at room temperature for 10 min, followed by addition of $CaCl_2$ to a final concentration of 50 mM and incubation at 40 °C for 5 min. **Fig. 2** shows that at low concentration of the extract, both enzymes were totally co-precipitated with the polymer-dye complex. With increase in concentration, the amount of PK precipitated was drastically reduced ultimately reaching nearly zero at an extract amount equivalent to 65% of the total volume. At this point, about 70% of LDH was precipitated.

For enzyme isolation, affinity precipitation was carried out at two different concentrations of the muscle extract - one at which maximal percentage of both LDH and PK were bound, and the other as an overload where LDH was predominantly bound. The precipitate was washed once with the above buffer containing $CaCl_2$, and then treated twice each with buffer containing 0.1 M KCl and 0.5 M KCl respectively, for the desorption of enzymes. After 0.5 M KCl treatment, 100 mM $CaCl_2$ was used during precipitation. At low concentration of the extract, (equivalent to 36 U LDH and 8.4 U PK per ml respectively), PK was desorbed in the presence of 0.1 M KCl with a yield of 70 %, along with 9% of LDH. Further treatment with 0.5 M KCl resulted in the recovery of 73% LDH and 5%PK. The increase in specific activities of PK and LDH in the two eluates were 4.4 and 7 fold respectively.

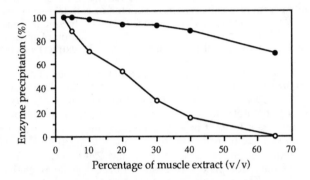

Figure 2 *Coprecipitation of lactate dehydrogenase (•) and pyruvate kinase (o) with Eudragit-Cibacron blue as a function of muscle extract load.*

When a high load of muscle extract (equivalent to 217 U LDH and 51 U PK per ml respectively) was used, major amount of PK was left in the supernatant after thermoprecipitation because of preferential binding of LDH. The bound LDH was desorbed directly with 0.5 M KCl buffer with a yield of about 70% and purification factor of about 10 (**Table 1**). Only 0.6% of the PK was present with the purified LDH. The specific activity of the purified enzyme was comparable with that (290 U/mg) of the commercial LDH (Sigma) determined using similar assay conditions.

Table 1 *Affinity thermoprecipitation of LDH at high concentration of muscle extract*

Stage	Volume	LDH			Total activity of PK
		activity	total activity	specific activity	
	ml	U/ml	U	U/mg	U
Starting (original extract vol. was 72 ml)	120	217	26 016	24	6 060
Supernatant after precipitation	116	46	6 336	ND	4 942
Washing	120	17	2 088	ND	348
Eluate 1	56	285	15 938	242	39
Eluate 2	30	68	2 034	204	3

ND: not determined

2.2.2 Affinity isolation of ADH from homogenate of Saccharomyces cerevisiae.
The binding of a number of proteins to triazine dyes is promoted in the presence of metal ions. The application of these interactions to protein purification by chromatography on a matrix with immobilized dye and by precipitation using bis-dyes has been reported.[15]

Isolation of ADH by metal ion promoted interaction with Eudragit-Cibacron blue and subsequent thermoprecipitation was studied. *S. cerevisiae* cells suspended in equal amount of 20 mM Tris-HCl buffer, pH 7.5 were broken by high pressure homogenization. The total protein content and ADH activity of the cell homogenate was 33 mg and 197 units per ml respectively. The separation of cell debris was facilitated by flocculation with 0.25% (w/v) polyethyleneimine (PEI) and subsequent centrifugation at 3000 x g for 10 min. This treatment also resulted in precipitation of nucleic acids and some proteins. The ADH activity in the clear supernatant (PEI-extract) was increased 1.5 fold.

Incubation of PEI-extract with Eudragit-Cibacron blue showed almost insignificant binding of ADH to the dye. Studies were performed to determine the effect of presence of metal ions on the enzyme binding to Eudragit-Cibacron blue (0.3% w/v). It seemed that Cu^{2+} and Zn^{2+} ions promoted the enzyme-dye interaction, but in case of Cu^{2+}, ADH was irreversibly inactivated. In Zn^{2+} promoted binding, increasing amount of enzyme was precipitated with increase in metal ion concentration. Upto 30-40 U of ADH per ml could be totally precipitated in the presence of 1 mM Zn^{2+} ions. The release of the enzyme from the affinity complex was best achieved by metal chelating agent, iminodiacetic acid (IDA). The other chelators, EDTA and imidazole formed precipitates with Eudragit-dye and were, therefore, unsuitable for desorption.

Affinity binding was carried out by mixing the PEI-extract with 0.3% (w/v) Eudragit-Cibacron blue and 1.5 mM Zn^{2+} ions in 20 mM Tris-HCl buffer, pH 7.5 in a volume yielding about 50 U ADH/ml, and incubating for 10 min at room temperature. This was followed by addition of 50 mM $CaCl_2$ and raising the temperature to 40 °C as before, for precipitation of Eudragit bound affinity complexes. Nearly all the enzyme was co-precipitated. After washing the precipitate once, the enzyme was desorbed by treatment twice with the buffer containing 200 mM IDA for 30-60 min at room temperature and finally the polymer-ligand was precipitated and recycled.

Table 2 shows the recovery and purification of ADH from PEI extract of about 600 g (packed weight) yeast cells. The affinity precipitation step gave an enzyme recovery of 50-55% with a purification factor of about 9. The binding capacity of the ligand and the purification of the enzyme was seen to be maintained during the reuse. The same procedure done at a smaller scale (25 ml) provided similar purification of the enzyme but the recovery was much higher (80%).

Table 2 *Isolation of ADH by affinity thermoprecipitation*

Stage	Sample	Volume ml	Activity U/ml	Total activity U	Specific activity U/mg
First cycle	PEI-extract	1000	265	262 890	14
	Eluate 1	500	209	104 435	135
	Eluate 2	500	52	26 210	120
Second cycle	PEI-extract	1000	280	280 630	15
	Eluate 1	500	236	117 935	134
	Eluate 2	490	77	37 539	99

Prior to PEI treatment, ADH in the yeast cell homogenate had specific activity of 6 U/mg.

Optimization of the process for larger volumes with regard to heating for thermoprecipitation, and continuous centrifugation should improve the yields. SDS-

PAGE profile (not shown) of the purified sample was essentially similar to that of commercial ADH (Sigma, 400 U/mg), except for a significant contamination in the form of hexokinase (subunit molecular weight of 51 000), which is also known to bind to Cibacron blue in the presence of metal ions.[15]

3 INTEGRATED AQUEOUS TWO-PHASE EXTRACTION AND AFFINITY PRECIPITATION

Although affinity precipitation can be used for protein isolation from relatively crude samples, prior removal of cell debris, particulate matter etc. is advantageous. This step is fairly easily done by partitioning in aqueous two-phase systems with minimal requirements of energy. Selective extraction by using affinity ligand modified with one of the phase forming polymers in two-phase system is also being worked out as a procedure for rapid isolation of proteins. Eudragit has been shown to partition to top PEG rich phase in a two-phase system; this partitioning is strongly promoted by increasing potassium phosphate concentrations. Thus, employing Eudragit bound ligand in a two-phase system makes it possible to design affinity extraction on the same grounds as using PEG bound ligands, followed by recovery of the polymer bound affinity complexes by precipitation for subsequent desorption of the protein. This concept of integration of affinity precipitation with extraction in aqueous two-phase system combines the features of gentle and rapid separation with the possibility of having a better control on the ligand and easy recovery of the target protein from the phase polymers by affinty precipitation. Purification of Protein A directly from the cell homogenate of recombinant *Escherichia coli* using Eudragit bound immunoglobulin G in PEG/Reppal two-phase system has been studied earlier.[16] This procedure provided high yields of the 26 fold purified protein.

3.1 Eudragit-Cibacron blue in aqueous two-phase system for isolation of muscle LDH

The integration approach was extended to the isolation of LDH from muscle extract in a two-phase system comprising of 6% (w/w) PEG and 8% (w/w) Dextran T250 in 50 mM potassium phosphate buffer, pH 7.6.[8] The volume ratio of the top to bottom phase was 1.35. In the absence of Eudragit-Cibacron blue, LDH partitioned predominantly into the bottom phase.Addition of increasing concentrations of the polymer-dye led to increase in the partition of the enzyme to the top phase. At Eudragit-dye concentration of 0.05%, increase in log K of the enzyme was about 2.45.

Affinity extraction of LDH directly from the crude extract (10% w/w) in the two-phase system containing 0.18% Eudragit-dye, followed by precipitation of Eudragit bound affinity complexes by reducing the pH to 5.1 and subsequent desorption of the enzyme from the complex in the presence of 0.5 M NaCl at pH 7.0 and finally precipitating the Eudragit-dye, yielded 5 fold purified enzyme. However, partitioning in two-phase system devoid of ligand prior to affinity partitioning, and subsequently washing the top phase containing the affinity complex with a fresh bottom phase before precipitation step yielded a much purer enzyme (with similar purity as obtained above by affinity thermoprecipitation) (**Table 3**). It seems from the studies that the number of extraction steps required for good purification is very much dependent on the nature of the crude feed and also of the ligand. The yield of the pure enzyme was 50%. The major loss of the enzyme activity took place in the precipitation step probably because of denaturation caused by low pH.

As an alternative to precipitation, the enzyme was also desorbed by addition of a salt phase (11 % w/w potassium phosphate) to the top phase containing the affinity complex. About 70% of LDH activity was recovered in the salt phase with a relatively lower specific activity (199 U/mg) than in the above case. The enzyme could be further purified to a specific activity of 297 U/mg by adsorption of the contaminants to DEAE-cellulose.

Table 3 *Isolation of LDH from muscle extract by using Eudragit-Cibacron blue in PEG/Dextran two-phase system*

Method	Specific activity (U/mg protein)	Purification factor
(1) Muscle extract and Eudragit-dye added to two phase system; Eudragit-dye-enzyme complex precipitated by reducing pH of PEG phase to 5.1; LDH desorbed with 0.5 M NaCl.	108	5.2
(2) Same as (1), except fresh bottom phase was added for "washing" once before precipitation.	53	7.3
(3) Muscle extract added in two phase system without Eudragit-dye; the top phase replaced by fresh PEG phase with Eudragit-dye; rest same as (1)	181	8.6
(4) Combination of (2) and (3)	248	11.8
(5) Same as (4) except instead of precipitation LDH desorbed directly from top phase by addition of salt	199	9.4

Specific activity of LDH in the crude extract was 21 U/mg.

Eudragit appears to be a promising candidate as a ligand carrier for affinity isolation of proteins. Further optimization studies are required on large scale use and recycling of Eudragit bound ligands.

REFERENCES

1. R. Kaul and B. Mattiasson, Bioseparation, 1992, 3, 1.
2. H.A. Chase and N.M. Draeger, J. Chromatogr., 1992, 597, 129.
3. M. Hayet and M.A. Vijayalakshmi, J. Chromatogr., 1986, 376, 157.
4. M. Fujii and M. Taniguchi, Trends Biotechnol., 1991, 9, 191.
5. D. Guoqiang, R. Kaul and B. Mattiasson, Bioseparation, 1993, 3, 333.
6. M.N. Gupta, D. Guoqiang, R. Kaul and B. Mattiasson, Biotechnol. Tech., 1994, 8, 117.
7. C.R. Lowe, M. Glad, P-O. Larsson, S. Ohlson, D.A.P. Small, T. Atkinson and K. Mosbach, J. Chromatogr., 1981, 215, 303.
8. D. Guoqiang, R. Kaul and B. Mattiasson, J. Chromatogr., in press.
9. D. Guoqiang, A. Lali, R. Kaul and B. Mattiasson, communicated.
10. D. Guoqiang, M.A.N. Benhura, R. Kaul and B. Mattiasson, communicated.
11. Worthington Enzyme Manual, Worthington Biochemical Corpn., Freehold, New Jersey, 1977.
12. P.K. Smith, R.I. Krohn and E.K. Hermanson, Anal. Biochem., 1985, 150, 76.
13. U.K. Laemmli, Nature, 1970, 227, 680.
14. G. Tsamadis, N. Papageorgakopoulou and Y. Clonis, Bioprocess Eng., 1992, 7, 213.
15. P. Hughes, In 'Protein-Dye Interactions : Developments and Applications', M.A. Vijayalakshmi and O. Bertrand eds., Elsevier, London, p. 207, 1989.
16. M. Kamihira, R. Kaul and B. Mattiasson, Biotechnol. Bioeng., 1992, 40, 1381.

Cadmium Biosorption and Toxicity to Laboratory-grown *Xanthomonas campestris*

M. I. Kefala, L. V. Ekateriniadou,[a] A. I. Zouboulis, K. A. Matis, and
D. A. Kyriakidis[a]

LABORATORY OF GENERAL AND INORGANIC CHEMICAL TECHNOLOGY;
[a]LABORATORY OF BIOCHEMISTRY, DEPARTMENT OF CHEMISTRY,
ARISTOTLE UNIVERSITY, GR-540 06 THESSALONIKI, GREECE

1 INTRODUCTION

Removal of metal ions from aqueous solutions by biosorption is a widely studied process and has many applications[1]. Removal of cadmium from industrial wastewaters using biosorption is of particular interest, due to its high toxicity[2,3]. Several microorganisms have been found to uptake effectively relatively high concentrations of cadmium from aqueous solutions[4].

Gram⁻ bacteria immobilize cadmium mainly in the cell envelope; it is suggested that phosphoryl and carboxyl groups of lipopolysaccharides may play a significant role in the cadmium entrapping[5,6]. The difference in the cell wall and membrane structure of Gram⁺ and Gram⁻ strains may resulted in differential affinity of cadmium biosorption to their cell components[6,7].

Bacterial extracellular polymers have been also studied for the capture and the subsequent removal of toxic metal ions from aqueous solutions[8,9]. The most important extracellular polymers are polysaccharides. *Xanthomonas campestris*, a Gram⁻ bacterium, produces xanthan, an extracellular polysaccharide widely used in several biotechnological applications[6].

The biosorption of cadmium by *X. campestris*, as well as by xanthan has been studied. A comparison was attempted also between *X. campestris* and three other strains, namely *E. coli* (Gram⁻, which do not produce extracellular polysaccharide), *B. subtilis* (Gram⁺, which also do not produce extracellular polysaccharide) and *L. mesenteroides* (Gram⁺, which produces dextran, another extracellular polysaccharide).

2 MATERIALS AND METHODS

2.1 Bacterial Strains

Xanthomonas campestris (ATCC 13951), *Escherichia coli* XL1, *Bacillus subtilis* (ATCC 6633) and *Leuconostoc mesenteroides* (ATCC 14935) were used in this study.

2.2 Media and Growth

E. coli and *B. subtilis* were grown in Luria-Bertani broth at 37° C and pH 7.0. *L. mesenteroides* was grown in a medium containing K_2HPO_4 0.5%, $MgSO_4$ 0.01%, yeast extract 0.25% and sucrose 10%, at 26° C and pH 6.7 and *X. campestris* was grown in

Luria-Bertani broth at 28 °C and pH 6.8. Cells were collected at the end of logarithmic phase.

2.3 Toxicity of cadmium

The toxicity of cadmium to the bacterial growth was estimated by the Minimal Inhibition Concentration (M.I.C.) with successive dilutions from 250 to 5 µg/ml Cadmium (prepared from $CdCl_2.2\ 1/2H_2O$).

2.4 Biosorption Assay

Cells after preliminary washing (twice) with 10 mM bis-tris at pH 7.0, were suspended at a final concentration of 0.5 mg cells(dry weight)/ml in a solution containing 25 µg/ml cadmium and 10 mM bis-tris at pH 7.0. After incubation at 28° C for 20 min, the suspension was centrifuged at 5,500 g for 10 min. The remaining cadmium concentration in the supernatant (final) was determined in triplicate samples with atomic absorption spectrophotometry (AAS). Biosorption (uptake) of cadmium was defined as the difference between the initial and the final cadmium concentrations[10,11].

2.5 Cellular Distribution of Cadmium

The distribution of biosorbed cadmium into Gram⁻ bacteria was determined by suspending the cells in distilled water and sonicating them until complete disruption. In the case of Gram⁺ bacteria, cells were disrupted by sonication, overnight incubation with 10 mg/ml lysozyme and resonication until complete disruption. The mixture was centrifuged at 12,500 g for 30 min[12]. The pellet (cell wall and cytoplasmic membrane) and the supernatant (cytoplasmic fraction) were digested with a boiling mixture of $HNO_3:HClO_4$ (4:1), solubilized with 6 N HCl and analysed for cadmium.

2.6 Cadmium Removal by Xanthan

Commercial xanthan gum was dissolved in 10 mM bis-tris (pH 7.0) containing a predefined cadmium concentration and incubated at 28 °C for 20 min under gentle shaking. Xanthan was separated by precipitation using 3 or 4 volumes of isopropanol and centrifugation at 12,500 g for 10 min[13]. The pellet was lyophilised, digested as mentioned above and analysed for cadmium, in order to determine the xanthan loading. All assays were performed in duplicate.

3 RESULTS AND DISCUSSION

Cadmium toxicity has been assessed on four different bacterial strains as is shown in Table 1. The Gram⁻ bacteria *X. campestris* and *E. coli* showed higher resistance to cadmium (M.I.C. 125 µg/ml and 80 µg/ml respectively), than the Gram⁺ strains *B. subtilis* and *L. mesenteroides.*

Table 1 *Minimal Inhibitory Concentration of cadmium to the bacterial growth*

Strain	MIC (µg/ml)
X. campestris	125
E. coli	80
B. subtilis	40
L. mesenteroides	10

Cadmium uptake by *X. campestris* was found to proceed rapidly, with maximal amounts biosorbed during the first 15-20 min (Figure 1). Cadmium biosorption at different media was also examined. The effects of different pH values in either 10 mM bis-tris (for pH 5.0-7.1) or 10 mM tris (for pH 7.1-9.1) on cadmium biosorption were also presented in the same Figure as inset. Best biosorption results were observed at pH values around 8.5. It was finally decided to use 10 mM bis-tris at pH 7.0 in the following experiments, since at that pH value the bacteria remain intact and cadmium precipitation was not observed.

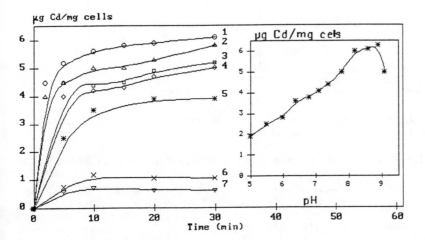

Figure 1. *Kinetics of cadmium uptake by X. campestris cells using initial cadmium concentration 5 µg/ml. Various conditions were used : (1) 0.25 M sucrose at pH 8.3, (2) 10 mM tris - 0.25 M sucrose at pH 8.3, (3) water at pH 8.3, (4) 10 mM tris at pH 8.3, (5) 10 mM bis-tris at pH 7, (6) 50 mM tris - 50 mM NaCl at pH 7, and (7) 10 mM tris - 0.25 M sucrose at pH 7. Inset: Cadmium biosorption at different pH values, at the initial cadmium concentration (5 µg/ml), 10 mM bis-tris at pH 5-7 and 10 mM tris at pH 7-9.1.*

In order to find the maximum uptake of cadmium, *X. campestris* cells, collected from the late exponential growth phase, were suspended in solutions containing increasing cadmium concentrations. It was observed that the maximum uptake was in the range of 20 µg Cd/mg cells, when the initial cadmium concentration in the solution was at least 50 µg/ml (Figure 2).

When cadmium uptake was studied in the four strains, *X. campestris* presented better ability for cadmium biosorption than the other strains (Table 2). Cadmium distribution in the cell components showed again that *X. campestris* immobilized higher cadmium concentration in the cell wall-membrane fraction (35%) than the other three bacteria. Interesting enough, cadmium located in the cytosolic fractions was found to be much more (65-85%) than that biosorbed in the cell wall membrane fraction (12-35%) in all bacteria used.

Measurements of cadmium removal by xanthan, either by increasing the xanthan concentration (Figure 3), or by increasing the cadmium concentration in the solution (Figure 4), showed clearly a noticeably high ability of this polysaccharide to remove cadmium.

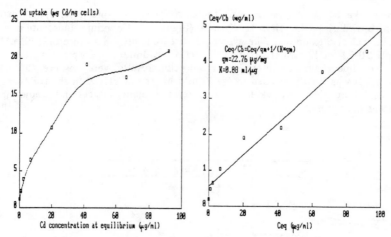

Figure 2. *Equilibrium sorption isotherm of cadmium by Xanthomonas campestris and Langmuir transformation of the isotherm.*

Table 2 *Uptake and cellular distribution of cadmium by different bacterial strains*

Strains	μg Cd/mg cells (dry weight)	% biosorbed Cd in cell wall-membrane	cytosol
X. campestris	13.6	35	65
E. coli	6.6	12	88
B. subtilis	13.5	26.5	73.5
L. mesenteroides	11.3	11.3	77

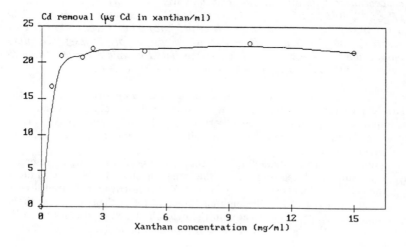

Figure 3 *Cadmium removal using different xanthan concentrations; 25 μg/ml initial Cd concentration.*

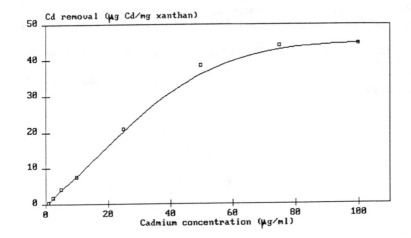

Figure 4 *Cadmium removal using constant xanthan concentration (1 mg/ml), but different initial cadmium concentrations.*

It is generally accepted that living cells can exhibit a wider variety of metal accumulation mechanisms than dead biomass[6,14]. Many microorganisms are able to immobilise relatively high amounts of cadmium in their cells and can also develop several strategies of resistance against toxic cadmium concentrations[2,15].

In this study we have clearly shown that *X.campestris*, a non pathogenic microorganism, and its extracellular polysaccharide xanthan could potentially be used as cadmium biosorbents for the detoxification of aqueous solutions. Since *X. campestris* can use whey as substrate, a byproduct of diary-industries, biomass of cells and xanthan could be easily produced in large scale for an effective removal of cadmium from wastewaters.

4 REFERENCES

1. G. M. Gadd and C. White, *TIBTECH*, 1993, **11**, 353.
2. M. H. Rayner and P. J. Sadler, "Metal-Microbe Interactions", Ed. R. K. Poole and G. M. Gadd, Oxford University Press, 1989, Vol. 26, Chapter 3, p. 39.
3. J. Remacle, "Biosorption of Heavy Metals", Ed. B. Volesky, CRC Press, 1990, Chapter 2.7, p. 293.
4. G. M. Gadd, "Microbial Control of Pollution", Ed. J. C. Fry, G. M. Gadd, R. A. Herbert, C. W. Jones and I. A. Watson-Craik, Cambridge University Press, 1992, p.
5. J. Remacle, "Biosorption of Heavy Metals", Ed. B. Volesky, CRC Press, 1990, Chapter 2.1, p.83.
6. C. L. Brierley, J. A. Brierley and M. S. Davidson, "Metal Ions and Bacteria", Ed. T. J. Beveridge and R. Doyle, John Wiley and Sons, Chichester, 1989, Chapter 12, p. 359.
7. I. C. Hancock, "Trace Metal Removal from Aqueous Solution", Ed. R. Thompson, Royal Society of Chemistry Special Pubs. 61, 1986, p. 25.
8. J. A. Scott and A.M. Karanjkar, *Biotech. Lett*, 1992, **14(8)**, 737.
9. A. B. Norberg and H. Persson, *Biotech. Bioeng*, 1984, **26**, 239.
10. P. Bauda and J. C. Block, *Environ. Techn. Letters*, 1985, **6**, 445.

11.   J. Remacle, I. Muguruza and M. Fransolet, *Wat. Res.*, 1992, **26(7)**, 923.
12.   F. Hambuckers-Berlin and J. Remacle, *Fems Microb. Ecol.*, 1990, **73**, 309.
13.   F. Garcia-Ochoa, J. A. Casas and A. F. Mohedano, *Separ. Sci. & Tech.*, 1993, **28(6)**, 1303.
14.   M. N. Hughes and R. K. Poole, "Metals and Micro-organisms", Chapman and Hall, 1989, Chapter 7, p. 303.
15.   M. N. Hughes and R. K. Poole, "Metals and Micro-organisms", Chapman and Hall, 1989, Chapter 6, p. 252.

Lincomycin Processing with Centrifugal Extractors *CENTREK*

G. I. Kuznetsov, A. A. Pushkov, A. V. Kosogorov, and M. I. Semenov

SCIENTIFIC AND INDUSTRIAL ASSOCIATION NIKIMT, 127410, MOSCOW, RUSSIA

Solvent extraction is widely used for separation and purification of biologically active substances. A number of specific problems are involved in an extraction process. They are connected with the formation of stable emulsions, considerable losses of products due to their decomposition, the use of large volumes of highly volatile expensive and toxic solvents.

The best way to solve these problems seems to be the use of single-stage centrifugal extractors, that are characterized by high separation factor and high specific capacity. A series of centrifugal extractors (*CENTREK*) with capacity from 0.02 to 20 m^3/h was developed. Technical specifications of the extractors are given in table. The extractors are made of stainless steel, titanium, zirconium and fluoroplastics.

Table. *CENTREK Centrifugal Extractors*

Parameter[*]	Type of Extractor								
	EC33	EC80	EC125	EC200	EC250		EC320		EC400
d, mm	33	80	125	200	250		320		400
n, s	50	50	50	50	50	25	50	25	25
Q, m^3/h	0.025	0.40	1.50	5.0	10.0	5.0	20.0	10.0	20.0
V_s, l	0.020	0.33	1.25	4.2	8.4	8.4	16.8	16.8	36.0
V_m, l	0.025	0.40	1.50	5.0	10.0	5.0	20.0	10.0	20.0
P, kW	0.040	0.25	0.70	2.2	7.5	1.5	10.0	3.0	10.0
l, mm	90	250	310	400	490	490	570	570	880
b, mm	105	250	310	400	490	490	570	570	880
h, mm	320	660	880	920	1100	900	1600	1500	1900
m, kg	6	42	105	230	340	260	980	914	1100

* d - rotor diameter; n - rotating speed; Q - capacity for the solutions: 1.1M TBP in kerosene - 2M HNO_3, when the ratio of flows of organic to aqueous = 1, emulsion type "oil in aqueous" and phase entrainment up to 0.05% vol.; V_s, V_m - holdup volumes of separating and mixing chambers; P - electrodrive power; l, b, h - length, width, height; m - mass.

The extractor (Figure 1) operates as follows: feed solutions are fed to a mixing chamber 1 where they are mixed by stirrer 2. The emulsion formed is transferred by screw conveyer 3 to rotor 4 where it is separated under the action of centrifugal force in separating chamber 5. Separated liquids are directed into circular collectors 6 of a fixed body 7 and then gravitate out of the extractor. The rotor unit together with electric gear 8 is easily removed and replaced by a new one, in case of need.

In distinction from differential centrifugal extractors, widely used in pharmaceutical industry, the simpler design of the *CENTREK* extractor permits to control of hydrodynamic and mass transfer. In particular, depending on process requirements, it is possible to change mixing intensity (by changing stirrer speed or stirrer diameter) and phase contact time (by changing mixing chamber volume).

In lincomycin extraction process (Figure 2) the extractors EC 125 (stage 7) and EC 250 (stages 1-6) are used. Intensive mixing provides a mass transfer efficiency of 93 - 96 % for each extractor. The extractor capacity is decreased considerably in comparison with the capacity in the table, but the capacity can be increased if one operates with higher phase entrainments.

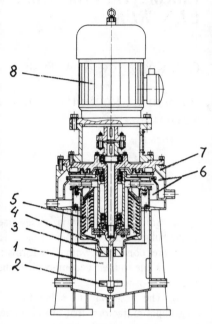

Figure 1. Centrifugal extractor *CENTREK*

The fraction of valuable component not extracted is:

$$1 - \varepsilon = \frac{\lambda - 1}{\lambda^{n+1} - 1}, \tag{1}$$

where n - number of stages, $\lambda = m \cdot R$ - extraction coefficient, assumed to be constant (m - distribution coefficient, R - organic:aqueous flow ratio).

Volumetric flows of organic and aqueous phase are:

$$Q_o = Q_o^c + Q_{aq/o}^s + Q_{aq/o}^e, \tag{2}$$

$$Q_{aq} = Q_{aq}^c + Q_{o/aq}^s + Q_{o/aq}^e, \tag{3}$$

where Q_o^c, Q_{aq}^c - organic and aqueous flows (without phase solubility and entrainment); $Q_{aq/o}^s, Q_{o/aq}^s$ - flows of aqueous phase dissolved in the organic phase and organic phase in the aqueous phase; $Q_{aq/o}^e, Q_{o/aq}^e$ - flows of aqueous phase entrained by the organic phase and organic phase by the aqueous phase.

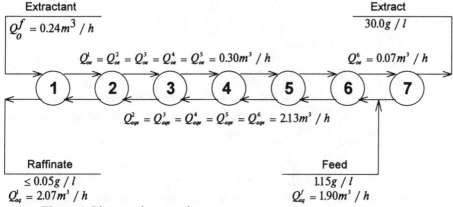

Figure 2. Lincomycin extraction process.

Assuming that phase entrainment is constant one can get for liquids which are soluble in each other

$$\lambda = \frac{m \cdot Q_o^c + Q_{aq/o}^e}{Q_{aq}^c + m \cdot Q_{o/aq}^e},$$ (4)

Distribution coefficient for lincomycin is $m=17$. The results for the lincomycin extraction process calculation taking into account the volumetric fraction of the entrained phases are shown on Figure 3.

On the basis of these results the operating conditions were established: entrainment of organic phase into the aqueous phase 0.02 and the entrainment of aqueous phase into the organic phase up to 0.2. This allows an increase in capacity of feed solutions from 1.5 to 2.0 m³/h without deterioration of technological performance. The first stage (Fig.2) was operated without entrainment of organic phase into raffinate and stage 7 was used to purify the extract entrained from the stage 6 aqueous phase. The phase entrainment is controlled by changing of interface position in the separating chamber[1,2].

In lincomycin processing partly dissolved liquids in each other are used: butylalcohol (butanol) - organic phase and lincomycin solution - aqueous phase. Butanol solubility in water

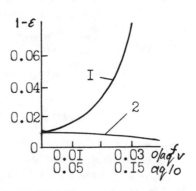

Figure 3. Calculated dependence of the fraction of lincomycin not extracted $(1-\varepsilon)$ from volumetric fraction of organic phase entrained in aqueous one $f_V^{o/aq}$ (1) and aqueous phase in organic one $f_V^{aq/o}$ (2)

$$k_{o/aq} = \frac{Q_{o/aq}^s}{Q_{o/aq}^s + Q_{aq}^c} = \frac{Q_{o/aq}^s}{Q_{aqe}} \approx 0.091 \tag{5}$$

water solubility in butanol

$$k_{aq/o} = \frac{Q_{aq/o}^s}{Q_{aq/o}^s + Q_o^c} = \frac{Q_{aq/o}^s}{Q_{oe}} \approx 0.172 \tag{6}$$

where Q_{aqe}, Q_{oe} - aqueous and organic flows without phase entrainment.

The mutual solubility causes considerable change of volumetric flows. From the balance of butanol and water flows it follows:

for extreme stages

$$Q_{aq}^f = Q_{aqe}^1 \cdot (1 - k_{o/aq}) + Q_{oe}^6 \cdot k_{aq/o}, \tag{7}$$

$$Q_o^f = Q_{oe}^6 \cdot (1 - k_{aq/o}) + Q_{aqe}^1 \cdot k_{o/aq}, \tag{8}$$

where Q_{aq}^f, Q_o^f - feed flows of aqueous and organic phase, respectively. Thus the flows from cascade are:

$$Q_{aqe}^1 = \frac{Q_{aq}^f \cdot (1 - k_{aq/o}) - Q_o^f \cdot k_{aq/o}}{1 - k_{aq/o} - k_{o/aq}}, \tag{9}$$

$$Q_{oe}^6 = \frac{Q_o^f \cdot (1 - k_{o/aq}) - Q_{aq}^f \cdot k_{o/aq}}{1 - k_{aq/o} - k_{o/aq}}, \tag{10}$$

One can calculate intermediate flows from a balance on the butanol and aqueous flows for the end stages, e.g. for first one:

$$Q_o^f + Q_{aqe}^2 \cdot k_{o/aq} = Q_{oe}^1 \cdot (1 - k_{aq/o}) + Q_{aqe}^1 \cdot k_{o/aq}, \tag{11}$$

$$Q_{aq}^e \cdot (1 - k_{o/aq}) = Q_{aqe}^1 \cdot (1 - k_{o/aq}) + Q_{oe}^1 \cdot k_{aq/o}. \tag{12}$$

Taking into account (9):

$$Q_{aqe}^2 = Q_{aqe}^{l+1} = \frac{Q_{aq}^f \cdot (1 - k_{aq/o})}{1 - k_{aq/o} - k_{o/aq}}, \tag{13}$$

$$Q_{oe}^1 = Q_{oe}^l = \frac{Q_o^f \cdot (1 - k_{o/aq})}{1 - k_{aq/o} - k_{o/aq}}, \tag{14}$$

where i = 1-5 - stage number.

Figure 2 shows that due to mutual solubility the organic:aqueous flow ratio $R = Q_{oe} / Q_{aqe}$ changes from 0.146 in stage 1 to 0.14 from stages 2 to 5 finally up to 0.033 in stage 6 when starting with a feed solution ratio of 0.126.

Knowing the distribution coefficient (m = 17) and volumetric flows calculated from equations (9,10,13,14) and using equations for the mass balance of lincomycin it is possible (using the method of successive approximations) to calculate its concentration in different stages of a cascade. Figure 4 shows the calculation results for the optimal organic:aqueous feed ratio $R_f = Q_o^f / Q_{aq}^f$, under these ratios the lincomycin concentration in the extract and raffinate are found within the given limits when its concentration in feed aqueous solution changes.

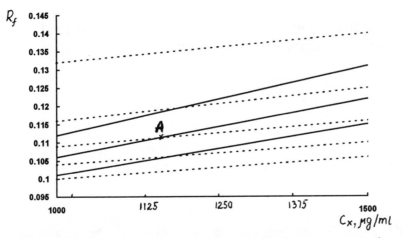

Figure 4. Optimal calculated parameters of centrifugal extractors operation.

Setting lincomycin concentrations in raffinate (µg/ml) 1-10, 2-20, 3-30, 4-40, 5-50 and one in extract - 6-25000, 7-30000, 8-35000.

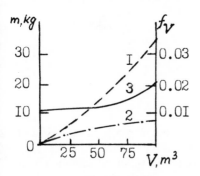

Figure 5. Dependance of solids mass (m) passed through the extractors (1), solids mass (m) accumulated in extractors (2) and fraction ($f_V^{o/aq}$) of organic phase entrained into aqueous one (3) from volume (V) feed solution passed through the extractors.

When the lincomycin concentration in the feed aqueous solution is 1150 µg/ml and its concentration in the extract is 30000 µg/ml (point A on Fig.4) the calculated concentration in raffinate will be ≈30 µg/ml and the feed flow ratio - R_f =0.11. Really, to fulfil these conditions the process was carried out under R_f =0.126 (see Fig.2). The lack of correlation between experimental and calculated data can be explained by the fact that in the model the entrained phases are not taken into account.

A specific feature of antibiotics processing is the presence of a solid phase in the feed aqueous solutions, that affect centrifugal extractor operations. In lincomycin processing the solution contains up to 5 g/l solids. A large part of the solid phase left the extractors with raffinate, but some fraction is accumulated in rotors being the reason for increased entrainments (Figure 5).

References

1. G. I. Kuznetsov, A. V. Kosogorov, A. A. Pushkov and L. I. Shklyar, ISEC'88, Conf. Papers, Moscow, 1988, **2**, 245.
2. G. I. Kuznetsov, A. A. Pushkov and S. M. Belyakov, Atomic Energy, 1986, **61**, (1), 23 (*in Russian*).

Building an Expert System to Assist the Rational Selection of Large Scale Protein Purification Processes

E. W. Leser* and J. A. Asenjo

BIOTECHNOLOGY AND BIOCHEMICAL ENGINEERING GROUP,
DEPARTMENT OF FOOD SCIENCE AND TECHNOLOGY, THE UNIVERSITY OF
READING, PO BOX 226, READING RG6 2AP, UK

This paper discusses the building and use of a knowledge based expert system for selection of protein recovery, separation and purification processes . It describes a rational approach that uses fundamental databases of characteristics of protein molecules to simplify the complex problem of choosing high resolution separation methods for multicomponent mixtures. This rationale is based on equations that describe heuristics and the behaviour of proteins in chromatographic separations. This paper discusses these equations and shows the interaction of the expert system with databases.

1 INTRODUCTION

The use of Expert Systems for solving Chemical Engineering problems is very well established[1] and their application to biotechnology problems is becoming more frequent[2]. Downstream processing for proteins, mainly due to the changes brought with the use of recombinant technology, contains many unsolved problems. The development of systems to deal with the question of protein purification started some years ago[3,4,5]. The use of commercial software proved to be suitable too[6,7,8]. The creation of an expert system to help in the design of large scale production of proteins has been well documented[9,10].

The system described here has been implemented using Nexpert Object, a commercial shell from Neuron Data Inc. The main target is to build an integrated system of objects and rules that leads to the same solutions that an expert, dealing with this kind of problem, would propose. The knowledge building must, therefore, be constructed using small bits of information, linking expert knowledge elements in a consistent structure to lead the reasoning through the search space in a logical direction. The analysis of the experts' heuristics and their implementation as a knowledge base is the bottleneck of the whole development. Heuristics may have some variation among experts and the knowledge engineer may have to handle either incomplete or ambiguous information. The proposed solution includes all unit processes from the end of fermentation to obtention of the final product. The present prototype has its knowledge

---

* on leave from Fundação Oswaldo Cruz, Brazil

base almost finished and part of its development is based on experimental data[11,12].

The first part of the process is the group of operations which obtain the product in solution starting from the fermented broth. It may include preconditioning, cell separation, cell disruption, cell debris removal and concentration. The second part includes the high resolution purification processes used to isolate and purify the product to the predefined levels of quality. The prediction of the sequence of high resolution purification steps represents the biggest challenge to the biochemical engineer. There are no models, nor sufficient data on contaminants found in the commonly used expression systems, and tradition in process scale-up is, basically, to repeat and amplify the procedures developed at the research laboratory scale.

In its present form the knowledge base contains around 200 rules and its construction was based on previous work[8,13,14], the authors' expertise and advice from industrial experts. The consultation consists of a dialogue between the programme and the user and the result is a set of suggestions with a sequence of operations and processes to achieve the desired level of purity for the product. This solution will be written and included in the programme's memory.

2 RECOVERY AND SEPARATION

The system starts asking questions to define the product: name and expression system. It will check in its memory whether there are any previous solutions for its downstream processing, and the user can decide whether he accepts that solution or continues with the search for an alternative one. The next steps ask for product utilization, stability (thermal and pH) and localization. The system also considers the pathogenicity of the expression system. This will subsequently define the requirements for cell separation. The choice between centrifugation and filtration depends on many parameters[15]. The next step is the analysis of cell disruption where, usually, only two options are considered: high pressure homogenizer or bead mill. However, it also leaves to the user the possibility of choosing non-mechanical methods: enzymatic or treatment with solvents, whenever adequate. Proteases as well as nucleic acids and cell debris need to be removed after the disruption of the cell wall. If the intracellular product is manufactured in *E. coli*, high expression of heterologous proteins will usually result in their accumulation as insoluble inclusion bodies[16]. If the intracellular product is manufactured in yeast, the protein is often present in homogeneously particulate form, typically 30-60 nm particles as in virus-like particles (VLPs)[17]. It is necessary to process these product aggregates into the native protein by solubilizing and refolding. The processing of intracellular particulate recombinant proteins is an important aspect of downstream processing and satisfactory methods exist for large scale separation, solubilization, and refolding of the particulate proteins and the system presents its suggestions to the user. The product is now in solution, ready to go on to the high resolution purification steps. If the concentration of the product needs to be increased (and this is decided by the user), a dewatering step is necessary to reduce the total volume and, therefore, the scale of purification stages. Preconditioning is the elimination of any suspended material that could cause fouling of the separation columns and may also be required[18].

3 ISOLATION AND PURIFICATION

Based on the previous definition[19] of a 'selection separation coefficient' (SSC) that can be used to characterize the ability of the separation operation to separate two or more proteins, it is possible to state that:

$$SSC = DF \cdot \eta \cdot \theta \qquad [1]$$

where   DF....deviation factor
                                       η......process efficiency
                                       θ......concentration factor

Initially, for the sake of simplification, a linear relationship among SSC and each one of the other variables has been assumed, as shown in **Eq. [1]**. A more accurate correlation can be used to show the effect of each variable on the SSC and this is being investigated in more detail.

It should be possible to use information on thermodynamic properties of protein product and main contaminant proteins to predict the performance of a particular separation. To do this it is necessary to find the differences in physicochemical properties that determine the separation behaviour of the different proteins in the system. The calculation of the deviation factor (DF), which allows for the differences between the product and each contaminant for a particular physicochemical or thermodynamic property (e.g. molecular weight, hydrophobicity or charge density at a particular pH) follows:

$$DF = \frac{||Protein\ Value| - |Contaminant\ Value||}{Max(|Protein\ Value| \wedge |Contaminant\ Value|)} \qquad [2]$$

The efficiency factor for the separation process in exploiting this difference in a specific property has to be included in the evaluation. Based on previous hypothetical definitions[8,13,14], confirmed by experimental results[11], the relative values used are:

Chromatographic Process	Efficiency (η)
Ion Exchange	0.70
Hydrophobic Interaction	0.35
Size Exclusion	0.20

The concentration factor will affect the selection criteria, since the contaminants in higher concentrations have to be removed first and this is a common rule-of-thumb that guides an expert.

$$\theta = \frac{Concentration\ of\ Contaminant\ Protein}{Total\ Concentration\ of\ Contaminant\ Proteins} \qquad [3]$$

To find the best sequence of high resolution purification processes, the system uses data on the main contaminants of the expression system employed in the production and compares them with those of the product. It calculates the SSC value for each of the selected properties and detects the maximum value. This maximum corresponds to a certain property and a chromatographic process will be associated with it. For example, if the maximum corresponds to molecular weight, the chosen process will be size exclusion chromatography. If it corresponds to charge density at a particular pH, the system will point to ion-exchange chromatography under the chosen conditions.

A procedure was developed to define the reduction of contaminants depending on their SSC value in the separation step. It has been assumed that the concentration of **each** contaminant will be reduced according to the following expression:

$$C_{n(i)} = C_{n(i-1)}(1 - \frac{\Delta}{100} \frac{DF_n}{DF_{max}}) \qquad [4]$$

It means that the concentration (C) of contaminant **n** after the stage **i-1** will be reduced proportionally to the value of the corresponding deviation factor (DF_n). The contaminant with the maximum value of SSC will be reduced to $\Delta\%$ of its initial concentration. The value of Δ was arbitrarily set equal to 99.

A particular situation will refer to affinity chromatography because the choice of this method is usually predefined by the user. The proposed solution is the system asking the user whether this method will be employed and if the answer is **yes** the system will reduce the concentration of all contaminants to 10% (or an experimentally measured value) of its initial value unless this technique is evaluated experimentally.

The programme continues the selection of processes and reduction of contaminants concentration until it reaches the desired level of purity for the product. If the product use is therapeutic, other steps are necessary to remove traces of contaminants according to specific regulations. The expert system will suggest the adding of a final polishing stage. If dimers of the product are present, polishing would correspond to size exclusion chromatography.

4 DATA USED FOR PROCESS SELECTION

The determination of physicochemical properties of contaminants, in main culture vehicles (*E. coli*, hybridoma and Chinese hamster ovary cell - CHO), in a systematic generation of data banks for process selection is presently being carried out[20,21]. It has also been recently demonstrated that charge density (net charge/molecular weight), determined over a range of pH values using an electrophoretic technique (titration curve), is the major factor affecting protein behaviour in ion exchange chromatography. With minor deviations this parameter can be used to predict the chromatographic behaviour of a protein[11].

To use the deviation factor, **Eq.[2]** means to exploit the differences between the charge densities of the proteins being compared, namely the product and each of the main contaminants. To adapt this logic to real titration curves, for cation-exchange chromatography values corresponding to pH values smaller than the pI will be equal to the real values of charge density in the curve. For those equal or bigger than the pI, the charge densities will be assumed to have a very small positive value. If they were equal to zero, this could lead to a mathematical indetermination. For anion- exchange chromatography, charge densities corresponding to pH values smaller or equal to the pI will be assumed to have very small positive values; and for pH values bigger than the pI, the charge density values used will be equal to the absolute value of those indicated by the curve.

The consequence of this transformation is that the system, when the chosen process is ion-exchange chromatography, will suggest whether to use an anionic or cationic matrix and the actual pH at which to carry out the separation. The calculated values of DF will be those given by **Eq. [2]**, which were used to predict separation performance. **Table 1** shows an extract of database, used to develop this part of the knowledge base, built from experimental data[11]. **Table 2** shows how the system acts on the mixture of proteins after each separation step.

TABLE 1: *Molecular weight, hydrophobicity and charge density values along a pH scale from 4.0 to 8.5 (extract) for four model proteins. Adapted from Watanabe[11]*

			Charge density (cm/sec.volt.Da)				...		
			pH 4.0	pH 4.0	pH 4.5	pH 4.5		pH 8.5	pH 8.5
protein	molwt (Da)	hydroph (*)	cation	anion	cation	anion		cation	anion
cont1	43,800	1.055	9.0	0.0001	4.0	0.0001		0.0001	5.0
cont2	78,500	1.085	1.6	0.0001	1.3	0.0001		0.0001	2.0
cont3	24,500	1.370	5.8	0.0001	1.0	0.0001		0.0001	10.0
product	71,000	0.930	3.8	0.0001	2.0	0.0001		0.0001	4.0

cont(i).... contaminant (i)

(*) Hydrophobicity units expressed as: $2-((\% \text{ Buffer B}?100) \times [(NH_4)_2SO_4])$

$[(NH_4)_2SO_4] = 1.5M$

2 was considered the maximum concentration (M) of $(NH_4)_2SO_4$ to be used.

TABLE 2: *Concentration (in g/litre and %) of four model proteins at the initial stage and after each separation step, as a result of the Expert System process selection rationale*

	Initial State		1st step	2nd step	3rd step	After 4th step	
protein	Conc (g/l)	Conc (%)	Conc (g/l)	Conc (g/l)	Conc (g/l)	Conc (g/l)	Conc (%)
cont1	10.0	39.2	0.1	0.09	0.09	0	0
cont2	5.0	19.6	4.25	0.04	0.04	0.03	0.3
cont3	0.5	2.0	0.5	0.5	0	0	0
product	10.0	39.2	10.0	10.0	10.0	10.0	99.7

In this example, the desired level of product purity was set to 99% and the system calculates four separation stages. However, as a selected process appears more than once, the final indication suggests only three processes, as shown in **Figure 1**.

At present a data bank of contaminants present in *E. coli*[21] is being implemented into the system.

5 CONCLUSIONS

The complex problem of selection of an optimal sequence of processes to purify proteins from multicomponent mixtures can be simplified. The utilization of a rational approach based on the exploitation of fundamental physicochemical and thermodynamic properties of protein molecules should lead to coherent and optimized decisions. To help accomplish this task, the use of contemporary computer techniques, supported by symbolic computation proves to be an extremely helpful tool.

To achieve a concentration equal to 99% three steps are needed:
1) cation exchange chromatography at pH 5.5
2) cation exchange chromatography at pH 5.0
3) anion exchange chromatography at pH 5.0

Figure 1   *Final text displayed to the user at the end of a consultation for suggesting a high resolution purification sequence by the expert system*

These techniques are being developed with the aim of optimizing solutions for large scale production applications and also for finding optimal and simplified separation schemes in the research laboratory. The authors used simplified models to describe the behaviour of high resolution separation processes to define how the contaminants are eliminated after each chromatographic step. More elaborate versions of these models are presently being investigated. This will result in a more accurate and scientifically sound basis to select separation and purification operations in a rational manner with a minimum number of steps.

6 ACKNOWLEDGEMENTS

Financial assistance by Conselho Nacional para o Desenvolvimento Científico e Tecnológico (CNPq), Brazil, European Science Foundation (ESF) Process Integration Programme, as well as the EBEN Network of the EC for support of E.W. Leser is gratefully acknowledged.

7 REFERENCES

1. G. Stephanopoulos and M. Mavrovouniotis, Comp.Chem.Eng., 1988, 12, V.
2. K.P. Clapp and G.J. Ruell, BioPharm., 1991, February, 23.
3. S. Wacks,'Design of protein separations and downstream processes in biotechnology' M.Sc. thesis, Columbia University, New York, USA, 39pp.
4. D.I.C. Wang, 'Separations for Biotechnology', M.S. Verral and M.J. Hudson (eds.), Ellis Horwood Ltd., Chichester, U.K., 1987, p.17.
5. S. Wheelwright and J.A. Asenjo, 'Downstream processing, recovery and purification of proteins. Handbook of principles and practice', K.H. Kroner and N. Papamicahael. Carl Hanser, München. (In press).

6.   I.J. Purves, 'The testing and development of an expert system for selection and synthesis of protein purification processes'. M.Sc. thesis, City University, London, 1990, 202 pp.

7.   J.A. Asenjo, 'Separations for biotechnology II', D.L. Pyle (ed.), Elsevier, New York, p. 11.

8.   J.A. Asenjo and F. Maugeri, 'Frontiers in Bioprocessing II' P. Todd, S.K. Sikdar and M. Bier (eds.) American Chemical Society, Washington, 1992, p. 358.

9.   E.W. Leser and J.A. Asenjo, J.Chrom., 1992, 584, 43.

10. E.W. Leser and J.A. Asenjo, Mem.Inst.Oswaldo Cruz, in press.

11. E. Watanabe, S. Tsoka and J.A. Asenjo, Ann.NY.Acad. Sci., 1994, 721, 348.

12. P. Woolston, M.J. Wharam, Kearns, M.J. and J.A. Asenjo, 'A database for the selection of large scale protein purification processes'. Presented at the Recovery of Biological Products VI, Interlaken, Switzerland, 1992.

13. J.A. Asenjo, L. Herrera and B. Byrne, J.Biotechnol., 1989, 11, 275.

14. J.A. Asenjo, J. Parrado and B.A. Andrews, Ann.NY.Acad Sci., 1991, 646, 334.

15. S.-M. Lee, J.Biotechnol., 1987, 11, 103.

16. J.F. Kane and D.L. Hartley, Trends in Biotechnol., 1988,6, 95.

17. F. Müller, D. Brühl, K. Freidel, K.V. Kowallik and M. Ciriacy, Mol. andGen. Genetic, 1989, 207, 421.

18. J.A. Asenjo and I. Patrick, 'Protein Purification Applications. A Practical Approach', E.L.V. Harris and S. Angal (eds.) Oxford University Press, Oxford, 1990, p.1.

19. J.A. Asenjo, 'Separation Processes in Biotechnology', J.A. Asenjo (ed.), Marcel Dekker, New York, 1990, p.11.

20. R. Turner, B.S. Baines and J.A. Asenjo, 'Physico-chemical database developments for baculovirus-produced proteins. The rational design of large scale protein purification', Separations for Biotechnology III, University of Reading, UK, 1994.

21. P. Woolston, personal communication.

Strategies in the Scale-up of the Anion-exchange Chromatography of Hen Egg-white Proteins

Peter R. Levison, Stephen E. Badger, David W. Toome, and Russell M. H. Jones

WHATMAN INTERNATIONAL LTD., SPRINGFIELD MILL, MAIDSTONE, KENT ME14 2LE, UK

1 INTRODUCTION

Ion-exchange chromatography is a widely used technique in the downstream processing of commercially important biopolymers. Ion-exchange chromatography involves the exchange of solute ions of like charge from a solid support bearing the opposite charge[1]. Low pressure ion-exchange media are traditionally based on polysaccharide supports including cellulose, agarose and dextran[1-3]. Dependent on the requirements for the ion-exchange step and process economics the separation can be carried out using either batch stirred tanks or column techniques. The features and benefits of these techniques have been reported elsewhere for the separation of hen egg-white proteins on Whatman DE52[4] and DE92[5] anion-exchange celluloses. Furthermore the chromatographic process can be used as a positive step whereby the target adsorbs to the medium and is selectively desorbed, for example the purification of restriction enzymes from bacterial cell lysate using Whatman P11[6] or as a negative step whereby the contaminants adsorb to the medium and the target passes through the medium as an unretained fraction, for example the purification of immunoglobulin G from goat serum? At first glance a positive step may appear to be the obvious route due to the selectivity offered by ion-exchange adsorption/desorption. However a negative step offers advantages particularly when the feedstock is relatively pure, since the minor contaminants will bind with valuable protein, binding capacity is not being used up on the major components ie the target. Furthermore while the unretained target peak does not undergo a concentration process, it remains in the same mobile phase throughout, rather than being subjected to a change in ionic strength and/or pH for desorption which may necessitate a diafiltration step in order to present the target in a suitable mobile phase for subsequent downstream processes.

Column chromatography is routinely used in large-scale ion-exchange processes and columns are available using either axial or radial flow modes of liquid flow. These two modes of operation have been compared previously[8,9]. In order for a column chromatographic process to be economically viable, it is important to optimize throughput at the required degree of product purity. Throughput in the column process is affected by capacity, resolution and flow rate all of which are interrelated. Column flow rate affects the overall process time and in circumstances where the ion-exchange step is the rate-limiting factor in the total downstream process then flow rate becomes a key driving force in the throughput calculation. Cellulose is a polysaccharide with a macroporous structure which, in its non-regenerated form, offers very high protein capacity and rapid adsorption/desorption kinetics[1], features which influence the first two throughput factors. Flow rate has been a limitation of traditional microgranular ion-exchange celluloses and in order to address this limitation, Whatman International Ltd have adopted a custom manufacturing strategy[10,11] where one opportunity for product manipulation is that of flow rate. Generally there is a trade-off between flow rate and protein capacity, and the process flow rate requirements should influence selection of the appropriate chromatographic medium. In this study we compare the process-scale chromatography of hen egg-white proteins on DE52, DEAE-cellulose/D529 and Whatman Express-Ion™ Exchanger D at flow rates of 30, 75 and 150cm/h respectively, where process flow requirements would influence which of these three media types would be used.

2 EXPERIMENTAL

Egg-White Feedback Preparation

The whites of 600 size 2 eggs (Barradale Farms, Headcorn, England) were separated and diluted with 0.025M-Tris/HCl buffer, pH 7.5 (Merck, Poole, England) to give a 14% (v/v) suspension. The suspension was treated with a total of 22kg of the weak fibrous anion-exchange cellulose, Cell Debris Remover (CDR) (Whatman International Ltd, Maidstone, England) in a batch mode to give a clear solution (200l) containing 8-9 mg/ml protein.

Dynamic Capacity Determination

Egg-white feedstock was applied to columns (15.5 cm x 1.5 cm i.d.) of DE52, DEAE-cellulose/D529 and Express-Ion D (Whatman International Ltd., Maidstone, England) previously equilibrated with 0.025M-Tris/HCl buffer, pH 7.5 at flow rates of 2, 2 and 4 ml/min respectively until the absorbance of the column effluent at 280nm was similar to that of the feedstock. Non-bound material was removed by washing with 0.025M-Tris/HCl buffer, pH 7.5 (100 ml) and bound material was eluted with 0.025M-Tris/HCl buffer, pH 7.5 containing 0.5M-NaCl (200 ml).

Process-Scale Chromatography

DE52 (25 kg) was equilibrated with 0.025M-Tris/HCl buffer, pH 7.5 to give a final slurry concentration of 30% (w/v). The slurry was pump-packed into a PREP-25 column (46 cm x 45 cm i.d., Whatman International Ltd, Maidstone, England) at a pressure of ~10 psi. The egg-white feedstock (200 l) was loaded onto the column and non-bound material removed by washing with 0.025M-Tris/HCl buffer, pH 7.5 (100 l). Bound material was eluted using a linear gradient of 0-0.5 M-NaCl in 0.025M-Tris/HCl buffer, pH 7.5 (200 l). Flow rate was maintained at 0.8 l/min (30 cm/h) throughout. The study was repeated using 25 kg of DEAE-cellulose/D529 and Express-Ion D operated at flow rates of 2 l/min (75 cm/h) and 4 l/min (150 cm/h) respectively.

Table 1 Typical Properties of DE52, DEAE-Cellulose/D529 and Express-Ion D

Property	Value		
	DE52	DEAE-Cellulose/D529	Express-Ion D
Moisture content (%)	73.8	70.2	66.0
Regains Free base form	2.49	2.18	1.88
(g/dry g) Hydrochloride form	2.99	2.30	1.92
Small-ion capacity (mequiv/dry g)	1.03	1.08	0.98
Column packing density (dry g/ml)	0.16	0.21	0.21
Protein capacity+ (mg/dry g)	745	510	290
Protein capacity+ (mg/ml)	118	105	61
Flow rate 50cm H_2O/cm	163	367	425
(cm/h) 75cm H_2O/cm	234	523	614

+ BSA, 0.01M-sodium phosphate buffer, pH 8.5.

3 RESULTS AND DISCUSSION

In order to address the flow rate limitations of our traditional ion-exchangers, we developed the products DEAE-cellulose/D529 and Express-Ion D Typical properties of these media are summarised in Table 1. The key features of the products are improved flow rate associated with an apparent reduction in protein binding capacity.

Protein capacity is a key parameter influencing throughput and by its very nature is entirely application dependent, and will be influenced by several factors including pH, conductivity, pI, molecular mass, nature of the contaminants and temperature. In order to determine the dynamic protein binding capacity of an ion-exchanger, a breakthrough curve would be obtained under conditions representative of those to be used for preparative chromatography. The breakthrough curve for DEAE-cellulose/D529 using the egg-white feedstock (8.8 mg/ml) is show in Figure 1. Similar curves for DE52 and Express-Ion D have been reported elsewhere[12,13]. Following salt-elution, the bound material was shown to be ovalbumin by FPLC analysis[13] and the maximum dynamic protein binding capacity for this system can be determined (Table 2). The data clearly indicate a reduction in protein capacity across the series of anion-exchange celluloses. In each of these experiments only ~50% (w/w) of the total ovalbumin present in the feedstock bound to the ion-exchangers. This clearly is an inefficient process strategy but gives an indication as to the ultimate scale of the chromatographic step.

Table 2 Capacity of DE52, DEAE-Cellulose/D529 and Express-Ion D following 100% breakthrough loadings of 10% (v/v) hen egg-white in 0.025M-Tris/HCl buffer, pH 7.5.

Medium	Protein Capacity (mg/ml column volume)
DE52	158
DEAE-Cellulose/D529	96
Express-Ion D	84

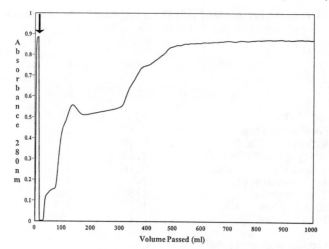

Figure 1 Absorbance profile of column eluate during a saturation loading of DEAE-cellulose/D529 with 8.8 mg/ml hen egg-white proteins using 0.025M-Tris/HCl buffer, pH 7.5. The absorbance of the feedstock is identified by the arrow.

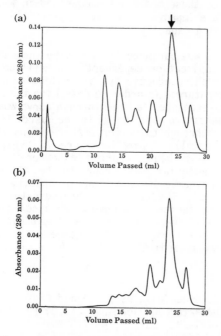

Figure 2 FPLC analysis using Mono Q of (a) egg-white feedstock and (b) salt-eluted material following process scale chromatography on DEAE-cellulose/D529. The ovalbumin peak is identified by the arrow.

Table 3 Protein Capacities of DE52, DEAE-Cellulose/D529 and Express-Ion D following process scale chromatography of hen egg-white in a 25l column

Medium	Feedstock total protein (g)	Total protein adsorbed (g)	Maximum theoretical capacity (g)	% of maximum capacity utilized
DE52	1703	1321	4013	32.9
DEAE-Cellulose/D529	1853	1466	2438	60.1
Express-Ion D	1856	1413	2134	66.2

In order to compare the three media we carried out a process-scale study where 25l columns of each medium were loaded submaximally with the egg-white feedstock. The results for this study are summarised in Table 3. The data demonstrate that under similar chromatographic conditions protein adsorption is very efficient for each medium at each flow rate used.

For each medium tested the binding efficiency for the adsorption process (percentage of available protein binding to the medium) was approximately 77% (w/w). Hen egg-white contains 63.8% (w/w) ovalbumin[14], the most acidic component of the mixture. The binding efficiencies of 77% suggest that 100% binding of applied ovalbumin should have occurred. This was borne out by FPLC analysis[13] of chromatographic fractions. These data are represented in Figure 2 for DEAE-cellulose/D529 and have been reported elsewhere for DE52[15] and Express-Ion D.

The diffusion kinetics of each medium facilitate the efficiency of protein adsorption and clearly the flow rate limitations of DE52 can be alleviated with the use of products such as DEAE-cellulose/D529 or Express-Ion D. It is likely that DE52 would bind significantly more protein than the other grades but at the expense of flow rate, which may, for the reasons discussed above, be critical. For the purposes of this study, flow rate variation within grades was not deemed relevant since the various grades cover the typical operating flow rate range for process-scale ion-exchange chromatography of proteins.

The data reported in this study demonstrate that careful media selection is necessary as part of the scale-up strategy and capacity/flow-rate considerations must be examined, at an early stage to enable scale-up to proceed smoothly and give the desired productivity. Given that column bed height remains constant and packing densities of the bed are maintained, it is our experience that scale-up effects are minimal provided that linear flow rate is maintained throughout.

REFERENCES

1. P.R. Levison, "Cellulosics: Materials for Selective Separations and Other Technologies", J.F. Kennedy, G.O. Phillips and P.A. Williams, eds., Ellis Horwood, Chichester, 1993, p.25.

2 E.F. Rossomando, Methods Enzymol., 190, 182, 309.

3 D. Friefelder, "Physical Biochemistry", Freeman, San Francisco, 2nd edn., 1982, p.249.

4   P.R. Levison, "Preparative and Production Scale Chromatography", G. Ganetsos and P.E. Barker, eds., Marcel Dekker, New York, 1993, p. 617.

5   P.R. Levison, S.E. Badger, D.W. Toome, M.L. Koscielny, L. Lane and E.T. Butts, J. Chromatogr., 1992, 590, 49.

6   J.M. Ward, L.J. Wallace, D Cowan, P Shadbolt and P.R. Levison, Anal. Chim. Acta., 1991, 249, 195.

7   P.R. Levison, M.L. Koscielny and E.T. Butts, Bioseparation, 1990, 1, 59.

8   L. Lane, M.L. Koscielny and E.T. Butts, Bioseparation, 1990, 1, 141.

9   P.R. Levison, S.E. Badger, D.W. Toome, E.T. Butts, M.L. Koscielny and L. Lane, "Upstream and Downstream Processing in Biotechnology III", A. Huyghebaert and E. Vandamme, eds., Royal Flemish Society of Engineers, Antwerp, 1991, p.3.21.

10   P.R. Levison, "Biotechnology International 1992", Century Press, London, 1992, p.211.

11   P.R. Levison, S.E. Badger, E.T. Butts and A.L. Khurana, 'Proceedings of the 9th International Symposium on Preparative and Industrial Chromatography 'PREP 92", SFE, Nancy, 1992, p.301.

12   P.R. Levison, "Preparative and Process-Scale Liquid Chromatography", G. Subramanian, ed., Ellis Horwood, Chichester, 1991, p.146.

13   P.R. Levison, S.E. Badger, D.W. Toome, M. Streater and J.A. Cox, J.Chromatogr., 1994, 658, 419.

14   W. Bolton, "Biochemists Handbook", C. Long, E.J. King and W.M. Sperry, eds., Spon, London, 1971, p.764.

15   P.R. Levison, S.E. Badger, D.W. Toome, D.Carcary and E.T. Butts, Inst. Chem. Eng. Symp. Series, 1990, 118, 6.1.

Kinetics of Protein Extraction Using Reverse Micelles: Studies in Well-mixed Systems and a Liquid–Liquid Spray Column

G. J. Lye,[a] J. A. Asenjo, and D. L. Pyle

BIOTECHNOLOGY AND BIOCHEMICAL ENGINEERING GROUP, DEPARTMENT OF FOOD SCIENCE AND TECHNOLOGY, UNIVERSITY OF READING, PO BOX 226, READING RG6 2AP, UK

[a]DEPARTMENT OF CHEMICAL ENGINEERING, IMPERIAL COLLEGE OF SCIENCE, TECHNOLOGY, AND MEDICINE, LONDON SW7 2BY, UK

1.   INTRODUCTION

Reverse micelles are surfactant aggregates in organic solvents which are able to selectively solubilise protein molecules. In Winsor II systems[1], the reverse micellar phase exists in equilibrium with an excess aqueous phase and it is known that the pH and ionic strength of this conjugate phase influences the partition of protein molecules between both phases. This paper summarises investigations into the factors governing protein mass transfer kinetics for liquid-liquid extractions carried out both in well-mixed systems and in a laboratory scale spray column.

2.   THEORY

Extraction in Well-Mixed Systems

Consider a system in which protein transfer occurs across an interfacial area, A, between two well-mixed liquid phases. If it is assumed that the two-film theory of mass transfer applies, the following expression can be derived[2] for protein transfer from phase 1:

$$\frac{C_1(t)}{C_1(0)} - \beta = (1-\beta)\exp-(\alpha t) \qquad [1]$$

where,
$$\alpha = \frac{K_L A}{V_1}(1+mV_r) \quad \text{and,} \quad \beta = \frac{mV_r}{1+mV_r} \qquad [2], [3]$$

Equation [1] thus relates the change in protein concentration in phase 1, $C_1(t)$, with time, t, to the overall mass transfer coefficient, K_L, the partition coefficient, m, and the physical parameters of the system.

Semi-Batch Extraction in a Liquid-Liquid Spray Column

The following model was used to describe protein extraction in a spray column in which the dispersed phase (either aqueous or organic) was passed in the form of discrete droplets through a stagnant continuous phase. If it is again assumed that film theory applies the following equation can be derived[2]:

$$\frac{C_{aq}(t)}{C_{aq}(0)} = \exp[-\gamma(1 - \exp-(\omega z))t] \tag{4}$$

where,

$$\gamma = \frac{Q_d}{mV_c} \quad \text{and,} \quad \omega = \frac{K_L S_d}{V_d U_t} \tag{5, 6}$$

Equation [4] relates the change in continuous phase protein concentration, $C_{aq}(t)$, with time, t, to the dispersed phase flow rate, Q_d, column height, z, and the droplet hold-up, V_d, interfacial area, S_d, and velocity, U_t. In this case, where the aqueous phase is continuous, the individual film mass transfer coefficients, k_{aq} and k_{rm}, are related to K_L by the following equation:

$$\frac{1}{K_L} = \frac{1}{k_{rm}} + \frac{1}{mk_{aq}} \tag{7}$$

3. MATERIALS & METHODS

Reverse micelle phases consisted of the anionic surfactant, AOT, in isooctane, the main protein investigated being lysozyme (pI 11.1, 14.3 KDa). For both forward and backward extraction in the well-mixed systems (total volume typically 50ml), the two phases were rapidly mixed in order to create a dispersion. Samples of this dispersion were taken over time and protein concentrations assayed in both phases; the rapid phase separation meant that protein concentrations did not change significantly after sampling. For forward extractions in the spray column (2.5cm i.d., height 30cm) experiments were performed over dispersed phase flow rates of 0-6ml/min, droplet sizes and dispersed phase hold-up being measured photographically. Protein concentrations in the emerging droplets and continuous phase were measured over time and for both systems studied, protein mass balances were generally better than 100±10%. Precise details of all experimental techniques have been given elsewhere[2].

4. RESULTS & DISCUSSION

Extraction in Well-Mixed Systems

The influence of pH and ionic strength on the forward and backward extraction of lysozyme is summarised in **Table 1**. Results such as these may be interpreted in terms of electrostatic interactions occurring between charges on the protein's surface and the electrical double layer created by the surfactant head groups[3]. In the case of the forward transfer step, both the rate and extent of protein extraction are seen to decline as pH and ionic strength are increased. This is because the magnitude of attractive electrostatic interactions decreases as the pH approaches the protein's pI, and because of increased Debye screening of the charges at higher ionic strengths. Consequently, recovery of protein from the micellar phase is seen to be favoured at high pH and high ionic strength.

Quantitatively, forward transfer is seen to be very fast with equilibrium being achieved in a matter of seconds, $K_LA= 43.4 \times 10^{-7}$ m^{3}s^{-1} at pH 7 for example. For back transfer, however, the rate is slightly slower if brought about by changes in pH ($K_LA= 11 \times 10^{-7}$ m^{3}s^{-1} at pH 12) or over 100 times slower if brought about by alteration of ionic strength ($K_LA= 9.5 \times 10^{-9}$ m^{3}s^{-1} at 2.0M KCl). Similar results were obtained for experiments using ribonuclease-a[2] although the rapid back extraction rates at pH>pI were not observed. The major draw-back of this type of experiment is that the interfacial area for mass transfer is unknown and hence only values of the combined mass transfer coefficient K_LA (m^{3}s^{-1}) can be reported to indicate the total mass transfer rate. Using a correlation relating K_L to energy input for rigid droplets[2] and also K_L values taken from experiments performed in stirred cells[4], it was estimated that the specific interfacial area could be as great as 10^4 m^{2}m^{-3}.

FORWARD TRANSFER (0.2M)			BACKWARD TRANSFER (1.0M)		
pH$_a$	m (C$_{aq}$/C$_{rm}$) (t=600 secs)	K$_L$A x10^7 (m^{3}s^{-1})	pH$_a$	m (C$_{rm}$/C$_{aq}$) (t=600 secs)	K$_L$A x10^7 (m^{3}s^{-1})
7.2	0.012	43.4	6.7	9.930	0.0
9.1	0.025	9.4	8.6	0.121	2.4
9.8	0.030	2.2	11.1	0.007	15.8
11.3	2.163	0.3	11.8	0.006	5.5
11.9	4.150	0.3	12.2	0.006	11.0

FORWARD TRANSFER (pH 7)			BACKWARD TRANSFER (pH 7)		
IS (M)	m (C$_{aq}$/C$_{rm}$) (t=1200 secs)	K$_L$A x10^7 (m^{3}s^{-1})	IS (M)	m (C$_{rm}$/C$_{aq}$) (t=3600 secs)	K$_L$A x10^9 (m^{3}s^{-1})
0.2	0.013	42.7	1.0	11.40	0.0
0.3	0.014	31.9	1.5	1.27	4.8
0.4	0.014	14.8	2.0	0.08	9.5
0.5	0.019	0.9			
1.0	9.044	0.2	1.5M KBr	0.13	41.5

Table. 1. Effect of (Top) pH and (Bottom) ionic strength on the equilibrium partition coefficients and combined mass transfer coefficients for the forward and backward extraction of lysozyme at V$_r$=1. The initial lysozyme concentration in the protein laden phase was 1 mg/ml. The reverse micelle phase contained 50mM AOT in isooctane whilst the aqueous phase consisted of KCl adjusted to the desired pH and ionic strength. Values of K$_L$A were calculated using equation [1].

The experiments reported above were repeated at a phase volume ratio of 5 ($V_r = V_{aq}/V_{org}$) for both lysozyme and ribonuclease-a[2]. While the partitioning behaviour was qualitatively similar at the two volume ratios, quantitatively it was found that forward transfer rates were generally higher at $V_r = 5$; for lysozyme, $K_LA = 92.5 \times 10^{-7}$ $m^3 s^{-1}$ at pH 7 for example. This is thought to be due to a change in the nature of the dispersion from one in which the organic phase is continuous at $V_r = 1$, to a continuous aqueous phase at $V_r = 5$. In order to further investigate this variation of K_LA with V_r, experiments were performed as outlined in **Table 2**.

Vol. Ratio (V_r)	Volume RM Phase (ml)	Volume AQ Phase (ml)	$[AOT]_{rm}$ (mM)	$[AOT]_{ov}$ (mM)	AOT_{ov} (mg)	m (C_{aq}/C_{rm})	$K_LA \times 10^7$ ($m^3 s^{-1}$)
1	15	15	50	25	330	0.019	0.9
1	5	5	50	25	110	0.025	0.3
5	5	25	50	8.3	110	0.101	0.3
5	5	25	150	25	330	0.002	1.1

Table. 2. Effect of phase volume ratio and surfactant concentration on the forward transfer of lysozyme. The reverse micelle phase contained the required concentration of AOT in isooctane whilst the aqueous phase consisted of 0.5M KCl, pH 7, and had an initial protein concentration of 1mg/ml. Values of K_LA were calculated using equation [1].

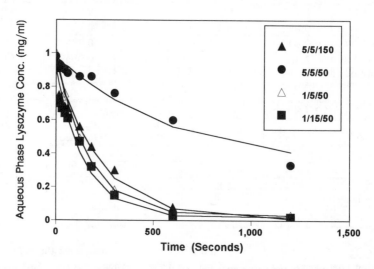

Fig. 1. Effect of phase volume ratio and surfactant concentration on the rate of lysozyme extraction from an aqueous phase of 0.5M KCl. In the legend, x/xx/xx represents V_r/volume RM phase/$[AOT]_{rm}$ respectively. See **Table 2** for experimental conditions. Solid lines are calculated from equation [1].

The results of these experiments at intermediate extraction rates, as depicted in **Fig. 1**, clearly show a reduced extraction rate in the system 5/5/50. From **Table 2**, it is clear that although the surfactant will be entirely solubilised in the organic phase under the conditions used[5], the system 5/5/50 is the one in which the *overall* concentration of surfactant, $[AOT]_{ov}$, is lowest. This would suggest that for a given energy input, lower values of $[AOT]_{ov}$ reduce the interfacial area for mass transfer and hence the extraction rate is decreased.

Retention of biological activity was also determined for both lysozyme and ribonuclease-a after undergoing the forward and backward extraction process, at $V_r=1$, under a range of conditions[2]. The enzymes were found to retain approximately 80% and 100% of their specific activity respectively. Activity retention was reduced by a further 10% (on average) when the forward transfer occurred at $V_r=5$. This was thought to be a consequence of multiple protein occupancy of each micelle at higher volume ratios[2] which would promote protein-protein interactions.

Semi-Batch Extraction in a Liquid-Liquid Spray Column

For the forward transfer of lysozyme in the spray column, the influence of pH (7-10), ionic strength (0.2-0.4M), and both the nature (aqueous or organic) and flow rate (0-6ml/min) of the dispersed phase were investigated. Values of K_L were of the order $1-8 \times 10^{-7}$ ms^{-1} and, as was found in the well-mixed systems, decreased with increasing pH and ionic strength.

Using existing correlations (e.g. Skelland & Cornish, Ruby & Elgin)[2] for predicting k_c and k_d values from knowledge of the droplet size, velocity and the physical properties of the two phases, it was possible to calculate K_L values from equation [7]. This was done for cases in which the dispersed phase droplets were assumed to be either stagnant or circulating. However, since it was only possible to operate the column over a limited range of dispersed phase flow rates and because values of m were similar for each set of experimental conditions investigated, only average values of K_L are reported in **Table 3**.

DISPERSED MICELLAR PHASES		DISPERSED AQUEOUS PHASES	
Stagnant	Circulating	Stagnant	Circulating
5.05	15.8	2.1	7.2

Table 3.   Predicted average values of the overall mass transfer coefficient K_L $\times 10^7$ (ms^{-1}) over the range of flow rates investigated for aqueous-reverse micellar, two-phase systems. Values calculated from equation [7] using correlations[2] for k_c, k_d and protein diffusion coefficients together with experimentally determined values of m and the physical properties of the phases.

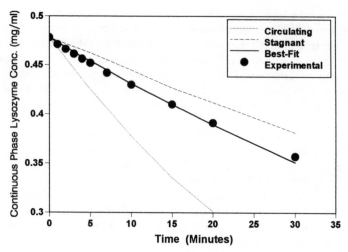

Fig. 2. Predicted and experimental protein extraction from the continuous (aqueous) phase at a dispersed (organic) phase flow rate of 4.3ml/min. Lines calculated from equation [4] using predicted values of K_L from **Table 3**. Aqueous phase consisted of 0.2M 90% KCl / 10% potassium phosphate buffer (pH 7) whilst the organic phase, which was pre-equilibrated with the aqueous phase (minus protein) before the extraction, consisted of 50mM AOT in isooctane.

A comparison of experimental and predicted extraction profiles is shown in **Fig. 2**; in this case the best-fit value of K_L, according to equation [4], was 5.5×10^{-7} ms^{-1} which suggests that stagnant drop correlations provide the most suitable predictions of mass transfer rates. This would be consistent with a study of droplet hydrodynamics which indicated that stagnant drops were formed regardless of which phase was dispersed (typical equivalent droplet diameter was 1.3mm)[2]. **Table 3** also predicts higher K_L values when the micellar phase is dispersed which was again consistent with experimental findings. Given that the dispersed phase droplets have stagnant interiors, this phenomenon is a consequence of the lower viscosity[2] and dynamic nature[6] of the micellar phase allowing diffusion/transport of the solubilised protein away from the interface.

5.   CONCLUSIONS

The work summarised in this paper shows that while pH and ionic strength control the equilibrium distribution of protein in reverse micellar systems, the partitioning kinetics can also be influenced by the operating conditions of the contacting device in which this process occurs (e.g. volume ratio, mixing speed, dispersed phase flow rate etc.). In the two systems studied here, it appears that the forward transfer step is limited by diffusion in the aqueous boundary layer film. In contrast, the back transfer process would appear to be limited by an interfacially controlled mechanism. Such considerations must be borne in mind for the design of reverse micellar extraction equipment and highlight the importance of this type of research.

The authors would like to thank the SERC for financial support of this work (CASE award for GJL).

6. NOMENCLATURE

Greek Symbols:

A	Interfacial area (m^2)		
C	Protein concentration (mg/ml)	α	Model parameter (s^{-1})
k	Film mass transfer coefficient (ms^{-1})	β	Model parameter
K_L	Overall mass transfer coefficient (ms^{-1})	γ	Model parameter (s^{-1})
m	Equilibrium partition coefficient ($=C_1/C_2$)	ω	Model parameter (m^{-1})
Q_d	Dispersed phase flow rate (m^3s^{-1})		
S_d	Droplet surface area (m^2m^{-3})	**Subscripts/Superscripts:**	
t	Time (s)		
U_t	Droplet terminal velocity (ms^{-1})	aq	Aqueous phase
V	Volume (m^3)	c	Continuous phase
V_r	Phase volume ratio ($=V_{aq}/V_{org}$)	d	Dispersed phase
V_d	Droplet hold-up (m^3m^{-3})	rm	Reverse micelle phase
z	Height (m)	1	Phase one
		2	Phase two

7. REFERENCES

1. P.A. Winsor, Trans. Faraday Soc., 1948, 44, 376.
2. G.J. Lye, Ph.D. Thesis, University of Reading, 1993.
3. K.E. Goklen & T.A. Hatton, Sep. Sci. Technol., 1987, 22, 831.
4. M. Dekker, Ph.D. Thesis, Wageningen Agricultural University, 1990.
5. R. Aveyard, B.P. Binks, S. Clark & J. Mead, J. Chem. Soc., Faraday Trans. 1, 1986, 82, 125.
6. P.L. Luisi & L.J. Magid, Crit. Rev. Biochem., 1986, 20, 409.

Extraction of Macrolide Antibiotics Using Colloidal Liquid Aphrons (CLAs)

G. J. Lye and D. C. Stuckey

DEPARTMENT OF CHEMICAL ENGINEERING AND CHEMICAL TECHNOLOGY, IMPERIAL COLLEGE OF SCIENCE, TECHNOLOGY, AND MEDICINE, LONDON SW7 2BY, UK

1.   INTRODUCTION

It is an accepted fact in biotechnology, that the downstream separation of dilute, labile products from complex mixtures often constitutes a significant fraction of the overall cost of a product, and at times may be as high as 70%. Because of this, industry is under constant pressure to develop new and economically more efficient downstream separation methods. One novel and promising technique which has been developed recently is the use of colloidal liquid aphrons (CLAs) for the pre-dispersed solvent extraction (PDSE) of molecules which are oil soluble.

CLAs have been postulated to consist of a solvent droplet encapsulated in a thin aqueous film which is stabilised by the presence of a mixture of non-ionic and ionic surfactants as shown in **Fig. 1**[1]. When polyaphrons are dispersed in an aqueous phase they form spherical droplets of 3-10μm diameter; the aqueous 'soapy' shell and highly charged surface[2] are thought to provide a large energy barrier to coalescence thereby stabilising the droplets[1]. Solute extraction is driven by a favourable partitioning of the target molecule between the feed solution and the oil core of the CLA, whilst the extremely large interfacial area (due to the small droplet size) allows equilibrium partitioning to be achieved in a matter of seconds. This technique has been shown to be suitable for the efficient extraction of a number of solutes from dilute solution[1-3]. Antibiotic purification is one area where PDSE may be applied in order to reduce purification costs for both new and existing processes. The advantages of this technique over conventional solvent extraction include reductions in the following; capital costs of mixer-settler units, power requirements for solvent dispersion, solvent inventories, and possibly solvent toxicity if used with direct broth extraction[2].

In this work, erythromycin was chosen as a model compound as it is produced commercially by large-scale fermentation of *Streptomyces*, and is currently purified by solvent extraction[4]. Erythromycin-A is a macrolide antibiotic of molecular weight 734Da consisting of a 14-membered oxygenated ring to which two sugar moieties are attached. The molecule has a pK_a of 8.8 which allows the aqueous-organic partition of the molecule to be controlled by manipulation of the extraction/stripping solution pH. The objective of this work was to carry out a preliminary investigation on the influence of parameters such as pH, ionic strength, phase ratio, feed concentration and mixing speed on both the extraction and stripping of erythromycin to and from CLAs using an aphron formulation which we know to be comparatively stable[5].

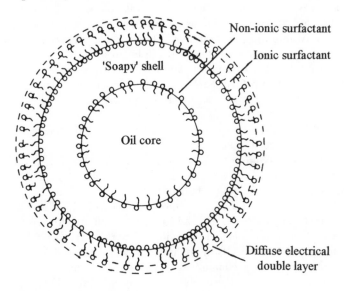

Fig. 1. Schematic structure of an aphron as proposed by Sebba[1].

2. MATERIALS & METHODS

Polyaphron phases were formulated from; decan-1-ol (99%, Aldrich), a non-ionic surfactant of the alcohol ethoxylate type (Softanol 120, Honeywill & Stein), and the anionic surfactant sodium dodecyl sulphate, SDS (Sigma). Erythromycin (Sigma, 98%) and the salts used for preparation of carbonate, phosphate and acetate buffers (BDH, 'AnalaR' grade) were of the highest purity available. Mobile phases for HPLC were prepared from mixtures of acetonitrile and methanol (Rathburn) containing organic phase modifiers of ammonium phosphate and tetramethyl ammonium phosphate (BDH). Sulphuric acid (1.84 s.g.) was from Fisons. HPLC grade water was used throughout this work (Purite).

Polyaphron phases were prepared by dropping the organic phase (1% w/v Softanol 120 in decanol) from a burette into a well-stirred (≈800rpm), foaming aqueous solution containing 0.5% w/v SDS - typically 1.5ml. The organic phase was added at an average flow rate of ≈0.5ml/min until the desired phase volume ratio of 9 was reached (PVR= V_{org}/V_{aq}), and a creamy-white polyaphron phase was obtained[5].

Equilibrium experiments were performed in test-tubes, the polyaphrons being dispersed by repeated vortex mixing over 30 minutes using a typical volume of $\approx$5ml. Kinetic experiments were carried out in a 250ml glass beaker mixed by a 21x7mm magnetic bar. Starting from an initial volume of $\approx$100ml, 5ml samples of the dispersion were taken at various time intervals, the CLAs being rapidly separated from the feed/stripping solution by a 0.22μm Swinnex membrane filter (Millipore). Details of erythromycin concentration, pH, ionic strength, mixing speed, CLA:feed ratio and CLA:stripping solution ratio will be given in the results section. For all experiments the glassware was tightly covered to prevent losses by splashing or evaporation, and no pH change was measured during the extraction/stripping steps. All experiments were performed at ambient temperature.

HPLC analysis was performed using a Waters (5μ, spherical), C18 reversed-phase column fitted to a Perkin Elmer, Series 4, system. Sample injection volume was 150μl, and erythromycin was detected by UV absorption at 215nm. Elution occurred isocratically at 0.8ml/min using a mobile phase of acetonitrile, methanol, 0.2M ammonium phosphate buffer (pH 6.5), 0.2M tetramethyl ammonium phosphate and water at a volume ratio of 30:30:5:10:25. This mobile phase had been optimised by Cachet *et al*[6], and gave a retention time for erythromycin of 10.1 minutes (maximum standard deviation of peak area quantification was 6%). Routine quantitative analysis of erythromycin was performed using a colourimetric assay based upon the reaction of sugar moieties on the erythromycin molecule with 18N H_2SO_4 to yield a yellow product absorbing at 479nm. The method is described by Dannielson *et al*[7] and allows detection to levels of 1-10μg/ml (maximum standard deviation was 4%). Particle size measurements were performed using a light scattering technique (Malvern M3.0) in which the polyaphrons were dispersed in water. Results are presented as Sauter mean diameters, the normalised standard deviation being 4%.

3.   RESULTS & DISCUSSION

Equilibrium Partitioning of Erythromycin

The influence of pH on the distribution of erythromycin between an aqueous phase and CLAs is shown in **Fig. 2**. Extraction experiments were carried out at a CLA:feed ratio of 1:100 v/v, the initial aqueous erythromycin concentration being 0.5g/l, whilst for stripping experiments, the phase ratio was 1:11 v/v and the initial concentration of erythromycin in the CLAs was 10g/l. For comparison, data is also shown for partitioning between water and pure decanol[3]. The partition coefficient, K is defined as the erythromycin concentration in the CLA phase divided by that in

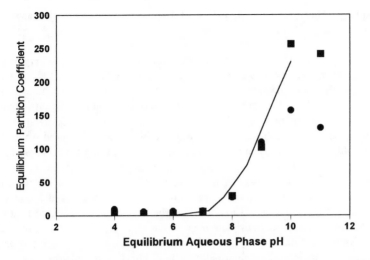

Fig. 2. Effect of pH on the distribution of erythromycin between a 20mM buffered aqueous phase and CLAs for (●) extraction and (■) stripping steps - results are the average of three determinations. Solid line represents data for water-decanol partitioning.

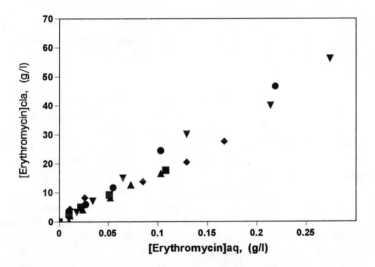

Fig. 3. Langmuir plot of the influence of erythromycin concentration on extraction at pH 9.5 from a 50mM buffered aqueous phase at CLA:feed ratios (v/v) of (▲) 1:50, (◆) 1:100, (■) 1: 150, (●) 1:200 and (▼)1:300. Maximum normalised standard deviation was 10%.

the aqueous phase (C_{cla}/C_{aq}); values of C_{cla} were calculated by mass balance because of the small volume of the aphron phase used.

As with conventional solvent extraction, the partitioning of erythromycin showed a strong dependence on pH, changing markedly around the pK_a. At values of $pH<pK_a$ the molecule exists in a charged state and can, therefore, be expected to reside preferentially in the aqueous phase. Extraction should thus be carried out at a $pH\geq pK_a$, and stripping when $pH\leq pK_a$. The large values of K determined for extractions at high pH are a consequence of the high phase ratio used which allows concentration of erythromycin in the CLAs up to levels of 30g/l in this case. Except for experiments at the highest pH values, **Fig. 2** shows no significant difference for aqueous-organic or aqueous-CLA partitioning. This suggests that the presence of surfactants, used in CLA formulation, have little influence on the *equilibrium* distribution of erythromycin. The reason for the decline in the values of K at pH 11 for both the extraction and stripping steps is unclear. It is not due to saturation of the CLAs with erythromycin (see **Fig. 3**), and quantitative agreement between erythromycin concentrations determined by both HPLC and colourimetric methods suggests that no change in molecular structure had occurred. Indeed, only at pH 4 and below did the HPLC trace suggest any structural degradation of the erythromycin.

The influence of aqueous phase ionic strength from 0.02-1.0M was also studied for extraction at pH 9.5 and stripping at pH 7.5 (other conditions as above - results not shown). Compared to the influence of pH, ionic strength had less effect on erythromycin partitioning with K values only varying by ≈50 over the range investigated. At higher ionic strengths, charges on the molecule will be 'screened' thus extraction was found to be enhanced whilst stripping of erythromycin from the CLAs was reduced.

With regard to stability of the dispersed CLAs, it was observed that in some cases a clear solvent layer was present on the surface of the dispersion when allowed to stand for a few minutes. This also coincided with a change in the appearance of the dispersion over the extraction/stripping time from milky-white to grey which suggested coalescence of the CLAs since larger droplets would scatter less light. The coalescence/breakage of the CLAs was most pronounced at pH<6 and ionic strengths >0.2M, which corresponds well with measurements we have made on the half-lives of various aphron formulations dispersed under various conditions. These results will be reported elsewhere, though it should be noted that in all cases the CLA half-life is considerably longer than the time required for extraction/stripping equilibrium to be achieved (see **Fig. 4**).

The influence of phase volume ratio and erythromycin concentration on extraction at pH 9.5 is shown in **Fig. 3** plotted according to the Langmuir model. It is seen that saturation of the CLAs with erythromycin is not reached up to absorbed concentrations of 60g/l but, because of the poor solubility of erythromycin in the aqueous feed solution at this pH, it was not possible to determine values for q_{max} and K_m for the system. From **Fig. 3**, it is also clear that in order to minimise erythromycin levels in the feed stream after extraction, either low CLA:feed ratios or multi-stage processes should be employed.

Kinetics of Erythromycin Partitioning using CLAs

Figure 4 demonstrates the extremely rapid extraction rate of erythromycin using CLAs (extraction occurred at a CLA:feed ratio of 1:100 w/w, and a mixing speed of 600rpm). The rate of the stripping step (at a CLA:stripping solution ratio of 1:6 w/w, a mixing speed of 600rpm and an initial concentration of erythromycin in the CLAs of 6g/l) was comparatively slower for experiments carried out over the pH range 5-9, near equilibrium extraction being achieved within 2-3 minutes (data not shown). Calculated values of the equilibrium partition coefficient for erythromycin at each pH were, however, similar to those presented in **Fig. 2** for both the extraction and stripping processes.

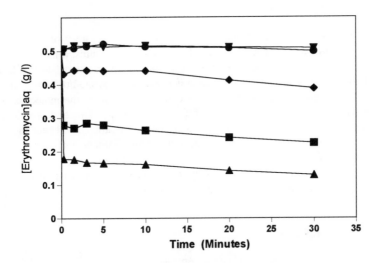

Fig. 4. Extraction rate of erythromycin from a 20mM buffered aqueous phase using CLAs at pH (▼) 6, (●) 7, (◆) 8, (■) 9 and (▲) 10. Maximum normalised standard deviation was 8%.

From **Fig. 4** it is seen that near equilibrium extraction is achieved within 15 seconds, and it is highly likely that this is achieved even faster since this is the minimum time in which it is possible to disperse the phases, take a sample and then separate the sampled phases using a Swinnex filter. These rapid kinetics are a direct consequence of the enormous interfacial area created using PDSE. Particle size measurements showed this aphron formulation to produce CLAs having a mean diameter of $4\mu m$, and a size distribution between $2-11\mu m$ (analysis of the scattering pattern suggested that these droplets were spherical); this results in a specific interfacial area for mass transfer of around $15,000m^2/m^3$ of feed solution in this case. The subsequent gradual extraction of erythromycin between 1-30 minutes, as at pH 10 for example, is due to the small driving force for mass transfer approaching equilibrium since the concentration of erythromycin in the CLAs is already around 32g/l after the first few seconds; the equilibrium value being 37g/l. Experiments were also carried out (data not shown), under the same conditions as above, in which the mixing speed of the dispersion was varied between 0-1000rpm. In the case of unmixed systems, the polyaphron phase was initially dispersed by a gentle swirling action after which no more mechanical mixing occurred. For extraction at pH 10, and stripping at pH 6, the difference measured for the process kinetics over this range of rpm was minimal. This suggests that the mass transfer process was interfacially rather than diffusion controlled which is likely considering the three surfactant layers across which the erythromycin molecule has to transfer (see **Fig. 1**).

4.   CONCLUSIONS

The results presented here are a preliminary investigation into factors influencing PDSE using CLAs. For poorly water soluble solutes, it has been shown that this is a very promising technique with considerable potential in the field of downstream processing. For erythromycin we have demonstrated that the extraction and stripping steps are extremely fast, and that the partition of the molecule can be manipulated through careful control of the system pH. The high solubility of erythromycin in the CLAs allows operation at high phase ratios, and hence the ability to greatly concentrate the molecule, whilst the strong influence of pH may allow some selectivity to be engineered into the process. Future work will investigate more precisely the influence of surfactants on erythromycin mass transfer rates and whether or not the process is interfacially controlled.

5.   REFERENCES

1.   F. Sebba, "Foams and Biliquid Foams - Aphrons", Wiley & Sons, Chichester, 1987.
2.   D.C. Stuckey, K. Matsushita, A.H. Mollah & A.I. Bailey, in "Proceedings 3rd International Conference on Effective Membrane Processes", BHR Publishers, London, 1993.
3.   A.H. Mollah, K. Matsushita, D.C. Stuckey, C. del Cerro & A.I. Bailey, submitted to Biotechnol. Bioeng., 1993
4.   S. Omura & Y. Tanaka, in "Biotechnology - Vol. 4", Eds. H.J. Rehm & G. Reed, VCH Publishers, 1986.
5.   K. Matsushita, A.H. Mollah, D.C. Stuckey, C. del Cerro & A.I. Bailey, Colloids Surfaces, 1992, 69, 65.
6.   Th. Cachet, I.O. Kibwage, E. Roets, J. Hoogmartens & H. Vanderhaeghe, J. Chromatogr., 1987, 409, 91.
7.   N.D. Danielson, J.A. Holeman, D.C. Bristol & D.H. Kirzner, J. Pharm.Biomed. Anal., 1993, 11, 121.

Lysozyme Refolding in Reverse Micelles: Initial Studies

S. Mall, S. M. West, J. B. Chaudhuri, and D. R. Thatcher[a]

SCHOOL OF CHEMICAL ENGINEERING, UNIVERSITY OF BATH, CLAVERTON DOWN, BATH BA2 7AY, UK

[a]ZENECA PHARMACEUTICALS, MERESIDE, ALDERLEY PARK, MACCLESFIELD, CHESHIRE SK10 4TG, UK

1. INTRODUCTION

The expression of recombinant proteins in *E coli* often results in the formation of intracellular inclusion bodies[1]. These insoluble aggregates are misfolded protein molecules held together by hydrophobic forces[2]. The production of recombinant proteins as inclusion bodies is advantageous in that the protein is present as a solid in a highly purified and concentrated state. However, in order for the protein to regain activity, it must be solubilised and then correctly refolded into its native tertiary structure.

Inclusion bodies are solubilised using strong denaturants such as guanidine hydrochloride (GuHCl) or urea and misformed disulphide bonds are broken under reducing conditions. The unfolded protein is refolded to the native, active state by removal of the denaturant, achieved by dilution with an appropriate buffer. Parallel to the refolding reaction is an unwanted, irreversible aggregation reaction which is second order or higher, whereas refolding is a first order reaction. Since the aggregation reaction dominates, the denatured protein concentration must be kept as low as possible to minimize aggregation and thus promote refolding. This has been shown by a study on the competition between aggregation and renaturation of lysozyme[3]. Refolding is completed by reforming the disulphide bonds by manipulating the redox environment[4].

An alternative method to reduce aggregation during refolding is to use reverse micelles[5] (Fig. 1). These are aqueous phase droplets surrounded by a layer of surfactant molecules suspended in an oil phase. The denatured protein molecules can be isolated from one another in the micelles. This reduces the likelihood of intermolecular reactions leading to aggregation, since typically there are more reverse micelles than protein molecules and the probability of there being more than one protein molecule per reverse micelle is therefore small.

Hagen *et al*[5,6] have shown that denatured ribonuclease A can be successfully transferred into AOT/isooctane reversed micelles and refolded to give an overall yield of 50%. They refolded the protein by combining 3 to 5 ml of the reversed

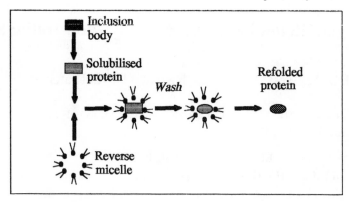

Figure 1.   Schematic diagram of protein refolding using reverse micelles

micelle phase containing the denatured protein with an equal volume of 0.1M KCl for 15 minutes. This step was repeated with the resulting organic phase and a fresh aqueous solution to remove additional denaturant. After ten such contacting stages, a ten-fold decrease in denaturant concentration was obtained. There have been very few further attempts to refold other protein molecules in reversed micelles.

We are studying the refolding of lysozyme using an AOT/isooctane reversed micelle system. In this paper we present our initial results using native and denatured pure lysozyme.

2. MATERIALS AND METHODS

Materials

Hen egg-white lysozyme (52000 units/mg), AOT (Sodium bis-2-ethylhexyl sulphosuccinate) and *Micrococcus lysodeikticus* cells were obtained from Sigma Chemical Co. Ltd., Poole, Dorset. Isooctane (2,2,4-trimethylpentane) was of AnalaR grade and purchased from BDH Chemicals, Merck Ltd., Poole, Dorset. Guanidine hydrochloride (GuHCl) was purchased from Aldrich Chemicals Co Ltd., Gillingham, Dorset. Karl-Fischer titrants were obtained from Fisons Scientific Equipment, Loughborough.

Experimental Methods

Forward Transfer. The forward extraction of lysozyme involved mixing equal volumes of an organic phase (50mM AOT/isooctane), with an aqueous phase (1mg/ml lysozyme containing different potassium chloride (KCl) concentrations in a phosphate buffer at a particular pH) for 1hr. This was achieved in tightly stoppered vials for protein transfer into the organic phase to occur.

The forward extraction of lysozyme in various denaturant (GuHCl) concentrations were performed in a similar fashion to above, except 400mM

AOT/isooctane was used and no salt was present in the buffer.

Inactivation of Lysozyme in GuHCl. Lysozyme (0.2mg/ml) was incubated in 0.06M potassium phosphate, pH6.2 in the absence and presence of increasing concentrations of GuHCl at 20°C for 15min. Enzyme activity was assayed in the same concentration of GuHCl and expressed relative to a control sample from which GuHCl was omitted.

Reactivation of Lysozyme

Hen egg-white lysozyme was denatured and reduced following the procedure of Goldberg *et al*[3] except 6M GuHCl was used as the denaturant. After dialysis the enzyme solution was stored at -20°C or at 4°C if it was to be used immediately. Renaturation-reoxidation of the enzyme solution was initiated by adding a sample to renaturation buffer (0.1M Tris-HCl, pH8.2, 1mM EDTA, 3mM reduced glutathione and 0.3mM oxidised glutathione), mixing with a Vortex mixer for 15s and incubating at 40°C. Enzyme activity was assayed at several time points and expressed relative to native lysozyme measured at the same concentration.

Analytical Methods

Enzyme and Protein Assays. Lysozyme concentrations were determined spectrophotometrically at 280nm, using an absorbance for 1mg/ml of 2.63 (cell path length=1cm)[7], for both the aqueous and organic phases after the forward extraction of the protein. An absorbance value of 2.37 was used for the reduced enzyme[7]. Lysozyme activity was determined at 25°C by following the decrease in absorbance at 450nm of a 0.25mg/ml *Micrococcus lysodeikticus* suspension in 0.06M potassium phosphate, pH6.2. The assay volume was 1ml and one unit of activity corresponds to an absorbance decrease of 0.0026 per minute[3].

Water Content in the Organic Phase The water content of the organic phase was measured using the Karl Fischer technique (Mettler DL37Coulometer). Water content is expressed as w_o, which is the molar ratio of water to surfactant.

3. RESULTS AND DISCUSSION

Characterisation of Protein Transfer

Studies were made on the transfer of native lysozyme (1 mg/ml) into a reverse micelle system comprising 50 mM AOT in isooctane. Forward transfer results were obtained as a function of aqueous phase pH at a concentration of 0.1M KCl (Fig. 2). During the forward transfer protein was observed to precipitate at the aqueous/organic phase interface. By measuring the residual protein concentration in the aqueous phase and the concentration in the organic phase the mass of protein precipitated at the interface was quantified (Fig. 2).

The large amount of precipitate formed between pH 2-4 shows that very little protein partitions into the organic phase. The protein remaining at the interface is most likely in the form of a protein-surfactant precipitate. The extent of precipitate

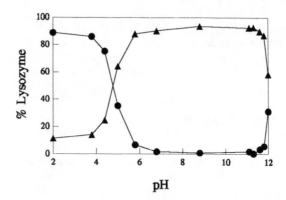

Figure 2.   The effect of pH on forward transfer and precipitation. (▲ = transfer; ● = precipitation)

formation has been determined by a mass balance. The figure illustrates that the maximum amount of protein transferred (90%) occurs over the pH range 7-11. After pH 11 a sharp decline in the transfer of protein may be observed, which coincides with the isoelectric point of lysozyme (pI=10.9). When the pH of the medium is less than the pI of lysozyme, the protein has a net positive charge and since the surfactant headgroups of AOT possess negative charges, the protein is attracted into the micelle. However, as soon as the pH of the medium exceeds the pI, the protein is repelled from the micelle due to the unfavourable electrostatic interactions and a decrease in transfer is observed.

A similar study was carried out for the back transfer of lysozyme (results not shown). It was found that maximum protein recovery occurred at pH 11, around the isoelectric point, where 90% of the protein is recovered. After the back transfer step 80% of the lysozyme specific activity was retained relative to the starting activity in the aqueous solution.

Folding and Unfolding Characteristics of Lysozyme in Aqueous Solution

In order to establish the denaturant concentration at which activity is lost, inactivation studies on lysozyme were carried out using GuHCl (Fig. 3). Lysozyme was inactivated at low concentrations of GuHCl with less than 5% activity remaining at 0.3M.

The amount of denaturant used to inactivate lysozyme is much lower than the concentrations required to significantly alter the protein structure or to solubilise inclusion bodies. Fluorescence studies (unpublished data) have shown that there is little change in the structure up to concentrations of 4.0M GuHCl but at higher concentrations the structure changes significantly. The observation that inactivation occurs at much lower GuHCl concentrations than large changes in enzyme structure is currently under investigation.

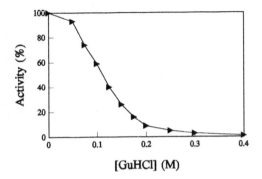

<u>**Figure 3.**</u>   **Inactivation of lysozyme with guanidine hydrochloride.**

In order to assess the refolding performance denatured and reduced lysozyme was renatured and reoxidised by dilution and incubation in renaturation buffer at a concentration of 0.015 mg/ml (Fig. 4).

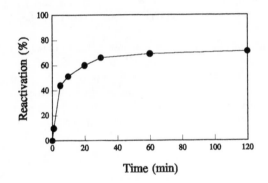

<u>**Figure 4.**</u>   **Reactivation of lysozyme.**

Lysozyme can be refolded to regain about 70% of its native activity. The initial recovery of 50% activity is achieved within 10 minutes. The results obtained agree with previously published data[8].

Transfer of Denatured Lysozyme into Reverse Micelles

Inclusion bodies require high concentrations of denaturant to be solubilised. It is necessary to transfer the denatured protein at high GuHCl concentrations into the reverse micelles. Investigations were made into the conditions under which denatured (non-reduced) lysozyme transfers into reverse micelles (Fig. 5). The protein was solubilised in phosphate buffer, at pH 7 containing either KCl or GuHCl.

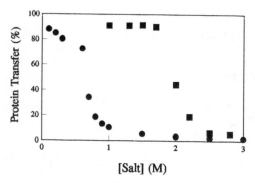

Figure 5. The effect of GuHCl concentration on the forward transfer of lysozyme. (● = KCl, ■ = GuHCl)

This figure shows a similar trend for lysozyme transferred using either KCl or GuHCl, but the decline in the amount of protein transfer occurs at much lower salt concentrations with KCl. The graph indicates that at a concentration of 1M GuHCl 90% of the protein is transferred into the micelle, whereas at 1M KCl only ~10% is transferred. The graph also shows that there is a limit to the amount of protein that can be partitioned inside the micelle as the GuHCl concentration increases.

The reason for this decrease in protein transfer is due to the electrostatic shielding of surfactant head-groups in the reverse micelle by the cations from the salt. These ions reduce the repulsion between individual headgroups of the AOT molecules causing the micelle to shrink in size and which are not therefore able to accommodate the protein. In addition, there may be a screening of charges between the protein surface and the surfactant headgroups. The effect of salt type and concentration on micelle size is shown in Figure 6.

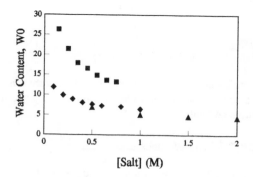

Figure 6. The effect of salt concentration on micelle size. (■ = NaCl, ◆ = KCl, ▲ = GuHCl)

Wo is related to the water pool size inside the micelle and gives an indication of the size of protein that may be accommodated. The smaller cations have more water of hydration associated with them, this gives rise to larger micelles (Fig. 6).

4. CONCLUSIONS

1. Lysozyme is a suitable model protein for this study as it may be transferred into and from AOT-isooctane reverse micelles retaining 80% of its activity.
2. In aqueous solution lysozyme loses the majority of its activity at 0.2M guanidine hydrochloride, major structural changes occur at concentrations of about 4.0M GuHCl. This implies that the denaturant concentration must be reduced to less than 0.2M to recover activity.
3. Denatured lysozyme may be refolded to recover about 70% of its activity by dilution in aqueous solution.
4. The transfer of denatured lysozyme into reverse micelles is limited to a concentration of 1.8M GuHCl. This results from the smaller size of the micelle that results from the guanidinium cation.

Current work is investigating the use of reverse micelles formed from non-ionic surfactants to increase the denatured protein and guanidine concentrations that may be solubilised in the micelle.

5. ACKNOWLEDGEMENTS

The financial assistance of Zeneca Pharmaceuticals is gratefully acknowledged, as is that of the BBSRC in the form of a CASE studentship to S Mall. This work is partially funded by the BBSRC (Biotechnology Directorate Grant GR/J19801).

6. REFERENCES

1. J. L. Cleland, 'Protein Folding In Vivo and In Vitro', ACS Symposium Series, 526, 1993,
2. A. Mitraki, and J. King, Bio/Technol., 1989, 7: 690.
3. M. E. Goldberg, R. Rudolph and R. Jaenicke, Biochem., 1991, 30, 2790.
4. R. Hlodan, S. Craig and R. H. Pain, Biotechnol. Gen. Eng. Rev., 1991, 9, 47.
5   A. J. Hagen, T. A. Hatton, and D. I. C. Wang, Biotechnol. Bioeng.., 1990, 35, 955.
6. A. J. Hagen, T. A. Hatton, and D. I. C. Wang, Biotechnol. Bioeng.., 1990, 35, 966
7. V. P. Saxena and D. B. Wetlaufer, Biochem., 1970, 9, 5015
8. B. Fischer, I. Sumner and P. Goodenough, Arch. Biochem. Biophys., 1993, 306,   183

An HPLC Technique for the Study of Adsorption of Amino Acids on Functionalized Resins

M. Martínez, J. L. Casillas, F. Addo-Yobo, C. N. Kenney, and J. Aracil

UNIVERSIDAD COMPLUTENSE, FACULTAD DE CIENCIAS QUÍMICAS,
DEPARTAMENTO DE INGENIERÍA QUÍMICA, MADRID 28040, SPAIN

1 INTRODUCTION

Advances in molecular biology, immunology and biochemical engineering have provided a wide range of different biochemical product that are or may become of great commercial value. Amino acids and related substances are often the main constituents in may of these products, which may be used as pharmaceutials or medical diagnostic reagents. Many of these biochemical compounds are produced in biochemical reactors, and often the solution in the reactor is a complex biochemical mixture with the substances of interest present only in very small concentration.

The efficient isolation and purification of these products is essential for commercial success. For this reason, downstream recovery represents a large portion of the product price and this expense may be a major production cost. Often, very high degrees of purity as well as high product yields could be obtained in a single purification step because the competitive edge of a given manufacture may depend on the optimisation of the downstream processing.

Adsorption processing using neutral non-polar macroporous resins is commonly used for the separation of biochemical products (antibiotic, amino acids and low molecular weight peptides) both on an analytical scale and in industrial manufacturing processes[1,2]. From a practical viewpoint an ideal method for study of adsorption processes for costly products should contact small quantities of adsorbate with an adsorbent for a short time. These are the conditions which are routinely attainable in analytical High Pressure Liquid Chromatography (HPLC).

Although liquid phase chromatography is commonly used industrially for adsorbent screening and to obtain preliminary kinetic and equilibrium data, the application of this technique appears to be much less developed than gas chromatography.

As the need for higher degrees of purification increases, modification of traditional separation methods to increase the selectivity of the process has become important. In order to increase the efficiency and selectivity of the commercial adsorbent a number of modified Amberlite XAD-2 resins have been obtained by chemical modification. Previous work[3,4] has led to successful preparation of polystyrene derivatives which have a more efficient separation of important antibiotics, eg. Cephalosporin C, and dipeptide Aspartame than the non-functional resin XAD-2 and XAD-4 from the impurities present in the fermentation broth.

In this study, due the simplicity of reverse phase chromatography, this technique has been used for the evaluation of the kinetics and equilibria of adsorption of seven representative amino acids (DL-Alanine, DL-Aspartic Acid, DL-Proline, DL-Serine, DL-Tryptophan, DL-Tyrosine, DL-Valine) on two modified Amberlite XAD-2 resins. The reason for selecting these amino acids is that the effect of a gradual increase in molecular size on pore diffusivities and equilibrium constant can be systematically evaluated.

2 EXPERIMENTAL APPARATUS AND PROCEDURES

2.1. Apparatus and Materials

The experimentals were carried out in a HPLC system, which consists of an Perkin-Elmer LC-100 high pressure liquid pump; a Rheodyne Type 7125 sample injection valve with a 20 μl sample loop; a 20x0.22 cm id stainless steel adsorption column; and a Perkin-Elmer LC-75 UV wavelength spectrophotometric detector.

The Amberlite XAD-2 was supplied by Rohm and Hass Co. The XAD-2-I and XAD-2-II resins were obtain by bromination and ethylbromination respectively of Amberlite XAD-2.

The eluent of controlled pH (3.0) and ionic strength (0.1 M) were prepared by dissolving known quantities of analytically pure Na_2HPO_4, KH_2PO_4 and NaCl in distilled deionised water in proportions specified by Perrin[5]. The flow rates studied were 0.5 to 1.5 cm^3/min and the temperature range was 293 to 313 K. High purity grade amino acids were supplied by Sigma Chemical Company. The resin temperature was controlled by a column oven.

2.2. Operation Procedures

Prior to each experimental run, the column was eluted overnight with buffer solution at a chosen pH. The temperature of the oven and water bath was raised to the desired point and the column and preheat tube were immersed in the oven and in the bath. The response of the on-line UV spectrophotometer to an eluted peak was monitored on a chart recorder, digitized and stored on a computer.

3 THEORETICAL

One of the simplest methods of characterizing the adsorption properties of a system is by measuring the chromatographic response. The Chromatographic response curves can be analyzed by the method of moments.

For a column packed with macroporous adsorbent and for a linear isotherm, the retention volume or the mean retention time is a measure of the equilibrium adsorptive capacity while the dispersion of the response peak arises from the combined effects of axial diffusion and finite mass transfer resistance. The mean retention time and the dispersion of the response peak are related to the first and second moments of the chromatographic response peaks by[6]:

$$\mu = \frac{\int_0^\infty C\,t\,dt}{\int_0^\infty C\,dt} = t_0 + \frac{L}{v}\left[\,1 + \frac{(1-\varepsilon)}{\varepsilon}\,k\,\right] \qquad (1)$$

where, t_0, the dead time, was estimated by analyzing the NaNO$_3$ peak as described above:

$$t_0 = \frac{L}{v} [\, \varepsilon + (1 - \varepsilon)\, \varepsilon_p \,] \tag{2}$$

and

$$k = \varepsilon_p + (1 - \varepsilon_p)\, k_{ads} \tag{3}$$

$$\sigma^2 = \frac{\int_0^\infty C\, (t - \mu)^2\, dt}{\int_0^\infty C\, dt} \tag{4}$$

The column efficiency may be evaluated by the height equivalent to a theoretical plate (HETP). Here, HETP is expressed in terms of moments[6] as:

$$\mathbf{\mathit{HETP}} = \frac{\sigma^2}{\mu^2}\, L \tag{5}$$

and

$$\frac{\sigma^2}{2\,\mu^2} = \frac{D_L}{v\, L} + \frac{v}{L}\,(\frac{\varepsilon}{1-\varepsilon})(\frac{R_p^2}{15\,\varepsilon_p\, D_p} + \frac{R_p}{3\, k_f} + \frac{r_c^2}{15\, k\, D_c})\, (\, 1 + \frac{\varepsilon}{(1-\varepsilon)\, k}\,)^{-2} \tag{6}$$

In a liquid phase system the axial dispersion contribution (D_L/vL) is essentially independent of velocity[7] so it is clear that a plot of HETP against velocity should be linear with intercept 2 D_L/vL and slope given by eqn (6). For a species which is either insignificantly adsorbed ($k\rightarrow0$) or very rapidly adsorbed ($D_p\rightarrow\infty$), it follows from eqn (6) that the HETP will be given simply by 2 D_L/vL and should thus be essentially independent of either velocity or temperature. The magnitude of the external mass transfer resistance may be estimated from Wakao's correlation[8]. Since the particle size is small the Reynolds number is always low over the entire range of velocity used in these experiments. At low Reynolds Number, the external mass transfer coefficient is given, approximately, by, $k_f\, R_p/D_m \approx 1$ ($N_{SH}\rightarrow2.0$), clearly, if $D_p \ll D_m \approx D_c$, and consequently, $R_p/3\, k_f \approx r_c^2/15\, k\, D_c \ll R_p^2/15\, \varepsilon_p\, D_p$ and the external resistance to mass transfer and the intracrystalline diffusional resistance will be negligible in comparison with the macropore diffusional resistance. With these approximations, eqn (6) may be written as:

$$\frac{\sigma^2}{2\,\mu^2} = \frac{D_L}{v\, L} + \frac{v}{L}\,(\frac{\varepsilon}{1-\varepsilon})(\frac{R_p^2}{15\,\varepsilon_p\, D_p})\, (\, 1 + \frac{\varepsilon}{(1-\varepsilon)\, k}\,)^{-2} \tag{7}$$

and the macropore diffusional time constant can be found, unambiguously, from the slope of plot of HETP versus fluid velocity.

4 RESULT AND DISCUSSION

A summary of physical properties modified resins are given in table 1.

Table 1: Physical properties of the Modified Resins.

RESIN	SURFACE AREA ($m^2\ g^{-1}$)	AVERAGE PORE VOLUME ($cm^3\ g^{-1}$)	AVERAGE PORE DIAMETER (Å)	%HALOGEN
XAD-2-I	168.2	0.48	118.3	30.48
XAD-2-II	168.6	0.50	119.3	26.17

The first and second moments were evaluated from the experimental chromatographic response curve by numerical integration.

The bed voidage was determined in two ways from the pressure drop by pumping buffer through the bed at flowrates within the range 0.1 - 5.0 ml min^{-1} (Re=1.0), and from the response for NaNO$_3$. The pressure drop was monitored as a function of the flowrate. When plotted according to the Carman-Kozany relationship the results showed that the bed voidage was 0.63 and 0.63 for XAD-2-I and XAD-2-II. Columns which yielded porosities outside this range were rejected. Furthermore, the porosities of the wet particles (ε_p) were estimated from these values and the dead time using eqn(2). Experiments in which the bed was bypassed showed that the contribution of the interconnecting pipe work to t_0 and peak spreading were negligible and eqn(2) was valid.

Since the NaNO$_3$ molecule is small enough to penetrate freely into the pores of the resins, there cannot be significant diffusion or adsorption ($k_{ads}=0$). According to eqn.(1), the plot of the first moment of the response against L/εv should be linear through the origin with a slope equal to bed voidage (ε). For the columns used in the present studies the void values determined in this way are 0.63 for XAD-2-I and 0.63 for XAD-2-II resins.

Thus, under conditions of axial dispersion control, one would expect to find a constant HETP. Since NaNO$_3$ molecules are small, penetrate freely and rapidly with no significant mass transfer resistance, the HETP values shows that HETP is approximately constant independent of fluid velocity and temperature over the experimental range. One may conclude that, mass transfer resistance must be negligible and HETP is determined by axial dispersion. The HETP values obtained are 0.40 for XAD-2-I and 0.58 for XAD-2-II.

4.1. Evaluation of Thermodynamic Parameters

Equilibrium adsorption constants (k_{ads}), summarized in Tables 2 and 3, were derived from the gradients of the plots of μ/L against reciprocal superficial velocity for all amino acids in the different adsorbates, using Eqn (1).

The k_{ads} value decrease, in all modified resins, with increasing temperature, showing that sorption equilibrium is more favourable at low temperatures. This implies that the adsorption of amino acids is an exothermic process. The differences in the values of k_{ads}, obtained for all amino acids examined in the different adsorbates may be explained in terms of compound molecular sizes and physico-chemical and polarities of the resins. Using all modified resins, a regular increase was observed in the equilibrium constants with the molecular size. The higher k_{ads} value that is observed for DL-Serine

Table 2: Thermodynamic and Kinetic Parameters of Amino Acids using XAD-2-I.

Aminoacid	Temperature (K)	Capacity Factor	k_{ads} (g/g)	$-\Delta H$ (KJ mol^{-1})	D_p (10^9 m^2 s^{-1})	D_p/R^2_p (10^3 s^{-1})
DL-Alanine	298	1.46	3.47	8.54	0.19	0.09
	308	1.40	3.14		0.20	0.10
	318	1.33	2.79		0.24	0.12
DL-Aspartic Acid	298	1.51	3.76	17.53	1.08	0.53
	308	1.42	3.28		1.18	0.58
	318	1.26	2.40		1.49	0.73
DL-Proline	298	1.39	3.11	2.87	0.353	0.17
	308	1.37	3.01		0.355	0.17
	318	1.35	2.89		0.358	0.18
DL-Serine	298	1.62	4.36	10.08	0.37	0.18
	308	1.53	3.88		0.38	0.19
	318	1.43	3.37		0.39	0.19
DL-Tryptophan	298	14.81	75.66	34.54	52.8	26.0
	308	10.88	54.40		58.1	28.6
	318	6.60	31.27		66.3	32.7
DL-Tyrosine	298	2.22	7.58	6.44	0.362	0.17
	308	2.11	7.03		0.367	0.18
	318	2.00	6.43		0.375	0.18
DL-Valine	298	1.59	4.20	12.28	0.72	0.35
	308	1.48	3.59		0.74	0.36
	318	1.38	3.07		1.04	0.51

Table 3: Thermodynamic and Kinetic Parameters of Amino Acids using XAD-2-II.

Aminoacid	Temperature (K)	Capacity Factor	k_{ads} (g/g)	$-\Delta H$ (KJ mol^{-1})	D_p (10^9 m^2 s^{-1})	D_p/R^2_p (10^3 s^{-1})
DL-Alanine	298	2.09	6.27	10.50	5.59	2.75
	308	2.01	5.88		6.79	3.34
	318	1.78	4.79		8.82	4.34
DL-Aspartic Acid	298	1.82	4.99	16.12	0.48	0.24
	308	1.59	3.86		0.49	0.24
	318	1.47	3.31		0.50	0.25
DL-Proline	298	2.31	7.34	22.19	4.98	2.45
	308	2.03	5.95		5.27	2.60
	318	1.65	4.16		5.57	2.74
DL-Serine	298	1.47	3.26	15.18	2.16	1.06
	308	1.38	2.86		3.55	1.75
	318	1.25	2.21		3.74	1.84
DL-Tryptophan	298					
	308					
	318					
DL-Tyrosine	298	9.19	9.19	11.44	5.45	2.68
	308	8.76	8.76		6.46	3.18
	318	6.85	6.85		7.14	3.52
DL-Valine	298	5.75	5.75	17.90	2.29	1.13
	308	4.06	4.06		2.60	1.28
	318	3.65	3.65		3.80	1.87

compared with other amino acids, which has a higher molecular size can be explained by the hydroxyl group of this amino acid. If the k_{ads} values for all amino acids are compared, excluding DL-Serine, in both resins it observed that have a much higher k_{ads} values in XAD-2-II than XAD-2-I. The results may be explained in terms of the pore diameter of the resins.

4.2. Evaluation of Kinetic Parameters

According to Eqn (7), when the controlling mass transfer resistance is macroporous diffusion and the axial dispersion term is either small or constant, the plots of HETP against interstitial liquid velocity should be linear with slope and intercept **which** are related to the mass transfer resistances and the axial dispersion coefficient respectively. The intercept occurred at a value of 0.40 for XAD-2-I and 0.58 for XAD-2-II, the same value obtained with the experiments using $NaNO_3$. The time constant and diffusivity constant derived from the slope are summarized in Tables 2 and 3. The derived diffusivity values are in the range of 10^{-8} to 10^{-9} m^2/s and increase with temperature, as expected for a hindered diffusion process. From the results it is observed that the sequence of diffusivities does not seem to follow the sequence of molecular size.

5 CONCLUSION

The application of liquid chromatography using a column packed with modified divinylbenzene-styrene resins as a means of measuring adsorption equilibrium constants and diffusivity constants in liquid phase systems has been demonstrated. The amino acids are all adsorbed, but there are significant differences in both the equilibrium constants and diffusivity constants in the three columns. It can be explained in terms of the physico-chemical properties of the resins and the polarities and molecular sizes of all the aminoacids tested.

6 REFERENCES

1. W. Voser, *J. Chem. Tech. Biotechnol.*, 1982, **32**, 109.
2. F. Addo-Yobo, N. K. H. Slater and C. K. Kennney, *Chem. Eng. J.*, 1988, **39**, B9
3. J. L. Casillas, J. Aracil, M. Martinez, F. Y. Addo-Yobo and C. K. Kennney, *Separations for Biotechnology*, 1990, **2**, 285.
4. J. L. Casillas, F. Y. Addo-Yobo, C. K. Kennney, J. Aracil and M. Martinez, *J. Chem. Tech. Biotechnol.*, 1992, **55**, 163.
5. D. D. Perrin, *Aust. J. Chem.*, 1963, **16**, 572
6. D. M. Ruthven, "Principles of Adsorption and Adsorption Processes", Wiley-Interscience, New York, 1984.
7. O. Levenspiel and K. B. Bischoff, "Advances in Chemical Engineering", Academic Press, New York, vol. 4, 1963, p 95.
8. N. Wakao and T. Funazkri, *Chem. Eng. Sci.*, 1978, **33**, 1375.

7 NOTATION

C: Adsorbate concentration in mobile phase, mmol g^{-1}
D_c: Intracrystalline diffusivity, cm^2 s^{-1}
D_L: Axial dispersion coefficient, cm^2 s^{-1}
D_P: Pore mass transfer coefficient, cm^2 s^{-1}
D_m: Solid-phase diffusion coefficient, cm^2 s^{-1}
k: Apparent constant
k_{ads}: Adsorption equilibrium constant, (g solute)(g resin)$^{-1}$.
k_f: External fluid film mass transfer coefficient, cm s^{-1}
L: Length of packing in column, cm

R_p: Radius of resin particle, cm
r_C: Crystal radius, cm
t: Time, s
v: Interstitial flow velocity, cm s^{-1}
GREEK LETTERS
ε: Bed porosity
ε_p: Porosity of the particle
μ: First moment
σ^2: Second moment

Solvent Selection Criteria for Protein Extraction Using Reverse Micellar Systems

H. B. Mat and D. C. Stuckey

DEPARTMENT OF CHEMICAL ENGINEERING AND CHEMICAL
TECHNOLOGY, IMPERIAL COLLEGE OF SCIENCE, TECHNOLOGY, AND
MEDICINE, LONDON SW7 2AZ, UK

1 INTRODUCTION

The application of reverse micellar systems for protein extraction has attracted considerable interest recently due to its ability to selectively solubilise proteins from an aqueous phase or from the whole broth and, in most cases, maintain their activity. The driving force for this transfer is either; electrostatic or hydrophobic interactions between the polar headgroups of surfactant aggregates and protein surface charge, a size exclusion effect, or a combination of all of these[1].

From past work[2] it is known that the solvent used in the formulation of reverse micelles has a significant effect on surfactant aggregation and structure, as well as on properties such as size and solubilisation capacity. However, most of the data on these effects are very fragmented and, with respect to protein partitioning, very little work has been done. Hence, the <u>objective</u> of this work was to examine the influence of a variety of solvents on the extraction of a protein with an anionic reverse micelle phase. In this paper, general guidelines for solvent selection for the reverse micellar extraction of α-chymotrypsin with the commonly used surfactant, sodium bis(2-ethylhexyl)sulfosuccinate (AOT) will be presented. The experimental techniques used were standard ones, and due to the lack of space are reported elsewhere[3]. It is hoped that this data will provide a firmer basis for the rational selection of a solvent phase.

2 INFLUENCE OF SOLVENT

In conventional liquid-liquid extraction the selection of solvents is an important part of process development, and high selectivity and capacity are of primary importance together with other parameters such as price, availability, toxicity and process compatibility[4]. In general, this selection often represents a compromise between the desired physicochemical properties of the system such as selectivity and capacity, and economic considerations such as cost and toxicity.

Solvent selection criteria for the liquid-liquid extraction of bioproducts are far more restrictive than those of conventional processes due to the fact that most bioproducts, such as proteins, are very sensitive to the solvent used. In the case of protein extraction using reverse micelles additional criteria have to be considered, and these include: whether or not the chosen solvent enables reverse micelles to form; the influence of the solvent on surfactant partitioning between the aqueous and solvent phases, and; the degree of water solubilisation (Wo = [water]/[surfactant]). These criteria ultimately influence the degree of protein extraction and stripping and, in the discussion below, data in these areas will be presented with the aim of drawing some general conclusions.

However, before discussing the data it would be useful to be able to classify the solvents based on a single physical property. There are numerous physical parameters that can be used for this such as; dielectric constant, dipole moment, and solubility in water, donor and acceptor number, and solubility parameter. Nevertheless, all these parameters only provide preliminary solvent extraction criteria since there is no direct relationship between them and solute selectivity[4]. In reverse micelle studies often the molar volume, solubility parameter, and log P are used, with the later two being used to explain the inactivation of enzymes and cells in an organic solvent[5]. Because of this, and other data obtained in our work (not shown), we have selected log P as the single best measure for solvent selection. Log P is defined as the logarithm of the partition coefficient of the solvent between octanol and water (log $P = \log([solvent]_{octanol}/[solvent]_{water})$).

2.1 Reverse Micelle Formation

It has been recognised that there are only a handful of solvents that form reverse micelles in comparison to the vast array available. As mentioned earlier, solvent type and structure affect the type of microemulsion formed, and its inherent physicochemical properties. Arkin and Singleterry[6] classified solvents into two groups; ones that are 'good', and ones that are 'poor' solvents for detergents. However, Magid[7] grouped solvents into: those forming micelles due to solvophobic interaction; those solvents that promote electrostatic interactions and as a result form reverse micelles, and; those in which aggregation does not occur. These three classes of solvents can be represented by the scheme below based on solvent polarity[8].

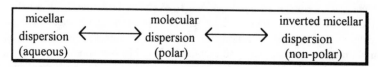

| micellar dispersion (aqueous) | ⟷ | molecular dispersion (polar) | ⟷ | inverted micellar dispersion (non-polar) |

On the basis of the above scheme, we can limit solvent selection only to those solvents that have nonpolar properties, and these include simple aliphatics, cycloaliphatic and aromatic hydrocarbons. Some chlorinated hydrocarbons such as carbon tetrachloride[9], chloroform[10] and trichloroethylene[1] were also found to have been used to prepare reverse micelles in the literature. Solvents, particularly those which have a hydrogen bond donor or acceptor, even if they are classified as a nonpolar solvent, for instance dioxan[11]or ethylacetate, are not good solvents for reverse micelle formation since the formation of hydrogen bonds seems to affect micellar aggregation, and prevents extensive association. Generally, we can conclude that the more non-polar the solvent is, the more likely the surfactant is to aggregate to form reverse micelles. Table 1, from literature studies, shows the range of solvents used to form reverse micelles, and it can be seen that most of them are aliphatics. Even though the number of solvents that form reverse micelles is small, at present it seems that there is no clear basis in choosing one solvent over another, especially in the area of protein extraction.

Table 1: Range of Solvents Used in the Literature

No.	Solvents	Log P[12]	Boiling Point(°C)[13]	Solubility[13]
1	2,2 dimethylbutane	3.3*	49.7	0.0018
2	Cyclohexane	3.2	80.7	0.0055
3	Hexane	3.5	68.7	0.0012
4	Heptane	4.0	98.4	0.0004
5	Octane	4.5	125.7	6.6×10^{-7}
6	Isooctane	4.5	99.2	0.0002
7	Decane	5.6	174.1	52ppb
8	Dodecane	6.6	216.3	3.7ppb
9	Tetradecane	7.6	253.7[15]	-
10	Hexadecane	8.8	287.0[15]	-
11	p-xylene	3.1	138.4	0.0156
12	Benzene	2.0	80.1	0.1791
13	Carbon tetrachloride	3.0	76.6	0.0770
14	Toluene	2.5	110.6	0.0515
15	Chloroform	2.1	61.2	0.8150

Solubility refers to solubility of the solvents in an aqueous phase in % (w/w)
* calculated according to reference 14.

2.2 Surfactant Partitioning

We have observed that solvent type and structure affect AOT partitioning, however, the extent of this partitioning also depends on the salt type and concentration (Figure 1)[16]. The difference in AOT partitioning into an aqueous phase for KCl and NaCl is due to

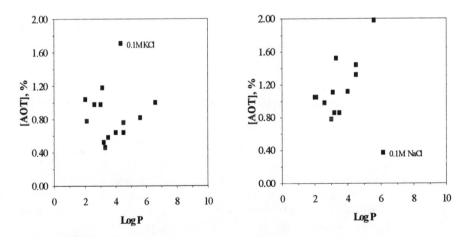

Figure 1: Effect of salt type on AOT partitioning into an aqueous phase for various solvents

differences in the hydrophilic-lipophilic balance (HLB) of the surfactant. NaCl promotes a higher HLB than KCl because Na has a larger radius of hydration than K, and hence results in a reduced screening affect on the surfactant headgroup. In the process development of reverse micelles for protein extraction, partitioning of the surfactant to the feed solution or to the aqueous product phase after back extraction is highly undesirable. This is not only because the amount of surfactant in the organic phase will decrease over time, but because of the more acute problem of product contamination.

2.3 Protein Extraction and Stripping

From our experimental work we have identified two classes of solvents: firstly, those for which water solubilisation, (C1-W) and protein transfer, (C1-T) significantly increased with decreasing polarity of the solvent (increasing log P), which corresponds to a series of simple aliphatic hydrocarbons, and; secondly, those for which the water solubilisation capacity (C2-W) significantly increased (at the 95% confidence interval) and protein transfer (C2-T) significantly decreased with decreasing polarity (Figure 2)[16]. In class 1 solvents increased protein partitioning was due to an increase in Wo which

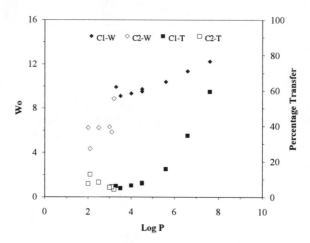

Figure 2: Solubilisation behaviour of α-chymotrypsin and water at 0.25M KCl and pH 5.5

indirectly reflects an increase in the size of the reverse micelles with decreasing solvent polarity. However, for class 2 solvents, increases in protein partitioning may also be due to the size of the reverse micelle formed. Decreases in Wo may be due to solvent penetration into the surfactant monolayer which restricts the amount of water solubilised. More polar solvents are expected to penetrate more deeply into the surfactant monolayer, and as a result, the water pool radius is less than the actual reverse micelle radius. When proteins partition into the reverse micelle water pool the interface is pushed outward to accommodate the protein molecules.

We also studied the backward transfer for simple aliphatic solvents, and the experimental results are given in Figure 3[16]. It was found that for the forward transfer of α-chymotrypsin under favourable conditions (0.1M KCl and pH 5.5), the extent of backward transfer at 1.0 M KCl and pH 7.5 increased with decreasing polarity of the solvent. It is possible that this increase in backward transfer is due to an increase in the α-chymotrypsin concentration inside the reverse micelle solvent phase. The activity of the recovered protein was measured using a bioassay, but no correlation was found between log P and protein activity recovered.

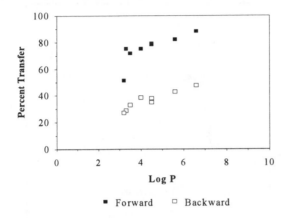

Figure 3: Percentage of α-chymotrypsin transfer for various solvents.

2.4 Water Solubilisation

We studied water solubilisation in the reverse micelle phase using a phase transfer technique for two commonly used systems; AOT/Solvent/KCl and AOT/Solvent/NaCl, and the results are presented in the form of a water solubilisation phase diagram shown in Figure 4[16]. The transition from one phase to another depends on the salt type and concentration, and the polarity of the solvent. For instance, all the solvents listed in Table 1 form reverse micelles in 1.0M KCl. However, by decreasing the salt concentration, some of the solvents were found to form bicontinuous microemulsions or micelles. In some cases, even though some solvents formed reverse micelles, they also formed a stable emulsion at the phase boundary which needed one to two weeks to separate.

In the literature on forward transfer, 0.05-0.1M KCl and NaCl are often used to make up a protein solution at the appropriate pH using solvents that have a log P between 3.3 to 6.6. These conditions produce reverse micelles large enough to solubilise the protein. In the case of backward transfer, higher salt concentrations of 1.0M KCl and NaCl are used in order to "salt out" the protein from the reverse micelle due to the formation of smaller micelles. Leodidis and Hatton[17] suggested that salts that form larger reverse micelles, such as sodium and calcium, seem to be better for forward transfer. In contrast, salts that form smaller sizes should be better for the backward process. However, in extracting a protein from a fermentation broth these conditions are usually not achievable since the ionic composition of the broth is determined by the quantity and type of salts in the nutrient media. Since most of the work reported in literature deals solely with the extraction of protein from a pure aqueous solution, Kamani compared the extent of α-chymotrypsin transfer to a 50mM AOT/isooctane reverse micelle system from either an aqueous solution of 0.1M KCl, or a filtered broth. He found that protein extraction from both systems were identical[18]. Therefore, it appears that the schematic phase diagram in Figure 4 will hold for broths as well as pure aqueous solutions. However, for the AOT/Solvent/NaCl system which forms larger reverse micelles (higher Wo), one would expect that the protein extraction efficiency would be higher than for KCl.

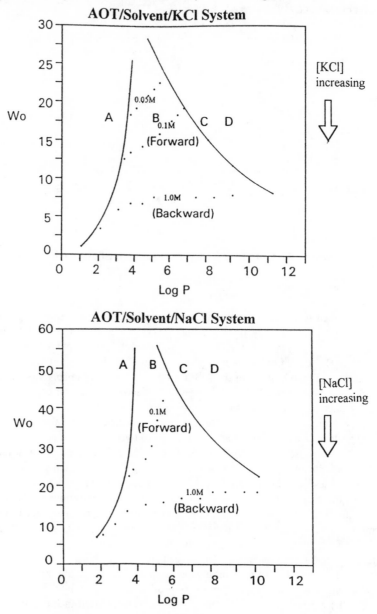

Figure 4: Schematic representation of a water solubilisation phase diagram[16].
(A, Stable emulsion; B, Reverse micelles; C, Bicontinuous
microemulsion; D, Micelles).([AOT]=50mM)

3 SOLVENT SELECTION CRITERIA

From the data presented, it can be seen that using log P as a selection criteria with Figure 4 will enable solvents to be chosen which will form reverse micelles as opposed to stable or bicontinuous emulsions, or micelles. In addition, depending on the salt type used eg NaCl or KCl, and its concentration, a preliminary estimate can be made of the Wo to be expected and hence the degree of protein extraction and stripping.

4 CONCLUSIONS

From the data presented above, preliminary solvent selection criteria based on log P and a water solubilisation diagram have been presented. These criteria specify that under certain physical conditions solvents with a log P between 3.3 and 6.6 provide a good compromise in terms of surfactant partitioning, solvent boiling point, solvent solubility in H_2O, toxicity and protein partitioning.

REFERENCES

1. T. A. Hatton, 'Surfactant Based Separation Processes' (Ed. by J.F. Scamehorn and H. Hatwell), Marcel Dekker Inc., New York, 1989.
2. H. F. Eicke, 'Topics in Current Chemistry', Springer-Verlag, New York, 1980.
3. H.B. Mat and D.C. Stuckey, 'Solvent Extraction In The Process Industries' (Ed. by D.H. Longsdail and M.J. Slater), Vol. 2, Elsevier, London,1993, 848.
4. M.J. Hampe, Ger. Chem. Eng., 1986, 9, 250.
5. L.E.S. Brink and J. Tramper, Biotechnol. and Bioeng., 1985, 27, 1258.
6. L. Arkin and C.R. Singleterry, J. Colloid Sci, 1946, 4, 537.
7. L. Magid, 'Solution Chemistry of Surfactants'(Ed. by K.L. Mittal), Vol.1, Plenum Press, New York, 1979.
8. A. Kitahara, T. Kobayashi and T. Tachibana, J. Phys. Chem., 1962, 66, 363.
9. K. Kono-no and A. Kitahara, J. Colloid interf. Sci, 1971, 35, 636.
10. R.C. Little and C.R. Singleterry, J. Phys. Chem., 1968, 68, 2709.
11. H.F. Eicke and H. Chrsten, J. Colloid interf. Sci, 1974, 46, 417.
12. C Laane, S. Boeren, K. Vos and C. Veeger, Biotechnol. Bioeng. 1987, 30, 81.
13. J.A. Riddick, W.B. Bunger and T.K. Sakano, ' Organic Solvents: Physical Properties and Methods of Purification', 4th. Edition, Wiley-Interscience, New York, 1986.
14. R.F.Rekker,'The Hydrophobic Fragmental Constant',Vol.1, Elsevier, New York,1977.
15. R. C. West, 'CRC Handbook of Chemistry and Physics', 63rd. Ed., CRC Press, Inc., Boca Raton, Florida, 1982.
16. H.B. Mat, Ph.D Thesis, in preparation.
17. E.B. Leodidis and T.A. Hatton, Langmuir, 1989, 5, 741
18. R. Kamani, Protein Extraction Using Reverse Micelles, M.Sc. Thesis, Imperial College of Science, Technology and Medicine,1993.

Metal-affinity Separations of Model Proteins Having Differently Spaced Clusters of Histidine Residues

V. Menart, V. Gaberc-Porekar,[a] and V. Harb

LEK, PHARMACEUTICAL AND CHEMICAL COMPANY, 61107 LJUBLJANA, SLOVENIA

[a]NATIONAL INSTITUTE OF CHEMISTRY, 61115 LJUBLJANA, SLOVENIA

1 INTRODUCTION

Demands for easy and low cost purifications of proteins are ever more increasing. Since the first report of Porath et al.[1] on the application of immobilized metal ions for protein separations, there have been many reports[2] of successful application of this principle for purification of naturally occurring, as well as for genetically engineered, proteins.

In engineering a protein for purification, two most general conditions must be fulfilled : namely, it must retain its biological properties and new amino acid residues must be accessible on the surface of the protein. In cases where the 3D structure is not known, this can be achieved most easily and directly by adding short purification tags to the N- or C-terminus of the protein. In the last few years poly-histidine tags added for separation on Ni(II)--nitrilotriacetic acid (NTA) columns seem to be the most popular approach for easy isolation of newly expressed proteins[3-6]. However, the attractiveness of more classical Metal--iminodiacetic acid (IDA) is not reduced, probably because it offers more variations in optimizing separation procedures. In addition, more specific approaches are possible if a 3D structure for the protein is available.

Our main objective was to test different possibilities for the introduction of basic amino acid residues, particularly histidines, in tumour necrosis factor alpha (TNF) molecule considering standard approaches as well as specific ones arising from structural and geometrical properties of the TNF trimer. TNF analogues having improved affinity to metal chelates should enable easy one step purification of some other TNF analogues currently under study.

2 MATERIALS AND METHODS

TNF structure was obtained from the Brookhaven Protein Data Bank[7]. Analysis of 3D structure was performed using program INSIGHT II (Biosym Technologies Inc.) and an IRIS graphical terminal.

TNF and analogues were produced mainly in *Escherichia coli* strain NM522. Pre-chromatographic steps included precipitation of crude bacterial extracts by polyethyleneimine (partial removal of nucleic acids and reduction of viscosity), followed by precipitation of a protein fraction with ammonium sulfate. Proteins were dissolved in phosphate buffered saline pH 7.1 to final concentrations usually not exceeding 2 - 3 mg of protein per milliliter of buffer.

All separations were performed at a flow rate of 1 ml per min, at room temperature. A Chelating Superose HR 10/2 column and standard FPLC instrument (two pumps P-500, UV-1 monitor, 2 ml sample loop) from Pharmacia were used. Eluates were collected as 1 ml fractions and analysed by SDS-PAGE. Proteins were visualized by silver staining. An imidazole gradient was applied for elution of TNF analogues because the stability[8] of TNF is known to be reduced below pH 5. The composition of buffers in all separations was essentially the same and imidazole gradients were formed combining buffers A and B (Figures 3, 4).

Buffer A : 0.2 M NaCl ; 0.02 M K-phosphate, pH 7.1
Buffer B : buffer A + 50 mM or 100 mM imidazole (pH corrected to 7.1)

The biological activity of crude bacterial extracts and purified proteins was determined by an in vitro cytotoxicity assay using L929 cell line[9].

3   RESULTS AND DISCUSSION

3.1 Possibilities for introduction of histidines in TNF

Native TNF monomer has only three histidine residues. Two of them (His15 and His78) are almost completely buried. Only His73 is at least partially accessible (Figure 1, e). TNF is considered to be a compact trimer[7] in solution and three surface accessible histidines per trimer seem to be responsible for weak retention on Cu(II)--IDA column, as already reported[10]. However, rather low affinity forces are not enough to enable one step purification from complex mixtures e.g. bacterial extracts.

Taking into account the basic structural and geometrical features of the TNF trimer, it was possible to determine a few regions where introduction of histidines (Figure 1) might substantially improve affinity to metal chelates as Metal--IDA or Ni(II)--NTA. Extension at the C-terminus is less attractive because the C-terminus is not on the surface, and changes at the C-terminus are known to affect or even destroy biological activity of TNF[11]. Introduction of histidines scattered all over the surface was not considered as a practical solution.

In contrast, clustering at the N-terminus and especially on exposed loops around the 3-fold axis seemed to be more useful.

While adding poly-histidine tags (e.g. 6His) at the N- or C-terminus is the usual approach, introduction of histidines inside the polypeptide chain is less common. However, it should be noted that affinity tags could be eventually removed after purification, if needed, while histidines inside the polypeptide chain remain in the protein and can affect biological activity.

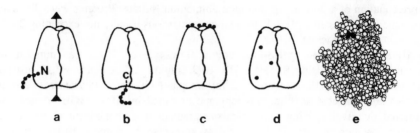

Figure 1 : *TNF trimer schematically presented with clusters of histidines (•) on N-terminus (a), C-terminus (b) and exposed loops (c) or scattered (d); the only surface accessible histidine in native TNF is shown in black (e).*

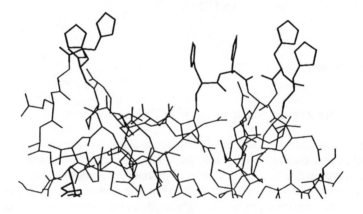

Figure 2 : *Structure of upper part of TNF trimer ; Glu107 and Gly108 are replaced by histidines in all three subunits. Surface accessibility and clustering in a plane perpendicular to 3-fold axis is evident.*

3.2 Extension of N-terminus

Short polypeptide sequences of histidines and arginines differing in the number and position were added to the N-terminus in order to compare the behaviour of equal number of histidines located on flexible peptide chains and those located on moderately rigid loops . Unfortunately, it turned out that in our case the very basic N-terminal extensions were quite susceptible to proteolytic degradation, at least in *E. coli* NM522 and some other strains used for the cytoplasmic expression of TNF analogues. As determined by Western blotting, two or three protein species of similar length always appeared in isolates, and because of this micro-heterogeneity, such cases were not analysed further.

3.3 His107 and His108

Much more useful results were found by replacing Glu107 or Gly108 with histidines such that both amino acid residues were located on exposed loops. Single mutations are multiplied due to the 3-fold axis of symmetry, so that the histidines are arranged circularly in a plane perpendicular to the axis of symmetry (Figure 2).
Both mutants were easily purified on Cu(II)--IDA in a single step. Somewhat lower affinity of His107 to Cu(II)--IDA in comparison to His108 was attributed to local chemical environments.   Otherwise, imidazole rings from histidines in both mutants seem to be geometrically very similar supposing that each imidazole ring can freely rotate and adopt the most appropriate orientation.

3.4 His107His108

If both histidines are present simultaneously, the affinity to Metal--IDA significantly increases which reflects in longer retention times. It was possible to isolate the pure His107His108 analogue on the Metal--IDA column using any of the metal ions tested (Cu, Ni, Zn, Co). Assuming that the concentration of imidazole necessary to elute the protein from the column is a good measure of affinity, it is interesting to notice that the order of elution was Co > Zn > Cu > Ni (Table 1). As reported[2], in most cases the order of elution for Cu and Ni is usually reversed. Our results (Table 1) refer to the concentrations of imidazole taken at maximum of the chromatographic peak and represent the mean value of the first three runs after loading the Chelating Superose column with an appropriate cation. On Cu and Ni the retention times are fairly constant, but in the case of Zn and Co, the retention times and the corresponding imidazole concentrations, necessary to elute the His107His108 analogue, tend to decrease with repeated use of the column, however, the resolution is not lost.

Table 1 : *Elution of His107His108 from Metal--IDA; imidazole concentrations were taken at maximum of each peak; all separations were done under identical conditions with linear imidazole gradient from 0 to 100 mM in 40 minutes (chromatograms not shown).*

	Metal ion			
	Co	Zn	Cu	Ni
Imidazole concentration [mM]	28	40	62	65

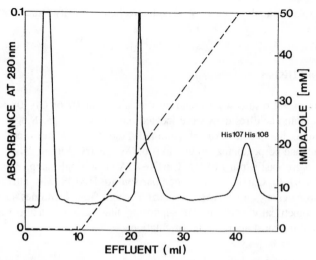

Figure 3 : *Standard purification of His107His108 from bacterial extract on Chelating Superose Cu(II)--IDA.*

We also tried to separate a composed mixture of purified TNF and its histidine analogues. By using the same imidazole gradient as in Figure 3, all the components were separated, but after optimization of the gradient much better resolution was achieved (Figure 4).

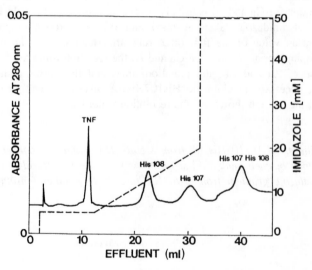

Figure 4 : *A mixture of approximately equal amounts (0.1 mg) of purified TNF and three histidine analogues separated on Cu(II)--IDA, using an optimized imidazole gradient.*

3.5 Conclusions

1) Replacement of Glu107 or Gly108 in TNF with histidine substantially increases the affinity to Cu(II)--IDA in comparison to native TNF.

2) The affinity appears to be additionally increased in double mutant His107His108. Isolation of a pure mutant was feasible in all Metal--IDA combinations (Co, Zn, Cu, Ni).

3) In all cases it was possible to isolate histidine analogues from crude *E.coli* extracts in a single step to achieve > 95 % purity only by changing imidazole gradient and holding other parameters (pH, salt concentration) constant. More than 90 % of the protein was usually recovered. The biological activity of histidine analogues described was completely retained.

4) We conclude that clustering of histidines in the plane perpendicular to the 3-fold axis enables cooperative action of most if not all histidines. Consequently, more specific and stronger interaction is possible than would be with the same number of histidines scattered over the entire surface of TNF trimer.

These results and conclusions will aid other experiments in protein engineering for purification, especially in proteins having stable oligomeric structure. We currently use the above experience for easy isolation of other TNF analogues. Along with practical value for purification purposes described, TNF analogues of this kind can be used as a good model for a thorough quantitative study of the protein - metal-chelate interactions.

Acknowledgements

A skillful technical assistance of Mrs. J. Lenarcic and Mrs. A. Jesenko is gratefully acknowledged. This research was supported by Ministry of Science and Technology of Slovenia.

References

1. J.Porath, J.Carlsson, I.Olsson and G. Belfrage, *Nature*, 1975, **258**, 598.
2. F. H. Arnold, *Bio/Technology*, 1991, **9**, 151.
3. E. Hochuli, H. Döbeli and A. Schacher, *J. Chromat.*, 1987, **411**, 177.
4. M. C. Smith, T. C. Furman, T. D. Ingolia and C. Pidgeon, *J. Biol. Chem.*, 1988 **263**, 7211.
5. R. Janknecht and A. Nordheim, *Gene*, 1992, **121**, 321.
6. J. Pohlner, J. Krämer and T.F. Meyer, *Gene*, 1993, **130**, 121.
7. M. J. Eck and S. R. Sprang, *J. Biol. Chem.*, 1989, **264**, 17595.
8. K. Haranaka, E. A. Carswell, B.D. Williamson, J.S. Prendergast, N. Satomi and L.J. Old, *Proc. Natl. Acad. Sci. U.S.A.*, 1986, **83**, 3949.
9. N. Mathews and M.L. Neale, " Lymphokines and Interferons ", IRL Press, Oxford, 1987, Chapter 12, p. 221.
10. E. Sulkowski, " Protein Purification : Micro to Macro", Alan R. Liss, Inc., New York, 1987, Chapter IV, p.149.
11. K. Gase, B. Wagner, M. Wagner, L. Wollweber and D. Behnke, *FEMS Microbiol. Lett.*, 1991, **84**, 259.

Chaperonin-assisted Refolding of Mitochondrial Malate Dehydrogenase

Andrew D. Miller, Karim Maghlaoui, Guido Albanese, Dirkjan A. Kleinjan, and Claire Smith

IMPERIAL COLLEGE OF SCIENCE, TECHNOLOGY, AND MEDICINE, DEPARTMENT OF CHEMISTRY, SOUTH KENSINGTON, LONDON SW7 2AY, UK

INTRODUCTION

Chaperonins are a subclass of the molecular chaperones which are a ubiquitous, abundant and highly conserved group of proteins which assist protein folding in cells.[1,2] Chaperonins first came to attention because of their specific induction during the cellular response of all organisms to heat shock[3,4] but are now known to be constitutively and abundantly expressed in the absence of any stress. Of all the chaperonins currently characterised, the best known are the *Escherichia coli (E. coli)* chaperonins cpn60 (groEL) and cpn10 (groES). Both are structurally quite well characterised. GroEL is a 14 subunit homo-oligomer composed of two stacked rings of 7 subunits each[5] whilst groES most probably consists of 7 subunits arranged in a single ring.[6] Subunit molecular weights derived from the gene-derived aminoacid sequences of groEL and groES are 57 259Da and 10 368Da respectively.[7]

Until recently, little was known of the way chaperonins assist protein folding but recent studies on chaperonins groEL and groES have begun to probe the mechanism of chaperonin intervention in protein folding. In particular, a number of well-defined *in vitro* systems have been devised to study chaperonin-assisted refolding of well-characterised enzymes using the *E. coli* chaperonins. These include studies on the refolding of chemically denatured Rubisco,[8,9,10] pre-β-lactamase,[11,12] citrate synthase,[13] LDH,[14] rhodanase[15,16] and mouse DHFR.[15,17] However, inspite of these studies, much of the mechanistic detail of chaperonin involvement in protein-folding remains elusive. In particular, it has still not been established whether chaperonins are catalysts or not of protein folding.

In an effort to understand the underlying chemical mechanism of chaperonin-assisted refolding of proteins, we have also been developing a number of model *in vitro* refolding systems using the *E. coli* chaperonins. In the following paper we report results obtained whilst studying the chaperonin assisted refolding of homodimeric porcine mitochondrial Malate Dehydrogenase (mMDH). Porcine mMDH was chosen as the substrate for an *in vitro* refolding model system because the enzyme is structurally characterised,[18] the refolding pathway for the enzyme has already been studied in detail[19] and the enzyme activity is readily determined by simple spectrophotometric assay.

Results described in this paper show that the chaperonins should not be regarded as refolding catalysts but as passive binding proteins which inhibit non-productive folding and aggregation reactions without substantially altering or accelerating the productive folding pathway itself.[1]

EXPERIMENTAL

Enzyme and Protein assays

mMDH homodimer concentrations were calculated from $A_{280}1\% = 2.5^{20}$ and a subunit molecular weight of 35 000Da.[21] Concentrations of *E. coli* groEL and groES were evaluated using extinction coefficients at A_{280} of 2.38×10^4 M^{-1} cm^{-1} for groEL and 3.44×10^3 M^{-1} cm^{-1} for groES.[9]

---

[1] Abbreviations used: cpn60, chaperonin 60; cpn10, chaperonin 10; Rubisco, Ribulose-1,5-bisphosphate carboxylase; DHFR, Dihydrofolate reductase; LDH, Lactate dehydrogenase; AMP-PNP, Adenosine 5'-(β,γ-imino)triphosphate; β-ME, β-mercaptoethanol

MDH renaturing experiments

GroEL and groES were prepared for use by dialysis against 150mM sodium phosphate, pH 7.6, containing 2mM β-ME and 1mM EDTA at 4°C. A stock solution of mMDH (~2.5 mg/ml) was prepared in the same buffer in a similar way.

An aliquot of mMDH stock solution was diluted to an enzyme concentration of approx. 0.3 mg/ml (4.3μM dimer concentration) in 150mM sodium phosphate, pH 7.6, containing 20mM β-ME, 10mM EDTA and 6M guanidinium hydrochloride. This solution was then incubated at 20°C for 2 h so as to fully denature the mMDH. Renaturation of mMDH was performed by diluting denatured mMDH to a concentration of 10 μg/ml (143nM dimer concentration, 30-fold dilution) in a variety of different renaturing buffers and incubating the solution for several hours at 20°C. Renaturing buffers (1ml) were typically composed of 150mM sodium phosphate, pH 7.6, containing 20mM β-ME, 10mM MgCl$_2$, 10mM KCl, 2mM ATP and varying amounts of groEL and groES. Fixed aliquots (20μl) of renaturing mixtures were removed at defined times and mixed with aliquots of an assay buffer (980μl) preincubated at 30°C. Assay buffer consisted of 150mM sodium phosphate, pH 7.6, 2mM β-ME, 0.5mM oxaloacetate and 0.2mM NADH. The initial rate of conversion at 30°C of NADH to NAD$^+$ (determined by the initial decrease [AU/min.] in the A$_{360}$ of assay mixtures) was used as a measure of mMDH reactivation during refolding. The activity of the refolding mMDH was expressed as a percentage relative to the activity of a control sample of undenatured mMDH (143nM dimer concentration) incubated at 20°C in a buffer of 150mM sodium phosphate, pH 7.6, containing 20mM β-ME, 10mM MgCl$_2$, 10mM KCl, and 2mM ATP.

RESULTS AND DISCUSSION

The experimental conditions, under which the refolding of mMDH was investigated, were devised as a compromise between the conditions used by Viitanen *et al.*[9] in the studies on chaperonin-assisted refolding of Rubisco and those conditions used by Jaenicke *et al.*[19] in studying mMDH refolding in the absence of chaperonins. In initial experiments, it was apparent that the refolded yields of active mMDH were substantially higher in the presence of groEL and groES than in their absence. Indeed, chaperonin-assisted refolding of mMDH resulted in at least 90 % recovery of active enzyme as compared to approx. 30 % recovery from spontaneous refolding in the absence of chaperonins and ATP (**Figure 1**).

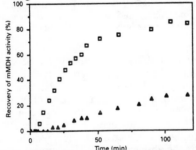

Figure 1. Time course of chaperonin-assisted (□) *and spontaneous refolding* (▲)*of mMDH. GroEL and groES homo-oligomer concentrations were 858nM and 1716nM respectively.*

The concentration of mMDH in the renaturing buffer during all these initial studies was 143nM (dimer concentration). Either increasing or decreasing this concentration failed to improve the yield of refolded mMDH in the presence of the chaperonins. Successful refolding of mMDH also required that the renaturation buffers contain phosphate and not Tris/HCl (in contrast with Viitanen *et al.* [9]) and that β-ME reducing agent be present throughout. A similar requirement for reducing agent was noted by Mendoza *et al.*[16] in their studies on rhodanase refolding.

A general scheme for the action of groEL and groES in refolding proteins has been outlined by Georgopoulos & Ang.[22] In this scheme, the unfolded protein binds to the large chaperonin groEL and then a combination of groEL-catalysed ATP hydrolysis and groES binding to groEL serve to promote protein refolding and/or subsequent release of the correctly folded protein from groEL. These studies on chaperonin-assisted refolding of mMDH closely support this general scheme. In the presence of groEL alone, without ATP present, there was no measurable recovery in mMDH activity with time at all, presumably because mMDH has all become bound to groEL (**Figure 2**). With groEL and ATP together, a small recovery (< 5 %) of mMDH activity was observed with time. However, when unfolded mMDH was combined with groEL and ATP followed by the subsequent addition of groES (after 45 min) a dramatic increase in the recovery of mMDH activity was observed. Indeed the recovery of mMDH activity paralleled, without loss in the yield of active mMDH, results obtained when unfolded mMDH was combined directly with groEL, groES and ATP together (**Figure 2**). Hence groEL, groES and ATP are *all* required for successful assisted refolding of mMDH.

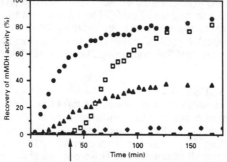

Figure 2. Chaperonin-assisted refolding of mMDH in the presence of groEL and groES (●), groEL alone (■), groEL with ATP (◆)and groEL with ATP followed by an addition of groES after 45 min. (□). Spontaneous mMDH refolding is also included (▲).Throughout, groEL and groES homo-oligomer concentrations were 858nM and 1716nM respectively and mMDH concentrations were 143 nM.

The reason that a modest amount of mMDH reactivation was observed with groEL and ATP alone may be because a small amount of the purified groEL is phosphorylated[23] so removing the requirement for groES to bring about the release of groEL bound protein. Alternatively groEL may contain a small amount of groES contamination. It has been shown that groES is not an absolute requirement for chaperonin-assisted refolding but apparently acts to improve the efficiency of the process.[11,13,15,17] However in the instance of chaperonin-assisted refolding of mMDH, groES must be considered an absolute requirement since the efficiency of assisted refolding is so low in the absence of groES (**Figure 2**).

Further studies with mMDH revealed that the chaperonin-assisted refolding was optimum when the groEL and groES homo-oligomers were used in at least a stoichiometric mole ratio with respect to each other (The preferred groEL:groES homo-oligomer mole ratio in all the experiments described in this paper was 1:2 which conferred a slight improvement over the 1:1 ratio). Furthermore, excess chaperonin was found to be essential for optimal yields of active mMDH with the best yields of refolded, active MDH obtained using a 6-fold molar excess of groEL homo-oligomer (together with a 12-fold molar excess of groES homo-oligomer) over mMDH homo-dimer. Further increases in the relative amounts of the chaperonins produced little further improvement. Similar observations have been made by Mendoza *et al.*[16] in their studies on rhodanase refolding. There is some debate[24] over the appropriate value of the extinction coefficient for groEL so it must be emphasised that calculations of groEL concentrations here are based upon the extinction coefficient determined by quantitative amino acid analysis.[9,12]

Investigations into the cofactor requirements of chaperonin-assisted refolding of mMDH revealed (**Table 1**) that K+ is not obligatory, as has been observed in other *in vitro* folding systems,[9] but ATP is essential. When ATP was substituted for by GTP or CTP, no active MDH was formed. A similar observation was made when ATP was substituted for by the nonhydrolysable ATP analogue AMP-PNP. Hence ATP hydrolysis is essential for chaperonin-assisted refolding of mMDH. This latter result differs from results obtained during studies on the refolding of LDH[14] and DHFR[17] which show that AMP-PNP is able to promote release of groEL-bound protein in an active form. At this stage there is no clear reason for the difference between these published observations and the results obtained here with mMDH.

Missing Component	Added Component	% Recovery of mMDH activity After 2 h incubation at 20°C
None	None	90
K+ ions	None	70
ATP	None	0
ATP	GTP*	0
ATP	CTP*	0
ATP	AMP-PNP*	0
ATP & groES	AMP-PNP*	0

*Nucleotides included at 2 mM concentration

Table 1. Influence of various factors on chaperonin-assisted refolding of mMDH.

In order to determine whether the chaperonins were changing the rate of mMDH refolding or not, a kinetic analysis of representative time course refolding data (see **Figure 1**) for chaperonin-assisted and spontaneous refolding of mMDH was carried out. In this analysis, an attempt was made to fit the observed refolding data using a simple kinetic model developed to explain the spontaneous refolding of mMDH by Jaenicke *et al.*[19] These authors have suggested that the spontaneous renaturation and reactivation of chemically denatured mMDH can be described *via* an irreversible consecutive transconformation-association process according to the minimum scheme in **Equation 1** (where N and N' are native enzyme and an inactive dimeric intermediate respectively; M and M' are two conformational states of catalytically inactive mMDH monomers).

$$2M \xrightarrow[\text{uni}]{k_1} 2M' \xrightarrow[\text{bi}]{k_2} N' \xrightarrow[\text{uni}]{k_{1'}} N \tag{1}$$

Jaenicke *et al.*[19] also suggested that the sigmoidal character of the kinetic traces for the time-dependent spontaneous recovery of mMDH activity (see **Figure 1**) indicated that the reactivation reaction could not be determined by a single rate-limiting step. Hence two alternative irreversible, consecutive reaction sequences, involving two rate-limiting steps, were proposed so as to describe the observed kinetic behaviour (**Equations 2 and 3**). In this instance, irreversibility only implies that the reverse reactions have rate constants much below those for the forward reactions.

$$2M \xrightarrow[\text{uni}]{k_1} 2M' \xrightarrow[\text{bi}]{k_2} N' \qquad (2)$$

$$2M' \xrightarrow[\text{bi}]{k_2} N' \xrightarrow[\text{uni}]{k_{1'}} N \qquad (3)$$

Although, both alternative pathways are plausible, Jaenicke *et al.*[19] found that the consecutive uni-bimolecular reaction process (**Equation 2**) was sufficient to describe both the sigmoidicity and the concentration dependence of the spontaneous reactivation process. Using a mathematical approximation of the consecutive uni-bimolecular reaction determined by Chien,[25] Jaenicke *et al.*[19] derived one set of kinetic constants, $k_1 = 6.5 \times 10^{-4}$ s^{-1} and $k_2 = 3 \times 10^4$ M^{-1} s^{-1}, which was sufficient to describe the reactivation behaviour of mMDH over the full range of concentrations used.

All our foregoing studies on chaperonin-assisted and spontaneous refolding of mMDH were performed under the same conditions of temperature and pH as the original refolding studies of Jaenicke *et al.*[19] Unsurprisingly therefore, representative time course data obtained for spontaneous refolding of mMDH agreed closely (**Figure 3**) with theoretical time course data obtained using both the mathematical approximation[25] for the consecutive uni-bimolecular reaction (**Equation 2**) and the previously determined values of k_1 and k_2 mentioned above. This theoretical data was calculated assuming an initial concentration of 94nM for the putative folding intermediate M (**Equation 2**) which when carried through the irreversible folding pathway (**Equation 1**) would result in a final concentration of native dimeric enzyme (N) corresponding to 47nM. Starting from an initial concentration of 143nM (dimer concentration) for the unfolded enzyme, 47nM corresponds to a 33 % recovered yield of active enzyme which agrees with the experimental results (**Figure 3**). However, of considerable surprise was that representative time course data obtained for chaperonin-assisted refolding of mMDH also agreed closely (**Figure 3**) with a theoretical time course calculated using the same mathematical approximation and values of k_1 and k_2 as above. The only variable altered was the initial concentration of M (**Equation 2**) which was increased to 260nM. Following the same argument as above, this would result in a final concentration of native dimeric enzyme (N) of 130nM which corresponds to a 90 % recovered yield of active enzyme, in agreement with the experimental results for chaperonin-assisted refolding of mMDH (**Figure 3**).

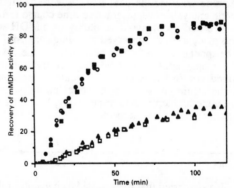

Figure 3. Curve fitting of representative time course data for chaperonin-assisted and spontaneous refolding of mMDH Two theoretical sets of time course data were calculated in which the concentration of monomeric folding intermediate M (Equation 2) was taken to be 260 nM (O) and 94 nM (▲) respectively. Rate constants of $k_1 = 6.5 \times 10^{-4}$ s^{-1} and $k_2 = 3 \times 10^4$ M^{-1} s^{-1}, obtained previously,were used to calculate both curves. Representative time course data for chaperonin-assisted (●, ■) and spontaneous refolding of mMDH (Δ,□) were plotted alongside the theoretical data for comparison.

Clearly the simple consecutive uni-bimolecular reaction scheme, used to describe spontaneous mMDH reactivation, applies equally well for the chaperonin-assisted reactivation. Moreover, a simple conclusion may be drawn from this result. The concentration of M represents the quantity of mMDH committed to entering the productive folding pathway leading to active enzyme. This quantity is three times higher for chaperonin-assisted refolding of mMDH than the spontaneous process. Hence it appears that the chaperonins, groEL and groES, are promoting mMDH refolding merely by increasing the effective concentration of productive folding intermediate M thereby increasing the flux of unfolded enzyme through the productive refolding pathway. There appears to be little or no alteration and/or acceleration of that pathway. How general is this conclusion?

Recent studies on the chaperonin-assisted refolding of citrate synthase[13] and pre-β-lactamase[11] have implied that refolding kinetics are unaffected by chaperonins and hence support the above conclusion. Moreover, recent work on chaperonin assisted refolding of pre-β-lactamase[12] has provided some evidence that the folding pathway of pre-β-lactamase is unaltered by the chaperonins. However by contrast, it has been reported that chaperonins cause a rate acceleration of Rubisco-refolding[9] and in other cases they have reportedly decreased the rate of protein refolding.[15,16] Hence it is not possible to completely generalise the above conclusion to the chaperonin-assisted refolding of all other proteins.

The characteristics of the chaperonins thus far described are *not* those of folding catalysts. However, chaperonins were found able to turnover at least six times in the following experiment. Denatured mMDH was diluted to a final concentration of 143nM (dimer concentration) in a renaturing buffer containing groEL homo-oligomer (143nM) and groES homo-oligomer (286nM). After 30 min, another mole equivalent of denatured mMDH was added to the renaturing buffer. Subsequently, four further mole equivalents of denatured mMDH were added to the renaturing buffer at 30 min intervals. **Figure 4** shows the stepwise increase of mMDH activity resulting from the first three additions of denatured mMDH. The final concentration of mMDH, following these six sequential additions was 858nM (dimer concentration). Final mMDH activity was approx. 40 % of that measured for a control sample of native mMDH at 858 nM (dimer concentration).

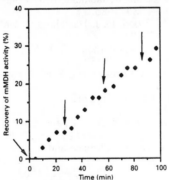

Figure 4. Observation of several cycles of chaperonin-assisted refolding of mMDH

Results with pre-β-lactamase[12] and homo-dimeric LDH[14] have shown that two polypeptide chains bind to the groEL homo-oligomer. Assuming the same is true of mMDH as well, then the refolding of mMDH observed in this experiment must arise from six sequential cycles of mMDH binding to groEL, refolding and release. In effect, each turn-over of the chaperonin-assisted refolding process is about 85 % efficient.

Langer *et al.*[26] have also demonstrated that groEL and groES may bring about successive rounds of protein refolding. In addition, they observed that groEL may act as a cellular catalyst of protein folding when working in collaboration with the *E. coli* chaperone DnaK and heat shock proteins DnaJ and GrpE. The results of Langer *et al.*[26] raise the possibility that groEL and groES may also increase the rate of mMDH refolding in the presence of DnaK, DnaJ and GrpE. With such a well characterised model refolding system as mMDH refolding, it should be relatively easy to put this possibility to the test.

ACKNOWLEDGEMENTS

We thank the Royal Society, the SmithKline (1982) Foundation, Roche Products Ltd, Merck Sharp & Dohme and the Wolfson Foundation for financial support.

REFERENCES

1. R. J. Ellis and S. M. van der Vies, *Annu. Rev. Biochem.*, 1991, **60**, 321.
2. M.-J. Gething and J. Sambrook, *Nature*, 1992, **355**, 33.
3. R. I. Morimoto, A. Tissieres and C. Georgopoulos, eds., 'Stress Proteins in Biology and Medicine', Cold Spring Harbor Laboratory Press, Cold Spring Harbor, NY, 1990.
4. M. J. Schlesinger, G. Santoro, and E. Garaci, eds., 'Stress Proteins: Induction and Function', Springer-Verlag, Heidelberg, 1990.
5. R. W. Hendrix, *J. Mol. Biol.*, 1979, **129**, 375.
6. G. N. Chandrasekhar, K. Tilly, C. Woolford, R. Hendrix and C. Georgopoulos, *J. Biol. Chem.*, 1986, **261**, 12414.
7. S. M. Hemmingsen, C. Woolford, S. M. van der Vies, K. Tilly, D. T. Dennis, C. P. Georgopoulos, R. W. Hendrix and R. J. Ellis, *Nature*, 1988, **333**, 330.
8. P. Goloubinoff, J. T. Christeller, A. A. Gatenby and G. H. Lorimer, *Nature*, 1989, **342**, 884.
9. P. V. Viitanen, T. H. Lubben, J. Reed, P. Goloubinoff, D. P. O'Keefe, and G. H. Lorimer, *Biochemistry*, 1990, **29**, 5665.
10. S. M. Van der Vies, P. V. Viitanen, A. A. Gatenby, G. H. Lorimer and R. Jaenicke, *Biochemistry*, 1992, **31**, 3635.
11. A. A. Laminet, T. Ziegelhoffer, C. Georgopoulos and A. Pluckthun, *EMBO J.*, 1990, **9**, 2315.
12. R. Zahn and A. Pluckthun, *Biochemistry*, 1992, **31**, 3249.
13. J. Buchner, M. Schmidt, M. Fuchs, R. Jaenicke, R. Rudolph, F. X. Schmid and T. Kiefhaber, *Biochemistry*, 1991, **30**, 1586.
14. I. G. Badcoe, C. J. Smith, S. Wood, D. J. Halsall, J. J. Holbrook, P. Lund and A. R. Clarke, *Biochemistry*, 1991, **30**, 9195.
15. J. Martin, T. Langer, R. Boteva, A. Schramel, A. L. Horwich and F.-U. Hartl, *Nature*, 1991, **352**, 36.
16. J. A. Mendoza, E. Rogers, G. H. Lorimer, and P. M. Horowitz, *J. Biol. Chem.*, 1991, **266**, 13044.
17. P. V. Viitanen, G. K. Donaldson, G. H. Lorimer, T. H. Lubben, and A. A. Gatenby, *Biochemistry*, 1991, **30**, 9716.
18. S. L. Roderick, and L. J. Banaszak, *J. Biol. Chem.*, 1986, **261**, 9461.
19. R. Jaenicke, R. Rudolph and I. Heider, *Biochemistry*, 1979, **18**, 1217.
20. E. M. Gregory, F. J. Jr. Yost, M. S. Rohrbach and J. H. Harrison *J. Biol. Chem.*, 1971, **246**, 5491.
21. C. J. R. Thorne and N. O. Kaplan, *J. Biol. Chem.*, 1963, **238**, 1861.
22. C. Georgopoulos and D. Ang, 'Seminars in Cell Biology: Molecular Chaperones', W. B. Saunders, Philadelphia, 1990, **1**, p. 19.
23. M. Y. Sherman and A. L. Goldberg, *Nature*, 1992, **357**, 167.
24. M. T. Fisher, *Biochemistry*, 1992, **31**, 3955.
25. J.-Y. Chien, *J. Am. Chem. Soc.*, 1948, **70**, 2256.
26. T. Langer, C. Lu, H. Echols, J. Flanagan, M. K. Hayer and F. U. Hartl, *Nature*, 1992, **356**, 683.

Mathematical Modelling and Computer Simulation of Continuous Aqueous Two-phase Protein Extraction

S. L. Mistry, J. C. Merchuk, and J. A. Asenjo

BIOTECHNOLOGY AND BIOCHEMICAL ENGINEERING GROUP,
DEPARTMENT OF FOOD SCIENCE AND TECHNOLOGY, UNIVERSITY OF
READING, PO BOX 226, READING RG6 2AP, UK

Abstract

A mathematical model is presented to describe the continuous, steady state operation of aqueous two-phase protein extraction. The model is based on the fundamental mass balances of the main components and phase equilibrium data in the form of equations for the binodial and the tie-line. SPEEDUP, a commercial simulation software, was used to solve the model. The model has been fitted to two sets of data i) Separation of thaumatin from *E.coli* homogenate proteins, and ii) Separation of α-amylase from *B.subtilis* supernatant. Model validation has been carried out for all of the experimental data. Simulation results showing the sensitivity to key process parameters, and the effect of process variables on performance are presented and discussed.

The model has now been extended to take into account phase separation and thus aspects of continuous processing. This model can be used to follow the effect of changes in key parameters which are essential to continuous operation in settlers, particularly input flow rates, phase composition and protein concentration.

1 Introduction

Aqueous two-phase systems (ATPS) exploit the incompatibility between aqueous solutions forming two phases[1]. The main difference between ATPS and other liquid/liquid extraction systems lies in the similarities of the physical properties, such as the density and polarity, between the two phases in equilibrium. This is due to the relatively high solubility of the basic components in both phases. They can be used to separate therapeutic and other proteins from contaminants and therefore represent an inexpensive and efficient initial downstream processing step. These systems can be relatively sensitive to their environmental conditions and hence, a mathematical model can be useful. It will describe the effect of key parameters on performance and will assist in process design, scale up. Numerous experimental data are available on phase equilibrium. However, no models for continuous ATPS have been reported in the literature until recently[2].

The main objectives for this work were to set up a steady-state mathematical model based on first principles and to use computer simulations to gain an insight on the system and the effect of process parameters on performance as well as to investigate the extension of this model to take into account rates of phase separation during continuous processing.

2 Modelling

2.1 Process Description

The process illustrated in Figure 1 shows the two stage separation system used. The assumptions made were that i)complete equilibrium occurred in each stage and ii)all the contaminants were lumped in one K_c. The first stage is the main separation step where *E.coli* lysate[3] with or without cell debris or Bacillus crude industrial fermentation supernatant[4] together with the protein product is entered. The protein product is partitioned to the top phase and then back extracted into a fresh bottom phase in stage 2. The top phase is recycled back into stage 1 to minimise PEG and product loss thus increasing the process yield in comparison to a single stage batch extraction.

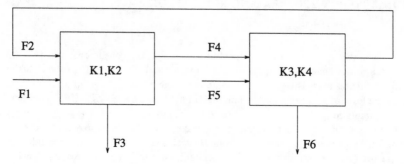

Figure 1: Flow Scheme of Process

2.2 Phase Equilibrium

To simulate the separation process, prediction of the phase diagram is essential. An exponential empirical curve (eq. 1) can be fitted to the experimental binodial data[2] :-

$$PEG_i = B_1 * exp(B_2 * PSAL_i) \tag{1}$$

For a binodial system of PEG4000/phosphate at pH7 $B_1 = 79\%(PEG)$ and $B_2 = -0.26(1/\%SALT)$.

2.3 Phase Separation

Figure 2 shows the set of equations representing the mass balances for stage 1. Similar equations can be written for stage 2[2]. The solution of the mass balances, together with the equations representing the binodial line, gives the composition of the phases leaving the stage, and are connected by a tie line of the form :-

$$PEG_i = C_1 * PSAL_i + C_2 \tag{2}$$

here C_1 is constant and C_2 varies for each tie-line.

3 Performance of Steady-State Model

3.1 Prediction of system performance

The model has been used to study the effects of input variables on system purity and yield over a range of conditions. The variables were increased/decreased until the model broke one of the following constraints :-

1. The system entered the homogeneous region, left of the binodial curve
2. The tie-line moved outside of the binodial curve, which is supported by experimental data, to the right.

An example of such a simulation is shown in Figure 3 [2].

Stage 1

$$F1 + F2 = F3 + F4 \tag{I}$$

$$F1^*PEG1 + F2^*PEG2 = F3^*PEG3 + F4^*PEG4 \tag{II}$$

$$F1^*PSAL1 + F2^*PSAL2 = F3^*PSAL3 + F4^*PSAL4 \tag{III}$$

$$F1^*PRO1 + F2^*PRO2 = F3^*PRO3 + F4^*PRO4 \tag{IV}$$

$$F1^*CON1 + F2^*CON2 = F3^*CON3 + F4^*CON4 \tag{V}$$

$$F1^*SALT1 + F2^*SALT2 = F3^*SALT3 + F4^*SALT4 \tag{VI}$$

$$SALT3 = SALT4 \tag{VII}$$

Figure 2: Mass Balance equations for stage 1

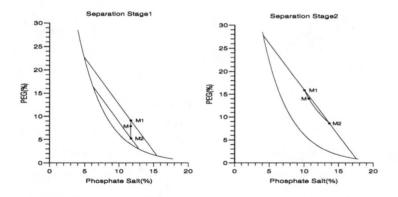

Figure 3: The effect of varying the PEG1 concentration entering stage 1

3.2 System Stability

To study system stability (defined by operating at a predefined distance from the binodial[2]), movements in the tie-line positions were studied for variations in PEG1, PSAL1, F5 and PSAL5 over their working range. If PEG1, PSAL1, F5 or PSAL5 are changed (which is how the manipulated variables would be changed in a real continuous system) this results in changes of the type shown in Figure 3 (M→M1, M→M2).

Figure 3 illustrates the effect of varying the PEG1 concentration entering stage 1. Increasing the PEG1 concentration (moving from M→M1) leads to a larger top phase and greater product separation in the first stage. This also leads to a smaller bottom phase in the second stage and lower product recovery. Moving from M to M2 represents decreasing the PEG1 conc. giving a smaller top phase in the separation. In the second stage the bottom phase becomes larger and product recovery is increased.

3.3   Purity and yield

The effect of varying four variables, PEG1, PSAL1, F5, and PSAL5, on purity and yield in the overall process were investigated[2]. An example is shown in Figure 4 in which the effect of changing the salt concentration entering the first stage (PSAL1) is shown. The purity and yield are fairly constant at high concentrations but there is a dramatic change at the lower range. In the region below PSAL1 concentration 7.5-8% the changes in purity and yield are very drastic. This can be attributed to the rapid decrease in tie-line length as we approach the binodial in stage 1. The high partition coefficient and minimal product loss (in the bottom phase of stage 1) are the principal reasons for the high yield (0.8-0.92) obtained for larger PSAL1. Further improvements in product purity could be obtained if a method for lowering the partition coefficient for contaminants is developed.

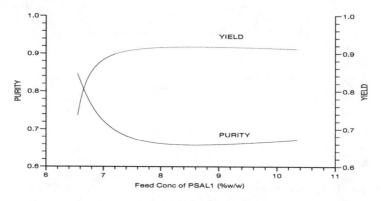

Figure 4: The effect of varying the PSAL1 concentration on purity and yield

4   Extended Model

This model differs from the previous one described above by incorporating :-
i)A cubic fitting for the binodial,

The binodial data of the Thaumatin system for a PEG4000/Phosphate system at pH7 was represented by a third order polynomial of the form :-

$$PEG_i = D_1 * PSAL_i^3 + D_2 * PSAL_i^2 + D_3 * PSAL_i + D_4 \qquad (3)$$

This form allowed a better fit of the experimental data than eq. 1.

ii)A function relating the partition coefficient to the concentration of NaCl instead of constant values of K, as used previously[2].

The partition behaviour of thaumatin[3] and α-amylase[4] was studied to show the effect of the salt concentration on the partition coefficient (K). Partitioning of contaminants could be modelled by a straight line as they were not a strong function of the concentration of NaCl.

$$LogK = E_1 * NaCl + E_2 \qquad (4)$$

The partition behaviour of the two proteins as a function of NaCl concentration in the system was more complex and was fitted to a Sigmoidal Boltzman curve of the form :-

$$LogK = \frac{G_1}{1 + exp\left\{\frac{(NaCl-G_3)}{G_4}\right\}} + G_2 \qquad (5)$$

The new cubic equation to model the binodial gave a better fit (see Figure 5) to the data than the original exponential fit. Also, the Sigmoidal Boltzman to express the value of the partition coefficient (Log K) of the thaumatin and α-amylase data gave very good fits (see Fig 6) and can now be used to model the partition coefficient instead of assuming it is constant in the separation and back extraction stages as in the simpler version of the model[2].

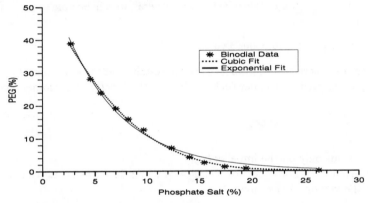

Figure 5: Comparison of Binodial Data to Cubic and Exponential Fits

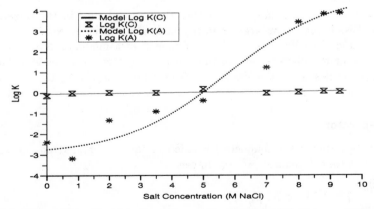

Figure 6: Comparison of α-Amylase(A) and Contaminant(C) Data and Model Fit

4.1   Modelling of Phase Separation

The steady-state model described above makes two important assumptions :-
i)total equilibrium is obtained in each of the contacting stages
ii)complete separation of the two phases can be obtained after each mixing/equilibration
stage.

These assumptions, do not take into consideration the performance of the equipment used for mixing and for separation. We will thus investigate how to relate performance and design information on the mixer and the separator of the two phases to the parameters of the model.

4.2   Mixer

The principle point here is that there is a minimum time required to complete mixing at the molecular scale (Micromixing). It will be assumed that the time required for macromixing, t_m, can be taken instead. In general, we will be able to find a function of the type :-

$$(N \bullet t_m) = f_1(\text{Re, Impeller Type}) \tag{6}$$

Efficient mixing requires turbulence and this establishes the lower limit for N. It has been widely reported that under such conditions, the function f_1 becomes a constant, I :-

$$t_m = \frac{I}{N} \tag{7}$$

where I depends only on the impeller type.

For a given total flow rate (for instance, [F1+F2] in stage 1), t_m can be used to calculate the volume of the mixer, V_{m1} :-

$$V_{m1} = \frac{J \bullet t_m}{[F1 + F2]} \tag{8}$$

The constant J takes into account that the required time in the mixer, which includes phase formation, is larger than the mixing time given by eq. 7.

In the case of on line mixers, the requirements of a minimum N is replaced by the election of the diameter of the tube (which together with the total flow rate establishes the Re). Once the diameter is fixed, a function relating t_m (or number of mixing elements) to flow rate will give the minimal volume of mixer required.

4.3   Separator

The separation after complete formation of the equilibrated phases has been achieved is a time-dependent process. In general :-

$$\left(\frac{dh^+}{dt}\right) = f_2(\Delta\rho, \sigma, \mu_h, \mu_L, d_d) = f_2(\text{Re,We}) \tag{9}$$

The mean droplet diameter is a function of the hydrodynamic conditions in the mixer. In a stirred mixer, it will be a function of Re, We and t_m :-

$$d_d = f_3(\text{Re,We,}t_m) \tag{10}$$

where the velocity of the droplets could be taken as proportional to N.

Integration of eq. 9 gives the profile of interphase height as a function of time and from it the time required for a complete separation, t_s, can be estimated as :-

$$t_s = f_4(\text{Re},\text{We}) \tag{11}$$

The volume of the separator required after the first mixer, V_{s1} can now be obtained from :-

$$V_{s1} = \frac{t_{s1}}{[F1 + F2]} \tag{12}$$

Equation 12 gives the maximal value of separator volume, since complete disappearence of the interface is assumed. It seems more logical to operate the separator at a steady-state where interface, top and bottom phases coexist. A volume balance will give :-

$$\left(\frac{dh^+}{dt}\right) * h_0 * A = F_1 + F_2 \tag{13}$$

If a certain value of h^+ (say 1/3) is adopted, this means that the separator will run continuously containing 1/3 bottom phase, 1/3 interface and 1/3 top phase. The total volume of the separator (h_0*A) can be obtained from eq. 9, and adopting h_0 such that the outgoing fluxes can be taken without interface, A, the cross sectional area can be obtained.

In this way, it is possible now to extend the simple steady-state model described above to take into account design variables for both the mixer and the separators. If, for instance, we take stage 1 in the original model[2], the expressions for the function f_1 can be found in the literature, f_2 and f_3 are presently being measured from experiments and from this we can obtain f_4. Taking eq. 7, 8, 11 and 12, and adding them to the original model[2] will allow the simulation of the complete system including the design and operation variables for the mixers and separators.

5 Acknowledgements

This work was partially supported by SERC to whom thanks are due.

Nomenclature

A	Cross sectional area of separator, [m²].
B_i	System constants for the exponential equation.
C_i	System constants for the tie-line equation.
CON_i	Concentration of contaminant in stream i, [kgm⁻³].
D_i	System constants for the cubic equation.
E_i	System constants for the contaminant partition coefficient equation.
F_i	Volumetric flowrate in stream i, [m³s⁻¹].
G_i	System constants for the protein partition coefficient equation.
h	Height of interphase, [m].
h^+	Relative height of interphase (= h/h_0).
h_0	Height of separator, [m].

I	Limiting value of f_1 at high Reynolds number.
J	Factor in eq. 8.
K	Partition coefficient.
N	Impeller velocity (rpm).
PEG_i	Concentration of PEG in stream i, $[kgm^{-3}]$.
PRO_i	Concentration of protein in stream i, $[kgm^{-3}]$.
$PSAL_i$	Concentration of phosphate salt in stream i, $[kgm^{-3}]$.
Re	Reynolds number.
$SALT_i$	Concentration of NaCl in stream i, $[kgm^{-3}]$.
t_m	Mixing time, [s].
t_s	Separation time, [s].
V_{mi}	Volume of mixer for stage j, $[m^3]$.
V_{si}	Volume of separator for stage j, $[m^3]$.
We	Weber number.
$\Delta\rho$	Density difference between phases, $[kgm^{-3}]$.
σ	Interfacial tension, $[Nm^{-1}]$.
d_d	Mean droplet diameter, [m].
μ_h	Viscosity of heavy phase, $[Nsm^{-2}]$.
μ_L	Viscosity of light phase, $[Nsm^{-2}]$.

References

[1] Albertsson P.A. *Partition of Cell Particles and Macromolecules.* J.Wiley, New York., third edition, 1986.

[2] Mistry S.L., Asenjo J.A., and Zaror C.A. Mathematical modelling and simulation of aqueous two-phase continuous protein extraction. *Bioseparation*, 3:343–358, 1993.

[3] Cascone O., Andrews B.A., and Asenjo J.A. Partitioning and purification of thaumatin in aqueous two-phase systems. *Enzyme Microb. Technol.*, 13:629–635, 1994.

[4] Schmidt A.S., Ventom A.M., and Asenjo J.A. Partitioning and purification of alpha-amylase in aqueous two-phase systems. *Enzyme Microb. Technol.*, 16:131–142, 1994.

[5] Sembria A., Merchuk J.C., and Wolf D. Characteristics of a motionless mixer for dispersion of immiscible fluids. i. a modified electroresitivity probe technique. *Chem. Eng. Sci.*, 41:1007, 1986.

[6] Sembria A., Merchuk J.C.and, and Wolf D. Characteristics of a motionless mixer for dispersion of immiscible fluids. iii. dynamic behaviour of the average drop size and dispersed phase hold up. *Chem. Eng. Sci.*, 43:373, 1988.

Direct Integration of Protein Recovery with Productive Fermentations

Philip Morton and Andrew Lyddiatt

BIOCHEMICAL RECOVERY GROUP, BBSRC CENTRE FOR BIOCHEMICAL
ENGINEERING, SCHOOL OF CHEMICAL ENGINEERING, UNIVERSITY OF
BIRMINGHAM, BIRMINGHAM B15 2TT, UK

1. INTRODUCTION

The effective cost of recovery and purification of biochemicals
manufactured by microorganisms in industrial fermentations may total
50–70% of market value. Economic realities confirm that consideration
of downstream processing should strongly influence (i) the design
strategies applied upstream to the assembly of microbial genomes,
(ii) the machineries of product expression, and particularly (iii)
the physical and biochemical manipulation of the source microorganism
and product in the bioreactor. In the context of (iii), the true
integration of protein product recovery with upstream processes of
fermentation, commonly reviewed as extractive bioconversion or
biotransformation (1), remains a coveted goal of bioprocess research.

1.1 Protein Recovery Integrated with Continuous Cultures

The integration of protein isolation with batch or continuous
fermentations has many process advantages. Thus, a simple reactor
(Figure 1a) operating at a productive steady state, and diluting at a
fixed rate, generates outputs which can be rapidly processed online.
Cells can be removed by continuous centrifugation, microfiltration or
passive sedimentation, and recycled if appropriate. Clarified streams
(product-rich) can then be subjected to the wide variety of unit
operations available for protein downstream processing. Rapid, online
processing bestows benefit upon product quality in respect of
bioactivity and molecular fidelity. This has particularly been shown
for direct recovery of anti-huIgG monoclonal antibodies from
continuous, serum-based culture of murine hybridomas in integrated,
fixed-bed immunoadsorption systems (2–5).

1.2 Protein Recovery Integrated with Batch Cultures

In contrast, integration of protein recovery with batch or fed-
batch fermentations offers more in the context of product yield and
quality. Continuous sequestration of product, either in situ or in an
external recovery loop with cell recycle, has many attractions.
Accumulating product concentrations are continually suppressed in the

reactor which promotes extracellular transport through increasing gradients of active or passive secretion. Natural feedback inhibition of biosynthesis/secretion will also diminish, leading to enhanced system productivity. Rapid separation of product from cells, broth or product antagonists encountered during time-dependent, accumulation in bioreactors commonly improves gross molecular yields. Direct recovery additionally will enhance overall product quality, as fore-shortened operations benefit the molecular and activity stabilities.

1.3 Potential Implementation of Extractive Biotransformation

Most publicised strategies are more applicable to micromolecules (organic acids/bases, alcohols etc; see 6) than macromolecular proteins. However, direct protein recovery can be operated within the reactor, or in an external loop. The former has advantages since the conditions of cell culture and product recovery can be readily maintained under central fermentation control. However, extraction agents (commonly a solvent or adsorbent) must be separated from cell biomass at termination of the productive cycle (see 1,6). In contrast, an external extraction loop can be designed and operated to achieve simultaneous separation and recycle of cells, together with selective product removal. Here, a major operating consideration is that time-based departures from physical and biochemical optima within the loop should not seriously hazard overall cell viabilities and productivity. In extremis, such operations may require an imposition of fermenter controls (temperature, pH, aeration, oxygen, mixing etc) in the loop. However, short residence times achieved by rapid circulation of broth through the loop will minimise detrimental effects. Protein may be recovered by pressure-driven transport across specialised membranes arranged in the reactor, or in a recirculating external loop (6,7). Membrane fouling must be minimised in such use of micro-filtration (M/F) or ultrafiltration (U/F) modules which only crudely fractionate broths rich in product. Maintenance of system sterility has been a significant bar to applications progress with animal cell cultures (described in 2-5). Solvent extraction can be applied directly in the fermenter, or an external loop, to achieve product recovery (see 1). However, both approaches require intermittent dispersion removal, phase separation and broth recycle (plus/minus cells).

2. PROTEIN RECOVERY THROUGH INTEGRATED ADSORPTION

Adsorption can be used directly for direct protein recovery by suspension contacting adsorbents within the reactor, or a contactor fed with clarified feedstreams from an external microfiltration loop. Inclusion of adsorbents in the reactor risks scouring of surfaces, mechanical attrition damage to particles, and requires a tripartite separation of cells, broth and particles at reaction termination. The treatment of a clarified stream in external fixed beds is necessarily complex, requiring the sterile return of two parallel streams (cell retentate from M/F and bed breakthrough) to sustain a batch reactor.

Direct contacting of whole broth in fixed beds is ruled out because they act as depth filters and rapidly blind with cells (note cell settler in Figure 1a). In contrast, sufficiently expanded, or

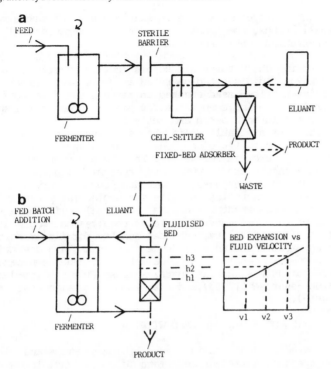

Figure 1 Schematic representation of protein recovery by selective adsorption integrated directly with (a) continuous culture of animal cells (2), and (b) batch culture of yeasts (17).

well mixed fluidised beds of selective adsorbents would theoretically allow the recirculation of cell suspensions with concomitant solute adsorption (see Figure 1b). Such a process should operate at many scales without severe pressure drop across the contactor. Direct fluidised bed adsorption of antibiotics from whole broths has been researched (8), and has found commercial application. However, operation with batch fermentations to specifically sequester product directly upon secretion, overcome feedback and other synthetic inhibitions, and increase productivity will be most effective with growth-related products.

2.1 Fluidised Bed Adsorption of Proteins

Adsorbent particles in a fluidised bed require properties of enhanced density and hydrodynamic behaviour which promote bed expansion in upward fluid flows without particle washout in stable operation (see Figure 1b). In a single-stage, recirculating fluidised bed, the particles are well-mixed and will adsorb solute to reach an equilibrium with continuously contacted feedstocks. This contrasts with expanded beds which yield a significant number of theoretical plates appropriate to true chromatographic separations (9,10). First principles suggest that expanded beds, which require lower operating fluid velocities but yield extended residence times, are best suited to fractionation of protein mixtures in single passes of suspension feedstocks (cell broths, homogenates, unclarified extracts) and <u>not</u> the direct integration considered here.

Despite lacking chromatographic sophistication (9-11), a single-stage recirculating fluidised bed operated at a higher fluid velocity is less sensitive to time-based changes in biomass and viscosity (11). A major constraint upon implementation for protein products in whole fermentation broths in both batch and fully integrated processes is a current shortage of available solid phases. Polystyrene adsorbents, successful in the direct, fluidised bed recovery of antibiotics (8), are unsuited to most protein recoveries. They commonly carry strong anion or cation exchange groups which, together with inherent hydrophobicity of the solid phase, exacerbate poor recovery. The microporous nature of particles inhibits macromolecular penetration and reduces working adsorbent capacities. The ideal solid phase for fluidised bed, protein recovery of proteins requires macromolecular accessibility and compatibility, enhanced density for fluidisation, physical and chemical resistance to crude feedstocks, and operational longevity. The most successful prototype materials have been ceramic-polysaccharide composites such as quartz-agarose (10), silica-dextran (11), kieselguhr-agarose (12,13),and titania or magnetite-agar or cellulose (14-16).

3. INTEGRATION OF FLUIDISED BEDS WITH BATCH FERMENTATIONS

Adsorbents operated as depicted in Figure 1(b) must withstand sterilisation treatments without detriment to operational longevity, and must not physically or chemically damage the recycled cells in use. Cell/debris interactions with the adsorbents should be minimal, whilst product adsorption should be as selective as possible (to maximise uptake with minimal particle inventory), and robust in the face of high or changing ionic strengths/ pH conditions in broths.

3.1 Microbial Test System

Fluidised bed development (11-17) required a suitable productive microorganism to validate experiments on the planned integration of protein recovery with fermentation. The key features of the chosen system of <u>Yarrowia lipolytica</u> (donated by Dr T W Young, University of Birmingham) are summarised in Table 1. In short, <u>Yarrowia</u>, grown in acidic pH and circumstances where simple nitrogen (ammonium sulphate) are replaced by protein substrates (eg whey protein), synthesises an acidic protease as a growth related product. It was proposed (11) that continual removal by direct adsorption of synthesised enzyme in an established population of <u>Yarrowia</u> would promote further synthesis (and diminish autolytic decay) where factors of oxygen and carbon supply, or accumulation of system antagonists, do not otherwise constrain the process.

3.2 Validation of Experimental Integration

The operation of a single-stage, recirculating fluidised bed of DEAE Spherodex (LS, 100-300 μm; 200% expansion) in an external loop

connected to a 2 l batch fermentation of Yarrowia has proved possible
because of the low ionic strength of fermentation broth (simple salts,
yeast nitrogen base, whey protein and glucose at pH 4.5). Adsorbent,
(25ml) in a 2,3 x 40 cm contactor, was fluidised at 0.5 cm/s. Bed
operation from zero-time extended the lag-phase and restricted cell
proliferation through cell and substrate interactions with the ion
exchanger. Activation of the fluidised bed in early exponential phase
(16 hours post inoculation) achieved immediate removal of free enzyme
in the broth without serious hindrance to biomass growth (17). Enzyme
biosynthesis was increased to replace adsorbed material, and system
productivity increased by more than 50% to 140,000 activity units (AU)
per gram glucose consumed. Quantitation of total product recovered
from both whole broth (by batch fluidised bed adsorption from
stationary phase cultures) and the integrated contactor revealed a
considerably enhanced yield of enzyme product per unit fermentation
time. Enzyme recovered from the integrated fluidised bed had specific
activities of 51,800 AU/mg protein, 40% greater than material
conventionally purified from stationary phase cultures (17). Rapid
adsorption achieved in fluidised bed also enhanced the molecular
integrity of the final product. Analysis by SDS-PAGE revealed a
minimum of autolytic breakdown products and a higher degree of
molecular purity than in product recovered by sequential microfiltration and fixed bed adsorption (11,17).

4. CORRECT SIZING OF ADSORPTIVE CONTACTORS

The sizing of the fluidised bed of anion exchanger proved critical
in the Yarrowia experimental system. For example, a maximum of 25 ml
of adsorbent can be operated at any one time in the external loop of
a 2 l fermentation. Increase of adsorbent capacity suppresses cell
growth and production – a consequence of using anion exchangers with

Table 1 Key characteristics of experimental vehicle used to study
extractive bioconversion of whey protein to protease product

* Microbial system	: Yarrowia lipolytica (oleaginous yeast)
* Morphology	: Budding and pseudo-hyphal forms – switched by growth conditions to yield varied rheological problems in downstream recovery
* Product	: Acidic protease (MW 38 kD, IpH 4.2) Synthesis induced by nitrogen deficiencies and pH < 6.0 Commercial potential in food and drinks manufacture
* Substrate	: Soluble protein (BSA or whey protein) plus simple salts, YNB, carbon source.
* Fermentation	: Batch, fed-batch or continuous cultures
* Biorecovery	: Anion exchange (DEAE) or bioaffinity (pepstatin) adsorption
* Generic links	: Product biosynthesis is feedback inhibited (by end-products of substrate proteolysis) Product is unstable through autolysis

affinities for negatively charged cell surfaces. (Cation exchangers
are not compatible with either the isoelectric characteristics or
operating pH of most protein recoveries.) The simple route to further
increase the enzyme productivity of the Yarrowia (and other) cultures
is to operate a manifold of fluidised bed loops of anion exchangers
(see Figure 2). These must be sized in a system-specific manner to
achieve enzyme adsorption without compromise to cell proliferation,
and may be sequentially brought on-stream as the preceding device
becomes product-saturated. Such a system could be controlled on a
time-based assumption of performance for each contactor, or using
continuous affinity HPLC or FIA monitoring of bed effluents (18-20).

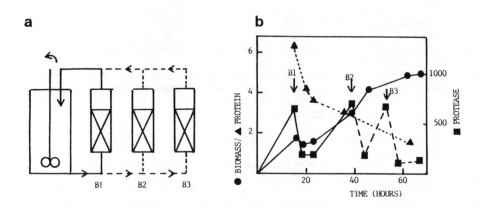

Figure 2   Proposed sequential operation of a manifold of fluidised
beds for integrated protease recovery from Yarrowia
cultures. (a) Outline flow-diagram. (b) System response to
sequential use of beds B1 to B3 (based on 11,17).

A generic solution to problems of cell/growth factor interaction
with anion exchangers may be the harnessing of specific interactions
commonly exploited in processes of bioaffinity chromatography. The
acidic protease of Yarrowia lipolytica is specifically inhibited by
pepstatin. Immobilisation of this peptide (<1.5 kD; 13) on suitable
solid-phases for use in integrated protease recovery is in progress.

5. CONCLUSION

The bulk properties of most proteins (size, lability, conformational
complexity/flexibility etc) are compromised in many of the processes
invoked for extractive biotransformation of fermentation substrates
into 'micromolecular' products (1,6). However, protein adsorption in
a single-stage, recirculating fluidised bed arranged as the external
loop for a batch fermentation appears to offer promise for integrated
recoveries of extracellular products. The general principle is

confirmed by preliminary data from the direct ion-exchange adsorption of acidic protease from batch fermentations of <u>Yarrowia lipolytica</u>. The enhancement of both product yield by avoidance of biosynthetic feedback inhibition, and product quality through shortened operation time-scales is very exciting. The development of bioselective or mild hydrophobic, fluidised bed adsorbents, not compromised by broth ionic compositions, may herald the generic application of this technology to other production systems where natural controls of product synthesis and inherent molecular stability compromise manufacture.

ACKNOWLEDGEMENT

We acknowledge the past support of PHM through an SERC BTD CASE Award in collaboration with D Thatcher at Zeneca Pharmaceuticals.

REFERENCES

1. B Mattiasson and O Holst (eds), 'Extractive Bioconversions', Marcell Dekker, Berlin, 1991.
2. J Rudge, M A Desai, S Shojaosadaty and A Lyddiatt. <u>In</u> 'Modern Approaches to Animal Cell Technology' (ed R E Spier and J B Griffiths), Butterworth, 1987, pps 556-575.
3. A Lyddiatt, M A Desai, J G Huddleston, J Rudge and S Shojaosadaty, <u>J Chem Technol and Biotechnol</u>, 1988, <u>45</u>, 47-60.
4. S B Mohan and A Lyddiatt, <u>Cytotechnology</u>, 1991, <u>8</u>, 201-209.
5. S B Mohan, S R Chohan, J K Eade and A Lyddiatt, <u>Biotech Bioeng</u>, 1993, <u>42</u>, 974-986.
6. A Freeman, John M Woodley and M D Lilley, <u>Bio/Technology</u> 1993, <u>11</u>, 1007-1012.
7. T I Linandos, N Kalogerakis and L A Behie, <u>Biotech Bioeng</u>, 1992, <u>39</u>, 504-510.
8. C R Bartels, G Kleiman, N Korsdun and D B Irish, <u>Chem Eng Prog</u>, 1958, <u>54</u>, 49-59
9. H A Chase and N M Draeger, <u>J Chromatography</u>, 1992, <u>597</u>, 129-145.
10. M Hansson, S Stahl, R Hjorth, M Uhlen and T Moks, <u>Bio/Technology</u>, 1994, <u>12</u>, 285-288.
11. P H Morton and A Lyddiatt. <u>In</u>: 'Ion Exchange Advances' (ed M J Slater) Elsevier Applied Science, 1992, pps 237-244.
12. C M Wells and A Lyddiatt. <u>In</u>: 'Separations for Biotechnology' (ed M J Verrall and M J Hudson), Ellis-Horwood, 1987, pps 217-224.
13. A W Sansome-Smith, Ph D Thesis, University of Birmingham, 1992.
14. N B Gibson and A Lyddiatt. <u>In</u>: 'Separations for Biotechnology 2' (ed D L Pyle), Elsevier, 1990, pps 152-161.
15. N B Gibson, Ph D Thesis, University of Birmingham, 1991.
16. N B Gibson and A Lyddiatt. <u>In</u>: 'Cellulosics; materials for selective separations and other technologies (eds J F Kennedy, G O Phillips and P A Williams), Ellis Horwood, 1993, pps 55-62.
17. P H Morton, Ph D Thesis, University of Birmingham, 1993.
18. H A Chase, <u>Biosensors</u>, 1986, 269-286.
19. S Shojaosadaty and A Lyddiatt. <u>In</u>: 'Separations for Biotechnology' (ed M J Verrall and M J Hudson), Ellis-Horwood, 1987, pps 436-445.
20. C Turner, N F Thornhill and N M Fish, <u>Biotech Techniques</u>, 1993, <u>7</u>, 19-24.

Production and Recovery of Recombinant Proteins of Low Solubility

Maria Murby, Thien Ngoc Nguyen,[a] Hans Binz,[a] Mathias Uhlén, and Stefan Ståhl

DEPARTMENT OF BIOCHEMISTRY AND BIOTECHNOLOGY, ROYAL INSTITUTE OF TECHNOLOGY, S-100 44 STOCKHOLM, SWEDEN

[a]CENTRE D'IMMUNOLOGIE ET DE BIOTECHNOLOGIE PIERRE FABRE, F-74 164 SAINT JULIEN EN GENEVOIS, FRANCE

1 SUMMARY

A novel strategy for the recovery of recombinant proteins with a strong tendency for aggregation is described. An extensive investigation of different expression strategies in *Escherichia coli* for the production of various fragments of the fusion glycoprotein (F) from the human respiratory syncytial virus (RSV) was carried out. Here, we describe the first successful production of unstable and poorly soluble RSV F-peptides. A fusion strategy was used in which the produced fusion proteins precipitated as inclusion bodies in the cell and efficient recovery was achieved by affinity chromatography. A chaotropic agent, guanidine hydrochloride, was present throughout the purification procedure which gave substantial amounts of full-length products. The strategy outlined here could be of interest for efficient recovery of heterologous peptides that form inclusion bodies when expressed in a bacterial host.

2 INTRODUCTION

A major consideration when setting up a production scheme for a recombinant protein, is whether the product should be expressed intracellularly or if a secretion system could be used to direct the protein out from the host cell. There are a number of advantages connected with the secretion strategy. First, the gene product will be less exposed to cytoplasmic proteases which might enable production of labile proteins.[1] Second, disulphide bond formation is enhanced in the oxidative environment outside the cytoplasm.[2,3] Third, the recovery of the recombinant protein is very much simplified since a large degree of the purification has been achieved through the secretion.[4] However, all proteins cannot be secreted even if an expression system with secretion signals is selected for the production. Proteins with a strong tendency to precipitate intracellularly and proteins containing hydrophobic transmembrane regions have been demonstrated to be extremely difficult to secrete.[5] Recent advances for *in vitro* renaturation of recombinant proteins from intracellular precipitates have greatly facilitated the recovery of proteins of low solubility, thus making intracellular production a more attractive alternative. Production by the inclusion body strategy has the main advantages that the recombinant product normally is protected from proteolysis and that it can be produced in large quantities. Levels of up to 50% of total cell protein content have been reported.[6]

A second important consideration is whether an affinity purification strategy could be utilised to simplify the recovery of the recombinant protein. A large number of gene fusion systems for the purpose of affinity purification have been described.[7,8] The staphylococcal protein A (SpA) system,[4] often in the form of the smaller derivative ZZ[9] as affinity handle has found extensive use in a number of hosts including bacteria, yeast, insect cells and CHO cells,[10] for the expression of a multitude of different recombinant proteins. In *E. coli*, the secretion signal of SpA has been found to direct the gene product to the periplasmic space[1] and in some cases also to the culture medium,[2,11] which thus greatly facilitates the recovery. For nonsecretable proteins, expression vectors for intracellular production have been developed.[5,12] The SpA or ZZ fusions are affinity purified using IgG-affinity chromatography, and yields between 90-100% are not unusual.[2,11] The high solubility of ZZ enables production of fusion proteins which remain soluble at very high concentrations within the *E. coli* cell and ZZ can also act as a solubilising fusion partner to improve the *in vitro* refolding properties for recombinant gene products.[13]

A related affinity purification system, with expression vectors for both intracellular[5] and secreted[14] production in *E. coli*, has been developed from the serum albumin binding region (denoted BB) of streptococcal protein G (SpG).[15] BB-fusions can be efficiently affinity purified on human serum albumin (HSA) columns.[15] A dual affinity fusion concept,[16] where proteins of interest are fused between the ZZ-domains and the BB-region, resulted in considerable stabilisation of labile gene products.[1] In addition, the two similar expression systems, based on ZZ and BB, respectively, have found use for the parallel expression of immunogenic peptides and proteins in vaccine research.[17] After affinity purification, the ZZ fusion protein may be used for immunisation and the corresponding BB fusion protein to analyse the induced antibody response to the fused peptide. In addition, the BB fusion protein can, after immobilisation to HSA-Sepharose, be used as ligand for affinity purification of peptide-specific antibodies.[14]

Respiratory syncytial virus (RSV) is the most important cause of lower respiratory tract infections in infants. Vaccine trials with inactivated virus have failed and attenuated RSV strains have been shown to be either overattenuated or connected with risks of reversion into virulence.[18] Since inhibition of virus replication and also protection against infection seem to be strongly correlated with serum antibodies, subunit vaccines have been suggested as the second generation vaccine candidates.[19] Of the ten virus proteins, most of the neutralising epitopes are attributed to the fusion (F) and the attachment (G) viral glycoproteins. The F protein shows very little antigenic variation between the two virus subgroups and has thus been selected as the target protein in several subunit vaccine approaches.[18,19] Reports of successful *E. coli* production of RSV F proteins are lacking although considerable efforts have been made.[20] This is probably due to that viral glycoproteins can be detrimental to the host as suggested by Steinberg *et al*[21] and/or to extensive degradation and difficulties in renaturation of precipitated gene products.

Here, we report an investigation of different expression strategies for the production in *E. coli* of a number of F proteins of varying size. The F proteins have been produced as fusions to the IgG-binding ZZ-domains of SpA and/or to the serum albumin binding region, BB, of SpG. Both secreted production and different inducible systems for intracellular expression have been analysed. Furthermore, we have studied how the properties of ZZ in combination with the use of chaotropic agents in moderate concentrations can be used to enable affinity purification of gene products that are prone to precipitation.

3 MATERIALS AND METHODS
mRNA purification and generation of gene fragments

mRNA was isolated from Hep-2 cells infected with RSV Long strain (ATCC VR-26) using oligo-dT coated Dynabeads (Dynabeads Oligo (dT)25, Dynal AS, Norway). Murine reverse transcriptase (Pharmacia Biotech, Sweden) was used for reverse transcription to create RNA-DNA hybrids of F-genes using F-specific primers (PF2, PF4 and PF6, see below for sequences). The RNA-DNA hybrids were used as templates in a subsequent PCR with F-gene specific primers containing restriction enzyme recognition sequences for further cloning.

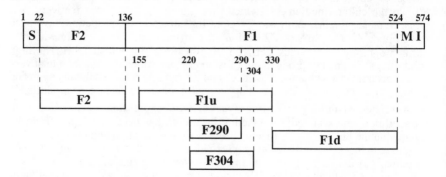

Figure 1. Schematic illustration of the RSV F protein. S represents the signal peptide and F2 the N-terminal part of F, generated by *in vivo* proteolytic cleavage at position 136. The C-terminal part, F1, contains a membrane anchoring region denoted MI. F2 and F1 are held together by disulfide bonds. Amino acid positions are presented above and the different F fragments investigated in this study are illustrated below.

The F2 gene fragment (Figure 1) was amplified with PF5 (5'-CCGAATTCCGCT TCTGGTCAAAACATCAC-3') and PF6 (5'-CCCGGATCCTCTTCTTTTCCTTTT CTTGC-3'), F1u (Figure 1) with PF1 (5'-CCGAATTCTAAGGTCCTGCACCT AGAAG-3') and PF2 (5'-GTTGGATCCTTCTTTTGTGTTGGTTG-3') and F1d (Figure 1) with PF3 (5'-GAAGGATCCAACATCTGTTTAACAAG-3') and PF4 (5'-CCCCTGC AGTTAGTCGACATTTGTGGTGGATTTACCAGC-3'). The obtained gene fragments were restricted with the appropriate enzymes, ligated into pRIT28[22] and sequenced according to Hultman *et al.*[23] To obtain the gene fragments F290 and F304 (Figure 1), a verified pRIT28F1u clone was used as PCR-template with primers RIT270 (5'-GGAATTCGGTGATAGAGTTCCAAC-3') and RIT271 (5'-CCAAGCTTAGT CGACGGACATGATAGAGTAAC-3') for F290 and primers RIT270 and RIT326 (5'-GGCTGCAGTTAGTCGACTGGTAATTGTACTACATATGC-3') for F304. The generated PCR fragments were subcloned into pRIT28 and sequenced as above.

Construction of expression vectors

The F2 and F1u gene fragments were isolated by *Eco*RI and *Bam*HI restriction of verified clones, and ligated into pEZZ308T,[24] pEBB318T, derived from pB1B2mp18[14] with the transcription and translation termination signals from pEZZ308T downstream of the *Hind*III site, and pRIT24.[16] The F1u fragment was also inserted into pUC19[25] using the same restriction enzymes. The obtained vectors were denoted pEZZF2T, pEBBF2T, pRIT24F2, pEZZF1uT, pEBBF1uT, pRIT24F1u and pUC19F1u, respectively. F1d was transferred to pEZZ308T and pUC19F1u via *Bam*HI and *Pst*I to obtain pEZZF1dT and pUC19F1. To construct the secretion vectors pRIT24F1d and pRIT24F1, an F1d fragment was isolated with *Bam*HI and *Sal*I and an F1 fragment (from pUC19F1) with *Eco*RI and *Sal*I and both fragments were inserted into pRIT24 restricted with the same enzymes. An F290 fragment restricted with *Eco*RI and *Hind*III, was inserted into pEZZ308T and pEBB318T restricted with the same enzymes to yield the vectors pEZZF290 and pEBBF290, respectively. An F304 fragment was inserted into the same vectors using *Eco*RI and *Pst*I, which yielded the vectors pEZZF304 and pEBBF304, respectively. The F290 and F304 fragments were both subcloned into pRIT24 via *Eco*RI/*Sal*I restriction to obtain the vectors pRIT24F290 and pRIT24F304.

Plasmids pEZZF2T, pRIT24Flu, pEZZF1dT, pUC19F1, pRIT28F290 and pRIT28F304 were all restricted with *Eco*RI and *Sal*I and the different F gene fragments were isolated and ligated into pUEXZZ (Erik Holmgren, personal communication) and pUEXZZBB previously restricted with the same enzymes. pUEXZZ is an expression vector derived from pUEX1[26] where the lacZ gene is removed and the lacI gene is partly deleted by Bal31 exonuclease treatment. ZZ is genetically fused in frame downstream of the remaining crolacI (44 amino acids) fusion peptide. Gene transcription is under control of the λ right promoter. pUEXZZBB is a dual affinity variant of pUEXZZ, where the BB domain has been fused downstream of the ZZ gene. The obtained intracellular expression vectors were denoted pUEXZZx and pUEXZZxBB, where x represents F2, Flu, Fld, F1, F290 and F304, respectively.

Plasmids pRIT28F290, pRIT28F304, pRIT28F2, pRIT28Flu, pRIT28F1d and pUC19F1 were restricted with *Eco*RI and *Hind*III and the resulting gene fragments inserted into pRIT44.[12] The constructed vectors encode ZZ-fusions where transcription of the gene fusions are under control of the *E. coli* trp-operon promoter and can be induced by tryptophan starvation or addition of β-indole acrylic acid (IAA).[12] A gene fragment, encoding a fusion protein consisting of one Z domain fused to the bovine pancreatic trypsin inhibitor (BPTI), was cut out from plasmid pKY481 (Sophia Hober, personal communication) by *Fsp*I/*Hind*III restriction and replaced by a similarly restricted fragment from pB1B2mp18, encoding BB. The obtained plasmid was named ptrpB1B2. Gene expression is under control of the *E. coli* trp-promoter and the vector confers kanamycin resistance. All the described F gene fragments were inserted into ptrpB1B2 using *Eco*RI/*Sal*I restriction except for the F304 fragment, which was inserted into ptrpB1B2 via *Eco*RI and *Pst*I restriction.

Expression and recovery of recombinant proteins

Shake flasks containing 100 ml Tryptic Soy Broth (TSB, 30g/l, Difco) supplemented with Yeast Extract (YE, 5g/l, Difco) and ampicillin (200 μg/ml) were inoculated with *E. coli* RRIΔM15[27] or RV308[28] cells harbouring the different expression vectors for secretion. The cells were grown over night at 37°C and harvested by centrifugation. Periplasmic material was released by osmotic shock and the fusion proteins were affinity purified on IgG-Sepharose® (Pharmacia Biotech, Sweden) or HSA-Sepharose as described by Ståhl *et al.*[14] The purified proteins were analysed by SDS-PAGE on the Pharmacia Phast system or on the Pharmacia Multiphore II (Pharmacia Biotech, Sweden), using 8-25 % and 12.5 % gels, respectively. The gels were stained with Coomassie Brilliant Blue 250.

RRIΔM15 or RV308 cells harbouring the heat-inducible or the IAA-inducible expression vectors were grown over night at 30°C. 100 ml TSB (30g/l) with YE (5 g/l) and ampicillin (200 μg/ml) was inoculated 1:100 (heat inducible) or 1:20 (IAA inducible) with the over night cultures and incubated at 30°C. Protein production was induced when OD (550 nm) reached 1.5. Transcription from the λ right promoter was induced by addition of 100 ml growth medium (56°C) and continued growth at 42°C for 90 minutes. The trp-promoter was induced by addition of IAA to a final concentration of 25 μg/ml and continued growth at 30°C for three hours. Cells were harvested by centrifugation, redissolved in cold TST (50 mM TrisHCl, pH 8.0, 200 mM NaCl, 0.05 % Tween 20, 0.5 mM EDTA) buffer and sonicated as described.[29] The supernatant after centrifugation was filtered (0.45 μm) and fusion proteins were isolated by affinity purification and analysed as described above.

Renaturation and recovery by a modified purification scheme

Insoluble material after sonication was pelleted by centrifugation and precipitated intracellular proteins were recovered by an initial solubilisation in 7 M guanidine hydrochloride (GuaHCl, Sigma) and 25 mM TrisHCl, pH 8.0 and incubated at 37°C during two hours. When cysteines were present in the fusion proteins, 10 mM DTT or 100 mM β-mercaptoethanol was also added. The solubilisation mixture was centrifuged and the supernatant was slowly pipetted into 100 ml of renaturation buffer containing 1 M GuaHCl, 25 mM TrisHCl, pH 8.0, 150 mM NaCl and 0.05 % Triton X100. The mixture was incubated at 4°C under slow stirring for 15 hours.

The renaturation mixture was centrifuged, filtered (0.45 μm) and applied to an affinity column containing 5 ml IgG-Sepharose® at 4°C at a low flow rate (0.5 ml/min). The IgG-Sepharose® had previously been pulsed separately with TSTG buffer (50 mM TrisHCl, pH 8, 200 mM NaCl, 0.05 % Tween 20, 0.5 mM EDTA and 0.5 M GuaHCl) and 0.3 M HAc, pH 3.3 containing 0.5 M GuaHCl. After sample loading, the column was washed with 100 ml TSTG and 25 ml 5 mM NH_4Ac, pH 5.5 containing 0.5 M GuaHCl. Bound proteins were eluted by 20 ml 0.3 M HAc, pH 3.3 with 0.5 M GuaHCl. The chaotropic agent was removed by dialysis against 2 l of 0.3 M HAc, pH 3.3, twice. After dialysis, the purified proteins were lyophilised and analysed as described above.

4. RESULTS AND DISCUSSION

Generation of gene fragments and construction of expression vectors

A number of gene fragments encoding various regions of the RSV F-protein were generated by reverse transcription of purified mRNA followed by PCR-amplifications using specific primers. The different F-gene fragments, sequenced using an automated solid-phase DNA sequencing method,[23] were found to display a few sequence variations as compared to the nucleotide sequence reported by López *et al.*[30] The F2 gene fragment showed an A to G nucleotide substitution at position 250 (nucleotide numbering as in López *et al*),[30] which results in an Ile to Met amino acid change. The same sequence appeared in the five individual clones sequenced. In the F1u gene fragment we found one T to A nucleotide change at position 980 in nine different clones, which led to a Thr to Ser modification of the encoded protein. In the F1d gene fragment we found three nucleotide changes in all eight clones sequenced, at positions 1338 (C to T, Ala to Val), 1354 (A to G, silent) and 1556 (C to A, His to Asn). Since the nucleotide changes occurred in all clones sequenced, one clone of each fragment was selected for further work and subcloned into the different expression vectors.

A number of expression vectors designed for secretion of the gene fusion products were constructed. Figure 2A gives an overview of the 14 different assembled expression vectors, with the encoded fusion proteins. The vectors encode F-peptides as single fusions to the affinity handles ZZ and BB, respectively, and also to both handles in a dual affinity fusion manner. In these vectors, the expression is under control of the SpA transcription and secretion signals.[29]

Two different types of expression vectors for intracellular production were constructed (Figure 2B and 2C), both with the advantage of inducible protein production. The first set of vectors (Figure 2B) took advantage of the heat inducible λ right promoter. Both ZZ and ZZBB fusions to all the generated F-peptides were constructed, all with an N-terminal extension (44 amino acids, denoted c) derived from the *E. coli* cro and lacI proteins. The second set of vectors for intracellular production (Figure 2C) were regulated by the tryptophan operon leader sequences and can thus be induced by tryptophan starvation or by addition of indole acrylic acid (IAA), which gives a more stringent control over the expression. All generated F-fragments were subcloned also into these vectors which encode both ZZ and BB fusions (Figure 2C). Altogether, 24 different intracellular gene fusions were assembled and characterised.

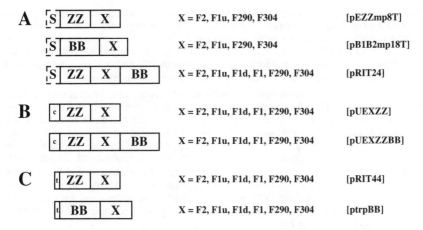

Figure 2. An overview of the 38 fusion proteins investigated in this study. **A.** Secreted gene products where S represents the SpA signal peptide, processed during secretion, ZZ, the IgG-binding domains derived from SpA, BB, the serum albumin binding region of SpG and X, the different F-fragments. **B.** Intracellular gene fusion products from the heat inducible vectors. **C.** Intracellular gene fusion products encoded from the tryptophan regulated vectors. Expression vectors used for the different constructions are shown within parenthesis (see Materials and Methods for details).

Expression of secreted gene products

When investigating the protein production and secretion in *E.coli* from the expression vectors designed for secretion of F-fusion proteins (Figure 2A), the results were rather discouraging. Although, these and related expression vectors have been shown to produce and secrete other target proteins with high yields,[2,11] no or very little product could be found and the degradation was extensive (data not shown). The dual affinity fusion strategy did however yield small amounts of product from the ZZF290BB and ZZF304BB fusions. From the medium fraction the quality of these proteins was somewhat better as compared to the same proteins recovered from the periplasm, which were partly degraded and associated with *E. coli* chaperones (GroEL and DnaK) (data not shown). These data indicated that the different F-fusions were secretion incompetent and proteolytically unstable.

Expression of intracellular gene products

Using the heat inducible intracellular systems (Figure 2B), the different F fragments could be expressed as ZZ and dual affinity fusions in the heat inducible systems (Figure 2B), but IgG- and HSA affinity purification of soluble cytoplasmic proteins revealed only small amounts of full length proteins (data not shown). Some degradation products were observed and large amounts of GroEL and DnaK were copurified with all the different F-fusions. The presence of the heat shock proteins indicates that the host cell is under stress, which could result from the increase in temperature during induction and/or that the F-fusion proteins exhibit properties that induce the stress response. When examining the insoluble fraction after sonication, we observed that a majority of the expressed fusion proteins was precipitated as inclusion bodies.

The inclusion bodies were solubilised using standard conditions[6] followed by renaturation as described in Materials and Methods. The fusion proteins were affinity purified by IgG-chromatography as described.[14] When washing buffer was applied to the IgG-Sepharose during affinity chromatography, the immobilised proteins precipitated on the columns (data not shown). The small amounts of the proteins that could be eluted from the columns were lyophilised, but after lyophilisation, the proteins were difficult to resolubilize (data not shown). These data indicated that the gene products were extremely proned to precipitate.

An alternative expression system was therefore investigated. With the tryptophan regulated vectors (Figure 2C) we could avoid the heat induced production of stress proteins and in addition, the N-terminal extension of unrelated amino acids, common for the ZZ and BB fusions, was shorter (eight amino acids originating from the tryptophan leader peptide). This is important since the ZZ-fusions will be used for immunisations and the corresponding BB-fusions for analysis of the antibody responses.[17] By using this system, all the different F-fragments could be expressed as ZZ-fusions. Despite the use of a lower temperature, the majority of the proteins were found as inclusion bodies.[31] When solubilising, refolding and IgG-purifying these proteins, we still had the problem with precipitation on the affinity columns when adding the washing buffer.

Alternative recovery scheme for fusion proteins with low solubility.

To circumvent the problems connected with the recovery of proteins with low solubility, we have developed an alternative purification process (Figure 3). After renaturation of precipitated material with 1 M GuaHCl (see Materials and Methods for details), the ZZ-fusion proteins were affinity purified using IgG-chromatography at 4°C in the presence of moderate concentrations (0.5 M) of GuaHCl in all chromatography buffers. The chaotropic agent was removed by dialysis and the purified fusion proteins lyophilized (Figure 3). Here, we have followed the production of three of the fusion proteins, tZZF2, tZZF1d and tZZF304 that by this strategy could be produced in reasonable amounts (Figure 4 and Table 1). SDS-PAGE analysis of insoluble material after sonication and in the pellet fraction after renaturation (Figure 4, lanes 1 and 2), clearly demonstrate that the fusion proteins were kept soluble during the refolding process. However, for tZZF304 residual amounts of the fusion protein could be found in the pellet after renaturation (Figure 4, tZZF304, lane 3). In Figure 4 (lanes 3), the IgG-purified fusion proteins, tZZF2, tZZF1d and tZZF304, can be observed. The different proteins recovered as full length products, are purified to high homogeneity. These results show that the problems associated with the expression of the F-proteins, degradation and insolubility, could be circumvented by this alternative purification strategy for intracellular products. To

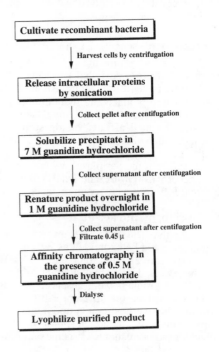

Figure 3. The alternative purification scheme used for recovery of the fusion proteins with low solubility.

analyse if complete elution was achieved, the IgG-Sepharose used for affinity purification was boiled under reducing conditions and run on SDS-PAGE followed by Western blot for analysis of SpA-derived protein. No IgG-binding proteins were visualised showing that the fusion proteins were kept soluble during the IgG-affinity chromatography when the alternative purification scheme was used (data not shown). Reasonable amounts of F-fusion proteins were recovered ranging from 20 to 50 mg/l (Table 1).

As alternative to GuaHCl the nonionic detergent Triton X100 was also demonstrated to keep the fusion proteins soluble when used in 0.01% concentrations in all buffers throughout the purification scheme (data not shown). For large scale production purposes a detergent might be a more attractive alternative since chaotropic agents are expensive and GuaHCl is extremely corrosive.

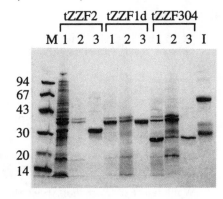

Figure 4. SDS/PAGE (12.5 %) analysis following the recovery of the fusion proteins tZZF2, tZZF1d and tZZF304. Lane M: molecular marker proteins with molecular weights in kDa to the left, Lanes 1: inclusion body fractions, Lanes 2: pellets after renaturation , Lanes 3: affinity purified products, Lane I: IgG, light and heavy chains.

Table 1. Fusion proteins recovered from inclusion bodies by the alternative purification strategy

Fusion protein	MW (kDa)	Precipitated fraction (%)	Yield (mg/l)
tZZF2	29.7	75	50
tZZF1d	37.2	100	20
tZZF304	25.2	75	40

Conclusions

This is the first reported thorough investigation where a secretion strategy is compared to intracellular expression for the production of related gene products. The intracellularly expressed fusion proteins were recovered from inclusion bodies and could be efficiently purified using affinity chromatography. This is also the first time IgG-chromatography has been performed in the presence of guanidine hydrochloride, to keep poorly soluble proteins in solution. Furthermore, we have demonstrated for the first time that RSV F-proteins can be produced in *E. coli* as full length proteins with good yields.

5 ACKNOWLEDGMENTS

This work has been supported by the European Biotechnology programme: "Human and veterinary vaccines" project N°920089. We thank Drs. L. Strandberg, P.-Å. Nygren, U. Öberg and H. Olsson for fruitful collaboration and Prof. M. Trudel for providing the RSV material.

6 REFERENCES

1. M. Murby, L. Cedergren, J. Nilsson, P.-Å. Nygren, B. Hammarberg, B. Nilsson, S.-O. Enfors and M. Uhlén, *Biotechnol. Appl. Biochem.*, 1991, **14**, 336.
2. T. Moks, L. Abrahmsén, B. Österlöf, S. Josephson, M. Östling, S.-O. Enfors, I.-L. Persson, B. Nilsson and M. Uhlén, *Bio/Technol.*, 1987, **5**, 379.
3. P. Carter, R. F. Kelley, M. L. Rodrigues, B. Snedecor, M. Covarrubias, M. D. Velligan, W. L. T. Wong, A. M. Rowland, C. E. Kotts, M. E. Carver, M. Yang, J. H. Bourell, H. M. Shephard and D. Henner, *Bio/Technol.*, 1992, **10**, 163.
4. M. Uhlén and T. Moks, *Methods Enzymol.*, 1990, **185**, 129.
5. A. Sjölander, S. Ståhl, K. Lövgren, M. Hansson, L. Cavelier, A. Walles, H. Helmby, B. Wåhlin, B. Morein, M. Uhlén, K. Berzins, P. Perlmann and M. Wahlgren, *Exp. Parasitol.*, 1993, **76**, 134.
6. R. Rudolph, *'Principles & Practice of Protein Engineering'*, J. L. Cleland and C. S. Craik, Eds., Hanser Publishers, New York, 1994, In press.
7. M. Uhlén, G. Forsberg, T. Moks, M. Hartmanis and B. Nilsson, *Curr. Opin. Biotechnol.*, 1992, **3**, 569.
8. E. Flaschel and K. Friehs, *'Biotechnology Advances'*, M. Moo-Young, and B. R. Glick, Eds., Pergamon Press Ltd., Oxford, England, 1993, 31.
9. B. Nilsson, T. Moks, B. Jansson, L. Abrahmsén, A. Elmblad, E. Holmgren, C. Henrichson, T. A. Jones and M. Uhlén, *Prot. Eng.*, 1987, **1**, 107.
10. P.-Å. Nygren, S. Ståhl and M. Uhlén, *TIBTECH*, 1994, In press.
11. M. Hansson, S. Ståhl, R. Hjorth, M. Uhlén and T. Moks, *Bio/Technol.*, 1994, **12**, 285.
12. K. Köhler, C. Ljungquist, A. Kondo, A. Veide and B. Nilsson, *Bio/Technol.*, 1991, **9**, 642.
13. E. Samuelsson, H. Wadensten, M. Hartmanis, T. Moks and M. Uhlén, *Bio/Technol.*, 1991, **9**, 363.
14. S. Ståhl, A. Sjölander, P.-Å. Nygren, K. Berzins, P. Perlmann and M. Uhlén, *J. Immunol. Meth.*, 1989, **124**, 43.
15. P.-Å. Nygren, M. Eliasson, E. Palmcrantz, L. Abrahmsén and M. Uhlén, *J. Mol. Recognit.*, 1988, **1**, 60.
16. B. Hammarberg, P.-Å. Nygren, E. Holmgren, A. Elmblad, M. Tally, U. Hellman, T. Moks and M. Uhlén, *Proc. Natl. Acad. Sci. USA*, 1989, **86**, 4367.
17. A. Sjölander, S. Ståhl, and P. Perlmann, *Immunomethods*, 1993, **2**, 79.
18. K. McIntosh and R. M. Chanock, *'Virology'*, 2nd ed., B. N. Fields, D.M. Knipe, et al. Eds., Raven Press, Ltd., New York, 1990, 1045.
19. G. L. Toms, *FEMS Microbiol. Immunol.*, 1991, **76**, 243.
20. A. Martin-Gallardo, K. A. Fien, B. T. Hu, J. F. Farley, R. Seid, P. L. Collins, S. W. Hildreth and P. R. Paradiso, *Virology*, 1991, **184**, 428.
21. D. A. Steinberg, R. J. Watson and W. M. Maiese, *Gene*, 1986, **43**, 311.
22. T. Hultman, S. Ståhl, T. Moks and M. Uhlén, *Nuleos. & Nucleot.*, 1988, **7**, 629.
23. T. Hultman, S. Bergh, T. Moks and M. Uhlén, *BioTechniques*, 1991, **10**, 84.
24. B. Löwenadler, B. Jansson, S. Paleus, E. Holmgren, B. Nilsson, T. Moks, G. Palm, S. Josephson, L. Philipson and M. Uhlén, *Gene*, 1987, **58**, 87.
25. C. Yanisch-Perron, J. Vieira and J. Messing, *Gene*, 1985, **33**, 103.
26. G. M. Bressan and K. K. Stanley, *Nucl. Acids Res.*, 1987, **15**, 10056.
27. U. Rüther, *Nucl. Acids Res.*, 1982, **10**, 5765.
28. R. Maurer, B. J. Meyer and M. Ptashne, *J. Mol. Biol.*, 1980, **139**, 147.
29. B. Nilsson and L Abrahmsén, *Methods Enzymol.*, 1990, **185**, 144.
30. J. A. López, N. Villanueva, J. A. Melero and A. Portela, *Virus Res.*, 1988, **10**, 249.
31. L. Strandberg and S.-O. Enfors, *Appl. Environ. Microbiol.* 1991, **57**, 1669.

Macro-Prep DEAE: A New Synthetic Weak Anion Exchange Support for Chromatography of Biomolecules

M. Navvab, L. Olech, C. Ordunez, M. Abouelezz, P. Tunón, and R. Frost

BIO-RAD LABORATORIES, 2000 ALFRED NOBEL DRIVE, HERCULES, CA 94547, USA

1 ABSTRACT

The Macro-Prep® DEAE weak anion exchange support is a new addition to a family of chromatographic supports developed for process chromatography. The retention times and the resolution are higher than currently available weak anion exchange supports. The material exhibits good flow properties; its resolving power is maintained at flow rates in excess of 1,500 cm/h. It has a dynamic binding capacity >35 mg of BSA/ml of support. The Macro-Prep support is stable to extended, repeated exposure to 1 M NaOH and a wide variety of organic solvents. The elution characteristics and the chemical and physical stability make the Macro-Prep DEAE support a valuable tool for purification and production of biomolecules.

2 INTRODUCTION

The Macro-Prep DEAE support has been developed to fill a need for a weak anion exchange support with higher selectivity than currently available supports. The derivative has been developed using a hydrophilic base bead (50 μm nominal bead size) produced by copolymerization of glycidyl methacrylate and diethylene glycol dimethacrylate (GMA/DEGDMA) base matrix. The macroporous epoxide activated base material is derivatized with diethyl amine groups. The unreacted epoxide groups are converted to diols to reduce non-specific interaction (**Figures 1 and 4**).

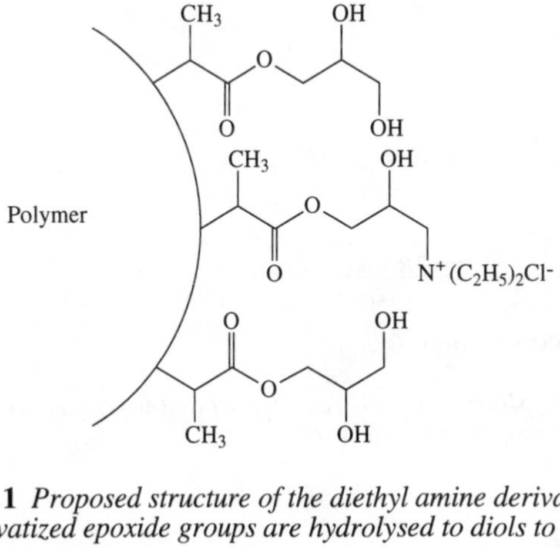

Figure 1 *Proposed structure of the diethyl amine derivatized Macro-Prep epoxide bead. Underivatized epoxide groups are hydrolysed to diols to reduce non specific interaction.*

The resulting diethyl amine derivatized support exhibits the properties essential for a process chromatography support: high dynamic protein binding capacity over a wide range of flow rates; very low non specific interaction; low operating pressure; negligible shrinking or swelling with changes in salt or pH; and it can be readily sanitized with commonly used reagents such as sodium hydroxide.

3 EXPERIMENTAL

3.1 Chemicals and Reagents

Macro-Prep DEAE support (Bio-Rad Laboratories, CA, USA), DEAE Sepharose Fast Flow from Pharmacia AB, Sweden, and Toyo-Pearl DEAE 650-M from Toso-Haas, Japan. All chemicals used were analytical grade. Sample proteins were obtained from SIGMA, weighed and diluted prior to use.

3.2 Chromatography

Chromatographic separations were performed using a low pressure chromatography Econo System (Bio-Rad Laboratories, CA, USA), or a Bio-Rad HRLC system, and A_{280} detection.

4 RESULTS AND DISCUSSION

4.1 Physical Characteristics

Chromatographic theory prescribes that a smaller particle gives better resolution but higher operating pressures, and conversely a larger particle gives lower resolution but also lower operating pressures. The theory also prescribes that a narrow particle size distribution gives better resolution and less band broadening. The nominal particle size of the Macro-Prep support is 50 μm (**Figure 2**).

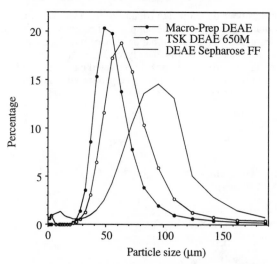

Figure 2 *Bead size distribution of the Macro-Prep DEAE support and two other weak anion exchange supports, measured using laser diffraction.*

Ideally the same bead should be used for both development and scale up work. After evaluating different size ranges a nominal particle size of 50 μm was chosen for the Macro-Prep base bead. The 50 μm size makes the bead small enough to give generally good resolution, but not so small that it necessitates high operating pressures (and more expensive equipment) for scale up work (**Figure 3**).

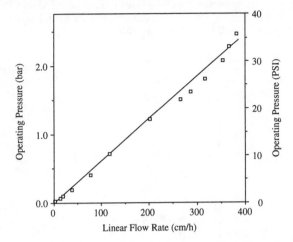

Figure 3 *Flow rate vs. operating pressure for the 50 μm Macro-Prep support. Column: Amicon Moduline® glass column, 14 cm i.d., 1/4" plumbing. Bed height: 17.4 cm. Pump: Cole-Parmer MasterFlex® pump, # 18 Norprene tubing, Easy-Load™ pump head.*

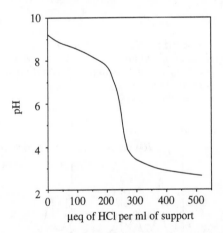

Figure 4 *Titration curve of the Macro-Prep DEAE support. Derivatization with diethyl amine produces a support with a single type of charged group, as revealed by the slope of the titration curve.*

4.2   Chromatography

The diethyl amine derivatized Macro-Prep bead exhibits distinctly higher retention times than do the DEAE Sepharose Fast Flow and the Toyo-Pearl DE650 M supports (**Figure 5**). The higher retention times result in greater available fractionation capability and therefore greater resolution. The mass transfer in the macro-porous rigid Macro-Prep bead is believed to account for the ability to maintain resolution (**Figure 6**) and higher dynamic binding capacity (**Figure 7**) with increasing flow rates .

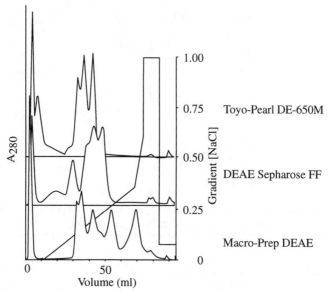

Figure 5 *The Macro-Prep DEAE support compared to DEAE Sepharose Fast Flow and Toyo-Pearl DE-650 M.   Column: 0.5 cm i.d., 2 ml of support.   Buffer: A: 50 mM Tris-HCl, pH 7.6. B: A + 1 M NaCl.   Sample: 100 μl of 9 mg/ml, myoglobin, conalbumin, ovalbumin, BSA and STI.   Flow rate: 153 cm/h (0.5 ml/min) .*

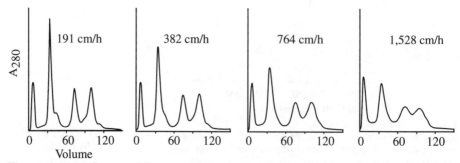

Figure 6 *The influence of flow rate on resolution. Column: 6.56 ml. Buffer: A: 50 mM Tris-HCl, pH 7.6. B: A + 1 M NaCl. Sample: 5 ml of myoglobin, conalbumin, BSA and STI, 3.65 mg/ml. Gradient: 0% B for 15 ml, 0-35% B in 85 ml, 35-65% B in 15 ml, 65-100% B in 15 ml.*

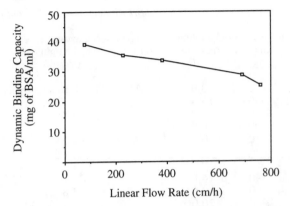

Figure 7 *Dynamic binding capacity for BSA of the Macro-Prep DEAE support. The support looses less than 30% of the capacity when the flow rate is increased from 75 cm/h to 760 cm/h. Column: 1 x 10 cm, 3 ml. Sample: 3 mg/ml of BSA, applied until the eluting buffer had an A280 of 50% of the incoming sample. Buffer: A: 10 mM Tris-HCl, pH 7.6. B: A + 1 M NaCl, pH 7.6.*

4.3 Chemical Stability

The Macro-Prep DEAE support is stable in a wide variety of buffers, salts and solvents commonly used in chromatography. It is stable to extended exposure to 1 M NaOH (**Figure 8**). It exhibits virtually no shrinking or swelling with changes in salt concentration or pH (**Table 1**). The shrinking and swelling can be completely eliminated if the support is packed in a column with a flow adaptor locked in place.

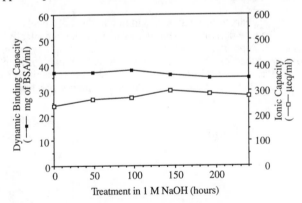

Figure 8 *Stability of the Macro-Prep DEAE support. Samples of support were taken at interval and tested for dynamic binding capacity as in fig. 7, and titrated for ionic capacity.*

Table 1 *Shrinking and swelling of the Macro-Prep support with changes in salt and pH.*

Time	Buffer	Linear Flow Rate (cm/h)	Operating Pressure (PSI)	Bed height (cm)	Change in bed height (%)
60 min.	A	0	0.0	19.0	0
30 min.	A + 1 M NaCl	390	13.4		
60 min.	A + 1 M NaCl	0	0.0	18.7	-1.6
30 min.	1 M NaOH	390	20.0		
48 hours	1 M NaOH	0	0.0	19.5	+2.5
30	A +1 M NaCl	390	14.8		
60 (rest)	A + 1 M NaCl	0	0.0	18.3	-3.8
20	1 M NaOH	390	20.0		
Over night	1 M NaOH	0	0.0	18.9	-0.5
30 min.	A +1 M NaCl	390	14.9		
60 min.	A + 1 M NaCl	0	0.0	18.2	-4.4
30 min.	Water	390	14.1		
90 min.	Water	0	0.0	18.8	-1
30 min.	1 M HCl	390	17.3		
60 min.	1 M HCl	0	0.0	18.5	-2.7

Buffer A: 20 mM Tris, pH 7.6. Measurements done with approximately 100 ml support in a 2.5 x 30 cm glass column, with the flow adaptor some distance from the bed.

5 CONCLUSION

The Macro-Prep DEAE support appears to be a valuable addition to the tools available for preparative protein isolation. The new derivative exhibits considerably higher selectivity than other commercially available supports. It gives high resolution, has high dynamic binding capacity and good flow properties; this combined with its ease of use, forming stable columns which flow well, and chemical stability which allows it to be sanitized and regenerated, should make it an excellent candidate for the development of separation processes.

Application of Charged Ultrafiltration Membranes in Continuous, Enzyme-catalysed Processes with Coenzyme Regeneration

B. Nidetzky, K. Schmidt, W. Neuhauser, D. Haltrich, and K. D. Kulbe

DIVISION OF BIOCHEMICAL ENGINEERING, INSTITUTE OF FOOD
TECHNOLOGY, UNIVERSITY OF AGRICULTURE, PETER-JORDAN STRASSE
82, A-1190 VIENNA, AUSTRIA

Selective and almost complete retention of the native nicotinamide-adenine-dinucleotide coenzyme has been achieved in a continuous reactor system equipped with derivatized ultrafiltration membranes carrying functional groups with a negative net charge. Employing this membrane technology for coenzyme retention, xylitol and gluconic acid could be continuously synthesized from xylose and glucose by a coupled, NAD(H) regenerating enzyme system.

1 INTRODUCTION

For the biotechnical utilization of dehydrogenases that require nicotinamide-adenine-dinucleotide as a coenzyme, two main prerequisites for process economics are (i) the regeneration of the coenzyme and (ii) its retention in the reaction system during continuous mode of operation. Several methods to recycle the oxidized or reduced form of the coenzyme such as coupled, enzymatic redox-reactions[1-3] or non enzymatic (electro)chemical[4] regeneration have been reported and thus the addition of only catalytic instead of stoichiometric amounts of the respective coenzyme was required. A suitable technology for continuous processes, however, has still to be developed.

Several concepts for the retention of coenzymes have been described, among which the most important apparently are: (i) the utilization of soluble, artificially enlarged modifications of the coenzyme derived from covalent coupling of NAD(H) to either polymers or the enzyme itself[5-8], (ii) dynamic recycling[9] and (iii) the application of specific membrane technologies[10-12] advantageously using the native form of the respective coenzyme. This could be especially important since several enzymes do not accept the modified form of the coenzyme[11]. It is concluded that, with achievable cycles of the native coenzyme of > 500,000, the costs of the coenzyme are no longer limiting for the total bioprocess[12].

Nicotinamide-adenine-dinucleotides are amphoteric molecules with acidic phosphate and basic aminogroups. Since, at pH values higher than 3, NAD(P)(H) carries a negative net charge, the electric repulsion forces between negatively charged functional groups introduced onto an ultrafiltration membrane and the coenzymes can be used to reject and to retain NAD(P)(H) inside the reaction system. Thus coenzyme losses can be prevented during continuous operation[10-12]. Following this concept, the enzymes and the coenzymes are specifically retained while low molecular weight substrates or products (provided that these do not carry negative charges) may unimpededly permeate the membrane. Here some new data on coenzyme retention using charged ultrafiltration membranes are presented and the application of this concept to produce xylitol and gluconate by a NAD(H)-linked enzyme system of aldose reductase (ALR) and glucose dehydrogenase (GDH) is demonstrated.

2 RESULTS AND DISCUSSION

Characterization of the reactor system

The retention of NAD(H) was experimentally determined in a convective, laboratory-scale reactor (total volume of 50 ml) which had a flat membrane configuration and was equipped with a magnetic stirrer. The ideality of the stirred tank reactor type was assured by analysis of the residence time distribution applying a pulse of tracer salt (NaCl) using conductivity measurement. Since the reactor was usually stirred at 3-6 s^{-1} and transmembrane fluxes were in the range of 10^{-4}-10^{-5} m $(s\ bar)^{-1}$, polarization of the membrane was avoided and the calculated retention coefficients are real.

The retention coefficient of a particular compound e.g. NADH is defined as:

$$R = 1 - \frac{C_{ft}}{C_r} \tag{1}$$

where C_{ft} and C_r are the coenzyme concentrations in the filtrate and reactor, respectively. R is the retention coefficient.

The initial concentration of NADH in the reactor, which is quantified by monitoring its characteristic absorption at 340 nm, decreases because of (i) incomplete retention and (ii) deactivation during continuous operation (without feed of fresh coenzyme). Taking into account both the deactivation of the reduced coenzyme which is assumed to be a first-order reaction, and the leakage through the membrane, a rate equation for the decrease of available NADH in the reaction system can be written as

$$\frac{dC_t}{dt} = -\frac{1}{\tau}(1-R)\,C_t - k_d C_t \tag{2}$$

After integration and rearrangement the retention coefficient R can be calculated as follows:

$$R = 1 + \tau k_d + \frac{\tau}{t}\ln\frac{C_t}{C_0} \tag{3}$$

with τ the residence time, t the reaction time, k_d the first-order deactivation constant of the coenzyme, C_0 the initial concentration of NADH in the reactor, and C_t its concentration after a reaction time t.

Characterization of the charged membrane

Among several types of membranes studied (**Table 1**) the most suitable properties - high retention coefficient for NADH, high permselectivity in order to avoid substrate or product accumulation inside the reactor and good transmembrane fluxes - are exhibited by the NTR type membranes of Nitto (Nitto Electric Industrial Company, Japan). NTR 7430, which is a modification of the commercially available membrane series 7400, was selected for further experiments because it showed the highest retention capacity for NADH (and somewhat less for NAD^+) while glucose or other aldose sugars were not significantly retained. Protein binding onto the membrane was also very low. This (composite) membrane has a supporting matrix of neutral polysulfone and is covered with a thin layer (0.2 - 0.5 µm) of sulfonated polyetherketone and polyethersulfone. Its charge density is approximately 1.5 mequiv.g^{-1} and it is resistant to temperatures up to 80 $^\circ$C and to organic solvent concentrations up to 20 % v/v. Utilizing the combined effects of size exclusion (molecular weight cut off = 0.7 - 1 kDa) and electrostatic repulsion, retention coefficients for NADH higher than 0.99 could be obtained.

Since NAD^+ and NADH are thermally inactivated (cf. **Equation (2)**) with NADH being quite unstable at pH values below 6.0 ($t_{1/2}$<30 hours at pH 6.0), in view of stabilizing the coenzyme[1] a possible antagonistic effect of additives on the retention of NAD(H) by the membrane has also to be considered. Thus, conditions for continuous operation need thus to be carefully balanced. It was found that retention coefficients are (i) increased in the presence of proteins (especially in the case of NAD^+), (ii) lowered with increasing concentrations of ionic compounds and, (iii) improved with increasing pH values.

NAD^+ is less well retained than NADH because of its lower negative net charge and its R-value may vary from approximately 0.85 to 0.93. Thus the performance of the reactor equipped with the charged membrane will depend on the ratio of NAD^+ to total coenzyme which in turn is governed by the activities of both enzymes.

Table 1   *Retention of pyridine nucleotide NADH and glucose by different types of charged ultrafiltration membranes*

supplier	type	transmembrane flux m (h bar)$^{-1}$	R-value NADH	R-value glucose
NITTO	SPA-I	5.5 10^{-1}	0.35	0.05
	NTR 7410	1.2 10^{-2}	0.73	0.05
	NTR7410x	7.0 10^{-3}	0.92	0.14
	NTR 7430	2.0 10^{-3}	>0.99	0.15
Amicon	Y05	2.6 10^{-3}	0.89	0.45
Desalination	DS5	5.1 10^{-4}	0.98	0.95
	G5	7.5 10^{-4}	0.85	0.70
Osmonic	SEPA-MX07	2.2 10^{-2}	0.95	0.70

To prevent leakage of NAD⁺, reaction conditions should be chosen such as to obtain a low f factor as defined in **Equation (4)**.

$$f = \frac{NAD^+}{NAD^+ + NADH}, \quad 0 \le f \le 1 \qquad (4)$$

Characterization of the coupled enzyme system

GDH from *Bacillus cereus*[13,14] was a commercial product (Amano, Japan) and its kinetic characterization has been reported by Schmidt[15]. ALR was partially purified from the cell extracts of *Candida tenuis* by a three-step procedure using ammonium sulfate-precipitation (30 % saturation), hydrophobic interaction- and anion exchange chromatography on phenylsepharose and Q-sepharose, respectively. By this procedure contaminating activities were removed and ALR was obtained 20-fold purified in yields of approximately 70 %.

Both ALR and GDH may utilize either NAD(H) or NADP(H). For economic reasons such as lower price and higher stability, NAD(H) was chosen as a coenzyme. The coupled reaction system investigated during this study is shown in **Figure 1**.

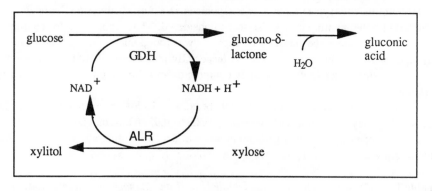

Figure 1 *NAD(H)-linked enzyme system of ALR and GDH for the conversion of xylose and glucose.*

ALR and GDH may operate in a rather broad pH range between 6.0 and 7.5 without appreciable decrease of activity. Both enzymes are stable for at least several days at room temperature. Their kinetic constants (K_m- and K_i-values) seem to be quite suitable since the affinity for the substrates is high and no evidence for strong product inhibition could be detected[15,16]. The activity for the reaction catalysed in the backward direction (xylitol to xylose and glucono-δ-lactone to glucose) was less than 2 % in the case of both enzymes at pH 7.0. Spontaneous hydrolysis of glucono-δ-lactone to gluconic acid (i) ensures that the reaction is irreversible but (ii) requires controlled addition of alkali to titrate the acid in order to maintain a constant pH value. The formation of gluconate is considered a disadvantage because this (negatively charged) compound is retained by the membrane to a significant extent (R=0.87 in the presence of protein). Gluconate thus slowly accumulates during reaction[17] and the concomitant increase in ionic strength in turn decreases the R-value for the coenzyme drastically and also inhibits ALR. However, the well known method of recycling the reduced form of pyridine nucleotide NADH using formate dehydrogenase[18,19] oxidizing formate to CO_2 may be difficult to apply in this particular case because formate is a strong inhibitor of ALR. A more than 5-fold reduction of the enzyme's activity is seen at a formate concentration of less than 10 mmol l^{-1} (W. Neuhauser et al., unpublished). The ALR from *C. tenuis* is an enzyme exhibiting a very broad substrate specificity[20] and in addition to xylose is also capable of reducing glucose (to sorbitol). Thus reaction conditions have to be chosen such as to avoid sorbitol formation by ALR as far as possible when using a substrate mixture of xylose and glucose.

Although K_m-values of ALR for xylose (0.25 mol l^{-1}) and glucose (> 5 mol l^{-1}) differ widely, it has been observed that when using a substrate mixture of xylose and glucose at pH values higher than 6.0, significant sorbitol synthesis by ALR occurred. Thus, in order to avoid any side product formation, reactions were carried out at a constant pH value of 6.0; since under these slightly acidic conditions NADH is quite unstable and since deactivation of NADH will be faster than the wash-out of NAD^+ from the reactor, a high f factor was chosen by employing a surplus of NADH oxidizing enzyme i.e. a ratio of 0.3 units GDH to each unit of ALR. The results of a continuous conversion experiment using a feed solution of 300 mmol l^{-1} xylose and glucose are shown in **Figure 2**.

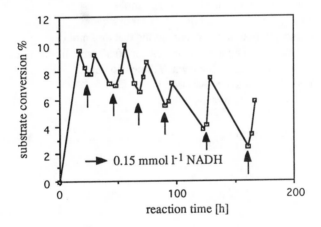

Figure 2 *Continuous conversion of xylose/glucose (300 mmol l^{-1} each) at 25 °C using 1.0 i.u. ALR ml^{-1} and 0.3 i.u. GDH ml^{-1} at an initial concentration of 0.25 mmol l^{-1} NADH(τ = 17 hours).*

Substrates and products were quantified by h.p.l.c with an Aminex HPX-87 C column and refractive index detection. A regular dosage of 0.15 mmol l^{-1} NADH was necessary to compensate for the decrease in the percentage of substrate conversion obviously caused by leakage of NAD^+. Based on the results of the discontinuous experiments, the theoretically calculated concentration of products (100 mmol l^{-1} equivalent to 30 % conversion) could not be attained which is due to lower real enzyme concentrations in the reactor than expected, probably caused by an unspecific adsorption of protein onto the membrane. Evidence for further time-dependent deactivation of the enzymes can be seen from the steady decrease of substrate conversion after dosage of fresh coenzyme.

3 CONCLUSIONS

Ultrafiltration membranes carrying negatively charged functional groups can be efficiently used for a selective retention of native nicotinamide-adenine-dinucleotide coenzymes during continuous operation (**Table 1**). The data in **Figure 2** demonstrate that the coupled enzyme system of ALR and GDH in combination with the charged membranes technology can be employed in a continuous process to synthesize xylitol and gluconate from mixtures of xylose and glucose. Further optimization of the efficiency of the enzymatic process described here is, however, necessary and currently attempted.

Nomenclature

- C_{ft} concentration of component in filtrate / mol l^{-1}
- C_r concentration of component in reactor / mol l^{-1}
- C_0 initial concentration of component in the reactor / mol l^{-1}
- C_t concentration of component in the reactor after reaction time t / mol l^{-1}
- f ratio of oxidized to total coenzyme / dimensionless
- k_d first-order deactivation constant / h^{-1}
- R   retention coefficient / dimensionless
- t   reaction time / h
- τ average residence time / h
- $t_{1/2}$ half life / h

References

1. H.K. Chenault and G.M. Whitesides, *Appl.Biochem.Biotechnol.*, 1987, **14**, 147.
2. M.-R. Kula and C. Wandrey, *Meth.Enzymol.*, 1987, **136**, 9.
3. K. Nakamura, M. Aizawa and O. Miyawaki, "Electroenzymology-Coenzyme Regeneration", Springer Vlg., Berlin/Heidelberg, 1988.
4. M.-O. Mansson and K. Mosbach. D. Dolphin, R. Poulson and O. Avramovic (Eds.), "Pyridine Nucleotide Coenzymes", J. Wiley & Sons, New York, 1987, pp. 217
5. S. Riva, G. Carrea, F.M. Veronese and A.F. Bückmann, *Enzyme Microb.Technol.*, 1986, **8**, 556.
6. K. Nakamura, H. Minami, I. Urabe and H. Okada, *J.Ferment.Technol.*, 1988, **66**, 267.
7. A.F. Bückmann, M. Morr and M.-R. Kula, *Biotechnol.Appl.Biochem.*, 1987, **9**, 258.
8. M. Persson, M.-O. Mansson, L. Bülow and K. Mosbach, *Bio/Technology*, 1991, **9**, 280

9. O. Miyawaki and T. Yano, *Biotechnol.Bioeng.*, 1992, **39**, 314.
10. M. Ikemi and Y. Ishimatsu, *J.Biotechnol.*, 1990, **14**, 211.
11. M.W. Howaldt, K.D. Kulbe and H. Chmiel, *Ann.N.Y.Acad.Sci.*, 1990, **589**, 253.
12. K.D. Kulbe, M.W. Howaldt, K.-H. Schmidt, T.R. Röthig and H. Chmiel, *Ann.N.Y. Acad.Sci.*, 1990, **613**, 820.
13. C.-H. Wong and D.G. Drueckhammer, *Bio/Technology*, 1985, **3**, 649.
14. C.-H. Wong, D.G. Drueckhammer and H.M. Sweers, *J.Am.Chem.Soc.*, 1985, **107**, 4028.
15. K. Schmidt, PhD Thesis, University of Stuttgart, 1993.
16. B. Nidetzky, W. Neuhauser, H. Schmidt, D. Haltrich and K.D. Kulbe, Abstract P112, presented at the "European Symposium on Biocatalysis", Graz, Austria, Sep. 12-17, 1993.
17. M.W. Howaldt, PhD Thesis, University of Stuttgart, 1988.
18. Z. Shaked and G.M. Whitesides, *J.Am.Chem.Soc.*, 1980, **102**, 7104.
19. K.H. Kroner, H. Schütte, W. Stach and M.-R. Kula, *J.Chem.Tech.Biotechnol.*, 1982, **32**, 130.
20. K.D. Kulbe, H. Schmidt, K. Schmidt and A.A. Scholze. J. Visser, G. Beldman, M.A. Kusters-van Someren and A.G.J. Voragen (Eds.), "Xylans and Xylanases", Elsevier Science Publishers, Amsterdam, 1992, pp. 565

Theoretical Description of the Distribution of Metal Ions in Aqueous Polymer Extraction Systems Used in Biotechnology

T. I. Nifant'eva, V. M. Shkinev, and B. Ya. Spivakov

V. I. VERNADSKY INSTITUTE OF GEOCHEMISTRY AND ANALYTICAL CHEMISTRY, KOSYGIN STR. 19, 117975 MOSCOW V 334, RUSSIA

1 INTRODUCTION

Two-phase aqueous polymer systems are used effectively for the separation of various biological materials[1,2]. Biopolymers and particles are usually transferred to a more hydrophobic polymer-enriched phase, which is produced, when water soluble polymer is salted out from a solution of different electrolytes or from a solution of another polymer. In our previous work it was shown that such systems could also be used for separation of metal ions[3-6]. The distribution coefficients D (which are the ratio of metal ion concentration in the polymer-enriched phase to that in the salt phase) vary over a wide range and depend on different parameters of the extraction system.

The partition of heavy metal ions during the separation of biological materials in two-phase polymer systems has not been widely discussed. Meanwhile, such metals are always present as impurities in $(NH_4)_2SO_4$ or other salts which are used at high concentrations (up to 40%) to form two-phase systems by salting out polyethylene glycol (PEG). Thus, there is a possibility of metal ion preconcentration in one of the phases such that the biotechnology product after the separation could contain some heavy metals. This may be important due to their toxicity and bioactivity.

This work presents the result of an investigation of the distribution of metal ions (Cu, Fe, Zn, Co, Eu) in the extraction system PEG 2000 - $(NH_4)_2SO_4$ - H_2O. The composition of the system and the experimental technique was described earlier[3,4]. Equations describing the dependence of the distribution coefficients upon pH, metal concentration and composition of extraction system are proposed.

2 RESULTS AND DISCUSSION

2.1 Extraction in the Absence of Organic Reagent

The following equilibria and the corresponding constants (with the charges omitted) are considered.
1. Complexation of PEG with metal ions

$$M + L \rightleftarrows ML; \qquad \beta_{ML} = \frac{[ML]}{[M] \cdot [L]} \qquad (1)$$

where L is PEG and M is metal ion.

2. Transfer of the complex into the PEG phase

$$ML \rightleftharpoons \overline{ML}; \qquad\qquad P_{ML} = \frac{[\overline{ML}]}{[ML]} \qquad\qquad (2)$$

where $\overline{ML}$ is the complex in the PEG phase.

3. Distribution of the polymer between the salt phase and the PEG phase

$$L \rightleftharpoons \overline{L}; \qquad\qquad P_L = \frac{[\overline{L}]}{[L]} \qquad\qquad (3)$$

4. Protonation of PEG

$$L + H \rightleftharpoons LH; \qquad\qquad K_L = \frac{[LH]}{[L] \cdot [H]} \qquad\qquad (4)$$

5. Distribution of the protonated form of PEG

$$LH \rightleftharpoons \overline{LH}; \qquad\qquad P_{LH} = \frac{[\overline{LH}]}{[LH]} \qquad\qquad (5)$$

6. Formation of sulphate complexes of the metal ion

$$M + i(SO_4) \rightleftharpoons M(SO_4)_i. \qquad\qquad \beta_{M(SO_4)i} = \frac{[M(SO_4)_i]}{[M] \cdot [SO_4]^i} \qquad\qquad (6)$$

7. Distribution of the sulphate complexes

$$M(SO_4)_i \rightleftharpoons \overline{M(SO_4)}_i. \qquad\qquad P_{M(SO_4)i} = \frac{[\overline{M(SO_4)_i}]}{[M(SO_4)_i]} \qquad\qquad (7)$$

8. Protonation of sulphate ion

$$SO_4 + H \rightleftharpoons HSO_4; \qquad\qquad \beta_{HSO_4} = \frac{[HSO_4]}{[H] \cdot [SO_4]} \qquad\qquad (8)$$

Based on eqs. (1)-(8) and assuming that [ML] and [LH] are small compared to the concentration of the other components in the extraction system, and that

$$\frac{\overline{M} + \sum_i \overline{M(SO_4)}_i}{M + \sum_i M(SO_4)_i} = bD_{SO_4} \qquad\qquad (9)$$

where D_{SO4} is the distribution coefficient for sulphate ion in the system, and b is an empirical constant, the following equation is obtained for D in the sulfate extraction system

$$D = bD_{SO_4} + \frac{P_{ML}\beta_{ML}\overline{L}_0}{\left(P_L + K_L P_{LH}H + P_{ML}\beta_{ML}M\right)} \cdot \frac{1}{\alpha_{MSO_4}} \tag{10}$$

where

$$\alpha_{MSO_4} = 1 + \sum_i \beta_i \frac{SO_4^i}{\alpha_{HSO_4}}$$

$$\alpha_{HSO_4} = 1 + \beta_{HSO_4}H$$

If the metal concentration is high and/or pH is low the following limiting case of eq.(10) is observed, in which D_{min} is no longer dependent on the metal ion concentration and pH

$$D_{min} = bD_{SO_4} \tag{11}$$

In order to confirm the applicability of eq. (11) for describing distribution coefficients in the region of their minimum values, the dependence of D on D_{SO4} in the systems of various composition was investigated: for Cu and Zn at high metal concentrations, for Fe, Zn and Co at low concentrations and low pH. From **Figure 1** it is seen that the relationship of D_{min} versus D_{SO4} in logarithmic coordinates gives a straight line with a slope equal to unity and log b value of 0.44.

The comparison of D at pH < 2.5 with the partition of the sulfate ion in the system (**Figure 2 a and b**) shows that the difference between log D and log D_{SO4} is approximately constant and has the value obtained for log b.

Thus, the extraction at low pH and high metal concentration depends only on the sulphate ion distribution coefficient, which itself depends on the $(NH_4)_2SO_4$ and PEG content in the system and could be evaluated from the phase diagram[3,4].

At low metal concentration and relatively high pH the other limiting case of eq. (10) is observed in which D_{max} is independent of these parameters, viz:

$$D_{max} = bD_{SO_4} + \frac{P_{ML}\beta_{ML}\overline{L}_0}{P_L} \cdot \frac{1}{\alpha_{MSO_4}} \tag{12}$$

In fact, the results presented in **Figure 2** reveal a region of constant D_{max} values at pH >4.

The dependence of D on the metal concentration exhibits two regions in which distribution coefficients are constant - D_{max} and D_{min}. For example, for Zn, Co and Cu this occurs at $C_M < 10^{-4}$ M and $C_M > 3 \cdot 10^{-2}$ M respectively.

If the total composition of the system is changed, the variables in eq.(12) are D_{SO4}, PEG concentration and the $\alpha_{M(SO4)}$ values. Taking into account the changes of these parameters the values of D_{max} for Zn, Fe and Eu were calculated. Good coincidence of the obtained results with experimental D for these metals (**Figure 3**) confirms the applicability of eq. (12) for the description of the extraction in the field of its maximum values.

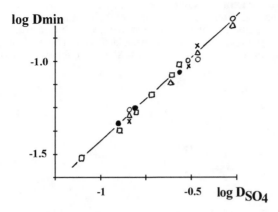

Figure 1 *Dependence of distribution coefficients in the field of their minimum
values on sulphate distribution in the extraction system:*
Co (o), Fe (△), Zn (×) - $C_M = 1 \cdot 10^{-5}$ M, pH 2.5
Cu (●), Zn (□) - $C_M = 3 \cdot 10^{-2}$ M, pH 5.0

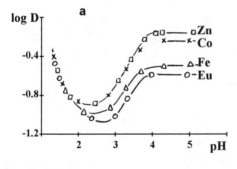

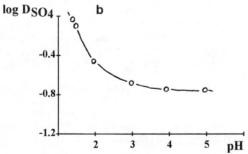

Figure 2 *Distribution of metals (a) and sulphate (b) and their
dependence on pH.* $C_M = 1 \cdot 10^{-5}$ M

2.2 Extraction of Metals with Organic Reagent

Let us consider now the distribution of metal ions in the presence of a water soluble organic compound ,in the most simple case - a monobasic acid HA. The equilibria in the system can be described by the previously derived eqs. (1)-(8) and the following:

$$M + A \rightleftarrows MA; \qquad\qquad \beta_{MA} = \frac{[MA]}{[M] \cdot [A]} \qquad\qquad (13)$$

$$MA \rightleftarrows \overline{MA}; \qquad\qquad P_{MA} = \frac{[\overline{MA}]}{[MA]} \qquad\qquad (14)$$

$$A + H \rightleftarrows HA; \qquad\qquad K_{HA} = \frac{[HA]}{[A] \cdot [H]} \qquad\qquad (15)$$

$$HA \rightleftarrows \overline{HA}; \qquad\qquad P_{HA} = \frac{[\overline{HA}]}{[HA]} \qquad\qquad (16)$$

At low metal ion concentration (which is the most common case) the following equation for D is obtained:

$$D = bD_{SO_4} + \frac{P_{ML}\beta_{ML}\overline{L}_0}{P_L + K_L P_{LH}H} \cdot \frac{1}{\alpha_{MSO_4}} + \frac{P_{MA}\beta_{MA}\overline{HA}_0}{K_{HA}P_{HA}H} \cdot \frac{1}{\alpha_{MSO_4}} \qquad (17)$$

If $K_L P_{LH}H > P_L$, then eq. (17) is reduced to the form

$$D = bD_{SO_4} + \frac{1}{\alpha_{MSO_4}H} \cdot \left(\frac{P_{ML}\beta_{ML}\overline{L}_0}{K_L P_{LH}} + \frac{P_{MA}\beta_{MA}\overline{HA}_0}{K_{HA}P_{HA}} \right) \qquad (18)$$

Let us consider the value $P_{MA}\beta_{MA}\overline{HA}_0/(K_{HA}P_{HA}) < P_{ML}\beta_{ML}\overline{L}_0/(K_L P_{LH})$. The experimentally determined curve showing the relationship of D versus pH (**Figure 4**) in the extraction of cobalt in the form of a thoron complex illustrates a special case of eq.(17) at various pH values. If pH < 2.5, a limiting equation for the distribution coefficient is obtained, namely, $D_{Co} = D_{SO_4}b$. When the pH value is increased, the condition $P_L > K_L P_{LH}H$ is fulfilled, and at pH > 4 a second region of constant D_{Co} values is obtained, which can be described by the eq.(12).

Finally, at high pH values, where the condition $A = \overline{HA}_0$ is met, the equation for D is the following:

$$D = bD_{SO_4} + \frac{P_{ML}\beta_{ML}\overline{L}_0}{P_L\alpha_{MSO_4}} + \frac{P_{MA}\beta_{MA}\overline{HA}_0}{\alpha_{MSO_4}} \approx \frac{P_{MA}\beta_{MA}\overline{HA}_0}{\alpha_{MSO_4}} \qquad (19)$$

i.e., a third region with constant D values is observed.

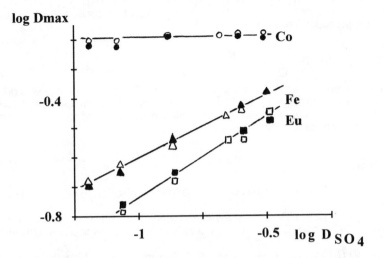

Figure 3 *Dependence of experimental (○,△,□) and calculated (●,▲,■) distribution coefficients in the field of their maximum values on the sulphate distribution in the system.* pH 5.0, $C_M = 1 \cdot 10^{-5}$ M

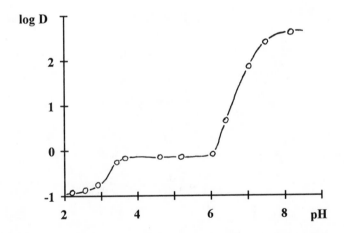

Figure 4 *Extraction of cobalt in the form of a thoron complex and its dependence on pH*, $C_R = 1.1 \cdot 10^{-2}$ M, $C_{Co} = 1 \cdot 10^{-5}$ M

In the cases where $P_{ML}\beta_{ML}\overline{L}_0/(K_L P_{LH}) < P_{MA}\beta_{MA}\overline{HA}_0/(K_{HA}P_{HA})$, the region of constant D values, corresponding to the maximum degree of blank extraction, may be missing in the pH dependence curves for $D^{3,4}$.

3 CONCLUSIONS

The equations proposed for the distribution coefficients of metal ions in the extraction system PEG-2000 - $(NH_4)_2SO_4$ - H_2O describe their dependence on different parameters and permit the selection of conditions for the predominant transfer of metal to a polymer or to a salt phase during biotechnology separations.

References

1. P.A.Albertsson, 'Partition of Cell Particles and Macromolecules', Wiley, New York, 3rd ed., 1986.
2.'Partitioning in Aqueous Two-Phase Systems. Theory, Methods, Uses and Application to Biotechnology', H.Walter, D.E.Brooks and D.Fisher (eds.), Academic Press, Orlando, Florida, 1985.
3. T.I. Zvarova (Nifant'eva), V.M. Shkinev, G.A. Vorob'eva, et al., *Mikrochim. Acta*, 1984, **3**, 449.
4. V.M. Shkinev, N.P. Molochnikova, T.I. Zvarova (Nifant'eva), et al., *J.Radioanal.Nucl.Chem.*, 1985, **88**, No. 1, 115.
5. T.I.Nifant'eva, V.M.Shkinev, B.Ya.Spivakov, Yu.A.Zolotov, *Z.Analyt.Kh.*, 1989, **44**, No.8, 1368.
6. T.I.Nifant'eva, V.M.Shkinev, B.Ya.Spivakov, Yu.A.Zolotov, *Dokl.AN SSSR*, 1989, **308**, No.4, 879.

Rapid Adsorbent Recycling for Continuous Separations: The Effects of Liquid Flow Rates on Apparatus Operation

Gordon W. Niven, Peter G. Scurlock, and Alison L. Gibb

DEPARTMENT OF BIOTECHNOLOGY AND ENZYMOLOGY, INSTITUTE OF FOOD RESEARCH, READING LABORATORY, EARLEY GATE, WHITEKNIGHTS ROAD, READING RG6 2EF, UK

1 INTRODUCTION

We are developing a novel strategy for carrying out continuous adsorptive separations. This has the potential to overcome some of the limitations of conventional methods for the purification of high-volume/low-cost products[1].

The majority of adsorptive separations techniques operate in batch mode and so their binding capacities are dependent on the surface area available. This is usually maximised by the use of highly porous adsorbent matrices such as those developed for column chromatography. The diffusion of solute into the matrix structure is a relatively slow process which increases the solid-liquid contact time which is required for maximum binding capacity, and therefore efficient use of adsorbent, to be achieved. Our approach does not involve the use of porous particulate matrices, but rather adsorption onto a non-porous planar surface. Although the resulting binding capacity is low, the contact time necessary between the solute and the adsorbent is much reduced. This enables continuous separations by the rapid recycling of adsorbent between adsorption and desorption environments.

The apparatus consists of a tank divided into four chambers arranged in-line (**Figure 1**). The adsorbent is in the form of a mobile continuous nylon belt which passes through each chamber. Peristaltic pumps deliver a flow of feedstock to the first chamber, wash medium to the second, eluent to the third and regeneration medium to the fourth. As the belt moves sequentially through each chamber, the target compound is transferred from the feedstock stream to the eluent in a continuous process. Although the total surface area of adsorbent is small, in a continuous process it is the area per unit time which is important to ensure a high throughput. This is facilitated by rapid adsorbent recycling.

This system offers several potential advantages for process scale separations. The low surface area enables a large amount of a biological compound to be purified using a relatively small amount of affinity ligand or antibody. As these can represent a major cost in process scale operations, this technique may have the potential to make highly selective separations feasible for a wider range of products. In addition, continuous processes can be monitored and computer controlled in real-time which maintains consistent product characteristics and aids quality assurance. Relative to porous separations matrices, the nylon belt is inexpensive and the system is less subject to fouling by particulate materials in the feedstock.

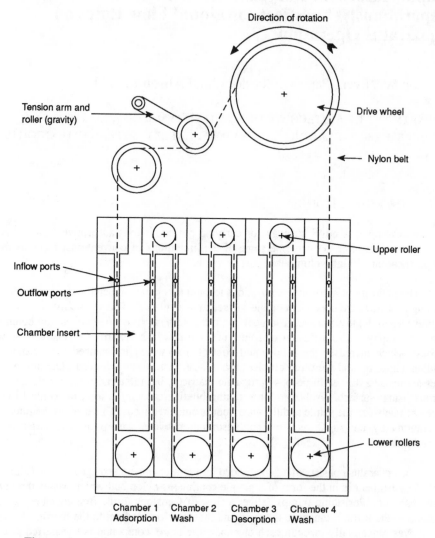

Figure 1   *Schematic representation of apparatus used for continuous separations*

The technical feasibility of this separations method was previously demonstrated by the isolation of trypsin from bovine pancreatic extract using soybean trypsin inhibitor (STI) as affinity ligand[1]. During 8 h continuous operation, it was estimated that 12 mg of trypsin was isolated using 0.1-0.3 mg of ligand of similar molecular weight, although this process was not optimised. This work showed that various factors influence apparatus performance, including target compound concentrations, the ligand recycle time, and liquid media flow rates. A thorough understanding of the complex interplay of these factors is required to develop a rational approach to process optimisation and apparatus design. This paper describes the use of Cu^{2+} transfer by chelation to investigate the effects of media flow rates on throughput and recovery and their interrelation. This chemical model system enabled the effects of operational parameters to be studied without interference from biological effects such as loss of activity and proteolysis which can occur in affinity systems such as trypsin/STI.

2 MATERIALS AND METHODS

The design and operation of the apparatus and the procedure for activating the nylon belt were described in detail elsewhere[1]. The chelating adsorbent was prepared by incubating the glutaraldehyde-activated belt (4 cm x 190 cm) in 20 mM iminodiacetic acid for 16 h at 25°C. With the belt fitted, each chamber contained 45 ml of liquid with a belt passage length of 30 cm. Unless stated otherwise, the apparatus was operated under the following conditions: belt speed, 1 m min^{-1}; feedstock, 2 mM $CuSO_4$ (chamber 1); wash, water, 10 ml min^{-1} (chambers 2 & 4); eluent, 50 mM ethylenediaminetetraacetic acid (EDTA) (chamber 3).

The concentration of Cu^{2+} was determined by measuring absorbence at 265 nm in the presence of 50 mM EDTA. A molar extinction coefficient of 3086 M^{-1}cm^{-1} was determined experimentally.

3 RESULTS AND DISCUSSION

During operation of the apparatus, a certain amount of liquid is transferred between adjacent chambers by the movement of the wet belt. This causes a small but continuous dilution of the contents of the adsorption and desorption chambers as liquid adhering to the belt is removed and replaced with wash media. This "carryover" is an important factor when determining the limitations of the separations system. It causes a concentration-dependent loss of material which places a limit on the achievable product concentration and results in a compromise between concentration and recovery. It may also effect the adsorption and elution conditions.

The volume of liquid carryover was estimated by measuring the transfer of Cu^{2+} from chamber 2 to chamber 3, in the presence of 50 mM EDTA to prevent transfer by chelation. The Cu^{2+} concentration in chamber 2 was maintained at 4.5 mM by means of a high flow rate (10 ml min^{-1}) and no flow of liquid was applied to chamber 3. Chambers 1 and 4 were flushed with water at 10 ml min^{-1}. A first order increase in Cu^{2+} concentration in chamber 3 was observed with time (**Figure 2**). The maximum level of Cu^{2+} would be obtained when the rates of Cu^{2+} transfer by carryover into, and out of, chamber 3 had equalized. From the Cu^{2+} concentration in chamber 2 (4.5 mM), and the initial rate of Cu^{2+} accumulation in chamber 3 (0.077 mM min^{-1}), it was estimated that the rate of liquid transfer due to carryover was approximately 0.77 ml min^{-1}.

To investigate the effect of feedstock application rate on apparatus performance, the apparatus was operated as described in the Materials and Methods section, using an eluent flow rate of 10 ml min^{-1}. Feedstock (2 mM $CuSO_4$) was delivered to the adsorption chamber at varying flow rates. The concentration of Cu^{2+} in the chamber outflow was monitored as the belt was cycled and the Cu^{2+} concentration at steady state was determined. There were three possible exit routes for feedstock material entering the adsorption chamber. It can be lost from the system via the chamber outflow, or by carryover into the waste stream of the wash chamber. Also, it can be adsorbed onto the belt and be transferred to the elution chamber. The loss by carryover is proportional to the concentration of Cu^{2+} in the adsorption chamber outflow and can be estimated from the carryover flow rate determined in the previous experiment. The apparent rate of transfer of Cu^{2+} via the belt is therefore the difference between the amount entering the chamber and the two sources of loss.

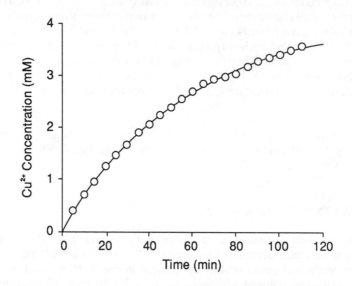

Figure 2   *Increase in Cu^{2+} concentration in chamber 3 as a result of carryover from chamber 2*

Figure 3 shows the effects of feedstock flow rate on the apparent Cu^{2+} transfer rate as determined by this method, and on efficiency which is the transfer expressed as a percentage of Cu^{2+} supplied. The apparent transfer rate increased with feedstock flow rate while the transfer efficiency decreased. These effects can be explained in terms of the concentration of Cu^{2+} in the adsorption chamber. This increased with feedstock flow rate from 0.13 mM at 0.5 ml min^{-1} to 1.47 mM at 15 ml min^{-1}. The amount of Cu^{2+} bound to the belt is dependent on its concentration in the adsorption chamber, in accordance with its adsorption isotherm. However, the quantity of Cu^{2+} lost through the chamber outflow and by carryover also increases with concentration. This resulted in a compromise between the rate of transfer and the efficiency. The maximum proportion of Cu^{2+} from the feedstock which could be bound to the belt under the conditions used was found to be 83%, at a flow rate of 0.5 ml min^{-1}. The apparent rate of transfer was 0.81 μmol min^{-1}. Transfer rates of approximately 6 μmol min^{-1} were possible at higher flow rates, but this resulted in lower transfer efficiencies of between 20% and 30%. The estimated loss by carryover ranged from 4% at 15 ml min^{-1} to 10% at 0.5 ml min^{-1}. The adsorption chamber outflow was therefore the major source of loss, particularly at higher flow rates.

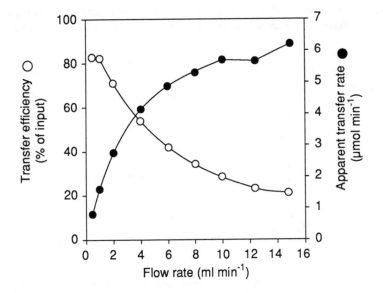

Figure 3 *The effect of feedstock flow rate on the rate of Cu^{2+} transfer ($\bullet$) and on transfer efficiency expressed as a percentage of Cu^{2+} entering the apparatus ($\circ$)*

Recovery of Cu^{2+} in the desorption chamber was investigated by monitoring the accumulation of transferred Cu^{2+} using different eluent flow rates. The flow of feedstock into the adsorption chamber was 4 ml min^{-1} which, according to the previous experiment gave an estimated transfer rate of 4.2 μmol min^{-1} at 54% efficiency. **Figure 4** shows the effects of eluent flow rate on the concentration of Cu^{2+} in the eluent and on recovery, expressed as a percentage of the amount of feedstock entering the apparatus. As expected, increased eluent flow rates resulted in decreased concentrations of Cu^{2+} in the eluent at equilibrium. Loss of Cu^{2+} from the desorption chamber by carryover consequently decreased with flow rate.

Carryover was the only source of loss from the desorption chamber, the outflow being the recovered product. The effect of eluent flow rate on recovery was therefore relatively slight compared with the effect of feedstock flow rate. The rate of Cu^{2+} recovery in the eluent at 10 ml min^{-1} was 4.0 μmol min^{-1}, compared to the transfer rate of 4.2 μmol min^{-1} estimated in **Figure 3**. The total loss during the desorption process was therefore less than 5%. The main issue concerning the optimisation of eluent flow rate is the compromise between product concentration and loss by carryover. In this particular experiment, it was observed that concentrations of between 0.4 mM and 1.9 mM could be obtained with a 8.5% variation in net recovery.

This model system has made it possible to analyse the performance of the continuous separations apparatus and to determine how certain variables influence its operation. The data generated by experiments such as those shown can be used to make rational decisions on how a process may be optimised. They also give insights into the operation of the apparatus which can be used to improve its design and construction. For example, efficiency will be improved by the inclusion of devices which reduce carryover.

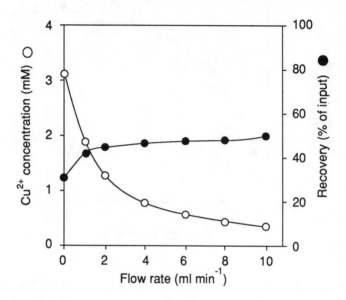

Figure 4 *The effect of eluent flow rate on product concentration (○) and recovery (●)*

It is apparent that the optimisation of the adsorption process is of fundamental importance as this had the greatest influence on system throughput and recovery. Data such as that presented in **Figure 3** makes it possible for the operator to decide how the compromise between throughput and recovery is to be balanced. For example, recovery may be of prime importance for the purification of an expensive therapeutic protein, while process rate may be most important if isolating a low value commodity from a waste stream. Both these parameters may be improved by the development of adsorbent belts with higher binding capacities. In the above experiments, the highest estimated belt binding capacity was 75 μmol m^{-2}. No attempts were made to improve this by investigating alternative belt materials or derivatization procedures. This is therefore an important area where further research is in progress. A major source of loss during the adsorption process was outflow from the adsorption chamber. Depending on the process stream, it may be possible to recover much of this material and return it to the feedstock flow after concentration.

The main product parameter which was determined by the eluent flow rate was product concentration. Carryover caused relatively little loss, other than at very low flow rates. Theoretically, maximum desorption and minimum loss are achieved at high flow rates but it is likely that maximum product concentration would be required for ease of subsequent processing and to reduce storage and transport costs. At low flow rates, carryover can cause dilution of the eluent, resulting in a reduction in desorption efficiency. It is therefore important to ensure that the eluent concentration is sufficiently high to compensate for dilution by carryover when low flow rates are used.

This novel approach to adsorptive separations offers the possibility of continuous high-resolution processes. This may complement continuous-flow production systems such as immobilized enzyme or cell bioreactors. It may also operate in-line with those low-resolution continuous separations methods which are currently in widespread use, such as centrifugation and filtration. With most product parameters dependent on liquid flow rates, automated monitoring and control can be used to maintain efficiency and ensure product quality.

The information obtained using this model system has enabled important conclusions to be made about apparatus function and process optimisation. This is currently being integrated into studies of protein separations. A 3 year project to further develop and assess this separation method is in progress in association with industry.

4 ACKNOWLEDGEMENTS

The multi-chambered tank was kindly provided by Mr Peter Wolstenholme of Biometra Ltd, Maidstone, UK. This work is supported by the European Union Agriculture and Agro-Industry Research Programme, Project Number PL920911.

This technique and apparatus is the subject of International Patent Applications.

1. G.W. Niven and P.G. Scurlock, *J. Biotechnol.*, 1993, **31**, 179.

Thiophilic Chromatographic Separations for Large Scale Purification of Whey Proteins

R. J. Noel, G. Street, and W. T. O'Hare

CHEMICAL AND BIOCHEMICAL ENGINEERING RESEARCH CENTRE,
UNIVERSITY OF TEESSIDE, MIDDLESBROUGH, CLEVELAND TS1 3BA, UK

1. INTRODUCTION

A solid-phase material suitable for large-scale chromatography should possess a high binding capacity for the target protein and have a high specificity of binding, in order to produce high purity products in a single step operation. The matrix and the attached ligand should be stable to cleaning solutions such as 0.5M sodium hydroxide. The ligand should be non-toxic and show a low rate of leakage from the matrix. Finally the ligand should be inexpensive to produce and immobilise onto the solid-phase matrix.

Thiophilic chromatography has been used to purify immunoglobulins from bovine and human colostral whey (1,2) and may have the potential to be suitable for large scale operation. The first thiophilic chromatography matrix (T-gel) was produced by reacting divinylsulphone-activated agarose with 2-mercaptoethanol. It has the chemical structure, P--OCH$_2$CH$_2$SO$_2$CH$_2$CH$_2$SCH$_2$CH$_2$OH where P represents the solid-phase. This small non-toxic ligand was shown to purify immunoglobulins from blood serum (3,4,5), tissue culture supernatants (6) and ascites fluid (7,8) in the presence of ammonium and potassium sulphate salts. High purity (>90%) separations of immunoglobulins can be achieved in a single step. Modifications to the chemical structure of the ligand have been made (7) in an attempt to improve the purity of immunoglobulins produced from a tissue culture supernatant.

The aims of this study are to determine the optimal conditions for the purification of immunoglobulins from cheese whey using T-gel and to evaluate the separations performed by modified T-gel ligands. The suitability of thiophilic ligands for process scale purifications from cheese whey are discussed in view of the results obtained.

2. MATERIALS AND METHODS

Pasteurised and skimmed cheese whey was obtained from Express Dairies Ltd.(Cumbria,UK), the pH was adjusted to 6.0 with sodium hydroxide before storage at -20°C. All chromatographic operations were performed at room temperature.

2.1 Preparation of Thiophilic Gels. Thiophilic gels were produced according to Knudsen,K.L. et al. (7). Divinylsulphone, 2-mercaptoethanol, 2-hydroxypyridine, 4-hydroxybenzoic acid and phenol were all obtained from Sigma Chemical Co. Ltd. 20g of suction-dried Sepharose 4B-CL (Pharmacia LKB Biotechnology) was added to 10ml 0.1M sodium carbonate buffer pH 11.0. The pH was adjusted to 11.0 with sodium hydroxide before the addition of 2ml divinylsulphone. The mixture was left overnight at room temperature and then washed with sodium carbonate buffer pH 11. The degree of divinylsulphone activation was determined by incubating 2 ml of gel, thoroughly washed with distilled water, with 2 ml 3M sodium thiosulphate for 24 hours followed by titration of the released sodium hydroxide with 0.05M HCl (10). The degree of activation routinely obtained was 100-300μmol/ml. 10g of divinylsulphone activated gel was added to 4ml of sodium carbonate buffer pH 11 and 1g of ligand dissolved in 2 ml of carbonate buffer. This reaction was allowed to proceed for 12 hours at room temperature. The number of vinyl groups remaining was determined as before to check all active groups had been substituted.

2.2 Column Separations. Analytical scale columns were made by packing 3 ml of the thiophilic matrix into 5 ml Gilson pipette tips plugged with glass wool. Each column was washed with 3 bed volumes (bv) of the appropriate concentration of $(NH_4)_2SO_4$ prior to sample loading. The columns were loaded with whey sample and washed with a further 3 bv of $(NH_4)_2SO_4$ solution. Proteins bound the column were eluted with 3 ml distilled water. 25ml columns (2.5cm x 5cm) were operated at linear flow rates of 2 m/h for breakthrough experiments.

2.3 Column Eluate Analysis. High performance gel permeation chromatography and SDS-PAGE were used to analyse the protein and peptide content of samples eluting from the T-gel. High performance gel permeation chromatography was performed on a BioSep-SEC-S2000 column 300 x 7.8mm (Phenomenex Ltd). A flow rate of 0.5 ml/min was used with 50mM sodium phosphate buffer pH 6.8. 20μl samples were loaded onto the column. Sodium dodecyl sulphate polyacrylamide gel electrophoresis (SDS-PAGE) with 12% acrylamide gels pH8.9 was performed using a Pharmacia GE-2/4 system. All gel samples were reduced with 2-mercaptoethanol prior to electrophoresis. Total protein analysis was performed by the total organic nitrogen or Kjeldahl procedure. A Kjeldahl factor of x6.38 was applied.

3. RESULTS

3.1 pH and $(NH_4)_2SO_4$ Concentration. The optimum concentration of $(NH_4)_2SO_4$ added to cheese whey in order to purify immunoglobulins on T-gel was found to be 0.8M. Immunoglobulins were not present in the eluate of T-gel until the final concentration of $(NH_4)_2SO_4$ in the whey reached 0.6M. The quantity of immunoglobulins purified by the T-gel increased as the $(NH_4)_2SO_4$ concentration of the whey was increased from 0.6-0.8M. Thereafter higher $(NH_4)_2SO_4$ concentrations served only to increase levels of α-lactalbumin (α-LA), thereby reducing the purity of the immunoglobulins.

When the pH of the cheese whey containing 0.8M $(NH_4)_2SO_4$ was varied between 5.5 and 8.0 no effect on the T-gel separation was observed. However, below pH 5.5 increased levels of contaminating α-LA and β-lactoglobulin (β-LG) were observed in the eluate of the T-gel.

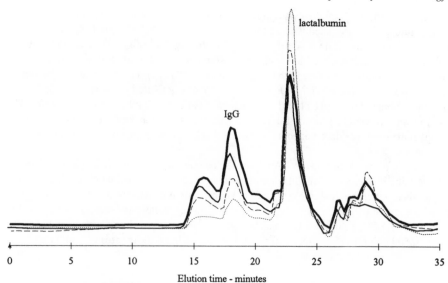

Figure1. High performance gel permeation chromatography of eluates from T-gel with increasing loadings of cheese whey applied to it - 1 bed volume (bv) dotted line (·······), 2bv dashed line (---), 3bv thin line and 4bv thick line. Elution times of pure proteins under the same chromatographic conditions are: human IgG 18 minutes, BSA 19 minutes, β-LG 21-22 minutes, α-LA 23-24 minutes.

3.2 *Column Capacity.* The purity of immunoglobulins eluted from T-gel was improved by increasing the column loading of cheese whey from 1 bv to 4 bv (**Figure 1.**) It can be seen that the level of contaminating α-LA continued to fall whereas the level of immunoglobulins in the eluate of the T-gel continued to rise. 10 bv of cheese whey can be applied to a T-gel column without any breakthrough of immunoglobulins being observed by gel permeation chromatography. The purity of immunoglobulins obtained after applying 10 bv of whey to T-gel is shown in **Figure 2a**. α-LA and β-lactoglobulin (β-LG) constitute less than 5% of the eluted protein. The gel permeation chromatogram shows a peak, absorbing significantly at 280 n.m., eluting between 14-16 minutes. This peak was analysed by SDS-PAGE and found not to contain any protein and is probably due to fat micelles present in cheese whey. The main immunoglobulin peak (18 minutes) was shown by SDS-PAGE (**Figure 2b.**) to contain IgG and a protein which is probably either lactoferrin or secretory IgA (sIgA). A 2.5cm x 5cm column of T-gel was treated to 50 IgG purifications, each purification cycle consisted of loading 10bv of cheese whey containing $0.8M.(NH_4)_2SO_4$. No change in flow rate (2m/h) or quantity of protein eluted per cycle (260 ± 5 mg) was observed.

3.3 *The Nature of the Thiophilic Ligand.* **Figure.3.** illustrates the effect on immunoglobulin purification when 2-mercaptoethanol is replaced by either phenol, 2-hydoxypyridine or 4-hydroxybenzoic acid in the production of the thiophilic ligand. After applying 4 bv of cheese whey containing 0.8M $(NH_4)_2SO_4$ the eluates from each gel were compared for immunoglobulin purity by gel permeation chromatography. 2-

mercaptoethanol-thiophilic gel (T-gel) possessed the highest capacity for immunoglobulins and was the least contaminated with α-LA.

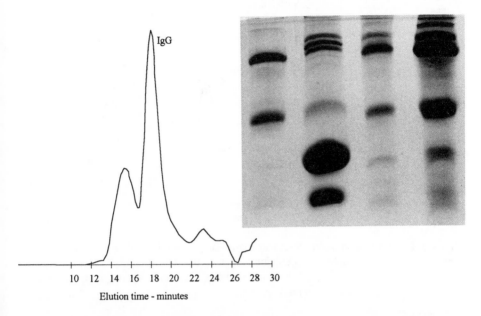

IgG

Elution time - minutes

Figure 2a. Analysis of immunoglobulin purity by high performance gel permeation and SDS-PAGE of an eluate of T-gel after application of 250 ml of cheese whey to a 25ml column at 2 m/h. **Figure 2b.** SDS-PAGE samples (from left to right); 1) pure humanIgG, 2) cheese whey, 3) 5µl T-gel eluate, 4) 25µl T-gel eluate.

The eluates from 4-hydroxypyridine-thiophilic gel contained similar quantities of immunoglobulins but contained higher levels of α-LA. Hydroxybenzoic acid and phenol-thiophilic gels both bound the highest levels of contaminants β-lactoglobulin (β-LG) was eluted from the phenol-thiophilic gel, whereas the α-LA.was eluted from the benzoic-acid thiophilic gel. SDS-PAGE analysis of the eluate from each gel revealed T-gel had the highest level of lactoferrin or sIgA

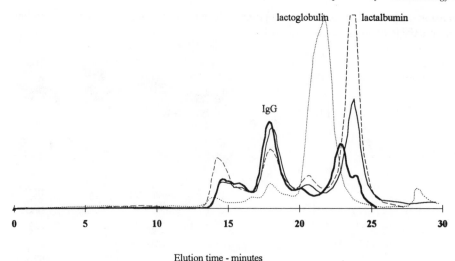

Figure 3. Eluates from different thiophilic matrices, analysed by high performance gel permeation chromatography. Each column contained the same quantity of ligand (40μmol); 2-mercaptoethanol (thick line), 2-hydroxypyridine (thin line), 4-hydroxybenzoic acid (dotted line) and phenol (dashed line).

4. DISCUSSION

The optimal conditions for the purification of immunoglobulins from cheese whey using T-gel (2-mercaptoethanol thiophilic chromatography) at room temperature were determined. The final concentration of $(NH_4)_2SO_4$ in the cheese whey should be 0.8M, the pH should be between 5.5 and 7.5. Increasing the quantity of whey loaded onto the column improves the purity of the eluted immunoglobulins as well as maximising the utilisation of the matrix. The capacity of the T-gel matrix was estimated to be 10mg IgG /ml matrix.

Over the same range of salt concentrations required for T-gel operation, Octyl-Sepharose purified only α-LA (results not shown). The use of H-gel (hydrophobic) and T-gel in series has been suggested for the purification of IgG and HSA from blood serum (11). Similarily, it is proposed that a column set-up consisting of octyl-derivatised gel followed by T-gel would purify α-LA and immunoglobulins respectively from whey.

No significant improvement in the purification of immunoglobulins was observed by changing the chemical structure of the thiophilic ligand. 2-hydroxypyridine-thiophilic gel showed a similar level of immunoglobulins in the eluate but had the disadvantage of a higher quantity of α-LA. This is in contrast to immunoglobulin purifications from tissue culture supernatants where the 2-hydroxypyridine derivatised gel gave an improved separation through decreased BSA contamination(5).

In conclusion, this simple single step chromatographic procedure run at high flow rate (2m/h) produced immunoglobulins of >90% purity and appeared to have good dynamic capacity. The T-gel ligand is simple and cost effective to produce, it is stable to repeated cycles of purifications. If any ligand leakage occurs from the solid-phase the small size of the ligand makes the separation away from the immunoglobulin products simple.

The requirement for large quantities of $(NH_4)_2SO_4$ (108g per litre) represents a significant expense but presents few practical difficulties. The product yield of immunoglobulins from whey has not yet been determined.

The 2-mercaptoethanol-thiophilic ligand can now be attached to various solid phase matrices in order to determine the most suitable for scale-up experiments.

REFERENCES

1.   T.W.Hutchens, J.S.Magnuson and T-T.Yip, <u>Pediatr Res</u>,1089,<u>26</u>, 623-628
2.   T.W.Hutchens, J.S.Magnuson and T-T.Yip, <u>J.Immunol Methods</u>,1990,<u>128</u>, 89-100
3.   J.Porath, F.Maisano and M.Belew, <u>Febs Lett</u>, 1985, <u>185</u>,306-310
4.   A.Lihme and P.M.H.Heegaard, <u>Anal. Bichem,</u> 1991,<u>192</u>, 64-69
5.   T.W.Hutchens and J. Porath, <u>Anal Biochem</u>, 1986, <u>159</u>, 217-226
6.   M.Belew, N.Juntti, A.Larsson and J.Porath, <u>J.Immunol Methods</u>,1987,<u>102</u>, 173-182
7.   K.L.Knudsen, M.B.Hansen, L.R.Henriksen, B.K.Anderson and A.Lihme, <u>Anal Biochem</u>, 1992, <u>201</u>, 107-177
8.   E.Juronen, J.Parik and P.Toomik, <u>J.Immunol Methods</u>,1991,<u>136</u>, 103-110
9.   B.Nopper,F.Kohen and M.Wilchek, <u>Anal Biochem</u>,1989,<u>180</u>, 66-71
10.   J.Porath,T.Laas,and J-C Janson, <u>J.Chromatography</u>, 1975, <u>103</u>, 49-62
11.   J.Porath and M.Belew, <u>Trends in Biotech.</u>,1987,<u>5</u>, 225-229

Development and Modelling of Continuous Affinity Separations Using Perfluorocarbon Emulsions

R. O. Owen, G. E. McCreath, and H. A. Chase

DEPARTMENT OF CHEMICAL ENGINEERING, UNIVERSITY OF CAMBRIDGE, PEMBROKE STREET, CAMBRIDGE CB2 3RA, UK

1 ABSTRACT

The aim of our work has been to develop techniques suitable for industry which can be used to extract proteins from particulate containing solutions (such as fermentation broths) on a continuous basis. More specifically, we have been concentrating on the design and development of novel contacting devices in which separations exploiting affinity chromatography can be carried out. Our processes consist of four stages required for the adsorption, washing, elution and regeneration of the affinity adsorbent being deployed. Two types of contactor are being investigated. The first consists of a series of mixer-settler type units in which adsorbent and process solution are contacted under well mixed conditions and are then separated. The second design enables the two phases to be contacted counter-currently, in a novel, falling bed type device, which is based on fluidisation of the adsorbent particles. The importance of modelling such systems is recognised, and preliminary ideas for achieving this for the latter design are outlined.

2 INTRODUCTION

Affinity chromatography is gaining popularity as a separation process in the biotechnology industry, particularly for recombinant protein purification. However, a comparatively large body of expertise still maintains that it is expensive, and somewhat cumbersome compared to traditional purification methods. This is largely because little research has been devoted to industrial scale applications, and plants used in industry are often simply larger versions of bench-scale equipment, which are operated in cyclical batch mode.

Our work has involved developing contactor designs aimed at industrial applications of affinity chromatography. These are capable of extracting proteins from solution on a continuous basis, and can operate in the presence of particulates commonly found in most fermentation broths, or preparations of disrupted cells. This is a considerable improvement on fixed bed extraction columns typically adopted in industry. The reason for this, is that such systems only cope clumsily with solutions containing particulates, and are additionally also time inefficient, since the need for column elution and regeneration only allows protein extraction to be carried out for a comparatively small fraction of the overall operating time. The new systems proposed consist of four stages arranged in series and in a loop, carrying out continuous adsorption, washing, elution and regeneration of the adsorbent material. A flowsheet of the four stage process is as shown in figure 1 below:

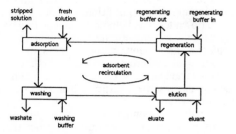

Figure 1. Schematic of four stage process.

3 DESIGN OF CONTACTORS

The first design requirement for each of the four stages is such that they should operate on a continuous basis and should cope with the presence of particulates in the process solutions. This is to avoid expensive centrifugation or filtration procedures prior to the extraction unit, and can be achieved by maintaining a sufficiently large voidage between adsorbent particles, hence allowing flow-through of particulate matter. Additionally, the four stages should enable easy separation of adsorbent material from its carrying phase, to ensure that it can be delivered easily from one stage to the next, and can be continuously recirculated round the entire system.

3.1 The PERCAS Unit

The first system to be investigated consisted of four mixer-settler units arranged in the loop as required from the diagram shown in figure 1. The mixing section of each stage is responsible for the actual extraction, washing, elution and regeneration steps of each stage, whereas the settler acts as a phase separator, removing the adsorbent particles from their carrying liquid. A detailed diagram of the overall flowsheet is shown in figure 2.

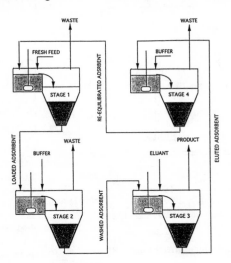

Figure 2. Flowsheet of the PERCAS unit.

The system is operated by supplying fresh feedstock and affinity adsorbent (from the regeneration stage, from which it is recycled) into the first mixing chamber where adsorption of the targetted protein onto the adsorbent takes place. The extent to which protein solution and adsorbent approach equilibrium depends upon the size of the mixing chamber and their respective flowrates. The volume of each mixing chamber is held constant by using a weir. This separates the mixer from the settler, which is where phase separation is then allowed to take place. The aqueous phase in the settler contains non adsorbed components and then flows to waste, whereas the adsorbent phase is pumped to the second mixing chamber. Here it is contacted with a wash buffer, which removes particulates or any other contaminants trapped on the adsorbent particles. The settler collects the flow over the weir from the mixing chamber, and again separates the waste aqueous phase, from adsorbent. Washed adsorbent is then pumped to the third mixing chamber, where it is contacted with eluant. This removes adsorbed proteins from their carrying adsorbent particles. Once again the settler separates the two phases, but in this case the aqueous phase is a product stream rather than being fed to waste. The fourth mixing unit accommodates exiting eluted adsorbent from the third stage, where it is re-equilibrated in adsorption buffer. After settling, the adsorbent is returned to the first stage and the whole process is repeated, with adsorbent continuously being pumped around this closed loop.

3.2 The Counter-Current Contactor

In this system, the concept of having four stages arranged in a loop with adsorbent recycle, is retained, but the individual stages operate on a different basis to that outlined above. A schematic diagram of the contactor proposed is as follows:

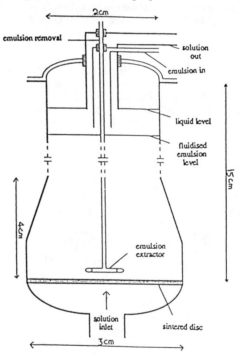

Figure 3. Schematic of counter-current contactor.

In this system, process solution enters each contactor at the base of the column, flows through the adsorbent and then exits the column at the top. Because of the upward flow of the entering process stream, each column behaves essentially like an expanded bed (e.g. Chase et al. 1992), in which the adsorbent particles are fluidised, but with minimal mixing of the solid and liquid phase. Importantly however, so as to ensure continuous, rather than batch operation, loaded adsorbent is constantly removed from the base of the bed (to be processed in the subsequent stage) and adsorbent from the previous stage is simultaneously supplied at the top. In typical expanded beds, each adsorbent particle remains stationary at a point governed by parameters such as liquid flowrate, particle dimensions etc. However, in this system, removal of adsorbent from the base of the bed, causes would-be suspended particles to fall with respect to the column, thus resulting in a downward plug flow of the adsorbing material. The continuous phase (process solution, eluant etc) is fed at the base of the column through the sintered disc and flows (also in the plug regime) upwardly against the falling particles. It is removed at the top of the column, just above the level of the expanded bed.

At steady state, the flow of the adsorbent in each of the four stages will be identical, and is determined by the degree of purification required in the adsorbing stage. The flowrates of eluant, washing medium and re-equilibration buffer to their respective columns are governed according to the flow-rate of adsorbent being treated.

The system is versatile, in that it can operate with a range of flowrates and extraction requirements. Being able to manipulate these, implies that the system must be able to cope with the columns containing beds of varying height and adsorbent inventory. Consequently, the conduits delivering and removing the reacting media to the column are made to be vertically adjustable, so their ends can be positioned correctly for effective operation. Additionally, they are designed to be concentric, so as to prevent re-circulation patterns from ensuing.

The expansion at the base of the column serves to drastically lower the degree of bed expansion at the point where adsorbent is removed. This ensures that the void fraction between adsorbent particles is low, so as to minimise loss of entering process solution exiting with newly loaded adsorbent. Additionally, submerging the adsorbent removal conduit in a volume of low void fraction adsorbent reduces the detrimental effect of the extraction duct on the plug flow of the liquid in the column proper.

4 ADSORBENTS USED FOR PURIFICATION

The adsorbent adopted for these processes consists of an emulsion of liquid perfluorocarbon droplets of average diameter 125 μm, which have been tightly coated with a layer of Cibacron Blue-derivatised poly vinylalcohol (PVA) (McCreath et al. 1993). Dyed PVA adsorbs to perfluorocarbon droplets due to its surface active nature, and the stability of the complex is then enhanced by crosslinking of the adsorbed polymer molecules coating the droplet surface. Because of the presence of the dye ligands, these novel, semi-liquid particles are capable of interacting biospecifically with certain proteins, enabling the emulsion to act as a true affinity adsorbent. A schematic diagram of the droplets is illustrated below:

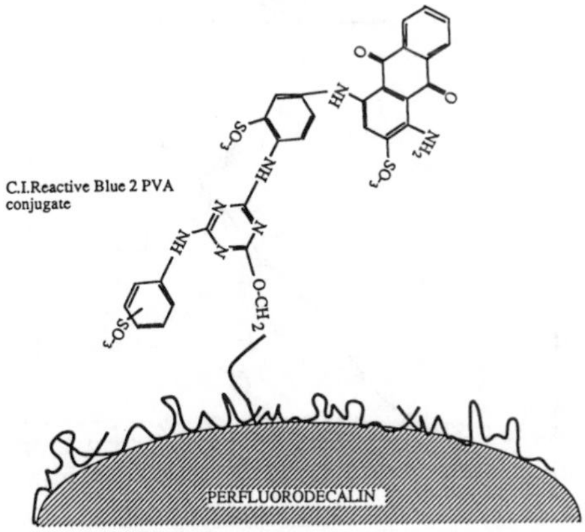

Figure 4 Schematic of emulsion droplet

The emulsion droplets are sufficiently dense for a falling-bed system (the density of the perfluorocarbon used in their manufacture is 1900 kg/m^3), and their semi-liquid nature allows for easy transportation by pumping, a desirable quality for continuous operation. Their chemical stability allows easy regeneration and their equilibrium adsorptive properties compare favourably with other commercial supports (e.g. for HSA (human serum albumin) the capacity $q_m = 30$ mg/ml; and the dissociation constant $K_d = 0.25$).

5 EXPERIMENTAL

5.1 Materials
HSA (fraction V) and BSA (bovine serum albumin) were purchased from Sigma Chemicals. Buffer salts were all purchased from Aldrich. Methods for the manufacture of affinity emulsions are descibed by McCreath et al. (1993). Protein assays were carried out using the Bradford method, and reagents for this were purchased from Pierce. Spectrophotometric analysis was carried out using a Shimadzu UV-160 spectrophotometer.

5.2 PERCAS
The ability of the PERCAS unit to extract protein on a continuous basis was then tested, with the aim to extract HSA from a 0.5 mg/ml single component solution flowing at a rate of 0.5 ml/min. Initially, PERCAS was made to reach hydrodynamic steady state by supplying stage 1 of the system with protein-free buffer solution. At steady state, the flowrates of the loading,washing, elution and regeneration streams were as follows:

> loading: 0.5 ml/min (20 mM phosphate buffer pH 5.0)
> washing: 1.06 ml/min (20 mM phosphate buffer pH 5.0)
> elution: 0.43 ml/min (20 mM phosphate with 0.5 M thyocyanante buffer pH 8.0)
> regeneration: 1.24 ml/min (20 mM phosphate buffer pH 5.0)
> adsorbent flowrate round the four stages: 0.76 ml/min

When steady state was reached, the extraction experiment was begun by substituting the buffer solution feed to stage 1 with protein solution. At this point, contact between HSA and adsorbent began to take place. Non-adsorbed HSA from the stage 1 settler was then fed to waste, whereas loaded adsorbent is fed onto stages 2, 3 and 4 for adsorbent washing, elution and regeneration respectively. Figure 5 shows the concentration profile of the product stream from stage 3. Concentration steady-state is reached after about 180 minutes from when protein solution was initially supplied. Figure 5 also shows the concentration profile of the stream exiting the adsorption stage, giving an indication of protein lost due to it not adsorbing in stage 1.

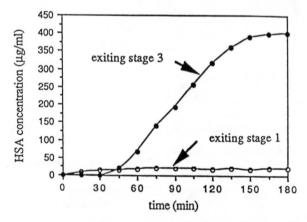

Figure 5. Concentration profiles of exiting streams from stages 1 and 3 respectively.

5.3 Counter-Current Contactor

Experiments were then carried out using an improved emulsion of significantly higher capacity (30 mg HSA/ml instead of 2 mg HSA/ml previously) to assess the performance of the counter-current contactor. As opposed to using an entire four stage unit for continuous purification, the performance of the single adsorption stage was investigated by passing a stock volume of fresh adsorbent through the contactor.

Initially the system was equilibrated by supplying buffer solution (20 mM phosphate pH 5.0) to the column in which 20 ml (settled volume) of adsorbent were being fluidised.The adsorbent was then made to flow downwards through the column by supplying and removing it to the top and bottom respectively at a rate of 2.2 ml/min. Because the adsorbent was supplied and removed at the same rate, the expanded bed height remained unchanged at 16.5 cm. The system reached hydrodynamic steady state, after which BSA solution (0.5 mg/ml, 5 ml/min) was used to replace the buffer solution feed. The exiting solution was collected in fractions and assayed continuously for protein concentration over a period of 140 minutes. The graph in figure 6 shows that the exiting solution was essentially protein free over this entire period.

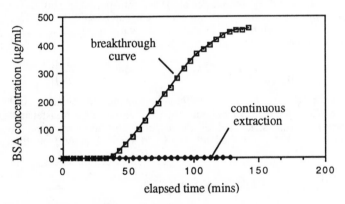

Figure 6. Concentration profiles obtained from column experiments.

For comparison, the counter-current contactor was then used as a simple expanded bed, in which there was no substitution of loaded adsorbent with fresh. Hence 20 ml adsorbent were fed into the column and were fluidised with 20 mM phosphate pH 5.0 buffer at 5.0 ml/min. As before, the bed expanded to a total height of 16.5 cm. After equilibration, the process feed was switched from buffer solution to a 0.5 mg/ml BSA solution (in the same buffer). Fractions were collected every four minutes to assay the exiting protein solution concentration as a function of time. This profile is also illustrated in figure 6.

6 DISCUSSION OF RESULTS

The experiments on the PERCAS unit demonstrate the feasibility of a truly continuous bioseparation procedure. The product stream from the elution stage delivered a 0.4 mg/ml HSA solution at a rate of 0.43 ml/min,thus delivering 0.172 mg/min protein. Comparison of this with the protein fresh feed rate to stage 1 of 0.255 mg/min, suggests an overall recovery of 67%. A further experiment which used blood plasma as a feed solution from which HSA was to be extracted, showed that 70% of all HSA subjected to PERCAS could be recovered in the product stream and was 91% pure (McCreath et al. 1993). These results highlight the success with which proteins can be purified on a continuous basis using PERCAS. Experiments using particulate containing feed solutions have also been carried out successfully, but these are described elsewhere (McCreath et al. 1993).

The counter-current column was also used successfully to extract BSA on a continuous basis, and the results showed that 100% of protein entering the column could be adsorbed onto the adsorbent. This shows improved extractive potential over what can be expected from CSTR type contactors. However, experiments involving an entire unit capable of adsorbing, washing, eluting and regenerating adsorbent counter-currently and continuously are still to be carried out.

At present, the use of fixed bed chromatographic columns is a ubiquitous technique used for protein purification in industry. Expanded bed adsorption (EBA) operates on a very similar principle, and the graphs in figure 6 illustrate the ineffectiveness of such systems convincingly. Firstly, it is clear that considerable protein losses are encountered if the column is operated past the breakthrough point, whereas the continuous extraction unit does not have such drawbacks. Additionally, if one considers the down-time to adsorption which results from the necessity to wash, elute and regenerate the columns, it is clear that a continuous counter-current device presents considerable advantages.

7 MODELLING OF COUNTER-CURRENT UNIT

If both the continuous and dispersed phases are infact transported in plug flow within the column, the conservation equations governing the system can be solved analytically, to yield an expression for the bed height, as a function of the various operating parameters. The form of the equation obtained is the following (Owen, 1993):

$$H = \frac{E}{\varepsilon A k_1}\left\{\frac{1}{S-T}\ln\left(\frac{C_{bout}-S}{C_{bin}-S}\right)+\frac{1}{T-S}\ln\left(\frac{C_{bout}-T}{C_{bin}-T}\right)\right\}$$

where H is the bed height, E is the adsorbent flowrate, ε is the bed voidage, A is the column area, and C_{bin} and C_{bout} are the inlet and outlet concentrations of the transferring species in the relevant process streams. k_1 and k_2 are respectively the forward and reverse rate constants for protein transfer to the adsorbent. S and T are obtained from the following expression:

$$S,T = \frac{1}{2}\left\{\left(C_{bout}-\frac{k_2}{k_1}+\frac{Eq_m}{L}\right)\pm\sqrt{\left(C_{bout}-\frac{k_2}{k_1}+\frac{Eq_m}{L}\right)^2+\frac{4k_2 C_{bout}}{k_1}}\right\}$$

In general, in modelling adsorption with reaction processes, overall rates of reaction are determined by considering intrinsic reaction rates and rates of mass transfer in series. In this model however, we fused these two phenomena, and described the overall rate of adsorption with a simple Langmuirian type expression, as suggested by Chase (1984). Consequently, the forward and reverse rate constants are not solely determined by intrinsic reaction rates, but quantify the effect of an amalgamation of resistances, including those to mass transfer. Additionally, it is anticipated that a Richardson-Zaki type relation can be used to obtain values for the void fraction ε, of the fluidised adsorbent.

REFERENCES

1. CHASE H.
   Journal of Chromatography, 297, p179, 1984

2. CHASE H., DRAEGER N.
   Journal of Chromatography, 597, p129-145, 1992

3. McCREATH G., CHASE H., PURVIS D., LOWE C.
   Journal of Chromatography, 629, p201-213, 1993

4. OWEN R.
   C.P.G.S Report (Internal Communication), University of Cambridge, 1993

Ligand-exchange of Nucleosides and Nucleic Acid Bases on aDAEG-sporopollenin and Kinetics in the Resin

E. Pehlivan, S. Yildiz, and M. Ersoz

DEPARTMENT OF CHEMISTRY, UNIVERSITY OF SELÇUK, 42079 KONYA, TURKEY

1 INTRODUCTION

Ligand-exchange resins have been employed during the past three decades with a great success in practically every branch of chemistry. They provide a powerful method for the separation of similar biochemical materials. The term "ligand exchange" applied to chromatography was first suggested by Helfferich[1,2] who demonstrated this technique by using a nickel-loaded carboxylic acid exchange for replacement of amines by elution with ammonia. Ligand-exchange reactions have also been studied extensively by Walton and his coworkers[3] who compared several resin types and coordinating metals in studies of ligand exchange between amine derivatives and ammonia. Recently, ligand-exchange chromatography has found a variety of applications, including the separation of amines, amino acids, nucleosides, nucleic acid bases, carbohydrates, and peptides[4-6]. The properties of metal-loaded resins of various types have also been studied by other investigators. New developments in ligand-exchange chromatography began with the use of resin with a stronger affinity for metals such as chelex-100, which contain iminodiacetate functional groups[7-9].

The mechanism of ligand-exchange is very similar to that of ordinary exchange, namely, different ligands have different affinities for the coordinating metal attached to the exchanger. Hence, their migration rates down a column are different and thus separation occurs. Ligand-exchange is a highly selective process and even very similar ligands, under proper conditions, may exhibit differences in the degree of formation of their metal-ligand complexes. The chemical behaviour of a ligand-exchange resin depends on the nature of the functional groups that are attached to the resin skeleton. An exchange of ligands takes place between the external solution and the coordination shells of the metal ions in the resin.

The purpose of the study was about the modification of Lycopodium Clavatum as a ligand exchanger and the utility of Cu^{++}, Ni^{++} and Co^{++} loaded aDAEG-sporopollenin as a solid support in the column chromatography for the separation of nucleosides and nucleic acid bases. Sporopollenin obtained from L. Clavatum, occurs naturally as a component of plant spore walls. It exhibits very good stability after prolonged exposure to mineral acids and caustics. It has important advantages over synthetic resins include the following: constancy of chemical structure, high capacity, chemical stability, uniformity of particle size and commercial availability [10,11].

The exchange reactions involving ions in solution and macromolecular resins are interesting. A number of recent reviews on this topic deal with general problems of ion exchange with its physicochemical aspects and ion exchange kinetics[12,13]. In the kinetics part, the ligand-exchange kinetics of nucleosides have been studied. The rate measurements have been carried out by a potentiometric technique. pH of the reaction mixture has been measured during the exchange process.

2   MATERIALS AND METHODS

Chemical and Reagents

Lycopodium Clavatum was purchased from BDH Chem. Corp., under the label Lycopodium. All other chemicals were purchased from Sigma and Fluka.

The synthesis of chelating resins were done in the following sequence:

i) Preparation of Diaminoethyl-sporopollenin (DAE-sporopollenin)

A suspension of sporopollenin from Lycopodium Clavatum in dry toluene containing 1,2 diaminoethane was mixed and refluxed for 9 hours[14]. The reaction is as follows

$$\text{(S)} + NH_2(CH_2)NH_2 \longrightarrow \text{(S)} - NH(CH_2)_2NH_2$$
$$\text{(DAE-sporopollenin)}$$

In the above equation, (S) indicates sporopollenin (L. Clavatum).

ii) Preparation of metal Loaded anti-Diaminoethylglyoximated-sporopollenin (Metal(II) Loaded aDAEG-sporopollenin)

Dried DAE-sporopollenin was placed in a reaction vessel and a slurry of monochloroanti-glyoxime was added to this resin. After the suspension was stirred for 15 hours at room temperature, the resin was washed with water and diluted acetic acid, and then water again[15]. The reaction can be considered as follows

anti-diaminoethylglyoxime-sporopollenin
(aDAEG-sporopollenin)

The product aDAEG-sporopollenin was treated with 1 M $CuCl_2$, 1 M $NiCl_2$ and 1 M $CoCl_2$ in different beakers to convert it into the metal(II) form. Excess $CuCl_2$, $NiCl_2$

and $CoCl_2$ were removed by repeated washings with water and the resin was allowed in 1 M ammonia overnight. The reaction is

Metal(II) loaded-anti-diaminoethyl glyoxime sporopollenin
(Me^{++} loaded aDAEG-sporopollenin)

iii) Chromatography

All experiments were carried out at room temperature. Columns were packed with aDAEG-sporopollenin resin. In the chromatographic set up, we used a peristaltic pump P1 (Pharmacia Fine Chemicals) and the chromatographic separations were followed by a continuously recording UV spectrophotometer (Shimadzu UV 160 A). Resin was packed in separate columns in distilled water. Each column was then washed with at least one total bed volume of eluant that we use in adsorption-desorption process in order to ensure that no metal leakage of the metal would occur during the actual experimental run later on. Samples were placed on the column with micropipettes. The eluting agent ammonia was pumped. Then the effluent was analysed.

iv) Kinetics

A known quantity of dry aDAEG-sporopollenin was placed in contact with a known volume of deionized water. Ligand-exchanger was stirred until a uniform slurry obtained. Then a known volume of 0.02 M cytidine was added to the suspension. As a ligand-exchange with the resin, the solution composition changes with time and this change was monitored by a pH meter (Orion, model SA-720).

The exchange reaction between water in the resin and cytidine, and the rate equation may be indicated as follows

$$\text{(S')} - H_2O + L \rightleftharpoons \text{(S')} - L' + L'' - H_2O$$

$$V = \frac{d[L'' - H_2O]}{dT} = k_{exp} \left[\text{(S)} - H_2O \right]^{n_1} [L]^{n_2}$$

In the above equation S' indicates Co^{++} loaded aDAEG-sporopollenin and L indicates ligands such as cytidine or guanine which will exchange with bound water of the resin, and L' and L'' are cytidine retained in the exchanger phase and the solution phase, respectively.

3   DISCUSSION AND RESULTS

Ligand-exchange is an useful technique for the separation of nucleic acid components which can form complexes with metal cations. Sporopollenin obtained from Lycopodium Clavatum is a suitable skeleton for ligand-exchange structure. It exhibits an excellent mechanical stability.

The functional groups in these resins were amino and oxime groups. Ethylenediamine complexes cannot be dissociate easily due to their stable structure. Glyoximes give stable strong complexes with transition elements. Transition elements are added to the exchanger as coordinating metals. Divalent metal ions, particularly strong complex formers such as copper-, nickel- or cobalt- are tightly bound to the resin. These counter-ions show a unique preference for sorbing nucleosides, amines, amino acids or other molecules that can act as ligands. The eluant attached to the resin matrix such as ammonia can be replaced by other ligands such as uridine, guanine, etc.

In this study, the ligand exchange chromatography of nucleosides and nucleic acid bases has been examined. Nucleic acid molecules possess a very complicated structure but they are capable of forming complexes with transition metal ions. Uridine and cytidine are separated on Cu(II)- aDAEG-sporopollenin by elution with 0.1 M ammonia solution, emerging from the column in the order in Figure 1. 0.5 M ammonia was used as eluant to desorb nucleic acid bases (Figure 2). Uracil lacking an amino group, is less strongly absorbed, whereas cytosine is strongly bound to the resin and emerged from the column later. Strong bases are more strongly held than weak ones. In every case, the elution order was the same. Uridine and uracil were eluted first, followed by cytidine and cytosine. During the separation process, ligands are distributed between the mobile and solid phase in accordance with thermodynamic stabilities of the sorption complexes formed.

Table 1   Retention volumes of nucleosides

Nucleosides	Peak	
	t (sec)	V (mL)
Uridine (0.02M, 30 μL)	220	13.7
Cytidine (0.02M, 60 μL)	480	30.0

The resins we obtained have remarkable chemical and physical properties. For instance, they are insoluble in concentrated acids, bases and salts. They have excellent thermal stabilities and their exchange capacities are high.

With regard to kinetics part, rate measurements were carried out by measuring the pH of the reaction mixture. The immobilized metal in the resin matrix was cobalt which is known to complex strongly with ligands. Measurement of the pH of the reaction mixture continued for an hour. It is clear, from the experimental data presented that the rate of cytidine binding is determined by the rate of chemical-exchange reaction between ligand and resin. Cobalt(II) cation in the resin matrix gives a chelate with the cytidine during the exchange reaction.

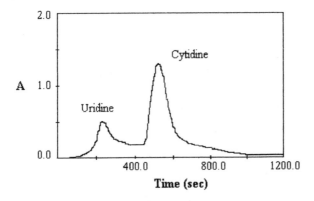

Figure 1   Elution of Nucleosides, 0.1 M NH$_3$, column (1x30 cm), V: 50 mL, λ: 280 nm, flow rate: 3 mL/min.

Table 2   Retention volumes of nucleic acid bases

Nucleic Acid Bases	Peak	
	t (sec)	V (mL)
Uracil (0.02 M, 30 μL)	1410	20.1
Cytosine (0.02 M, 30 μL)	2110	30.1

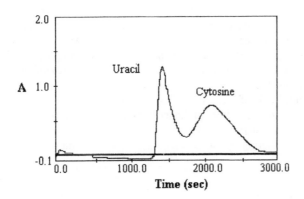

Figure 2   Elution of Nucleic Acid Bases, 0.5 M NH$_3$, column (1x30 cm), V: 40 mL, λ: 254 nm, flow rate: 0.85 mL/min.

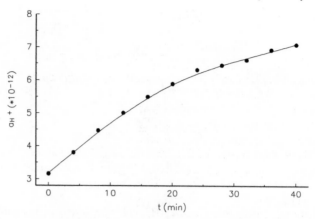

Figure 3 Dependence of a_{H^+} on the time, 0.5 g resin, 0.02 M, 12 mL cytidine

It can be seen from Figure 3 that the first experimental points fit very well into a straight line. It can be accepted that the rate at the beginning of the reaction, v_0 is the slope of the straight line v_0 value was calculated as 0.15 from Figure 3.

4   CONCLUSION

The novel chelating exchanger is formulated through the reaction of diaminosporopollenin with monochloro-anti glyoxime and is used for the access of transition metal ions. This chelating exchanger is compared with synthetic exchangers such as Chelex-100 and cellulose exchangers which have some disadvantages owing to halide salts and acidic solution leach metal ions from the resin. Another problem is their poor mechanical strength, low capacities, unfavourable chemical and physical properties. On the contrary, aDAEG-sporopollenin resin gave more satisfactory resolution even if the column size is small and metal leakage was not observed during elution and have strong chelating ability with various kinds of metal ions. This resin can be suggested as a chromatographic method for the rapid separation of ligands, and for partial resolution of mixtures of nucleosides and nucleic acid bases. Further studies are underway involving glyoximes and carboxyl groups of differing chelation strengths, selectivities and sorption properties.

REFERENCES

1. F. Helfferich, <u>Nature</u> , 1961, <u>189</u>, 1001.
2. F. Helfferich, <u>J. Am. Chem. Soc.</u>, 1962, <u>84</u>, 3242.
3. H. F. Walton and H. Stokes <u>J. Am. Chem. Soc.</u>, 1954, <u>76</u>, 3327.
4. K. Shimamura, L. Dickson and H. F. Walton, <u>Anal Chem. Acta</u>, 1967, <u>37</u>, 102.
5. B. Hemmasi and E. Bayer, <u>J. Chromatogr.</u>, 1975, <u>109</u>, 43.
6. J. Porath, <u>J. Chromatogr.</u>, 1988, <u>443</u>, 3.
7. G. Goldstein, <u>Anal. Biochem.</u>, 1967, <u>20</u>, 477.
8. J. Porath and B. Olin, <u>Biochem.</u>, 1983, <u>22</u>, 1621.
9. M. L. Antonelli, R. Bucci, and V. Carunchio, <u>J. Liquid Chromatogr.</u>, 1980, <u>3</u>, 885.
10. J. Brooks, M. D. Muir, and G. Shaw, <u>Nature</u>, 1968, <u>220</u>, 678.
11. J. Brooks and G. Shaw, <u>Nature</u>, 1968, <u>220</u> 532.
12. L. Liberti, 'In Mass Transfer and Kinetics of Ion-Exchange', Boston, 1983.
13. F. Mata and R. Villamanon, <u>J. Macromol. Sci.-Chem.</u>, 1981, <u>A16(2)</u>, 451.
14. E. Pehlivan and S. Yildiz, <u>Anal. Letters</u>, 1988, <u>21(2)</u>, 297.
15. S. Yildiz, E. Pehlivan, M. Ersöz and M. Pehlivan, <u>J. Chromatogr. Sci.</u>, 1993, <u>31</u>, 150.

Extractive Biotransformations Using Cells Hosted in Reverse Micelles

S. Prichanont, D. C. Stuckey, and D. J. Leak[a]

DEPARTMENT OF CHEMICAL ENGINEERING AND [a]DEPARTMENT OF BIOCHEMISTRY, IMPERIAL COLLEGE OF SCIENCE, TECHNOLOGY AND MEDICINE, LONDON SW7 2BY, UK

1 INTRODUCTION

The synthesis of optically active compounds from pro-chiral intermediates is, at present, one of the major focal points for target-oriented research within the pharmaceutical and agrochemical industries. This results from growing regulatory and public pressure over health concerns since the different enantiomers can exhibit different modes of action and toxicities. Because of their potential cost savings over chemical routes, biocatalysts are starting to be employed commercially to produce a range of optically active compounds, and epoxides are one such example. Epoxides are very useful compounds since their highly reactive nature makes them valuable as building blocks in organic syntheses. More importantly, all epoxides, with the exception of ethene oxide, are either chiral or can generate chiral products which are useful as precursors in the synthesis of optically active natural products or drugs[1]. However, microbial epoxidation processes catalysed by monooxygenases are not yet sufficiently well developed to be scaled up, and many problems remain unsolved. Firstly, most commercially important substrates have a low water-solubility, hence substrate limitation is unavoidable. Secondly, monooxygenases are sensitive to the amount of epoxide formed which leads to product inhibition. Thirdly, certain microorganisms also contain enzymes which can further metabolise epoxides, thus the efficient removal of products is critical in order to obtain a high overall productivity. Finally, cofactor regeneration is needed in order to replenish reducing equivalents used in the monooxygenase reaction. In order to address these problems, more effective microbial systems are being screened, and novel multi-phase reactor designs are being evaluated[2-5].

One novel technique which has the potential to address the problems mentioned above is the encapsulation of whole cells in water in oil(w/o) microemulsions stabilised by nonionic surfactants(In the literature these are referred to as "reverse micelles", despite the fact that we have measured the size of these aggregates using a Particle Size Analyser(Malvern), and found that they vary between 1-10 μm, which is considerably larger than a true reverse micelle). However, there is a total lack of information on microbial epoxidation in reverse micelles, although studies on whole cell immobilization in reverse micelles have been carried out by Luisi's group since 1985[6-9]. They found that microorganisms can be solubilized in selected organic solvents, and yet still maintain their viability. This solubilization results in a solution which is transparent, and thermodynamically stable over time[6]. In reverse micelle systems, cells are located inside water droplets, which are stabilised in the organic solvent by a layer of surfactants. A single reverse micelle can be viewed as a continuous bioreactor with the microorganism acting as a biocatalyst to which the substrate is constantly being supplied by thermodynamic equilibrium to the water pool of the micelle. Simultaneously, the product is continuously being stripped out of the cell environment due to its higher solubility in the bulk solvent. As a result, problems due to product inhibition, product degradation, and substrate limitation should be overcome in these systems.

The aim of this study was to assess the potential of carrying out a wholecell biotransformation in reverse micelles to produce an epoxide. The bioconversion of allyl phenyl ether (APE) to the epoxide, phenyl glycidyl ether (PGE - Fig.1), by *Mycobacterium* sp. strain M156[10], which contains an alkene monooxygenase enzyme, was investigated both in a conventional aqueous system and a novel reverse micelle system. The optimum conditions for the epoxidation in the aqueous system were found to be at 37^{0}C in 25 mM phosphate buffer (pH7.5). For the reverse micelle system of Tween 85/Span 80/n-hexadecane, the effects of substrate concentration, and W_0 ([water]/[surfactant]) were evaluated.

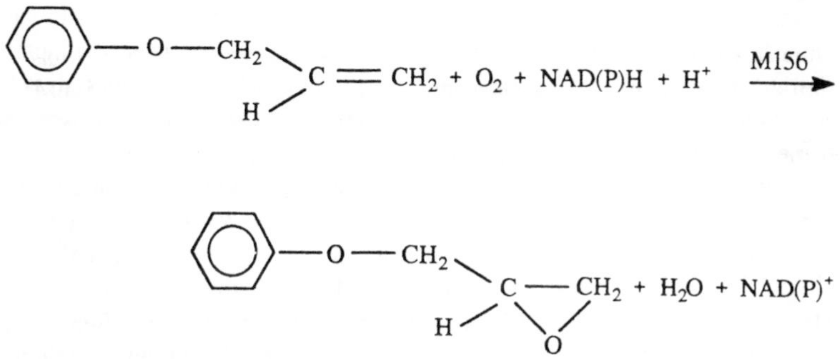

Figure 1 Epoxidation reaction of APE to PGE by *Mycobacterium* sp. strain M156

2 MATERIALS AND METHODS

Chemicals

Allyl phenyl ether (99%), 1,2-epoxyoctane (97%), (+,-)-1,2-epoxy-3-phenoxy propane (99%)(phenyl glycidyl ether), and 2-pentanol (98%) were obtained from Aldrich Chemical Co.Ltd. n-Hexadecane and Tween 85 were from Sigma Chemicals. Span 80 was from Fluka, and other chemicals were from BDH and of AnalaR grade.

Microorganism

Mycobacterium sp. strain M156 (NCIMB 40156) was isolated using 10%(v/v) propene as the sole carbon source in The Department of Biochemistry, Imperial College.

In order to obtain large amounts of bacteria, M156 was grown on propene in a 20L bioreactor (Chemap) containing 15L of a basal ammonia minimal salts medium[11] modified to contain 5 times the concentration of Zn and 2.5 times the concentration of Cu. Harvested cells were washed twice and resuspended in phosphate buffer (pH6.7), and then finally drop frozen in liquid nitrogen and kept as pellets at -70^0C.

Epoxidation Assay in Aqueous Systems

Frozen cells were thawed and suspended in phosphate buffer at pH7.5 (25 mM), washed twice and resuspended to an OD_{540} of 10 (2.5 mg dwt/ml). The cell suspension (4 ml) was placed in a series of identical 25 ml screw cap bottles, incubated in a 37^0C shaking water bath (280 rpm) for two minutes before adding APE to a final concentration of 2.47 mM, and incubated again. At appropriate time intervals, a bottle was selected from the series and then plunged into an icecold bath to stop cell activity. An internal standard (1,2 epoxyoctane) was added and vortex mixed. Components were extracted into diethyl ether and analysed by a gas chromatography (Philips PU4500) using an FID detector. The packed column used was 3% SP2100 on Chromosorb WHP 80-100 mesh. All experiments were done in triplicate and the standard deviation was found to be $\pm 3\%$.

Epoxidation Assay in Reverse Micelle Systems

Reverse micelles containing resting *Mycobacterium* sp. strain M156 were made by adding appropriate amounts of a cell suspension in pH7.5 phosphate buffer (25 mM) to 4 ml of Tween 85/Span 80 (20wt%, HLB 10 - weight ratio of Tween 85 to Span 80 = 5.7) in n-hexadecane (saturated with air). This was vortex mixed for 30 seconds, and then incubated in a 37^0C shaking water bath (280 rpm) for 2 minutes. APE was then added and the sample was incubated again for an appropriate time interval. An internal standard (2-pentanol) was added, and the cells spun down immediately to stop the reaction. The supernatant liquid was analysed by a gas chromatography (Shimadzu 14A) using an FID detector. The packed column used was 3% SP2100 on 80/100 Supelcoport. Concentrations quoted in the reverse micelle system refer to total solution volume in all cases. All experiments were done in triplicate, and the standard deviation was found to be $\pm 8\%$.

3 RESULTS AND DISCUSSION

Microbial Epoxidation in an Aqueous System

The microbial epoxidation of APE in an aqueous system was investigated over time under optimum conditions, as shown in Figure 2. It can be observed that at the initial time point shown in the graph, PGE was already detectable which indicates that this was not the actual starting time of the reaction since a sample could not be taken until approximately one minute after the reaction started. The increase in product, and the decrease in substrate, occurred at a constant rate for the first twenty minutes of the reaction, the initial PGE production rate being 18.3 nmol/mg dwt-min. Subsequently, the product was degraded, primarily by a biocatalysed reaction since the degradation was rapid and chemical degradation was found not to be very significant over this time (data not shown). It is known that under certain growth conditions M156 contains a hydrolase activity which can degrade the epoxide produced. The decrease in the consumption rate of the substrate, and the decrease in concentration of the product in the latter part of the reaction may be due to a combination of substrate exhaustion (K_s = 1.2 mM), product inhibition and product degradation. Within 100 minutes of initiating the reaction, no detectable product (PGE) was found in the reaction medium.

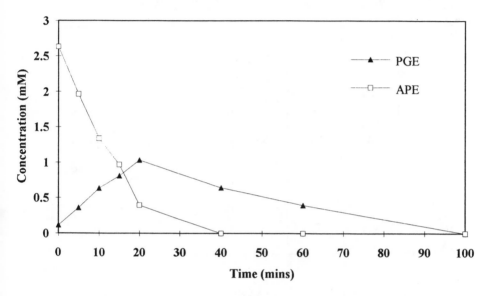

<u>Figure 2</u>   Appearance and disappearance of PGE and APE over time. The reaction was initiated by the addition of 2.47 mM APE at the optimum conditions: 37°C, pH7.5, 25 mM phosphate buffer, and 280 rpm shaking speed.

Parameters Affecting Microbial Epoxidation in Reverse Micelle Systems

Several experiments were performed in order to determine suitable conditions for microbial epoxidation in reverse micelles. These experiments are summarized in Table 1. At W_0 = 50, epoxidation could be detected with APE concentrations ranging from 15 mM upwards. In contrast, at W_0 = 18 epoxidation activity could not be observed under any conditions, either with varying substrate or surfactant concentrations. Moreover, epoxidation did not seem to occur until the W_0 value reached 35, corresponding to a 6.6% (vol/vol) water content. Since W_0 is a surrogate measure for micelle size, this may be due to the fact that the cells need to be surrounded by a pool of water molecules. As one might expect, it was found that the larger the W_0 value the less stable the system became. For the system with W_0 value = 18 a clear and stable solution could be maintained for at least 12 hours, while at W_0 = 50 this condition only lasted for 3 hours.

In conclusion, W_0 was found to be a vital parameter determining epoxidation in reverse micelles. The effect of surfactant concentration is not known, and more experiments are needed in order to assess this parameter.

Table 1 Effects of Different Parameters on Microbial Epoxidation in Reverse
Micelles: (a) at 20% (w/w) HLB 10 surfactant , and (b) at 30 mM APE
Conditions: 37°C, shaking speed 280 rpm.

Epoxida-tion	W_0=18	W_0=30	W_0=35	W_0=40	W_0=45	W_0=50
(a) APE(mM)						
2	NO	-	-	-	-	-
5	-	-	-	-	-	NO
10	NO	-	-	-	-	-
15	-	-	-	-	-	YES
30	NO	-	-	-	-	YES
50	NO	NO	YES	YES	YES	YES
75	-	-	-	-	-	YES
100	-	-	-	-	-	YES
(b) surfactant (wt%)						
10	NO	-	-	-	-	-
15	NO	-	-	-	-	-

Microbial Epoxidation in a Reverse Micelle System

Figure 3 shows PGE accumulation over time measured at $W_0 = 50$ with 5, 50, and 75 mM APE. In the latter two systems, PGE was found to slowly accumulate before reaching maximum levels. In contrast to the aqueous system, the product extracted from the reverse micelle systems was not subsequently further degraded. This appears to be due to the high solubility of PGE in n-hexadecane, causing it to be stripped from the cell environment. At 5 mM APE, no epoxidation was detected. This was probably due to an insufficient concentration of APE in the water pool resulting in a low reaction rate, since the partition coefficient of APE in n-hexadecane ($P = [APE]_{solvent}/[APE]_{water}$) was around 60, and for PGE was 7. The choice of solvent for these systems is difficult since it should ideally be nontoxic to the cell, have a low partition coefficient for the substrate, while a high partition coefficient for the product. Hexadecane met two of these three criteria. In addition, there should not have been any oxygen limitation since the solubility of oxygen is very high in the solvent, and the solvent was air saturated. Epoxidation started when 15 mM APE was supplied. These results are not shown here since the amount of PGE, detected in the first 90 minutes of initiating the reaction, was too low to give any accurate numbers. However, the maximum amount found was measured at around 0.16 mM, which is obviously less than that found when 50 and 75 mM of APE was supplied. From the four different APE concentrations mentioned above, the reverse micelle system supplied with 50 mM substrate seemed to give the best epoxidation results.

This was most likely due to the system reaching an optimum between a moderate substrate concentration in the water pool ($[S] \sim K_S = 1.2$ mM) with low inhibition, and a higher substrate concentration in the water pool ($[S] > K_S$) with stronger product inhibition. The initial epoxidation rate at 50 mM APE was found to be 3.76 nmol/mg dwt-min, which was almost five times less than that found in the conventional aqueous system. This was probably due to solvent inhibition of cell metabolism. As already mentioned, a small amount of PGE was also found at the initial time point, as shown in the graph. The production rate at this point seemed to be astonishingly

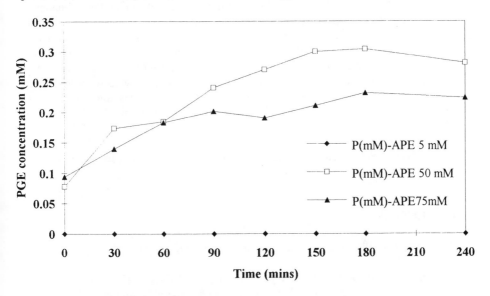

Figure 3 Epoxidation in a solution of (20 wt%,HLB10)Tween85/Span80/ n-hexadecane at various APE concentrations. Conditions: $W_0 =$ 50 (0.47 mg dwt/ml), 37^0C, shaking speed 280 rpm.

high compared to the 3.76 nmol/mg dwt-min rate found, if roughly one minute was allowed for the sample to be taken. This was a puzzling observation, but it occurred consistently with a variety of samples in triplicate, and may be due to very high initial rates of epoxidation. However, more work is needed to determine the real cause of this phenomenon.

From a theoretical standpoint, the accumulation of PGE was expected to increase at a constant rate over the whole reaction period since there should have been no substrate limitation, product inhibition, or product degradation. As can be seen from Figure 3, this did not occur. Instead, after about 180 minutes the PGE reached a steady concentration, and the reaction ceased, in contrast to the behaviour of M156 in the aqueous system. Based on the data available the reasons for this behaviour are not clear. Since the time to reach a constant PGE value was similar, it may be that cell viability was influenced by the reverse micelle system or by the high substrate levels which might eventually partition into, accumulate, and disrupt the cell membrane since APE was more soluble in the cell membrane than in the water pool. Attempts to recover the cells from the reverse micelle phase and test cell activity in an aqueous phase proved unsuccessful due to an inability to separate the cells from the surfactants. Another possible explanation for the data is that the monooxygenase was deactivated over time.

4 CONCLUSIONS

From this preliminary data the optimum conditions for aqueous phase epoxidation were established, and it was noted that after 20 minutes of PGE production a biological mechanism (probably a hydrolase) was responsible for product degradation. When this reaction system was transferred to a reverse micelle in n-hexadecane it was found that both W_0 and substrate concentration influenced epoxide production, and the "optimum" conditions were found to be at $W_0 = 50$, and $[APE] = 50$ mM. The reasons for this "optimum" behaviour were not clear, although a hypothesis was put forward, and more work is needed to clarify the underlying mechanisms. It appears at this point that "reverse micelle" encapsulation of cells may have some potential for the production of chiral epoxides.

ACKNOWLEDGMENTS   The authors would like to thank the support of a Royal Society Research grant and The British Council.

REFERENCES

1. D.J. Leak, P.J. Aikens, and M. Seyed-Mahmoudian, TIBTECH, 1992, 10, 256.
2. R.D. Schwartz and C.J. McCoy, Appl. Envi. Microb., 1977, 34, 47.
3. A.Q.H. Habets-Crutzen, L.E.S. Brink, C.G. van Ginkel, J.A.M. de Bont, and J. Tramper, Appl. Microb. Biotechnol., 1984, 20, 245.
4. O. Miyawaki and L.B. Wingard Jr., Biotechnol. Bioeng., 1986, 28, 343.
5. L.E.S. Brink and J. Tramper, Enz. Microb. Technol., 1986, 8, 334.
6. G. Haring, P.L. Luisi, and F. Meussdoerffer, Biochem. Biophys. Res. Comm., 1985, 127, 911.
7. N. Pfammatter, A.A. Guadalupe, and P.L. Luisi, Biochem. Biophys. Res. Comm., 1989, 161, 1244.
8. A. Hochkoppler and P.L. Luisi, Biotechnol. Bioeng., 1991, 37, 918.
9. N. Pfammatter, A. Hochkoppler, and P.L. Luisi, Biotechnol. Bioeng., 1992, 40, 167.
10. S.R. Rigby, C.S. Matthews, and D.J. Leak, Bioorg. and Med. Chem., 1994 (Accepted for publication).
11. R. Whitterburg, K.C. Phillips, and J.F. Wilkinson, J. Gen. Microbiol., 1970, 61, 205.

Isolation of Amphotericin B by Liquid Ion Exchange Extraction

M. J. Rees, E. A. Cutmore, and M. S. Verrall

SMITHKLINE BEECHAM PHARMACEUTICALS, BROCKHAM PARK, BETCHWORTH, SURREY RH3 7AJ, UK

1 ABSTRACT

Amphotericin B was produced by extracting *Streptomyces nodosus* whole broth at pH 10.5 with an organic solvent (preferably butan-1-ol) containing the liquid ion exchange reagent Aliquat 336. Amphotericin B precipitated from the organic phase in preference to the co-extracted amphotericin A, the precipitation being effected by a decrease in the pH of the organic phase. Careful washing of the crude product with aqueous organic solvent mixtures then neat solvent yielded a product of >90% purity directly from whole broth. It was discovered that this product was a different crystalline form to that normally found. However the product could be converted to the other crystalline form by re-crystallisation from a mixed solvent system.

2 INTRODUCTION

A supply of amphotericin B was required for chemical derivatisation studies. Because of the low solubility of the compound in most solvents (generally >1g/L (1) the use of published isolation procedures was difficult with the available facilities and a liquid ion exchange procedure was developed to overcome these problems.

Liquid ion exchange (LIX) extraction (2) is a process commonly used in the hydro-metallurgic industry which has found some applications within the pharmaceutical industry, for example the extraction of olivanic acids (3). The process is similar to ion pair extraction in which a hydrophobic ion pair is formed in the aqueous phase and subsequently extracted into a water immiscible solvent. However in the case of liquid ion exchange the hydrophobic extractant ion is supplied as an organic solvent solution and exchanges its counter ions with an ion in the aqueous phase. During the process a small amount of the hydrophobic extractant may transfer to the aqueous phase and form an ion pair with the product ion. The formation of the ion pair is dependent upon the target compound possessing a suitable ionisable functional group, eg. a carboxylic acid or an amine, if this group is ionised it can combine with a counter ion to form an ion pair. In the present work the amphoteric compound amphotericin B was converted to its anion at pH 10.5 and extracted with the cationic reagent Aliquat 336 (a commercial form of trioctylmethyl ammonium chloride).

3 EXPERIMENTAL

Amphotericin B is an antifungal compound poorly soluble in aqueous and organic solvents. The molecule possesses a system of 7 conjugated double bonds (see below)

AMPHOTERICIN B

which gives this compound a characteristic UV/VIS spectrum with a strongly absorbing chromophore eg at λmax 405nm. The co-produced compound amphotericin A is the 28,29 dihydro analogue. The compound also has two ionisable groups, a carboxylic acid and an amine. The carboxylic acid group lends itself to extraction with a cationic reagent.

Methods and Materials

- Aliquat 336 was obtained from Henkel Chemicals.
- Methanol and acetone bulk solvents were obtained from Charles Tennant & Co. Ltd.
- Butan-1-ol was obtained from Alcohols Ltd.
- Triethylamine was obtained from Rhone Poulenc Ltd.
- Amphotericin B standard was obtained from Sigma.
- All other chemicals were obtained from Fisons Chemicals Ltd.
- *S. nodosus* fermentation broth was grown on a complex medium and was harvested at 120 - 144hr.

Extraction Method

The broth was harvested and its pH adjusted to 10.5 using 7% aqueous sodium hydroxide solution. The broth was then mixed for 0.5h under continuous agitation with one half volume of ethyl acetate or butan-1-ol containing 7% w/v Aliquat 336. pH was maintained at 10.5 with further additions of sodium hydroxide solution. The phases were then separated by passage through a disk stack centrifuge (Westfalia SA-7). The aqueous phase was discarded and the organic retained and assayed, by U.V. absorbance at 405nm, to determine the amphotericin B content.   If the extracting solvent was butan-1-ol, 1% v/v ethyl acetate was added and the mixture stirred briefly then the extract was left to stand for 3-5 days whilst precipitation of the amphotericin B occurred. The amphotericin B precipitated in the form of spherulites which sank to the bottom of the holding vessel.   When the concentration of amphotericin B had fallen to <20% of its original level the crude product was recovered from the mother liquor.

Product Recovery

The precipitated amphotericin B was resuspended in the mother liquor by stirring. The product was then recovered by passage through a tubular bowl centrifuge (Sharples). The solid phase was recovered, resuspended in aqueous acetone and was once more recovered by passage through the Sharples centrifuge. The solids were then resuspended in acidic aqueous acetone/water, pH 5 with glacial acetic acid, and recovered again by centrifugation. The solids were removed, resuspended in acetone and recovered by filtration through a filter cloth, the cake was washed *in situ* with methanol then dried *in vacuo* at 30 °C.

4 RESULTS AND DISCUSSION

Initially ethyl acetate was used as carrier solvent but with higher broth titres it was found that the percentage of amphotericin B extracted using this solvent was decreasing as the titre increased. The explanation appeared to be that the extractant solution was becoming saturated with amphotericin (see Fig 1); the level at which this saturation occurred was approx. 4.2g/L of amphotericin B (amphotericin B has a solubility of only 0.3g/L in pure ethyl acetate). The concentration of LIX reagent was not limiting as a large molar excess was present. The carrier solvents were therefore re-appraised (see table 1). Butan-1-ol became the solvent of choice, with this solvent extractions of > 80% available amphotericin B were possible from a broth of titre 5.0 g/L equivalent to > 8g/L amphotericin B in the extractant.

The amphotericin B could be precipitated by adding glacial acetic acid; however this produced a very impure product. If the acid was added slowly the crystallisation was more controlled, which generated a purer product. One simple method of very slow pH adjustment was to add 1% v/v ethyl acetate to the butan-1-ol extract, the ester was hydrolysed by the dissolved alkaline aqueous phase in the extract resulting in the slow release of acid which neutralised the residual alkali and provoked the crystallisation of amphotericin B. As the pH of the extract approached neutrality the hydrolysis of the ethyl acetate stopped. The level of ethanol in the extract after 5 days was measured at ~ 0.1%. When ethyl acetate was use as the extracting solvent this process occurred spontaneously.

Amphotericin B precipitates from most solvent systems as a microfine suspension which is difficult to recover on a process scale. Also because of the large surface area and the amphipathic nature of the compound, impurities are readily adsorbed. Using the method described in this paper amphotericin B precipitated in the form of discrete spherulites (10-40 μm in diameter, see Figs 2a & b) these proved easy to recover by centrifugation or filtration, the spherulites appeared to be the result of slow crystallisation of the amphotericin B from the extract; rapid crystallisation by direct addition of acid produced an amorphous mass.

Figure 1

The Extraction of Amphotericin B with Ethyl Acetate + 7% Aliquat 336.

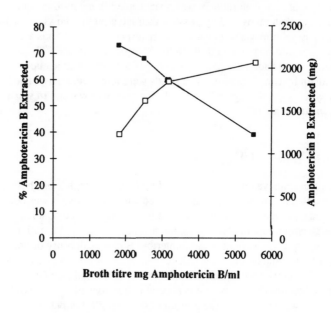

──■── % Amphotericin B Extracted.

──□── Amount of Amphotericn B Extracted.

Legend to Fig. 1.

1L batches of ***Streptomyces nodosus*** broth of varying titres of amphotericin B were extracted at pH 10.5 with 500ml of ethyl acetate containing 7% w/v Aliquat 336. The amount and % of total available amphotericin B extracted was calculated. With increasing broth titre the % of total available amphotericin B extracted decreased. The amount of amphotericin B extracted increased but reached a plateau at which level the extractant became saturated.

Table 1

Extracting Efficiencies of Solvents as % of Total Available Amphotericin B Extracted

Solvent	Absorbance 405nm	% Vol. of Solvent Recovered	% Total Amphotericin B Extracted
1)Hexane	0.142	80	6
2)Dichloromethane	0.343	85	15
3)Ethyl acetate	1.084	61	35
4)Toluene	1.327	64	44
5)M.I.B.K.	1.582	74	62
6)Butan-1-ol	1.602	85	72

1L batches of *Streptomyces nodosus* broth, 5.5 g/L amphotericin B, were extracted with 500ml of solvent containing 7% w/v Aliquat 336 at pH 10.5. The absorbance at 405nm, proportional to the concentration of amphotericin B, was measured.

Analysis of the product by NMR and HPLC showed the product to be >90%, amphotericin B. However comparison between this material and that produced by a previous method and that obtained from Sigma showed certain differences in IR spectrum and dissolution characteristics. Also in the subsequent chemical modification process the liquid ion exchange material possessed different characteristics to those seen with material from other sources, the liquid ion exchange material proved to be more difficult to solubilise in the reaction solvent. The liquid ion exchange polymorph, designated Form II could be converted to the other polymorph, Form I using an adaptation of a published recrystallisation method (4).

Infra-red Spectroscopy

Nujol mull infra-red spectroscopy (I.R.) permits the characterisation of and discrimination between the two different, crystalline, polymorphic forms of amphotericin B, termed Form I and II, because there are some prominent differences as well as several similar features between their I.R.spectra.

Thus the Form I polymorph exhibits a sharp, strong band at 1687 cm^{-1} and a strong absorption at 1550 cm^{-1}, due to the C=O stretch of the ester group and the C=C stretching mode of the heptaene chain respectively. Also this form shows a group of three medium intensity bands at 1211, 1189 and 1164 cm^{-1} arising from the C-O-C asymmetric stretch of the ester group. In addition there is a C-H deformation of the pyranose moiety, on the lower frequency side of the 853 cm^{-1} band, found as a spur at 838 cm^{-1} in the mull I.R of Form I.

Conversely the Form II polymorph possesses a mull I.R. spectrum in which the ester C=O stretching mode is found as a sharp, strong absorption centred at the appreciably higher frequency of 1709 cm^{-1} compared with the I.R of Form I. Furthermore the heptaene group gives a C=C stretch at the significantly higher frequency of 1576 cm^{-1} in the Form II than in the Form I.

The Form II I.R. also shows only two bands at 1196 and 1176 cm^{-1} whilst the third peak located at 1211 cm^{-1} in the Form II.R. spectrum is absent in that of Form II. In addition the medium intensity, sharp spur at 838 cm^{-1} in Form I is absent in the I.R. of Form II in which only absorption at 852 cm^{-1} occurs. There are also several other minor disparities between other absorptions discernible from comparing the I.R.spectra of the two forms. Thus these considered in conjunction with the described primary differences between the Nujol mull I.R.'s of Forms I and II afford an unequivocal fingerprint of and a means of differentiating between these two crystalline polymorphs.

Figure 2 (a)

Figure 2 (b)

5 SUMMARY

The sparing solubility of amphotericin B in organic solvents can be greatly enhanced using liquid ion exchange reagents. Very slow pH adjustment, achieved by the *in situ* hydrolysis of ethyl acetate resulted in the precipitation of a distinct crystalline form different from that obtained by other methods. The selectivity of the extraction combined with the controlled crystallisation conditions enabled product >90% pure to be obtained directly from fermentation broths.

We wish to thank Mr. J. Warrack and the Microscopy Section for their kind assistance.

References

1. Asher *et al* Analytical Profiles of Drugs. Academic Press. 1977. Vol. 6.
2. Science and Practice of Liquid-Liquid Extraction. Oxford Science Publications 1992. Vol. 2. p 222.
3. Hood, J. D., *et al* J. Antibiot. 1979, **32**, 295.
4. United States Patent. 4,902,789. Michel *et al* 1990.

Optimization of the Reverse Micellar Extraction of Horseradish Peroxidase by Response Surface Methodology

C. Regalado, J. A. Asenjo, S. Gilmour,[a] L. A. Trinca,[a] and D. L. Pyle

BIOTECHNOLOGY AND BIOCHEMICAL ENGINEERING GROUP, DEPARTMENT OF FOOD SCIENCE AND TECHNOLOGY, THE UNIVERSITY OF READING, PO BOX 226, READING RG6 2AP, UK

[a]DEPARTMENT OF APPLIED STATISTICS, THE UNIVERSITY OF READING, READING RG6 2FN, UK

1 INTRODUCTION

Recently there has been a tremendous increase in the volumetric yields of fermenters and the production of biochemicals in large quantities by cell culture. Continuous processes which can be scaled up for the separation of these diluted biological products are still of major concern in biotechnology.[1,2] Liquid-liquid extraction constitutes an attractive alternative for bioseparations since it is a well known technique which is readily scaleable and can be operated in a continuous basis. Reverse micelles can be applied as the organic solvent system for the solubilization and recovery of hydrophilic biomolecules, showing good selectivity with little denaturation or loss of biological activity.[3,4] The mechanism of this extraction process involves direct interaction between the reverse micelles generated by the surfactant in the organic phase and the biomolecule in the aqueous phase.[5] Current understanding of reverse micellar extraction is inadequate for *a priori* selection or prediction of the optimum operating conditions to maximise protein recovery. A true optimum can be more efficiently obtained using a factorial design rather than the traditional way of varying one factor at a time, while keeping others constant.[6]

Here we report the use of Response Surface Methodology (RSM) to obtain the combination of factors that maximise the recovery of a cationic isoenzyme of horseradish peroxidase (HRP) (E.C. 1.11.1.7). The significance of factorial interactions between continuous variables is obtained through RSM, and the responses are described in terms of a mathematical model[7]. HRP is a haem-containing glycoprotein having a pI of 8.8, and M_r of 41.6 kDa.[8] Peroxidase has been used as a combined antigen and immunological marker in biological systems to localise cellular constituents, for quantitative analysis, and as part of an assay for rapid detection of listeriae.[9,10] Enzyme recovery was assessed in three ways: total protein, specific activity and total activity. This study focuses on the effects of forward transfer conditions for reverse micellar solubilization of HRP in isooctane using the anionic surfactant AOT. Conditions used to recover the protein to another aqueous phase were kept constant throughout. Previous phase transfer experiments with HRP isoperoxidases[11] showed that pH, ionic strength, and surfactant concentration were the major factors affecting the extraction, and were therefore chosen as the variables for this study.

2 MATERIALS AND METHODS

2.1 Reagents

HRP type IX, AOT, guaiacol, potassium chloride and hydrogen peroxide, all at the highest available grades of purity, were purchased from Sigma (Poole, Dorset, England), and were used as supplied. Spectrophotometric grade isooctane was obtained from Aldrich (Gillingham, Dorset, England). De-ionized water was used to prepare all the aqueous solutions. All other chemicals were obtained from commercial suppliers at the highest available purity.

Isoelectric focusing and titration curve analyses carried out on HRP type IX show that it was a highly purified isoenzyme with an isoelectric point of around 8.7. However, from the silver stained gels, two additional faint bands of protein were observed at pI's 6.7 and 6.4 (data not shown)

2.1 Methods

Protein concentration was assessed using the BCA method, and checked against the absorbance of the haem group at 403 nm. HRP activity was assayed with guaiacol as the chromogenic substrate[12] by following the absorbance of the guaiacol oxidation products at 436 nm. Titration curve and isoelectric focusing analyses were carried out in a Phast System (Pharmacia), using the silver staining technique.

Aliquot samples were taken, as required by the experimental design, from a stock of concentrated HRP solution kept under refrigeration. Equal volumes of aqueous and organic phases were contacted at 25 °C, during forward and backward transfer until equilibrium was reached (5 and 15 min respectively). 40 mM buffers were used to control the required aqueous phase pH. In a typical experiment 1.5 ml of this buffered solution, adjusted to the appropriate ionic strength, containing 6.25 μM HRP, were mixed (using a magnetic stirrer), with 1.5 ml of AOT at the appropriate concentration dissolved in isooctane. The phases were separated after 1 min centrifugation, and assayed for protein content. Back transfer was accomplished by contacting equal phase volumes, using a fresh 0.1 M phosphate buffer (pH 8.0) also containing 1 M KCl.

2.3 Experimental Design

The variables used for the optimisation of the forward transfer conditions were pH, ionic strength, and surfactant concentration. It was observed that at pH $\geq$ 4.0 no protein partitioning occurred, while below pH 2.3 the enzyme activity was lost. It was also noticed that protein recovery after back transfer was seriously decreased when forward transfer ionic strength was $\geq$ 0.25 M. A surfactant concentration above 110 mM did not produce any increase in protein recovery, while, below 50 mM AOT, no reverse micellar phase was obtained.

Multifactor systems are often modelled by using a full second order polynomial design, which provides the estimation of all the main effects and two factor interactions independently of each other. This is a good approximation to the true model, especially in the region around the optimum, provided that all the three factor and higher

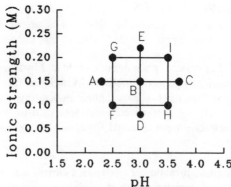

Figure 1. Combination of factors used for central composite design experiment.

interactions are negligible. Preliminary trial runs indicated a slight effect from the detergent concentrations, and because of the small amount of the protein samples, it was decided to use a modified model in which the quadratic term corresponding to surfactant concentration was eliminated. Thus the model employed here is:

$$R = A_0 + B_0 + A_1 x_1 + A_2 x_2 + A_3 x_3 + A_{11} x_1^2 + A_{22} x_2^2 + A_{12} x_1 x_2 + A_{13} x_1 x_3 + A_{23} x_2 x_3 \quad (1)$$

where R is the estimated response; x_1, x_2 and x_3 are the levels of the factors; A_0 is the intercept term; B_0 is the block effect. A_{12}, A_{13} and A_{23}, the interaction terms, measure how much the slope, with respect to one factor, changes as the other factor increases or decreases[13]. A central composite design experiment for two factors (ionic strength, and pH, each at 5 levels) and a third factor (surfactant at two levels) was employed to determine the ten parameters in equation (1). The experiment was divided into six blocks (here, separate days), where two blocks comprised eight runs, and the remainder had six. Blocks were assigned to days at random, while the order of each run within a block was also randomised. Figure 1 shows the treatment combinations used for each of the surfactant levels under study (50 mM and 110 mM). The axial points (A, C, D, E) were repeated twice at each surfactant level to increase the precision of the fitted model, and a total of 40 runs was carried out, including 16 centre points (B). The analysis of data was carried out using an IBM-compatible PC with SAS software.

3 RESULTS AND DISCUSSION

3.1 Protein Recovery

Figure 2 shows the response surface of the predicted protein recovery as a function of pH and ionic strength, for 50 mM AOT. The response surface is most sensitive to changes in pH, while the effect of varying ionic strength produces only weak changes. Factors significantly greater than zero at the 10 % level of confidence were considered to significantly contribute towards the fitted model for protein recovery:

$$R_1 = 68.25 + 0.38 B_0 - 18.90 pH_c - 0.94 IS_c - 0.62 S_c - 1.61 pH_c \cdot S_c$$
$$+ 3.20 IS_c \cdot S_c - 4.54 IS_c \cdot pH_c - 15.89 pH_c^2 - 2.87 IS_c^2 \quad (2)$$

where R_1 is the estimated protein recovery (%); $pH_c = (pH-3.0)/0.5$; $IS_c = $ (ionic strength -0.15)/0.05; and $S_c = (S-80)/30$, where S is the surfactant concentration. Maximum response was only slightly affected by surfactant concentration (77 % for 110 mM, and 74 % for 50 mM AOT).

The analysis of variance shows a high significance in the linear (excluding the surfactant), quadratic, and interaction terms shown in equation (1). The model gives a good fit to the experimental region, since the coefficient of determination is high (0.988) and the lack of fit is not significant (p = 0.19). Canonical analysis was carried out to remove all cross-product terms and to find a stationary point of the response surface (where the slope of the response surface is zero in all directions.[13]) The high negative pH eigenvalue indicates a strongly parabolic response surface, while the smaller negative eigenvalue obtained for ionic strength, indicates a slight curvature of the response surface, both opening downwards. The small, positive, eigenvalue obtained for the surfactant means that any maximum response for this factor will be found away from the stationary point. The combination of signs shown by the eigenvalues corresponds to a saddle stationary point, where more than one maximum can exist.

The contours of the response surface are plotted in figure 3. The maximum predicted protein recovery is located at the centre of the smallest ellipse, which corresponds to pH = 2.6 and ionic strength = 0.135 (standard error of the predicted value = 0.536). It must be noted that only two values of surfactant concentration were used in these experimental studies and thus no strong conclusions can be drawn by plotting surfactant concentration versus ionic strength or pH. However, due to the presence of a minimax (saddle) surface, it is theoretically possible to find an increased response at smaller surfactant levels than studied here. However, when 40 mM AOT was used it was not possible to produce a reverse micellar phase at ionic strength levels between 0.08 and 0.15 M. To test the goodness of fit of the model a normal probability plot of the residuals (observed minus predicted values) is shown in Figure 4, where about 95 % of the data lie within a ranked value of ± 2, and the actual deviations from a straight line are within experimental error. It can be concluded that the statistical model gives a good fit to the response surface. The protein recovery depended heavily on pH, while ionic strength and

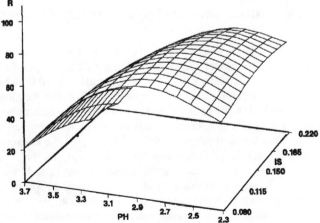

Figure 2. Response surface for HRP protein recovery (50 mM AOT).

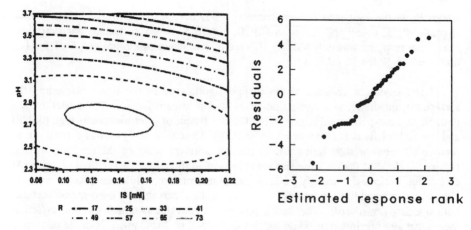

Figure 3. Contour plot of estimated
HRP protein recovery (50 mM AOT).

Figure 4. Normal probability plot for
HRP protein recovery (50mM AOT).

the block effect showed a slight influence on the predicted values. However, all the interactions are important for recovery optimisation.

3.2 Enzyme Activity Recovery

Equation (1), with R now equal to enzyme activity, gave a poor fit to the data. The variance of the residuals appeared to have more spread associated with the higher predicted values (heteroscedasticity of the error). In order to validate the model the response variable was transformed[7] in the form R^α, where $-1 \leq \alpha \leq 1$. The value of α producing a fitted model with the highest coefficient of determination (0.948), and the highest non-significant lack of fit (p = 0.38), was considered the best transformation, conditions that were met for $\alpha = -0.25$. From an analysis of variance (data not shown), the model shown in equation (3) was derived:

$$R_2 = (0.3068 + 0.0014 B_0 - 0.0015 S_c + 0.0011 pH_c - 0.0009 IS_c$$
$$- 0.0011 pH_c \cdot S_c - 0.0011 IS_c \cdot pH_c + 0.0002 IS_c \cdot S_c + 0.0008 IS_c^2 + 0.0067 pH_c^2)^{-4} \quad (3)$$

where R_2 is the enzyme activity recovery (%). The block effect was only significant during the first two days. This model was used to generate a contour plot (figure 5) for 50 mM AOT, which shows a maximum (R_{max} = 109 %) at pH 3.0 and 0.182 M in ionic strength (standard error = 0.576). Comparison with figure 3 shows that the predicted optimal conditions differ by 0.4 pH units and 0.044 M in ionic strength respectively. At 110 mM AOT a small but significant increase in maximum response is estimated. Because of the minimax surface obtained it might be thought to be possible to produce higher activity recovery by lowering the surfactant concentration region, but this is constrained by the restrictions imposed by reverse micellar solubilization (that is, its phase diagram). Regarding the increase in activity, Regalado et al.[11] found increased specific activity of a mixture of HRP isoenzymes after reverse micellar solubilization, because of selective isoenzyme extraction. From isoelectric focusing studies with silver staining two additional faint bands of protein appeared in the original sample at pI's 6.7 and 6.4, of which only

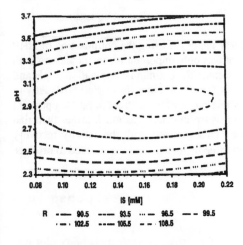

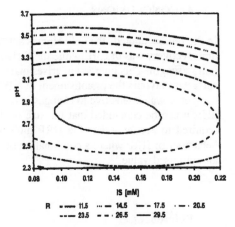

R ⋅— 90.5 --- 93.5 ⋅⋅⋅— 96.5 — - 99.5
⋅—⋅ 102.5 ---- 105.5 — ⋅⋅ 108.5

R ⋅—— 11.5 ⋅⋅⋅ — 14.5 —- 17.5 ⋅ — ⋅ 20.5
---- 23.5 - - ⋅ 26.5 —— 29.5

Figure 5. Contour plot of estimated HRP activity recovery (50 mM AOT).

Figure 6. Contour plot of estimated HRP units recovery (50 mM AOT).

the first appeared after reverse micellar extraction. This may account for the enhanced specific activity.

3.3 Total Units Recovered

In this case neither untransformed nor transformed RSM models gave a good fit to the observed set of data. The best fit gave a good coefficient of determination (0.981), but the lack of fit was just slightly significant (p=0.075). The analysis of variance showed that the surfactant and ionic strength concentrations do not significantly affect the estimated response, leaving the pH as the main linear factor responsible for overall recovery of units. The fitted second-order model showed two qualitative differences from that for protein recovery (equation 2), namely, the insignificant contribution by ionic strength and the pH-surfactant interaction to the estimated response. The fitted model is shown in equation (4):

$$R_3 = 29.26 - 0.52B_0 - 7.06pH_c + 0.30S_c - 0.04IS_c + 1.42IS_c \cdot S_c \qquad (4)$$
$$- 1.37IS_c \cdot pH_c - 0.39S_c \cdot pH_c - 7.85pH_c^2 - 1.49IS_c^2$$

where R_3 is the estimated value of the overall activity recovered. The block effect was significant during the first two days, as a result of decreased initial activity. Statistical analysis indicated that there are clear optimal values for pH and ionic strength, while the effect of surfactant concentration is to produce a ridge which slightly rises towards high levels, due to its interaction with IS. Figure 6 shows the contour plot at 50 mM AOT, where the optimum R_{max} = 30.4 estimated units recovered (corresponding to 78.8 % of original, standard error = 0.286) was found at pH 2.75, and 0.135 M ionic strength. As expected, the optimum lies between the optima for protein and activity recoveries, but closer to the former than the latter. It is interesting to note that the opposite effect was obtained for 110 mM AOT, mainly because of the higher ionic strengths required to obtain the maximum recoveries.

4 CONCLUSIONS

In the region studied, the only significant linear factor affecting the optimisation of HRP units recovery was the pH, while the surfactant was only important for its interaction with ionic strength. This interaction arises mainly because the optimum combination of pH and surfactant favours the establishment of a Winsor II system. Since the overall HRP activity recovered was insensitive to changes in the whole range of ionic strength (at the optimum pH), it can be concluded that the size of the reverse micelles was much larger than that required to accommodate the HRP molecule. This contrasts with the activity recovery behaviour of HRP where a combination of all three factors studied was needed to produce an optimised value.

References

1. K.E. Göklen and T.A. Hatton. Biotechnol. Prog. 1985. 1. 69.
2. R.S. Rahaman, J.Y. Chee, J.M.S. Cabral and T.A. Hatton. Biotechnol. Prog. 1988. 4. 218.
3. P.D.I. Fletcher, A.M. Howe and B.H. Robinson. J. Chem. Soc. Faraday Trans. 1. 1987. 83. 185.
4. R.Wolf and P.L. Luisi. Biochem. Biophys. Res. Commun. 1979. 89. 209.
5. P.D. Haaland, "Experimental Design in Biotechnology" Marcel Dekker, New York, 1989, Chapter 1, p. 9.
6. V.M. Paradkar and J.S. Dordick. Biotechnol. Bioeng. 1994. 43. 529.
7. G.E.P. Box and N.R. Draper, "Empirical Model Building and Response Surfaces", John Wiley and Sons, New York, 1987, Chapter 8, p. 268.
8. K.G. Paul and T. Stigbraund. Acta Chem. Scand. 1970. 24. 3607.
9. S. Avrameas and B. Guilbert. Biochimie. 1972. 54. 837.
10. E.T. Roysser and E.H. Marth, "Listeria, Listeriosis and Food Safety", Marcel Dekker, New York, 1991, Chapter 17, p. 223.
11. C. Regalado, J.A. Asenjo and D.L. Pyle. 1993. Proceedings of the VI European Congress on Biotechnology. Vol. I. p. MO307. Florence.
12. H.U. Bergmeyer, M. Graßl and H. Walter, "Methods of Enzymatic Analysis", 3rd edition, H.U. Bergmeyer, J. Bergmeyer and M. Graßl (ed.), Verlag Chemie, Weinheim, 1983, Vol. 2, Chapter 2, p. 267.
13. A. Khuri and J.A. Cornell, "Response Surfaces. Designs and Analyses" Marcel Dekker, New York, 1987, Chapter 5, p. 174.

Phase Recycling in Aqueous Two-phase Partition Processes: Impact Upon Practical Implementation of Protein Recovery from Brewery Waste

Marco Rito-Palomares, Jon G. Huddleston, and Andrew Lyddiatt

BIOCHEMICAL RECOVERY GROUP, BBSRC CENTRE FOR BIOCHEMICAL ENGINEERING, SCHOOL OF CHEMICAL ENGINEERING, UNIVERSITY OF BIRMINGHAM, BIRMINGHAM BI5 2TT, UK

ABSTRACT.

The paper presents experimental studies of phase recycling in two-stage processes. Comparison is made of the effect of salt addition(NaCl) upon phase recycling with that of the manipulation of tie-line length, pH and volume ratio without salt addition, for the recovery of bulk protein from brewers' yeast. The outline economics of a prototype process is presented.

1. INTRODUCTION.

Aqueous two-phase partition in systems comprising binary mixtures of hydrophilic polymers, or polymers and salts has exhibited many advantages in the handling of particles in feedstock suspensions, and for the recovery of macromolecules. Advantages of high water content, biocompatibility, low cost, and space-time yield support the extended use of this technique. However, poor understanding of the mechanistic basis of the partition, the limits of system selectivity, and ignorance concerning the practicality of phase recycling have compounded a negative view of the potential for economic operation, by the bioprocessing industries (1-4).

In a typical aqueous PEG-phosphate process created under appropriate conditions, the first stage or loading extraction yields a bottom phase containing cell debris and contaminants, and a top phase containing the intracellular protein. In the second stage or stripping extraction, manipulation of salt addition, system pH and tie-line length (TLL), can concentrate the intracellular protein in the bottom phase. Further processing of that phase by ultrafiltration yields a protein concentrate and salt solution as permeate (1,2).

In the last decade, the importance of aqueous two-phase systems (ATPS) in downstream processing operations for the recovery of intracellular products has widely increased (1-8). However, lack of a full consideration of phase recycling in the developed processes is common. In particular the practice of assuming the ease of phase recycling in modelling studies of ATPS should be interpreted with caution. The effects of phase recycling upon the recovery of intracellular products are still unknown for the majority of processes, and must be studied carefully and individually for each product or process. In particular, non-product solutes (salts, pigments, macromolecules) recycled with phase feedstocks may have significant and possibly cumulative impacts upon subsequent cycles of operation.

The present study reports experimental studies of phase recycling in two-stage processes. The effect of salt addition upon phase recycling, and the establishment of strategies for the design of a two-stage process for the recovery of bulk protein from brewers' yeast, are considered. In addition, the outline economics are compared for a prototype process operated at litre scale with, and without, phase recycle.

2. MATERIALS AND METHODS.

Waste lager yeast was donated by Bass Cape Hill Brewery Birmingham, UK and stored at -18°C before use. Polyethylene glycol (PEG) (nominal molecular weight of 1000 daltons) was obtained from Sigma, Poole UK. All other chemicals were of analytical grade.

Yeast disruption and characterisation of PEG-phosphate system.

Yeast cake was slurried (30% wet w/v) in 20mM potassium phosphate buffer pH 7.6 (buffer A) and disrupted at 15 l / h in a Dynomill (KDL 0.61) operated at 3200 rpm with 0.2 - 0.5 mm glass beads. Wet-milled Brewers' yeast was diluted two-fold in buffer A before use in ATPS experiments, due to the high concentration of cell debris. The bulk protein concentration was 2.5 ± 0.1 mg/g in 20g experimental systems (2). Binodal curves were determined at 25°C by turbidimetric analysis (8) at pH 8.9 for loading conditions, and at pH 6.5 for stripping conditions. Tie-lines were fixed by using a point of defined system composition and the known volume ratio (9), whilst tie line-lengths were estimated as described by Huddleston *et al* (8)

Loading extraction.

Systems (20g) were assembled on a fixed mass basis for convenience on a top-loading balance. Complete phase separation was achieved by low speed centrifugation at 2200 rpm for 20 min, 20°C. Acid and alkaline adjustments to working pH values were achieved without impact upon phase behavior by addition of small volumes of molar orthophosphoric acid or sodium hydroxide. Samples were carefully extracted from phases for determination of protein concentration (10) and subsequent estimation of protein partition coefficients (ln Kp) and practical bulk protein yield (%). The influence of neutral salt concentration (NaCl) upon loading and stripping extraction **(Figure 1 (a)** and **(b))** was investigated in 20g experiments, comprising 27%w/w PEG / 14%w/w phosphate, pH 8.9 and 14%w/w PEG / 12%w/w phosphate, pH 6.5 respectively.

Stripping extraction and recycling experiments.

The top phase from loading stages was taken forward to the stripping extraction. Fresh phosphate was added and the pH adjusted to 6.5. Concentrations of PEG and phosphate in top phases from loading and stripping extractions were estimated by the interception of tie-line with the binodal curve (9). Recycling experiments were carried out wherein the top phase at stripping was used again in a new loading extraction. The deficits of PEG, phosphate and wet-milled Brewers' yeast were adjusted with fresh materials, and the pH was re-established at pH 8.9 as before.

3. RESULTS AND DISCUSSION.

Phase recycling with addition of NaCl.

Addition of neutral salt (eg. sodium chloride) in two-stage aqueous two-phase systems is an important parameter which manipulates the partition of bulk protein between the phases. Usually low concentrations of salt are used in the stripping extraction, in order to switch the protein partition from top to bottom phase (1). However salt addition inevitably involves the inclusion of another chemical impurity and added cost to the separation process. A two-stage process for the recovery of bulk protein from brewers' waste was developed in our group (2). This process involved no neutral salt addition in the loading extraction and only a low concentration (0.3 moles / kg) of sodium chloride in the stripping extraction to yield an overall protein recovery of 34% without phase recycling.

Phase recycling in a two-stage process involving neutral salt addition at any stage must be studied with care. When the phase recycling experiments were attempted in the prototype system (2) the sodium chloride contained in the top phase from the stripping extraction was recycled into a further loading extraction. Such situations compromise the performance of subsequent cycles since the presence of low concentrations (less than 0.6 moles / kg) of sodium chloride showed significant negative effect upon the protein partition coefficient (ln Kp) and protein yield in the loading extraction (**Figure 1 (a)**).

In the stripping extraction, low concentrations of sodium chloride (0.2-0.4 moles/kg) improved the protein recovery (**Figure 1 (b)**). This implies that the concentration of sodium chloride of the top phase from the loading extraction must be diluted carefully through each cycle before the stripping extraction has been assembled. In addition, salt concentration above 0.4 moles/kg showed negative effects upon protein recovery (**Figure 1 (b)**) confirming the sensitivity of such systems to added solute. Phase recycling in processes with salt addition involves problems of manipulation, measurement and control of solute concentrations in the stages of extraction. The accumulation of salt (in addition to pigment and contaminants inherent to the process) after multiple recycling induced pronounced protein precipitation with associated product loss.

Phase recycling without NaCl.

In response to the above results the design of a two-stage process for the recovery of bulk protein from brewers' waste was attempted without salt addition. For the loading extraction, preliminary experiments with systems in the "wedge-shaped" zone in the top left of the phase diagram (2) were attempted. However, the cell debris occupied all of the bottom phase and part of the top phase, thereby limiting the quantitative recovery of soluble protein in the PEG-rich phase.

After that, the volume ratio (Vr) and the tie-line length (TLL) parameters were manipulated to maximise the recovery of bulk protein in the top phase. Systems at different tie-lines lengths were selected (see **Figure 2 (a)** and **Table 1**) and those with longer TLL and higher Vr showed better ln Kp and recovery (Yt) values (see **Table 1**). However, the system E (longest tie-line system) showed a decrease in ln Kp and Yt, when compared with system D, due probably to increased protein precipitation. In the systems studied, the maximum yield obtained was 75% i.e. system D.

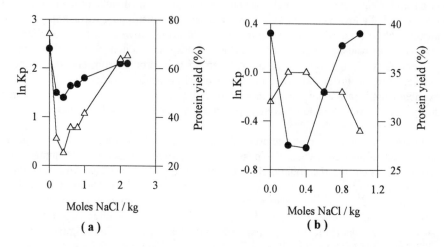

(a)   (b)

Figure 1.   Effect of NaCl concentration upon protein partition coefficient and practical protein
                   yield for yeast protein; (a) loading extraction, (b) stripping extraction.
Wet-milled brewers' yeast was directly partitioned in 27%w/w PEG / 14%w/w phosphate (pH 8.9) and, then in 14%w/w PEG / 12%w/w phosphate (pH 6.5) for loading and stripping extraction respectively. Partition coefficients (ln Kp; •) and protein yields, (Δ) in loading (a) and stripping (b) extractions represent that material recoverable from PEG-rich top phase and from phosphate-rich bottom phase respectively. The latter is expressed as a percentage of soluble protein loaded.

For stripping extraction, tie-line length and system pH were the important parameters manipulated to switch protein partition from the PEG-rich top phase to the phosphate-rich bottom phase. Systems along a simple tie-line close to the binodal (see **Figure 2 (b)** and **Table 1**) were selected, and the system with smallest Vr showed highest protein recovery (Yt) and partition coefficient (ln Kp) (**Table 1**). The maximum protein yield achieved from the phosphate-rich bottom phase, was 57%. However this value is practically increased by the subsequent recycling of the protein content in the PEG-rich top phase. The overall bulk protein yield was maximised for a two-stage process (without salt addition) at approximately 42% of the total protein loaded.

Preliminary experiments as described in Materials and Methods were also carried out for bovine serum albumin and clarified wet-milled brewers' yeast (data not shown), which demonstrated the feasibility of phase recycling with no significant effect upon ln Kp and protein yield. Phase recycling using wet-milled brewers' waste containing particles (cell debris) was then attempted. The loading extraction achieved very good cell debris removal and no significant effect upon ln Kp and protein yield through each cycle that was observed (**Figure 3 (a)**). The loading extraction is a robust system due to its deep position on the phase diagram. The values of protein yield and ln Kp were 79±5% and 2.55±0.15 respectively through 5 cycles of operation.

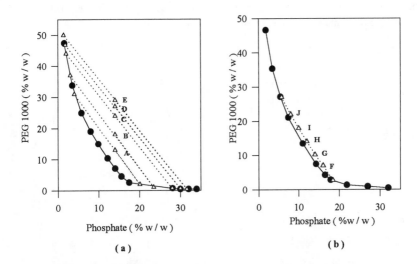

Figure 2. Binodal curves and experimental tie-lines of PEG / phosphate systems; (a) loading extraction, (b) stripping extraction.

Binodal curves (•) were estimated as described in Materials and Methods at pH 8.9 for loading (a) and at pH 6.5 for stripping extractions (b). Experimental phase volumes and PEG 1000 weight fraction for system A to J (Δ) are detailed in Table 1.

Table 1. Protein partition coefficient and protein yield of PEG / phosphate systems.

LOADING EXTRACTION				STRIPPING EXTRACTION				
System	Vr	WF	ln Kp	Yt(%)	System	Vr	WF	ln Kp	Yb(%)
A	0.63	0.13	0.50	56	F	0.20	0.07	0.30	57
B	1.10	0.18	1.00	62	G	0.41	0.10	0.00	55
C	1.40	0.24	2.40	68	H	1.10	0.14	-0.23	50
D	1.60	0.27	2.30	75	I	1.95	0.18	-0.63	47
E	1.70	0.29	2.10	69	J	4.20	0.22	-0.87	40

The volume ratio (Vr) of the top and bottom phases in 20g non-biological experimental systems at points in differents tie-line A to E in Fig. 2 (a) and at points along a single tie-line F to J in Fig. 2 (b) was estimated after phase separation in graduated centrifuge tubes and compared with the weight fraction (WF) of PEG 1000. Wet-milled Brewers' yeast was partitioned for loading extraction systems (A to E) and for stripping extraction systems (F to J) respectively. Protein yield in loading (Yt) and stripping (Yb) extractions, represents that material recoverable from PEG-rich top phase and from phosphate-rich bottom phase respectively, and are expressed as a percentage of total soluble protein loaded to systems.

The stripping extraction was demonstrated to be more sensitive with respect to the ln Kp than the loading extraction upon phase recycling, and the values of protein yield and ln Kp were 50±3% and 0.5±0.2 respectively (**Figure 3 (b)**) through 5 cycles. The sensitivity of the stripping extraction can be explained by the proximity of the system to the binodal curve, wherein small changes in system composition induce great change in the phase composition (9). However, after 5 cycles of operation, no gross negative effect was observed upon the total recovery of bulk protein in this study. The polymer (PEG 1000) and phosphate recycled in the experiments described were approximately 60%w/w and 20%w/w of the fresh material respectively (**Table 2**).

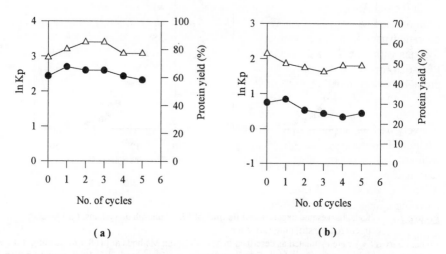

(a)   (b)

Figure 3.   Effect of polymer recycling upon protein partition coefficient (ln Kp) and practical protein yield (%); (a) loading extraction, (b) stripping extraction.
Protein partition coefficient (•) and practical protein yield (Δ) for loading (27%w/w PEG / 14%w/w phosphate, pH 8.9) and stripping extraction (7%w/w PEG / 16%w/w phosphate, pH 6.5) are expressed relative to the number of recycling. Protein yields(%) for loading (a) and stripping (b) extractions, represent that material recoverable from PEG-rich top phase and from phosphate-rich bottom phase respectively, and are expressed as a percentage of soluble protein loaded.

Table 2. Economic analysis of the prototype processes.

Process	Experimental system (g)	Solute re-used(%) PEG1000	Phosphate	Chemical used (g)	Bulk Protein yield (%)	Cost (£)
Without recycle	1000	0	0	420	43	10.4
With recycle	1000	60	20	284	42	7.8

The costs of chemical were obtained from Sigma and took no account of saving from bulk purchase. Chemical used represents the quantity of fresh PEG 1000 and phosphate used.

An outline economic analysis based solely upon chemical costs was undertaken for the prototype process operated at the one litre scale (**Table 2**). The polymer recycling implemented here achieved a gross cost reduction of 25%. Further options for consumable saving based upon the recycling of other phase components are currently under study.

4. CONCLUSION.

Polymer recycling in a two-stage process applied to the recovery of bulk protein from brewers' yeast involving salt addition must be viewed with caution. Salt addition caused problems of manipulation, measurement and control of solute concentrations and as a result the stability of the process was low. In the design of a two-stage process without salt addition, tie-line length (TLL), Vr and pH of systems were selected as the main practical operating parameters to manipulate the protein partition in the desired or target phase. Systems with long TLL, large Vr and high pH showed high recovery of protein from the PEG-rich top phase. On the other hand, systems with low values of TLL, Vr and pH all promote better protein recovery from the phosphate-rich bottom phase.

Phase recycling without salt addition showed no significant negative effect upon the recovery of bulk protein from brewers' yeast through at least 5 operational cycles. In addition, the process with polymer recycling showed an economic advantage over those without. However, the total effect of polymer recycling (particularly accumulation of contaminants) upon the recovery of specific intracellular products (eg. target enzymes), where the recovery is greater, must be examined with care. Further effects of polymer recycling upon specific target enzymes are under current study. The practicality of polymer recyling in two-stage ATPS shown in these preliminary results partially offsets negative economic views of the potential generic implementation of this technique.

ACKNOWLEDGEMENT.

The authors thank Consejo Nacional de Ciencia y Tecnologia (CONACYT-Mexican Government), for financial support for MRP.

REFERENCES.

1. Greve A and Kula M-R, J. Chem. Tech. Biotech., 1991, **50**:27-42.
2. Flanagan J A, Huddleston J G and Lyddiatt A, Biosep., 1991, **2**:43-61.
3. Hustedt H, Biotech. Lett., 1986, **8** (11) 791-796.
4. Hart R A and Bailey J E, Enz. Microb. Technol., 1991, **13**:788-795.
5. Cascone O, Andrews B A and Asenjo J A, Enz. Microb. Tech., 1991 **13**:629-635.
6. Papamichael N, Borner B. and Hustedt H, J. Chem. Tech. Biotech.,1992, **54**:47-55.
7. Mistry SL, Asenjo JA and Zaror CA, Biosep., 1993, **3**:343-358
8. Huddleston JG, Ottomar K, Ngonyani D and Lyddiatt A, Enz. Microb. Tech., 1991, **13**:24-32.
9. Albertsson P-A " Partition of Cells and Macromolecules", N.Y., 3nd ed. 1986.
10.Bradford MM, Anal. Biochem., 1976, **72**:248-254.

Virus Safety of Purified Biological Products for Therapeutic Use

P. L. Roberts

RESEARCH AND DEVELOPMENT DEPARTMENT, BIO-PRODUCTS
LABORATORY, DAGGER LANE, ELSTREE, HERTFORDSHIRE WD6 3BX, UK

1 INTRODUCTION

Bio-products for potential therapeutic use in man may be derived from a variety of human or animal sources. They may also be produced from cell culture, e.g. monoclonal antibodies, or be based on recombinant DNA technology. One aspect of product safety that is of current concern is that infectious agents may be present in such products. In this respect viruses are the major problem, as other types of microorganisms like bacteria and fungi are easy to detect and can be conveniently removed by well-established techniques such as membrane filtration. In addition, unconventional agents or prions that cause spongiform encephelopathy are relevant in specific situations where bovine material or certain human tissues are used, e.g. pituitary glands. In this article, methods that can be used for ensuring the safety of bio-products are considered together with relevant examples.

2 GENERAL VIRUS SAFETY STRATEGIES

There are two main approaches available for ensuring the virus safety of a bio-product. The first involves the testing of the start material, process intermediates and final product for the presence of viruses. The exact testing strategy will depend on the nature of the product and the production system involved. This method is limited by the virus detection range of the tests used. Also, it is of relatively low sensitivity in that, if only a small proportion of the final product units were contaminated, this would go undetected. Nevertheless, this approach is part of the virus safety testing strategy for all bio-products.

The second approach involves testing the ability of the purification system to inactivate and/or physically remove large amounts of added virus. The current opinion is that the total process should give a virus reduction value of >10 log with >5 log occurring in a single step.

Specific requirements will depend on the potential risks associated with the particular product. In order to achieve these values, additional specific virus-inactivating steps may need to be added to the basic process or standard process steps may need to be altered to maximise virus reduction. Where additions or modifications are needed, their effect on product yield, potency, immunogenicity, etc., will need to be tested. Virus reduction steps should be robust and reproducible. Re-contamination after a virus reduction step must be prevented and, for this reason, treatment in the final container is the ideal situation.

2.1 Evaluating Virus Reduction Steps

Regulatory guidelines that outline the approach that should be adopted with bio-products have been published[1]. The relevant step(s) in the manufacturing process should be tested, using a scaled-down system where appropriate. The process is challenged by the addition of high titres of relevant viruses whose removal/inactivation is then followed during the purification process, e.g. for plasma products, human immunodeficiency virus (HIV), hepatitis B and C; for hybridoma cell lines, retroviruses and Epstein-Barr virus. Other "model" viruses are also used in cases where there are technical, safety or other difficulties in using relevant viruses. In addition, the use of "model" viruses allows an expanded range of virus types to be tested. Virus is quantified by either physico-chemical methods, e.g. radiolabelling, immunochemical or nucleic acid detection, or by infectivity assay. Assaying for infectivity is the best approach, where practicable, because the other methods are not dependent on virus infectivity and thus can under-estimate virus reduction.

Virus reduction values, measured in terms of virus-infectivity, are determined by the addition of virus to an intermediate and then carrying out the purification process or step. The total infectious virus before and after treatment is then determined by infectivity assay using, preferably, a cell-culture system to ensure sensitivity and accuracy. For the particular virus, a log reduction factor is then calculated. The purification process can be tested either by the addition of virus to the crude start material and sampling each process intermediate or, where virus reduction is very high, by testing each stage individually. The log virus reduction values for all the purification stages are then used to estimate the reduction value for the total process by addition.

There are other aspects of virus reduction that may need to be considered for the process. For instance, the kinetics of virus inactivation should be followed where relevant. Where a manufacturing range is to be set for a process step, the effect of these parameters, e.g. temperature, pH, column age, etc., on virus reduction should be considered and tested. This is especially

important for a process step identified as a major virus
reduction step.

3 VIRUS REDUCTION METHODS

3.1 Virus Testing

The original source from which the product is derived
can be tested. This can involve for instance the medical
screening of each donor and can include a range of
biochemical tests including those for viruses. Tissue or
cell-cultures should be screened for viruses either by
isolating infectious virus or by detecting virus proteins
or virus genomes. Indirect methods based on the detection
of antibody to virus are used in some situations.

In the case of blood/blood-derived products[2], selection
criteria based on a questionnaire are used to exclude
individuals of high risk. Where paid donors are used,
extensive and regular medical testing may be involved.
Each donation is then tested for HIV-1/2 and hepatitis C by
a sensitive antibody detection method such as ELISA and for
hepatitis B surface antigen. Plasma is later pooled (500-
5000 donations) and fractionated/purified to produce
coagulation factors, albumin and immunoglobulin. The start
material, certain intermediates and the final product are
also tested for viruses.

With hybridoma and genetically engineered cell-
lines[3,4,5,6], the master cell-bank/working cell-bank are
extensively tested for a wide range of viruses of human and
animal origin using virus isolation and other techniques.
Tests for specific viruses and more wide-ranging methods
are included. Extended cell-banks, produced after a
manufacturing run, are also tested. Viruses of particular
concern include those that can be latent/non-cytopathic in
cell culture and those with oncogenic potential including
retroviruses. Relevant viruses may include mouse retroviruses, human retroviruses (e.g. HIV and human T-cell
leukaemia virus) and herpes viruses such as Epstein-Barr
virus and HHV-6.

3.2 Virus Removal and Inactivation

3.2.1 The Standard Purification Process. The
inactivation and/or removal of viruses may occur during
protein purification. Process steps which can be important
include precipitation, chromatography, freeze-drying, low
pH, ethanol, virus neutralisation and product storage. For
example, during the ethanol fractionation procedure used
for the production of immunoglobulins and albumin from
plasma, both virus inactivation and precipitation occur[7].
During the production of immunoglobulins, neutralisation by
antibody to virus takes place. Even such steps as storage
of the final liquid product at 4°C can lead to significant
virus inactivation.

Affinity chromatography using metal chelate or monoclonal antibody-based methods or ion-exchange chromatography have proved useful for virus reduction. Extended column washing and/or the use of several types of buffer may all contribute to this. Inactivation by the eluting agent, e.g. acid or thiocyanate, can also be important. For example, during the purification of human monoclonal antibodies for potential therapeutic use by affinity-chromatography on Protein-G Sepharose FF and acid-elution[8], virus reduction values were 6-7 log for herpes simplex virus and 5 log for Sindbis virus. In the case of the acid-resistant virus polio, virus reduction was only 3 log and was exclusively due to physical virus removal. During the purification of a high purity Factor IX concentrate (9MC/Replinine), a copper chelate affinity chromatography stage is used. In this process, a number of wash buffers, including one at a pH of 4.2, are sequentially applied to remove contaminants. This column produces good levels of virus reduction[9] and, together with a specific virus inactivation step using solvent/detergent, produces a product with a high level of safety (**Table 1**).

3.2.2 *Specific Steps*. Heat-treatment and the use of solvent/detergent are the two most commonly used specific virus inactivation methods. These were originally developed for blood products but can also be applied to other bio-products. Heat-treatment in the liquid state has long been used to treat albumin solutions in the final container. Caprylate is added as a product stabiliser which does not effect the virus. Treatment is for 10 hr at 60°C. Under these conditions, high levels of virus inactivation have been determined (**Table 2**). The inactivation rate is very rapid and in some cases occurs before the sample has reached temperature. Other products are bulk pasteurised during the production process. This method is used with antithrombin III where citrate is included as a stabilising agent but is then later removed

Table 1   *Virus Reduction During the Manufacture of Factor IX (9MC/Replinine)*

		Virus Reduction (log)		
Virus	Envelope	Solvent/ Detergent	Affinity Column	Total
Sindbis	+	≥7.1	6.9	≥14.0
VSV	+	>5.5	–	>5.5
HSV-1	+	>5.6	–	>5.6
Vaccinia	+	2.4	≥5.9	≥8.3
HIV-1	+	≥6.3	–	≥6.3
Polio-1	–	0.0	4.6	4.6

by chromatography. The presence of this stabilising agent had an insignificant effect on virus inactivation with ≥ 5 log inactivation of vaccinia virus occurring within 10 mins at 60°C for antithrombin III, with or without stabiliser, or albumin[10]. In both products, the inactivation of the HIV virus (≥ 5 log in 10 mins) was also demonstrated. Pasteurisation is also used with coagulation factor concentrates, however higher product losses occur.   In addition, the stabilisers that must be used, sugars and/or amino acids, can substantially decrease virus inactivation.

Dry heat treatment of freeze-dried products is carried out on the final product. Various temperature (60-80°C)/ time combinations have been used in the past, however treatment of coagulation factors at 80°C for 72 hr has become accepted[11].   Under these severe conditions, no transmission of hepatitis B, C or HIV has occurred. Virus inactivation studies (**Table 3**) have confirmed its effectiveness against a range of viruses including HIV[12]. However, it is less effective against extremely dry heat resistant viruses such as vaccinia virus and possibly human parvovirus.   Further studies have allowed process ranges, e.g. sugar level, residual water content, to be set for consistent virus inactivation.

Solvent/detergent, i.e. tri-n-butyl phosphate and non-ionic detergent, is probably the most widespread virus inactivation method in current use[13].   These chemicals act on the essential lipid envelope of viruses and render them non-infectious.   Although non-enveloped viruses exist, these are much less important in bio-products.   The chemicals need to be removed after treatment but this can usually be achieved by carrying out the procedure early in the purification process.   A new high purity Factor IX (9MC/Replinine) includes treatment with 1% TNBP and 0.3% Tween-80 at 22°C for 5 hr[14]. The inactivation of ≥ 6-7 log of a range of viruses was demonstrated, with substantial inactivation occurring within 30 mins (**Table 1**).   Vaccinia virus, a pox virus, was partially resistant to treatment. However, this unusual virus type is not known to pose a

Table 2   *Virus Inactivation in Albumin (Zenalb 4.5%) During Pasteurisation*

	Inactivation Time ≥ 5 log	
	Temperature (°C)	
Virus	56°C	60°C
Vaccinia	9	3
Polio-1	10	0
Sindbis	30	–
Vesicular Stomatitis	–	10

Table 3   *Virus Inactivation in Factor VIII (Intermediate Purity Product 8Y) by Dry Heat Treatment*

	Virus Inactivation (log)			
	Time (h) at 80°C			
Virus	4	8	24	72
Sindbis	-	3-5	>6	>6
Vaccinia	-	-	1-3	2-4
HIV-1	>5	>5	>5	>5

problem in bio-products. After treatment, the solvent/detergent is removed by metal chelate affinity chromatography, a step which further contributes to virus safety.

3.2.3 New Techniques. One potential approach to virus inactivation is the use of gamma-irradiation. This method would have the advantage that it can be carried out on the final container. Unfortunately, the limited data available would suggest that a dose necessary to inactivate virus would significantly effect the product[15]. Irradiation with UV is another potential approach, however this requires the product to be in a thin liquid film because of the poor penetrating power of UV[16]. Photodynamic methods involving the use of dye and visible light have proved promising, however the dye must be removed and small non-enveloped viruses are not inactivated[16]. Recently some manufacturers have developed membrane filters, i.e. Viresolve[17] (Millipore), Nylon 66 (Pall), or hollow-fibre depth filters such as Planoray[18] (Asahi) that are capable of removing viruses. The Viresolve membrane has been shown to remove viruses in a predictable manner based on size. These new filters are most effective with viruses of medium to large (>50 or >80 nm) size. The size of the protein product also needs to be taken into consideration when selecting a filter and yields across the process determined.

4   CONCLUSION

A number of techniques are available for the screening, removal or inactivation of viruses in bio-products. Although specific virus inactivation/removal techniques have been developed, virus reduction during the standard purification procedure can be substantial. It may thus be advantageous to consider virus reduction, as well as product purity, activity and yield, during product development. All the methods in use suffer from some limitation in terms of the range of viruses that are fully susceptible. In those situations where the complete elimination of all viruses is required, the combination of complementary virus reduction methods may be the only successful approach.

REFERENCES

1.   Committee for Proprietary Medicinal Products: Ad Hoc
     Working Party on Biotechnology/Pharmacy and Working
     Party on the Safety of Medicines: EEC Regulatory
     Document, *Biologicals*, 1991, **19**, 247.
2.   Committee for Proprietary Medicinal Products: Ad Hoc
     Working Party on Biotechnology/Pharmacy and Working
     Party on the Safety of Medicines: EEC Regulatory
     Document, *Biologicals*, 1991, **20**, 159.
3.   Committee for Proprietary Medicinal Products: Ad Hoc
     Working Party on Biotechnology/Pharmacy and Working
     Party on the Safety of Medicines: EEC Regulatory
     Document, *Biologicals*, 1991, **19**, 133.
4.   US Food and Drug Administration, Points to Consider in
     the Characterization of Cell Lines used to Produce
     Biologicals, 1987.
5.   US Food and Drug Administration, Points to Consider in
     the Manufacture and Testing of Monoclonal Antibody
     Products for Human Use, 1987.
6.   Committee for Proprietary Medicinal Products, *Tibtech*,
     1988, **6**, G5.
7.   J.J. Morgenthaler, Virus Inactivation in Plasma
     Products, ed. J.J. Morgenthaler, Curr. Stud. Hematol.
     Blood Transfus, Karger, Basel, 1989, **56**, 109.
8.   R.M. Baker, A-M. Brady, B.S. Cambridge, L.J. Ejim,
     S.L. Kingsland, D.A. Lloyd, P.L. Roberts, this volume.
9.   P.L. Roberts, C.P. Walker, P.A. Feldman, *Vox Sang*,
     1994, in press.
10. L. Winkelman, P.L. Roberts, *Brit J. of Haem*, 1989, **76**,
     Suppl, 35.
11. L. Winkelman, P.A. Feldman, D.R. Evans, Virus
     Inactivation in Plasma Products, ed.J.J. Morgenthaler,
     Curr. Stud. Hematol. Blood Transfus, Karger, Basel,
     1989, **56**, 55.
12. P.L. Roberts, A. McAuley, C. Dunkerley and L.
     Winkelman, *Thromb. and Haemo.*, 1991, **65**, 1163.
13. B. Horowitz, A.M. Prince, M.S. Horowitz, C.
     Watklevicz, Virological Safety Aspects of Plasma
     Derivatives, ed. F. Brown, Dev. Biol. Stand., Karger,
     Basel, 1993, **81**, 147.
14. P.L. Roberts, J.W. McPhee and P.A. Feldman, *Thromb.
     and Haemo.*, 1993, **69**, 1282.
15. H. Mohr, B. Lambrecht, H. Schmitt, Virological Safety
     Aspects of Plasma Derivatives, ed. F. Brown, Dev.
     Biol. Stand., Karger, Basel, 1993, **81**, 177.
16. B. Cuthbertson, K.G. Reid, P.R. Foster, Blood
     Separation and Plasma Fractionation, ed. J.R. Harris,
     Wiley, New York, 1991, 385.
17. A.J. DiLeo, D.A. Vacante, E.F. Deane, *Biologicals*,
     1993, **21**, 287.
18. S. Manabe, Animal Cell Technology: Basic and Applied
     Aspects, ed. H. Murakami, S. Shirahata, H. Tachibana,
     Kluwer Academic Publishers, Dordrecht, Netherlands,
     1992, 15.

Protein A Leakage from Affinity Adsorbents

S. D. Roe

BIOSEP, B353, AEA TECHNOLOGY, HARWELL, OXON OXI I ORA, UK

1 INTRODUCTION

The use of therapeutic proteins for the treatment of human illness has necessitated the removal of contaminants which may cause a variety of undesirable side-reactions, including antigenicity, pyrexia and transmissable diseases. The contaminant concentration in a product, below which an immune response should not occur, appears to be around 10 ppm[1], varying with therapeutic dose, contaminant and patient. Such contaminants may be derived from the source material or added to the product during upstream and downstream processing. The most important categories of contaminants are aggregated, fragmented and chemically modified product, contaminating protein, DNA, viral particles, pyrogens and ligands derived from chromatographic matrices.

The use of proteinacious affinity ligands such as Protein A and antibodies has required the use of a subsequent chromatographic step such as ion exchange or gel permeation to ensure ligand removal from the product stream. Although this approach has been widely accepted as a remedy to the problem of ligand leakage, it does not address the cause of such matrix deterioration. While several papers have appeared on the subject of ligand leakage and Protein A purification of IgG in recent years[2-5], BIOSEP research has aimed at developing a more fundamental understanding of the causes of such ligand loss. Initial work, reported here, has focused on the leakage of the most widely used multiple-point attached ligand, Protein A, using a commonly used matrix, Sepharose 4 Fast Flow (Pharmacia).

2 MATERIALS AND METHODS

Protein A Sepharose 4 Fast Flow (Pharmacia) was slurry packed into an HR 5/5 FPLC glass column (5 mm diameter) to give a 1 ml bed volume, bed height 5.1 cm, cross sectional area 0.196 cm^2. The bed was then settled by using a flow rate of 4 ml/minute (1224 cm/hr) for 1 minute, and subsequently operated at 0.2 ml/minute (61.2 cm/hr). All chromatography was operated using an automated FPLC system (Pharmacia). Pre-purified mouse polyclonal IgG (Sigma) in the absence of interfering contaminants, such as BSA, has been used as a protein source. This protein was purified without use of protein A chromatography and was prepared as a 0.5 mg/ml solution in Buffer A and filtered through a 0.22 micron Millex GS sterile filter (Millipore). The Protein A column was routinely stored at 5°C when not in use.

The following purification cycle was used:

(a) Bed equilibration for 3 bed volumes in 0.1M citrate, 0.1M sodium dihydrogen phosphate, 3M NaCl, pH 8.9 (Buffer A).
(b) Sample loading in Buffer A.
(c) Bed washing in Buffer A for 6 bed volumes.
(d) Elution in 0.1M citrate, 0.1M sodium dihydrogen phosphate pH 2.5 (Buffer B) for 10 bed volumes.
(e) Re-equilibration in Buffer A for 3 bed volumes.

Protein A was measured using ELISA kits obtained from Biotage Europe, Hertford, U.K. All samples were assayed in triplicate. Initial Protein A binding was carried out overnight at 5°C, allowing improved Protein A binding and increasing the assay sensitivity.

Following packing, the Protein A column was put through a purification cycle without sample loading to measure the concentration of Protein A leaking from the bed in the absence of IgG. Following this blank run the bed was loaded up with murine polyclonal IgG and the breakthrough measured. 0.37 mg/ml IgG was applied in buffer A. The bed was then eluted as described above and the leakage of Protein A measured throughout the entire purification cycle. A variety of bed storage conditions were then evaluated. The bed was put through a weekly routine of storage for, on average, 6 days, followed by operation for 2-3 purification cycles. The nature of the storage conditions was altered to determine the influence on ligand leakage. This routine was repeated for a total of 21 cycles over a period of 63 days. This cycling was then followed by a repeat capacity measurement, washing with guanidine hydrochloride and a final capacity measurement. During each cycle the bed was loaded with 8.0 mg of murine polyclonal IgG, with the IgG peak being collected during elution and assayed for Protein A by ELISA.

3 RESULTS

Leakage in the absence of IgG binding.

Figure 1 shows that following column packing, a high Protein A concentration of 90 ng/ml was measured in the initial two column volumes eluted, reflecting the release of interstitial, unbound or loosely bound ligand. Following this initial release the leakage concentration fell to below 2 ng/ml for the majority of the cycle, with a slight rise to 3 ng/ml during elution at pH 2.5 and regeneration to pH 8.9. An approximate 5% variation in the leakage concentration was found for each sample, assayed in triplicate. The Protein A leakage concentration in the absence of IgG was therefore very low, following release of loosely associated ligand.

Adsorbent capacity and leakage with IgG loading

The breakthrough of IgG showed a large degree of tailing with an outlet concentration still well below the applied IgG concentration. This indicated excessive pore diffusion limitation within the beads and prevented the accurate measurement of bed capacity without a very prolonged breakthrough study. However use of the BIOSEP Simulus programs[6] allowed prediction of the bed capacity as between 21 and 22.3 mg IgG/ml of bed. This capacity is well below the manufacturer's quoted capacity of 35 mg of human IgG/ml but compares favourably with the capacity of >18.5 mg/ml for murine IgG3 measured by Fuglistaller[3]. The dynamic capacity (10% of breakthrough) was measured to be 10.6 mg IgG. Subsequently 75% of the dynamic capacity was used, with 8 mg of murine polyclonal IgG applied to the bed during each cycle.

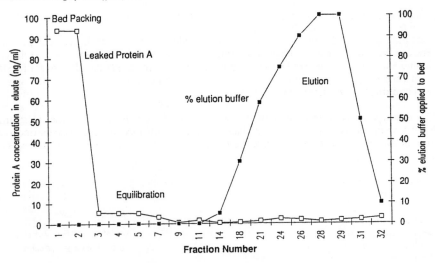

<u>Figure 1</u> Protein A leakage from Sepharose 4 Fast Flow without loaded IgG.

Subsequent elution of bound IgG from this bed and measurement of leaked Protein A (Figure 2) again showed an initial release of loosely bound Protein A during bed equilibration, followed by a low leakage concentration of approximately 5 ng/ml during bed loading and washing. However a significant increase in Protein A concentration to 73 ng/ml or 10.26 ng/mg IgG (10.26 ppm) was found in association with the eluted IgG. This compares favourably with the Protein A leakage measured by Francis et al[2] where a slightly higher Protein A leakage of around 14.4 ppm was measured in the first cycle. The leakage level remained slightly above average at 10 ng/ml after peak elution. This study indicates that although the low pH elution conditions should ensure disruption of the Protein A binding to the IgG Fc region, the Protein A still co-elutes with the IgG, to a concentration which does not correlate with the leakage levels expected without IgG binding.

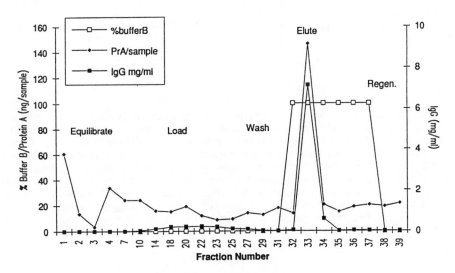

<u>Figure 2</u> Variation in leaked Protein A during a purification cycle on Protein A Sepharose 4 Fast Flow with murine IgG applied to measure the breakthrough.

Ligand leakage with adsorbent re-use.

For the majority of a bed's working life it is under storage between purification runs. Therefore the nature of the storage conditions may have a strong influence on the degree of ligand leakage. Any storage condition must prevent microbial contamination of the bed, preserve the integrity of the protein A and preserve the integrity of the matrix and spacer arm. The influence of storage conditions on leakage concentration was measured over 21 cycles during a period of 63 days. The results of the analysis of protein A leakage in eluted IgG following storage under a variety of conditions are shown in Figure 3. This shows that the storage conditions of the column have a strong influence on the concentration of leaked protein A found in subsequent eluted product.

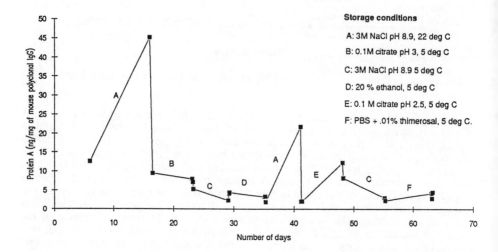

Figure 3 Protein A leakage expressed per mg of IgG eluted in successive purifications of murine polyclonal IgG on Protein A Sepharose 4 Fast Flow: the influence of storage conditions.

Adsorbent capacity after repeated use.

Following cycling of the bed for 23 consecutive purification cycles the capacity of the bed was re-measured. The bed was then washed in 3 column volumes of 6M guanidine hydrochloride, flow rate 0.2 ml/minute (61.2 cm/hr), re-equilibrated in pH 8.9 buffer for 10 column volumes and the bed capacity measured again. Table 1 shows that the bed capacity had increased from 21.7 mg/ml following column packing, to 29.3 mg/ml after 21 cycles. Following bed washing in guanidine hydrochloride, the bed capacity increased still further to 34.5 mg/ml. Both breakthrough curves also showed a marked reduction in the pore diffusion limitation found in the bed initially. It is interesting to note that the final bed capacity of 34.5 mg/ml compares favourably with the manufacturers quoted capacity of 35 mg of human IgG per ml.

Table 1 Increase in capacity of Protein A Sepharose 4 Fast Flow
with repeated use and after washing with guanidine hydrochloride.

Age of Column	Capacity (mg of murine polyclonal IgG per ml of bed)
Fresh column after initial blank cycle without IgG loading	22.3
After 21 cycles	29.3
After 22 cycles and guanidine hydrochloride wash	34.5

4 DISCUSSION

The data presented here and in other publications[2,3] indicate that Protein A "leakage" is highest at 10-15 ppm during the first few purification cycles and falls off steadily with bed re-use reaching a near constant concentration of below 5 ppm after 5 cycles. Much of this initially released material is probably the consequence of loosely associated or un-bound Protein A being present in the adsorbent as obtained from the manufacturer and its release may account for the apparent increase in bed capacity found in this study. On bed equilibration following removal from storage some "loosely associated" Protein A is also found in the column eluent. However, without binding of IgG the leakage rate is very low throughout the purification cycle (2-3 ng/ml) indicating that the binding of IgG is a pre-requisite to ligand displacement. With IgG binding the majority of ligand leakage occurs in a peak precisely associated with the emerging antibody. The concentration of ligand appears to correlate with the concentration of IgG eluting to give a constant Protein A: IgG ratio (or ppm measurement), the absolute value of which will vary according to the bed age and previous storage conditions.

The suggestion from this data is that the initial adsorbent as packed into the column has internal agarose pores which are not fully accessible to IgG, potentially due to a combination of a sub-optimal pore diameter and a super-optimal Protein A loading, causing pore blinding or steric hindrance. Some of this Protein A may be non-covalently or loosely bound and easily removed during the first few cycles of IgG binding and elution. Therefore as leakage of Protein A occurs with repeated use the internal agarose bead pores become more accessible to IgG binding, increasing the working capacity of the bed.

Although Protein A has a molecular weight of 42 Kd, the molecule is asymmetric with an apparent molecular weight closer to 100 Kd. The consequence is a high degree of pore diffusion resistance which has also been found in other Protein A adsorbents such as Eupergit[3]. The pore size for effective affinity chromatography, with minimal pore diffusion resistance, may be around 10 times the gel pore size which may be suitable for other chromatographic techniques (i.e. 100 nm as opposed to 10 nm). This report shows that 10-20 ppm of Protein A is removed during initial Protein A cycles and it is incorrect to assign this high concentration entirely to the release of bound ligand. Some of this released Protein A is most likely to be unbound in the adsorbent pores, non-covalently associated or covalently bound loosely, through one point of attachment: in all cases the ligand may be easily removed during the first few cycles of IgG binding and elution. Therefore as leakage of Protein A occurs with repeated use the internal pores of the agarose bead become more accessible to IgG binding, increasing the working capacity of the bed.

Bed cleaning with guanidine hydrochloride is already routinely practised and strongly recommended to remove non-specifically bound protein, sterilise the bed and remove endotoxins. A subsequent rise in bed capacity occurred despite the use of a clean feedstock of pure IgG in this study.

Incorrect storage conditions significantly increased the level of Protein A leakage into the product and into the first column volume emerging during bed equilibration at pH 8.9. Storage at room temperature (25°C) in 3M NaCl, pH 8.9 or in 0.1M citrate pH 2.5 caused the highest subsequent leakage rates. The storage conditions which gave least subsequent Protein A leakage on bed start up and into the product on elution were 20% ethanol, 3M NaCl pH 8.9 and PBS pH 7.2 + .01% w/v thimerosal, all at 5°C. Following incorrect storage, a high level of Protein A leakage was found in the first subsequent purification cycle. In following cycles the leakage level in the eluted IgG fell dramatically. The use of improper storage conditions did not appear to influence the subsequent pattern of leakage after the first cycle. Furthermore, the influence of improper storage on leakage appears to diminish with bed use, indicating that initial Protein A leakage during the first few cycles may be the result of the release of interstitial or non-covalently bound Protein A.

Given the results outlined above, two main mechanisms of ligand leakage are proposed, with initial release attributed to removal of interstitial and loosely bound Protein A. This may be followed by a slow release associated with binding of antibody, possibly as a result of nucleophilic attack at the high pH of binding.

REFERENCES

1. A.F. Bristow, "Protein Purification Applications" IRL Press, Oxford, Chapter 2, 1990, p. 29.

2. R. Francis, J. Bonnerjea, J., and C.R. Hill, "Separations for Biotechnology" Elsevier Applied Science, London, 1990, p. 491

3. P. Fuglistaller, Journal of Immunological Methods, 1989, 124, p. 171.

4. A.C. Kenney, 143-160, "Monoclonal Antibodies: Production and Application", Alan R. Liss Inc, 1989, p. 143.

5. D.S. Pepper, "Laboratory Methods in Immunology", CRC Press, Volume 2, 1990, Chapter 10, p. 169.

6. D.J. Wiblin, "Simulus: Theoretical and Technical Reference", BIOSEP, 1994

Simultaneous Biochemical Reaction and Separation Using a Novel Rate-zonal Centrifugation Process

S. J. Setford and P. E. Barker

DEPARTMENT OF CHEMICAL ENGINEERING AND APPLIED CHEMISTRY,
ASTON UNIVERSITY, BIRMINGHAM B4 7ET, UK

1 INTRODUCTION

In recent years, the continuing demand for cost reductions and safety considerations in biochemical production processes has led to the study of integrated bioprocess operations. The combination of the biosynthesis and separation stages of a process into a single unit operation is a good example of bioprocess integration. Such operations can be of further advantage in biocatalytic reactions exhibiting reversible, consecutive or product inhibition effects. In these cases, the simultaneous removal of product or inhibitor molecules from regions of high enzyme activity can improve product yields and reaction rates.

Chromatographic systems have been widely employed as combined reactor-separators, notably by Barker and co-workers,[1,2] who investigated the synthesis of B-512(F) dextran, a commercially important bacterial polyglycoside produced by the action of B-512(F) dextransucrase enzyme on sucrose. The formation of this polymer is a complex process, involving an insertion mechanism. The fructose by-product formed during the reaction acts as an 'acceptor' molecule, causing the release of growing dextran chains by releasing them from the enzyme molecules. This results in an increase in low mol. wt. dextran which has few commercial uses.

Barker and co-workers found that by exploiting differences in affinity between the eluent and column packing, and the various reaction species, it was possible to remove the fructose by-product, as it was formed, from the zone of biocatalysis. A 100% increase in the yield of high molecular weight dextran product (>160 000 Da) was recorded using this technique. Process integration was achieved by virtue of the simultaneous separation of product as it was formed.

Work in this laboratory[3,4] has shown that combined bioreaction- separation is also possible using a novel rate-zonal centrifugation process. Again, the biosynthetic conversion of sucrose to B-512(F) dextran and fructose was considered. By exploiting differences in the sedimenting properties of the dextransucrase enzyme and inhibitor fructose by-product in high centrifugal fields, a 100% increase in the yield of commercially important (12 000 - 98 000 molecular weight) dextran was recorded.[3] However, it was found that viscosity build-up in the bioreactor due to dextran formation had an adverse effect on system productivity.[4]

This paper describes the use of pre-synthesised acceptor molecules to reduce viscosity build-up in centrifuge vessels operated as combined bioreactor-separators. The effects of a number of important process variables on bioreactor performance in the presence of these pre-added acceptor molecules are also reported.

2 VISCOSITY AND ACCEPTOR ADDITION

Broths containing native dextran molecules form extremely viscous solutions and even gels at low concentrations. Broths containing 0.5% w/w native dextran have been found to have viscosities of around 5 mPas, equivalent to a 40% w/w sucrose solution, rising to 1000 mPas for 10% w/w broths.[3] This is due to the extended structure of the polymer which hydrogen-bonds with other dextran molecules to form extensive, cross-linked structures. A means of 'tailoring' the size of these molecules and hence decreasing medium viscosity was required. It was expected that these smaller structures would still be able to aggregate with other dextran molecules, forming more compact particles with good sedimenting properties, whilst making a significantly lower contribution to overall medium viscosity.

Alsop[5] has shown that the addition of small quantities of lower mol. wt. pre-synthesised dextran to dextransucrase reaction broths resulted in an overall decrease in the mean mol. wt. of synthesised dextran product. The pre-added dextran fractions acted as acceptor molecules, able to terminate the polymerisation process by releasing the growing dextran chains from the enzyme's active sites. Lower mol. wt. acceptors (~5000 Da average) yielded an overall increase in low mol. wt. dextran product (<12 000 Da) whereas intermediate mol. wt. acceptors (~20 000 Da. average) improved the yield of clinical range dextran (12 000 - 98 000 Da). These differences in product mol. wt. distribution can be simply explained by steric effects.

3 THE RATE-ZONAL BIOREACTION-SEPARATION PROCESS

The combined Rate-Zonal Bioreaction-Separation (RZBS) process is based on the rate-zonal centrifugation technique. A suspension of dextransucrase enzyme is carefully layered onto a pre-formed supporting sucrose substrate gradient, so that it forms a relatively narrow zone at the top of the gradient. The enzyme solution creates a negative gradient, but is prevented from premature sedimentation by the steep positive density gradient beneath it.

Under the influence of an applied centrifugal field, the enzyme molecules sediment into the substrate gradient at a rate dependent on their molecular or particle weight, size, shape and density. Reaction between the enzyme and substrate results in the formation of high mol. wt. dextran polysaccharide and low mol. wt. fructose monosaccharide. The polymer product and enzyme-polymer complexes sediment relatively rapidly and are thus simultaneously separated from the slowly sedimenting fructose molecules during the biosynthetic process.

A Beckman J2-MC centrifuge and JCF-Z zonal rotor, containing a Reorientating Gradient ('Reograd') core was used in these studies. Zonal rotors represent the only practical way to significantly increase the amount of material that can be handled using the rate-zonal centrifugation method. These rotors are generally 'bowl' shaped with capacities 50-100 times that of typical centrifuge tubes. The cylindrical cavity of the bowl is divided into sector-shaped compartments by vanes attached to a central core that slots into the rotor cavity; the rotor is enclosed by a threaded lid. A rotating seal assembly allows fluid to be pumped in and out of the cavity while the rotor is spinning. The RZBS process in a zonal rotor is shown in **Figure 1**.

4 EXPERIMENTAL DETAILS

The B-512(F) dextransucrase enzyme was produced and purified in the laboratory.[3] The activity of the enzyme was determined using a colourimetric method[3] and expressed in DSU cm^{-3}. 1 DSU is the amount of enzyme that will convert 1 mg of sucrose to product in 1h at 25°C and pH 5.2.

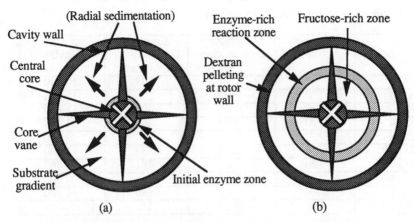

Figure 1 *The Rate-Zonal Bioreaction-Separation Process in a Zonal Rotor. (a) The enzyme is introduced to the vessel and initially forms an annular reaction zone on top of the sucrose gradient. (b) Under an applied centrifugal field, fructose, dextran and enzyme-dextran complexes sediment at different rates and are therefore separated.*

Prior to gradient loading, the rotor was spun at a relatively low speed (~1000 rev min^{-1}). The sucrose gradient (in 0.1 mol. dm^{-3} acetate buffer, pH 5.2, 25°C) was prepared using gradient forming apparatus and introduced, least dense end first, to the rotor cavity via an inlet port cut into the Reograd rotor core. The applied centrifugal force resulted in the gradient material forming an annular layer at the cavity wall. The increasing density of the gradient material entering the vessel displaced the lower density material away from the cavity wall and towards the outlet, located in the central rotor core. Loading continued until gradient material was seen exiting the system. A known volume of dextransucrase enzyme was then introduced to the vessel via the 'outlet' port, forming an annular zone bounded by the central core and the least dense end of the substrate gradient. This resulted in the displacement of an equivalent volume of dense gradient material via the cavity inlet port.

The application of an intensified centrifugal field resulted in the radial sedimentation of the reaction components towards the cavity wall. Dextran and dextran-enzyme complexes sedimented more rapidly and were thus separated from the fructose by-product as the process proceeded. The design of the rotor core allowed the relative order and distribution of the material in the cavity at the termination of a run to be retained in the effluent stream of the rotor, compressed air being used as the displacement medium. Gradient fractions were collected during the cavity unloading procedure and the exact volume of each fraction measured so that the radial position that it occupied in the bowl at run termination could be calculated. Samples were analysed for saccharide content using an Aminex HPX-87C column.

The run conditions used in these studies are given in **Table 1**. The rotor cavity had a total sample volume of 1.75 dm^3 and an enzyme volumetric loading of 60 cm^3 was used in all of these studies. Unless otherwise stated, a rotor speed of 15 000 rev min^{-1}, a run time of 6h and a sucrose gradient increasing in concentration from 15-25% w/w was used in all runs. Pre-added dextran T20 and T5 (~20 000 and ~5000 mean mol. wt) acceptors were dissolved into the gradient material prior to loading. Acceptor concentrations as low as 0.02% w/v were used, since initial studies had shown that significant decreases in broth viscosities could be expected at these levels. Column 5 lists the total enzyme activity loaded in each run, which equalled the product of the volume and activity of each loaded enzyme solution and hence had units of DSU. Experimental data corresponding to these runs is presented in **Table 2**.

Table 1 *Run conditions for the RZBS studies.*

Run number	Acceptor type	Acceptor concentration (% w/v)	Enzyme activity (DSU cm^{-3})	Total activity (DSU)
RG 1	-	-	183	10 980
RG 2	T20	0.02	215	12 900
RG 3[†]	T20	0.02	194	11 640
RG 4	T20	0.02	460	27 600
RG 5	T20	0.02	64	3 840
RG 6[††]	T20	0.02	185	11 100
RG 7	T20	0.05	181	10 860
RG 8	T20	0.20	155	9 300
RG 9	T5	0.02	196	11 760
RG 10	T5	0.20	159	9 540

[†] Run time: 3h   [††] Rotor speed: 10 000 rev min^{-1}.

Table 2 *Experimental results from RZBS studies.*

Run number	Mass of dextran sedimenting to wall (g)	Dextran sedimentation to wall per hour per unit total enzyme activity (10^5 g h^{-1} DSU^{-1})	Dextran sedimentation to lower 50% of gradient (g)	Gel dextran: fructose ratio
RG 1	5.64	8.56	11.34	1.61
RG 2	4.38	5.66	17.58	2.37
RG 3	1.53	4.38	4.71	2.77
RG 4	9.24	5.58	33.54	1.37
RG 5	1.86	8.03	4.42	3.26
RG 6	1.50	2.25	13.08	1.95
RG 7	3.42	5.25	9.30	2.25
RG 8	2.58	4.62	7.02	2.33
RG 9	4.26	6.03	15.54	2.28
RG 10	1.68	2.93	8.40	2.18

5 RESULTS AND DISCUSSION

5.1 Effect of Acceptor Addition

An important indicator of bioreactor performance was given by the mass of dextran that had pelleted at the cavity wall under a given set of run conditions (column 2, **Table 2**). In the absence of pre-added acceptor (run RG 1), 5.64g of dextran sedimented to the cavity wall; the pre-addition of 0.02% w/v T20 acceptor (run RG 2) decreased this value to 4.38g. A similar result was observed in the presence of 0.02% w/v T5 acceptor (4.26g). Therefore it was concluded that the pre-addition of acceptors led to a decrease in dextran sedimentation by reducing the mean mol. wt. of the dextran synthesised in the bioreactor.

However, when the distribution of dextran in the vessel was examined in more detail, a different conclusion was drawn. The values in column 4 of **Table 2** represent the mass of dextran recovered from the 'lower' 50% of gradient material recovered from the rotor, that is, the half of the gradient that occupied the zone adjacent to the cavity wall during centrifugation. Enhanced pelleted dextran values of 17.58g and 15.54g were recorded in the presence of 0.02% T20 and T5 acceptors respectively, compared with a value of 11.34g when no acceptor had been added. Thus in this context, overall dextran sedimentation was actually improved. **Figure 2** shows the effect of T20 acceptor concentration on dextran distribution within the bioreactor vessel.

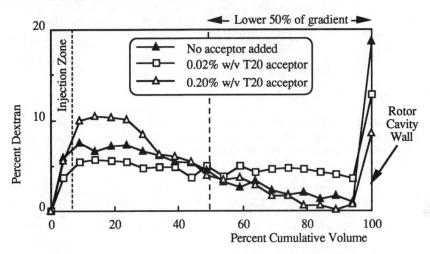

Figure 2 *Effect of acceptor concentration on dextran distribution.*

Poor dextran sedimentation was observed when 0.20% w/v concentrations of T20 and T5 acceptor were used (runs RG 8 and 10). Dextran sedimentation to the lower 50% of the gradient was measured at 7.02g and 8.40g in the presence of 0.2% T20 and T5 acceptor respectively, compared with 17.58g in the presence of 0.02% T20.

It would appear that the pre-addition of acceptor molecules resulted in a decrease in very high mol. wt. dextran, thus decreasing the yield of dextran product of a sufficient particle weight to sediment to the rotor wall, but an improvement in the rheological properties of the reaction medium such that there is an overall improvement in dextran sedimentation, particularly into the 'lower' half of the gradient.

5.2 Effect of Enzyme Activity

The effect of enzyme activity on bioreactor performance was evaluated at three different levels; 215, 460 and 64 DSU cm^{-3} (runs RG 2, 4 and 5 respectively). Run RG 4 yielded the highest mass of dextran recovered from the rotor wall recorded in all of the trials performed (9.24g). The recovery of dextran for the equivalent runs at 215 and 64 DSU cm^{-3} were 4.38g and 1.86g respectively. It was concluded that, as expected, an increase in enzyme activity resulted in a greater recovery of dextran from the rotor cavity wall.

However, when the sedimentation of dextran to the rotor wall per unit time and unit enzyme activity (column 3, **Table 2**) was calculated at the different enzyme activities, interesting data was obtained. Values of 5.58 x 10^{-5} g h^{-1} DSU^{-1}, 5.66 x 10^{-5} g h^{-1} DSU^{-1} and 8.03 x 10^{-5} g h^{-1} DSU^{-1} were recorded at enzyme activities of 460, 215 and 64 DSU cm^{-3} respectively. This data suggests that at low enzyme activities, viscosity build-up in the bioreactor is sufficiently retarded to allow a greater proportion of the dextran synthesised in a run to sediment to the rotor cavity wall. This would be a particularly important consideration in biotransformations involving expensive biocatalysts or valuable products. Dextransucrase enzyme activity appeared to have little effect on this factor for activities of between 215 and 460 DSU cm^{-3}.

A negative aspect of high enzyme activity was highlighted by the poor dextran:fructose ratio recorded in the recovered gel fraction (1.37:1). This was due to 'entrainment' effects.

High mol. wt. native dextran molecules in aqueous solution readily form extended, cross-linked structures by intermolecular hydrogen bonding, which entrap smaller molecules within their structure. Consequently, native dextran sedimentation to the rotor wall was accompanied by the sedimentation of the other reaction components, including fructose. High enzyme activities result in higher dextran conversions, hence a greater degree of dextran intermolecular bonding and fructose entrainment. A dextran: fructose ratio of 3.26:1 was recorded in the lowest enzyme activity run, the highest ratio recorded in this study. A compromise between bioreactor productivity and product separation is required.

5.3 Effect of Centrifugal Force

A measure of the effect of centrifugal force on reactor performance was obtained by comparing data from run RG 6 (rotor speed of 10 000 rev min^{-1}, corresponding to a maximum relative centrifugal field (RCF) of 9950 gee) with run RG 2 (15 000 rev min^{-1}, RCF: 22 385 gee). The ratio of the RCF between these runs was 2.25.

As expected, dextran sedimentation to the rotor wall was greater at the higher RCF: 4.38g in run RG 2 compared with 1.50g in run RG 6. The ratio of these values yielded a factor of 2.92, which approximately corresponded to the RCF ratio between the two runs. Thus, dextran accumulation at the rotor wall may be closely related to the applied RCF.

5.4 Effect of Run Time

In run RG 3, the run time of the reaction was set at 3 hours. The rate of dextran pelleting at the rotor wall per unit time and unit enzyme activity was recorded at 4.38×10^{-5} g h^{-1} DSU^{-1}, compared with a value of 5.66×10^{-5} g h^{-1} DSU^{-1} in the 'standard' run (RG 2). It was also found that the mass of dextran sedimenting into the lower 50% of the gradient per unit time and unit enzyme activity was approximately 1.7 times as high in the longer run. These preliminary findings indicated that dextran build-up at the rotor wall and in the lower portion of the gradient was not directly proportional to run time and that overall improved rates of dextran sedimentation can be achieved by careful selection of this factor.

The 3h run yielded a pelleted dextran:fructose ratio of 2.77:1, compared with a ratio of 2.37:1 for the equivalent 6h run. This result can be accounted for by a decrease in entrainment effects due to the shorter run time. Although short run times and lower enzyme loadings improve product separation, this is at the expense of productivity.

5.5 Alternative Rotor Core Design

Preliminary studies have indicated that improved bioreactor productivity can be achieved using a Beckman 'Continuous Flow' (CF) core in place of the Reograd core. The design of the CF core is such that the mean radius of rotation in this vessel is 8.13 cm as opposed to 5.75 cm for the Reograd core, giving a mean relative centrifugal force (RCF) value of 20 440 gee compared with 14 460 gee at a rotor speed of 15 000 rev min^{-1}. The volumetric capacity of the CF core is 660 cm^3.

The mass of dextran sedimenting to the rotor wall per unit time and unit enzyme activity was improved by up to 2.8 times when using the CF core and a low enzyme activity. For example, a dextran sedimentation value of 22.9×10^{-5} g h^{-1} DSU^{-1} was recorded with this equipment at an enzyme activity of 57 DSU cm^{-3} compared with a value of 8.03×10^{-5} g h^{-1} DSU^{-1} using the Reograd core and an enzyme activity of 64 DSU cm^{-3}. Comparable run conditions (rotor speed, gradient concentration and acceptor) were used for both runs.

Interestingly, when the enzyme activity was increased to 158 DSU cm^{-3} in the CF core, the dextran sedimentation value fell to 8.73×10^{-5} g h^{-1} DSU^{-1} compared with a value of 5.66×10^{-5} g h^{-1} DSU^{-1} at an enzyme activity of 215 DSU cm^{-3} using the Reograd core.

This data indicates that using the CF core and a low enzyme activity charge, a significant improvement in bioreactor productivity and performance can be achieved. This improvement can be attributed to decreases in rates of viscosity build-up and higher mean RCFs.

5.6 Scale Constraints and Economic Considerations

The intention of this work was to demonstrate an entirely novel method of process intensification: bioreaction *and* separation using centrifugal force as the separation mechanism. The dextransucrase reaction has proved a valid system for the evaluation of this process. Economic assessments of the system using this reaction scheme indicate that, as it stands, the process is uneconomic. Dextran production costs are approximately ten times higher using this method in comparison with the current method of dextran production.

A major problem is that since the system requires the generation of relatively high maximum centrifugal forces, the size of the rotor is restricted, in this case to a capacity of 1.75 dm^3. It appears possible, using established correlations, that rotors with volumetric capacities of 7 times greater than this figure, capable of generating equivalent centrifugal forces, can be constructed. Even with these modifications the process would remain uneconomic and research is currently on-going to find alternative systems that would benefit from this process. However, the dextransucrase reaction has sucessfully demonstrated for the first time that combined bioreaction-separation is possible using this novel technique.

6 CONCLUSIONS

This work has demonstrated that it is possible to improve the overall performance of a rate-zonal bioreactor-separator by the pre-addition of suitable acceptor materials to reaction broths. The experimental data indicated that the acceptor molecules decreased the yield of very high mol. wt. dextran, thus reducing the amount of dextran product of a sufficient particle weight to sediment to the rotor cavity wall. However, the improvement in the rheological properties of the solution resulted in an overall improvement in dextran sedimentation into the lower regions of the gradient.

Reducing the dextransucrase enzyme activity from 215 to 64 DSU cm^{-3} significantly increased the proportion of dextran synthesised in a run that sedimented to the rotor cavity wall, as well as improving product separation. This was due to the lower rate of viscosity build-up in the bioreactor medium. However, the low enzyme loading resulted in a decrease in the level of pelleted dextran and hence productivity.

Bioreactor productivity can be improved by using the CF rotor core. This core generates a mean centrifugal force 1.4 times greater than the Reograd core when spinning at the same rotational speed. When a low enzyme activity was injected into the bioreactor, the mass of dextran sedimenting to the rotor wall per unit time and unit enzyme activity was improved by 2.8 times using the CF core.

References

1. I. Zafar and P.E. Barker, Chem. Eng. Sci., 1988, 43(9), 2369.
2. G. Ganetsos, P.E. Barker and N.J. Ajongwen, 'Preparative and Production Scale Chromatography', Marcel Dekker, New York, 1992, Chapter 16, p. 375.
3. S.J. Setford, PhD Thesis, 1992, Aston University, Birmingham, UK.
4. S.J. Setford and P.E. Barker, 'The 1993 IChemE Research Event', The Institution of Chemical Engineers, Rugby, Warwicks., 1993, p. 28.
5. R.M. Alsop, 'Progress in Industrial Microbiology', Elsevier Scientific Publishers, Amsterdam, 1983, p.1.

The support of the SERC Biotechnology Directorate is gratefully acknowledged.

Downstream Separation of Chiral Epoxides Using Colloidal Liquid Aphrons (CLAs)

M. Rosjidi, D. C. Stuckey, and D. J. Leak[a]

DEPARTMENT OF CHEMICAL ENGINEERING AND CHEMICAL TECHNOLOGY; [a]CENTRE FOR BIOTECHNOLOGY, DEPARTMENT OF BIOCHEMISTRY, IMPERIAL COLLEGE OF SCIENCE, TECHNOLOGY, AND MEDICINE, LONDON SW7 2BY, UK

1. INTRODUCTION

The synthesis of enantiomerically pure compounds is of increasing importance to both the pharmaceutical and agrochemical industries. This is a result of growing regulatory and public pressure over health concerns in cases where the biological activity of a compound lies in a single enantiomer. This has stimulated research on various chemical and biological routes for the manufacture of useful optically active intermediates. One such class of intermediates are the epoxides which have utility in synthetic organic chemistry, readily undergoing nucleophilic substitution. However, chemical methods for the synthesis of chiral epoxides are relatively expensive, while in contrast, microbial systems have the ability to synthesise epoxides of high enantiomeric purity from a wide range of substrates with a potential for considerable cost savings [5,6].

Nevertheless, in many microbial biotransformations products may be dilute, labile and inhibitory and often tend to be difficult and expensive to separate from the broth. For poorly water soluble products solvent extraction is often the primary separation technique; however, this method has a number of drawbacks, namely: the high power requirement for solvent dispersion; the capital cost of mixer settlers; large solvent inventories, and; potential toxicity problems if the cells come into contact with the solvent[3]. One novel technique which has the potential to ameliorate these drawbacks is the use of polyaphrons in pre-dispersed solvent extraction (PDSE). Polyaphrons were initially described by Sebba[1], and are an oil (solvent) droplet encapsulated in an aqueous shell. Polyaphrons disperse easily in water to form colloidal liquid aphrons (CLAs). The CLAs are of the order of 1-10 microns in diameter thus facilitating a rapid equilibrium as a result of their enormous contact area[1,3,4].

Toxicity of a reaction product to the biocatalyst (microorganism) is often found in biotransformation processes, including the epoxidation of Allyl phenyl ether (APE) to Phenyl glycidyl ether (PGE) using *Mycobacterium* sp M156 [6]. In order to overcome this problem, rapid removal of PGE from the environment of the microorganism is essential, and in this work CLAs have been used to extract PGE from an aqueous phase. However, despite their numerous advantages, CLAs are very slow to separate from an aqueous solution due to the small phase density difference, and their size. To facilitate their separation colloidal gas aphrons (CGAs) have been used to float the CLAs, but this results in breakage of the aphrons making product recovery more difficult, and prevents them from being recycled[3]. In this work, the use of crossflow ceramic membrane microfiltration was proposed to separate CLAs from the aqueous phase in which they are dispersed. Although suitably formulated CLAs are difficult to break[1], they will probably show different characteristics from solid or other colloidal particles during membrane separation. The objective of this work, therefore, was to study the effect of parameters such as transmembrane pressure (TMP), feed concentration and feed flowrate on CLA separation. The stability of CLAs during the separation was also assessed based on size measurement.

2. MATERIAL AND METHODS

Reagents

Phenyl glycidyl ether (Aldrich, 99%); Allyl phenyl ether (Aldrich, 99%); Sodium dodecyl sulphate (SDS) (Aldrich, 98%); Softanol 120/Alcohol ethoxylates no.12 (BP); 1-Decanol (Aldrich, 99%) ; 2-Decanone (Aldrich, 98%).

Preparation of Polyaphrons

The method used for the preparation of polyaphrons was based on the principles described in previous works [3,4]. 1-2ml of 0.5% SDS solution was placed in a 100ml conical flask using a magnetic stirrer at 600rpm to provide adequate mixing conditions. 10-20ml of 1% Softanol 120 solution in an organic solvent was then dropped into the aqueous phase using a burette to regulate the solvent flowrate. This was necessary to ensure that all the solvent already dropped into the flask was converted to polyaphrons. The phase volume ratios (PVR) of polyaphrons formed was calculated from the volume ratio of solvent added to the aqueous phase. Correctly formulated polyaphrons have a white, creamy texture with gel like properties, and the best formulation yields polyaphrons which are stable over a period of months.

Extraction of PGE using CLAs

Polyaphrons made of 1% Softanol 120 in 2-Decanone and 0.5% SDS in distilled water were investigated for extracting PGE from an aqueous feed solution. This polyaphron had a PVR of 20, and a mean diameter of 6.9 microns. The polyaphrons, and a 5 mM PGE feed solution in distilled water at volume ratios (CLA : feed) of 1%, 2%, 3%, and 4%, were mixed in flasks with a stirrrer speed of 200 rpm. Samples were taken over a 0-120 second time period. Prior to being analysed, the CLAs were separated from the aqueous phase using a disposable microfiltration membrane with a pore size of 0.45 microns. The analysis of PGE was performed by employing Gas Chromatography (Philips, PU 4500) using a flame ionisation detector (FID). The packed column used was 3% SP2100 on Chromosorb WHP 80-100 mesh. Aqueous phase samples were first extracted into Diethyl ether prior to injection onto the column, and the standard deviation of this technique was $\pm$ 3%, and duplicate samples were always taken.

CLA Separation Using a Plate-Frame Inorganic Membrane

The membrane module employed was an "Ansep" crossflow filtration module obtained from Ceramesh Ltd. This module is a plate-frame device consisting of 4 membrane sheets, arranged in series, giving a maximum working membrane area of 0.042 m^2. The membrane sheets used had a pore size of 0.1microns, and consisted of a ceramic layer deposited on a metal mesh. The size of CLAs was determined using a Malvern Particle Size Analyser (2600 Series), which is based on the principle of laser diffraction, and gives a size distribution and an average diameter. The determination of CLA concentration was performed by optical density measurements at 640 nm on a Perkin Elmer UV/VIS spectrophotometer.

The influence of pressure. A schematic diagram of the experimental set-up is shown in Figure 1. 5l of a 2% CLA suspension was initially placed in the reservoir, which was gently agitated to maintain an homogenous CLA suspension. The CLAs were then passed repeatedly through the membrane module using a peristaltic pump (Watson-Marlow 502S) at a flowrate of 977ml/min (crossflow velocity = 0.55 m/sec, Re $\approx$ 400), the maximum

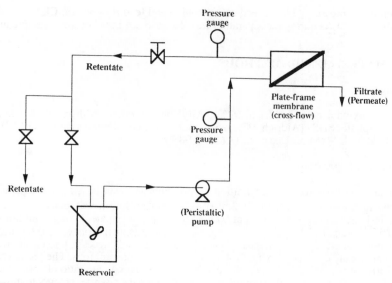

Figure 1: Schematic diagram of CLA separation using a plate-frame crossflow microfiltration.

flowrate which could be achieved. The initial inlet pressures to the filter were adjusted to 0.17, 0.33 and 0.65 bar, and the initial transmembrane pressure ($TMPs_{init}$) were 0.14, 0.29, and 0.60 bar respectively. The experiments were run until the CLA suspension was roughly concentrated by a factor of 10. Parameters such as filtrate flowrate (flux), transmembrane pressure (TMP), CLA concentration and CLA size were monitored during filtration.

Determination of CLA gel layer. The flowrate of the peristaltic pump was adjusted to 213ml/min (crossflow velocity = 0.12 m/sec, Re ≃ 100), a minimum flowrate, and the position of the valve opened fully so that the feed flowrate (tangential velocity), shear force, and transmembrane pressure drop occurring across the membrane surface was expected to be very low. This was done to avoid the possibility of damaging any fouling layer if it was formed. The first step of this experiment was conducted using distilled water which was repeatedly passed through the membrane module containing fresh membrane sheets at a constant flowrate of 213ml/min. Subsequently, 2l of a 2% CLA suspension was filtered until the CLA suspension was concentrated roughly by a factor of 10. Finally, fresh distilled water was again passed through the system. The whole experiment was monitored by measuring the flux of filtrate.

3. RESULTS AND DISCUSSION

Extraction of PGE using CLAs

Extraction of 5mM PGE in distilled water using various volume ratios (CLA : feed) of 1%, 2%, 3% and 4% at pH 5.86 is shown in Figure 2. It can be seen that the extraction occurred extremely rapidly, and near equilibrium was approached within 10 seconds. This shows that although the CLAs were encapsulated by a soapy film (thin film of surfactant solution), their enormous contact area could facilitate rapid mass transfer. In the experiment at a 1% volume ratio, 70% of the epoxide could be extracted from the aqueous solution. By increasing the volume ratio to 4%, 89% of the epoxide could be extracted.

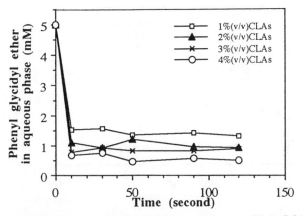

Figure 2: The extraction of PGE using CLAs at pH = 5.86. CLAs made of 2-Decanone; PVR=20;CLA mean diameter = 6.9 micron (the mean diameter of CLAs derived from the volume distribution).

<u>CLAs separation using a plate-frame inorganic membrane</u>

The influence of pressure. Figure 3 shows filtrate flux at the different initial inlet pressures. For the initial inlet pressure of 0.17 and 0.33bar, the filtrate flux was roughly constant during the filtration process. This is different from what is typically found with other substances, including colloidal particles, where the filtrate flux decreases in the early part of filtration as a result of membrane fouling (i.e. particles clogging the membrane pores and forming a gel layer on the membrane surface). This poses the question as to whether any CLA gel layer was formed on the membrane surface. For the higher initial inlet pressure of 0.65 bar, filtrate flux was initially constant (about 63 l m^{-2}hr^{-1}), but then decreased after 64 minutes when the CLA concentration in the retentate reached about 113 g/l. At the lower pressures, even when CLA concentrations in the retentate reached these levels no noticeable drop in the flux occurred. This shows that an increase in driving force across the membrane, i.e.an increase in transmembrane pressure, significantly influences the filtrate flux at high CLA concentrations, and the decline in flux was probably due to CLAs forming a gel layer on the surface, which was compressed at higher TMPs.

The transmembrane pressure drop of the module (TMP) was also monitored during the filtration. Figure 4 shows that TMP increases with time, which corresponds to an increase in the CLA concentration recycled to the membrane. When the initial inlet pressure was low, at 0.17bar, the TMP only increased slightly during the filtration. However, the rate of increase of TMP became higher during the filtration when the initial inlet pressure was set to higher values. This is probably due to the higher driving force across the membrane, and decreased CLA size (see Figure 5).

The CLA sizes measured during the filtration process are shown in Figure 5. It can be seen that the mean diameter of the CLAs decreases with time, and that the rate of change increases with the initial inlet pressure. Size distribution data (not shown) reveals a shift from larger CLAs (>10 microns) to a significant fraction of small ones (1-2 microns) in a bimodal distribution[7]. This bimodal phenomenon obviously does not reflect breakage and coalescence since this would result in larger CLAs being formed. To understand the phenomenon more clearly further experiments determining the concentration of surfactant in the filtrate will be performed in future work.

The concentration of CLAs measured during the filtration process is shown in Figure 6. As expected the CLAs could be separated, and a separation degree of 1.0 was obtained where the filtrate was clear (CLA free). It can be seen that CLAs can easily be concentrated

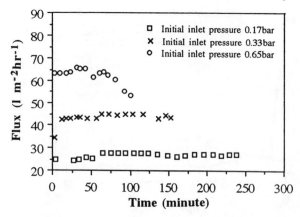

Figure 3: The variation of filtrate flux with time.

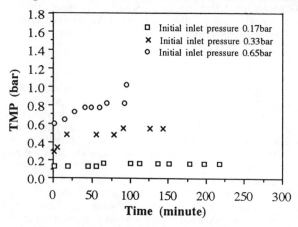

Figure 4: The transmembrane pressure (TMP) measured during filtration.

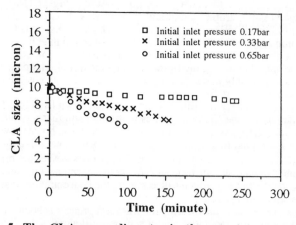

Figure 5: The CLA mean diameter in the retentate measured during filtration.
This mean diameter derived from the volume distribution.

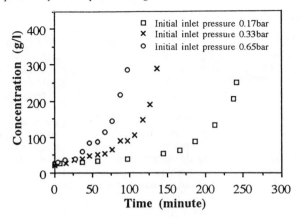

Figure 6: The CLA concentration in the retentate measured during filtration.

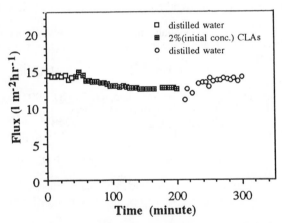

Figure 7: Determination of the formation of a CLA fouling layer by measurement of filtrate flux.

from 2% to 30% (w/v) under quite low pressures. More work is needed at higher pressures to determine whether separation times can be reduced.

Determination of CLA gel layer. Flux was monitored during the whole filtration experiment, and the results are shown in Figure 7. The constant filtrate flux in the first stage indicated that the system was free from impurities. When a 2% CLA suspension was passed across the membrane, the filtrate flux declined slightly and then seemed to remain constant In the third stage where clean distilled water was again employed, the flux gradually increased and reached the same level as at the start of the run. This indicated that flux reduction was caused by the formation of a reversible CLA "gel layer", and that when pure water was used this layer disappeared. No irreversible gel layer occurred which was expected since the solution did not contain any high molecular weight solutes.

4. CONCLUSIONS

The extraction of PGE using CLAs made of 2-Decanone occurred rapidly, and near equilibrium was approached within 10 seconds. At a volume ratio (CLA : feed) of 4%, 89% of the epoxide could be extracted. It was also apparent that CLAs could be separated and concentrated using a crossflow membrane (a plate-frame inorganic membrane), and at low TMPs the filtrate flux obtained was satisfactory (i.e. 63 $lm^{-2}hr^{-1}$ at 0.60bar TMP_{init}).

Although the size of CLAs changed during filtration this did not represent aphron breakage, and hence all of the solute in the solvent phase can potentially be recovered. During filtration at low TMPs, a reversible "gel layer" appeared to form on the surface of the membrane which could be removed by flushing through with water. Based on this preliminary assessment it appears that crossflow membrane separation of CLAs has considerable potential, and more work will be carried out to quantify the important parameters controlling the process.

REFERENCES

1. F. Sebba, "Foams and Biliquid Foams - Aphrons", John Wiley and Sons, 1987.
2. F. Sebba, "Surfactant-based Separation Processes", Mercel Dekker, New York, 1989, 91.
3. D.C. Stuckey, K. Matsushita, A.H. Mollah, and A.I. Bailey,"3rd International Conference on Effective Membrane Processes", Mechanical Engineering Publication Limited, London, 1993, Pub. no.3, 3.
4. K. Matsushita, A.H. Mollah, D.C. Stuckey, C.d. Cerro, and A.I. Bailey, Coll. and Surf., 1992, 69, 65.
5. D.J. Leak, P.J. Aikens and M. Sayed-Mahmoudian, TIBTECH, 1992, 10, 256.
6. M. Sayed-Mahmoudian, PhD Thesis, 1989, Imperial College.
7. L.E.S. Brink and J. Tramper, Biotech. Bioeng., 1985, 27, 1258.
8. L.E.S. Brink, J. Tramper, K.Ch.A.M. Luyben and K. Van't Riet, Enzyme Microb. Tech., 1988, 10, 736.
9. M.H. Lojkine, R.W. Field and J.A. Howell, Trans. IChemE, 1992, 70c, 149.
10.J.A.L. Santos, M. Mateus and J.M.S. Cabral, "Chromatographic and Membrane Processes in Biotechnology", Kluwer Academic Publisher, Netherland, 1991, 177.

Extraction of Valeric Acid from Aqueous Solutions Using Tri-n-butylphosphate

M. Olga Ruiz, Isabel Escudero, José L. Cabezas, José R. Alvarez,[a] and José Coca[a]

DEPARTMENT OF CHEMICAL ENGINEERING, UNIVERSITY COLLEGE BURGOS, 09008 BURGOS, SPAIN

[a]DEPARTMENT OF CHEMICAL ENGINEERING, UNIVERSITY OF OVIEDO, 33071 OVIEDO, SPAIN

1   INTRODUCTION

Liquid-liquid extraction is a suitable alternative for the recovery of organic acids from dilute aqueous solutions. Valeric acid (n-pentanoic acid) is industrially produced by oxidation of amyl alcohol or by fermentation processes and can be found as a subproduct in the manufacture of adipic acid. As pure adipic acid is required when used as the monomer in polymerisation processes, any valeric acid produced must be removed during the adipic acid purification step.

Physical extraction of carboxylic acids from aqueous solutions is not very efficient because of their low activity in aqueous solution, resulting in very low distribution coefficients. The extraction process can be highly improved by reactive extraction using extractants which can form ion-pair complexes or solvating species with the carboxylic acid. Organophosphorous compounds (Lewis bases) have been used for the extraction of carboxylic acids [1-4] based on the strong interaction of the organic acid with the extractant which allows the formation of solvation complexes. Aliphatic amines have also been used to extract carboxylic acids [1-3,5,6] based on the formation of ion pair complexes, the organic acid being extracted as an ammonium salt.

In this work, tri-n-butylphosphate (TBP) dissolved in kerosene as diluent was investigated for the recovery of valeric acid from aqueous solutions in a range of concentration of 0.1 to 40 g/L. Equilibrium data are reported at 25 and 50 °C and a simple model for the extraction with TBP is proposed. Data for the back-extraction of the acid from the organic phase are also reported.

2   EXPERIMENTAL

2.1 Materials

Chemicals for the extraction experiments were commercially available: Valeric acid (Aldrich, >99%), tri-n-butylphosphate (Fluka, >97%) and kerosene (Aldrich, 64% alkanes, 26% non-aromatic cycloparaffins, 9.8% aromatics and sulphur), and were used as supplied. Bidistilled water was used in all experiments.

2.2 Methods

The extraction solvent consisted of a mixture of TBP with kerosene as diluent at concentrations ranging from 5 to 50 vol %. Solutions of valeric acid were prepared with a concentration of 6 g/L. Isothermal equilibrium data were determined by contacting fixed volumes of the extractant and valeric acid solution in screw-capped flasks of 60 mL volume. After vigorous manual shaking and thermostating in a water bath, the phases were allowed to settle and samples of both phases were then taken and equilibrium compositions determined.

The aqueous phase acid concentration was determined by titration with standard sodium hydroxide solutions (0.01 or 0.1M) using phenolphthalein as indicator. The acid concentration in the organic phase was determined by a two-phase titration in order to strip the acid from the organic phase; 1 mL of the organic phase was mixed with 7 mL of ethanol and 3 mL of water and was titrated with a sodium hydroxide solution (0.01 or 0.1M) using phenolphthalein as indicator[7]. The TBP concentration in the organic phase was analysed following the method described by Ritcey *et al.* [8].The mass balance closure to the system was of the order of 2 mass %.

Back-extraction of valeric acid was studied at 25 and 50 °C by contacting 15 ml of the organic phase (TBP-kerosene 10 vol % or 30 vol %) loaded with valeric acid with the same volume of a sodium hydroxide solution. Valeric acid concentration was analysed in both phases following the aforementioned procedure.

3   RESULTS AND DISCUSSION

Reactive extraction of valeric acid (HA) using TBP yields a reaction complex (HATBP) which remains in the organic phase and may be represented by:

$$H_nA_{(w)} + TBP_{(o)} \Leftrightarrow (HATBP)_{(o)}$$

The degree of extraction, E, can be defined as:

$$E = 1 - \frac{C_{HA(w)}}{C_{HA(w)i}} \qquad (1)$$

The influence of extractant concentration on E was studied at 25 and 50 °C. Results are shown in **Figure 1** as percentage extraction, % E.

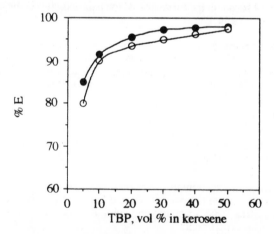

Figure 1 *Extraction of valeric acid using TBP in kerosene at 25 and 50 °C. Initial concentration:* $(C_{HA(w)i})$ *: 6 g/L.* ●*:25°C,* ○ *: 50°C.*

From data in **Figure 1** and practical considerations, two concentrations of the extractant were selected, i.e., 10 and 30 vol %. The effect of the extractant concentration and physical extraction by the diluent (kerosene) is shown in **Figure 2**.

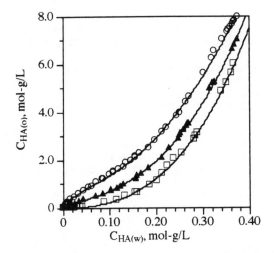

Figure 2 *Equilibrium isotherms for the extraction of valeric acid using TBP in kerosene at 25 °C. □: Kerosene, ▲: TBP 10 vol % and ○: TBP vol 30 % in kerosene*

3.1 Equilibrium data modelling

3.1.1 Physical extraction with kerosene. Experimental data on the physical extraction of valeric acid using kerosene show that the influence of temperature is negligible in the range of 25 - 50 °C. Taking into account the partial dissociation of the acid in the aqueous phase and its association in the organic phase at high concentrations[1], the distribution of valeric acid between water and kerosene can be described according to the following expressions:

(i) The dissociation of the acid in aqueous solution:

$$HA_{(w)} \Leftrightarrow H^+_{(w)} + A^-_{(w)} \qquad K_{HA} = \frac{\left[H^+\right]_{(w)}\left[A^-\right]_{(w)}}{\left[HA\right]_{(w)}} \qquad (2)$$

(ii) For the distribution of the undissociated acid between the aqueous (w) and organic (o) phases:

$$HA_{(w)} \Leftrightarrow HA_{(o)} \qquad P = \frac{\left[HA\right]_{(o)}}{\left[HA\right]_{(w)}} \qquad (3)$$

(iii) If the association of the acid in the organic phase is considered as a dimerization reaction:

$$2HA_{(o)} \Leftrightarrow (HA)_{2(o)} \qquad K_d = \frac{\left[(HA)_2\right]_{(o)}}{\left[HA\right]^2_{(o)}} \qquad (4)$$

The total concentration (analytical) of the acid in all its possible forms in the organic phase, $C_{HA\,(o)}$, can then be expressed as:

$$C_{HA(o)} = P\left[HA\right]_{(w)} + 2K_d P^2\left[HA\right]^2_{(w)} \qquad (5)$$

Kerosene is basically a mixture of non-polar compounds and, therefore, a poor degree of extraction is expected. When the concentration of the acid increases, it tends to form a stable dimer in kerosene. The parameters in Eqn. (5) can be estimated using the Levenberg-Marquardt algorithm and their values are given in **Table 1**.

There is poor agreement between the experimental and the predicted data when the formation of a dimer is assumed, as shown in **Figure 3**. If further association is considered, such as the formation of a trimer, then

$$(HA)_{2(o)} + HA_{(o)} \Leftrightarrow (HA)_{3(o)} \qquad\qquad K_t = \frac{\left[(HA)_3\right]_{o)}}{\left[(HA)_2\right]_{o)}\left[HA\right]_{(o)}} \qquad (6)$$

and the total acid concentration in organic phase can be expressed as:

$$C_{HA(o)} = P[HA]_{(w)} + 2K_d P^2[HA]_{(w)}^2 + 3K_d K_t P^3[HA]_{(w)}^3 \qquad (7)$$

NMR tests performed on samples of different valeric acid concentrations using a diluent consisting of 85% of deuterated cyclohexane and 15% of deuterated toluene cannot distinguish between free-acid, dimer and trimer. However, if the trimer formation is considered, and Eqn. (7) is used, a much better fit of the data is obtained, as shown in **Figure 3**.

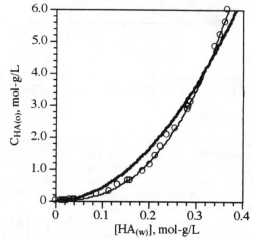

Figure 3 *Valeric acid extraction equilibrium data. Kerosene/Water at 25°C.*
━━ : *Dimer;* ──── : *Trimer*

3.1.2   Reactive extraction with TBP.   The extraction equilibrium of valeric acid using a strong solvating extractant such as TBP can be represented by the formation of a solvent-acid complex, which in general from may be written as[1]:

$$H_n A_{(w)} + m S_{(o)} \Leftrightarrow (H_n A)S_{m(o)} \qquad (8)$$

where m is the solvation number of the acid and n is the number of carboxyl groups of the acid. It has been reported that the solvation number of the acid is equal to the number of its carboxylic groups[4,9]. Wardell *et al.*[10] reported that for the extraction of acetic acid the ratio between the distribution coefficient (K_D) and the weight fraction of the phosphoryl compound in the solvent mixture was independent of the concentration of the phosphoryl compound. Such behaviour might correspond to simple, stoichiometric interaction between acetic acid and TBP, with no effect derived from consumption of the phosphoryl compound[11]. Valeric acid seems to form a complex with TBP at low acid concentrations as shown in **Figure 4**. There is a clear difference between the extraction using kerosene and kerosene/TBP systems. As the TBP content increases so does the degree of extraction.

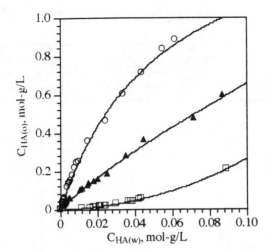

Figure 4 *Valeric acid extraction using kerosene and kerosene/TBP at low acid concentrations.* □: *Kerosene,* ▲: *TBP 10 vol % and* ○: *TBP vol 30 % in kerosene*

For valeric acid extraction using TBP, first the TBP-acid complex is formed, and then it modifies the polarity of kerosene, allowing the presence of free acid in the organic phase besides the acid complex, plus dimer or trimer molecules. The mathematical formulation of this behaviour may be expressed by the addition of two terms: One corresponds to the physical extraction, and the other accounts for the TBP-acid complex.

The formation of the TBP-acid complex can be formulated as follows:

$$HA_{(w)} + TBP_{(o)} \Leftrightarrow HATBP_{(o)} \qquad\qquad K_C = \frac{\left[HATBP_{(o)}\right]}{\left[HA_{(w)}\right]\left[TBP_{(o)}\right]} \qquad (9)$$

and

$$[TBP]_{(o)T} = [TBP]_{(o)} + [TBPAH]_{(o)} \qquad (10)$$

Combining Eqn (9) and Eqn (10), the concentration of TBP-acid complex is:

$$[HATBP]_{(o)} = \frac{K_C\,[HA]_{(w)}\,[TBP]_{(o)T}}{\left(1 + K_C\,[HA]_{(w)}\right)} \qquad (11)$$

and taking into account Eqn (7), the total concentration of valeric acid in the organic phase will be

$$C_{HA(o)} = P[HA]_{(w)} + 2K_d P^2[HA]_{(w)}^2 + 3K_d K_t P^3[HA]_{(w)}^3 + \frac{K_C\,[HA]_{(w)}\,[TBP]_{(o)T}}{\left(1 + K_C\,[HA]_{(w)}\right)} \qquad (12)$$

Parameters in Eqn. (12) depend on the extractant/diluent ratio, and the estimated values when kerosene is used do not apply. Kerosene, as diluent, changes its extracting power in the presence of TBP as a result of the modification of its polarity. **Table 1** shows the coefficients in Eqn. (12) obtained using the Levenberg-Marquardt algorithm to fit the experimental data. As the concentration of TBP increases, P increases, and K_d and K_t decrease. These results show that the acid does not tend to aggregate, as in the case of pure kerosene, because of the higher polarity of the solvent. The parameter K_C, for the TBP-acid complex remains constant at both TBP concentrations. **Figure 2** also shows the agreement between the experimental data (symbols) and the model (lines).

Table 1 *Curve fitting values of the partition coefficient, formation of dimer, trimer and TBP-acid complex*

Solvent	P	K_d	K_t	K_C
Kerosene	0.122	435.241	35.273	-
Kerosene-TBP10%	0.559	44.601	2.175	24.508
Kerosene-TBP30%	2.014	4.561	0.217	24.508

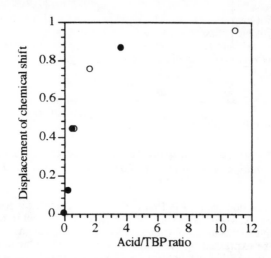

Figure 5 *Change of chemical shift for the phosphoryl groups with the ratio acid/TBP. O:TBP 10 vol % in kerosene; TBP, ●:30 vol % in kerosene*

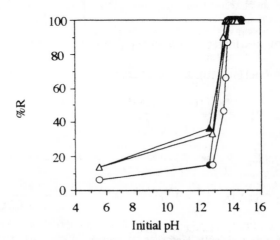

Figure 6 *Back-extraction of valeric acid from TBP/kerosene with NaOH solutions as a function of the pH of the stripping solution. ●:TBP 30 vol % at 25°C; O:TBP 30 vol % at 50°C; ▲:TBP 10 vol % at 25°C; △:TBP 10 vol % at 50°C*

NMR ^{31}P spectra were obtained with different concentrations of valeric acid in TBP at 10 and 30 vol% in kerosene. A single peak was observed for every sample analysed. A linear displacement of the chemical shift was observed when the ratio valeric acid/TBP increased from zero to one, as shown in **Figure 5**. At ratios greater than 2 there is no significant change in the chemical shift. Gatrone *et al.*[12] observed the same effect for the extraction of monocarboxylic acids using octyl(phenyl)-N, N-diisobutylcarbamoylmethylphosphine oxide (CMPO). They found a linear change in the chemical shift for the phosphoryl group until the available phosphoryl sites were occupied (acid/CMPO=1)

3.2 Back-extraction

Back-extraction of valeric acid was carried out by washing the organic phase with NaOH solutions. The degree of back-extraction (%R) increases with the pH of the stripping solution but it is not very sensitive to temperature, as shown in **Figure 6**.

Pure valeric acid was obtained when the sodium valerate solution was acidified with a strong mineral acid, as a result of its poor solubility in water.

NOMENCLATURE

C	total (analytical) concentration
HA	undissociated acid
H^+	proton
A^-	dissociated acid
S	solvating extractant

Subscripts:

(w)	aqueous phase
(o)	organic phase
i	initial
T	total

Symbols:

[ ]	concentration in moles per litre
K_{HA}	dissociation constants of the acid in the aqueous phase
P	partition constant between the organic and aqueous phases
K_d	dimerization constant of the acid in the organic phase
K_t	trimerization constant of the acid in the organic phase
K_C	equilibrium constant for the formation of the Acid/TBP complex
m	solvation number of the acid
n	number of carboxyl groups in the acid molecule

REFERENCES

(1) Kertes, A.S.; King, C.J. *Biotechnol. Bioeng.* **1986**, 28, 268-282.
(2) Schügerl, K. in *"Separations for Biotechnology"*, Chapter 20, Ellis Horwood Ltd., Chichester, England, 1987, pp. 260-269.
(3) Schügerl, K.; Degener, W. *Int. Chem. Eng.* **1992**, 32, 29-40.
(4) Hano, T; Matsumoto, M.; Ohtake, T.; Sasaki, K.; Hori, F.; Kawano, Y *J. Chem. Eng. Japan* **1990**, 23, 734-738.
(5) Puttemans, M.; Dryon, L.; Massart, D.L. *Anal. Chim. Acta* **1985**, 161, 381-386.
(6) San Martín, M.; Pazos, C.; Coca J. *J. Chem. Tech. Biotechnol.* **1992**, 54, 1-6.
(7) Bar, R.; Gainer, J.L. *Biotechnol. Progr.* **1987**, 3, 109-114.
(8) Ritcey, G.M.; Ashbrook, A.W. *"Solvent Extraction. Principles and Applications to Process Metallurgy. Part I"*, Elsevier, Amsterdam, 1984.
(9) Abbasian, K.; Degener, W.; Schügerl, K. *Ber. Bunsenges Phys. Chem.* **1989**, 93, 976-980.
(10) Wardell, J.M.; King, C.J. *J. Chem. Eng. Data* **1978**, 23, 144-148.
(11) Clack, G.A.; Gatrone, R.C.; Horwitz, E.P. *Solvent Extr. Ion Exch.* **1987**, 5, 471-491.
(12) Gatrone, R.C.; Howitz, E.P. *Solvent Extr. Ion Exch.* **1987**, 5, 493-510.

Aqueous Two-phase Extraction Using Colloidal Gas Aphrons in a Spray Column

S. V. Save* and V. G. Pangarkar

DEPARTMENT OF CHEMICAL TECHNOLOGY, UNIVERSITY OF BOMBAY,
MATUNGA, BOMBAY 400 019, INDIA

1.   INTRODUCTION

Aqueous two-phase systems (ATPS) offer a gentle environment for cells, cell organelles and biologically active proteins and partition can be exploited to effect separation otherwise difficult or impossible to achieve. Consequently, the major interest during the early applications of ATPS was concentrated in cell biology. The dramatic changes in production of proteins by genetic engineering and developments in enzyme technology have brought about renewed interest in protein separation processes and their quantitative description as a basis for scale-up. Therefore, major interest in application of ATPS moved away from cell biology and is presently concentrated in the following areas :

- fundamental analysis of phase separation and protein partitioning
- improvement in economy
- improvement in selectivity of extraction
- multistage operation

Solvent extraction technology with multistage operations is well established. Preliminary studies on aqueous two phase extraction using conventional contactors like Kuhni column, Graesser contactors or mixer-settler batteries has been reported in the literature[2].   Joshi et al.[3] extensively investigated various conventional contactors such as spray column, York - Schiebel column, plate column to effect aqueous two-phase extraction.

Recently Save et al.[4] developed a technique to intensify the mass transfer in aqueous two-phase extraction. This involved conversion of the dispersed phase into colloidal gas aphrons (CGA). CGA consists of individual aphrons which are made up of gas (air) entrapped in a soapy film stabilized by surfactant. The properties of CGA are widely reported[4-8]. This new technique offers the following advantages:

- lower disengagement time owing to higher phase density difference
- higher phase purity
- considerably higher interfacial area as compared to conventional contactors[4]
- higher utilization of costly polymers due to enhanced mass transfer rate thus increasing economic viability of aqueous two-phase extraction[4].

*present address: Department of Food Science and Technology, University of Reading, P.O.Box 226, Whiteknights, Reading RG6 2AP, England

Save *et al.*[4] reported the effects of surfactant type and its concentration, dispersed phase velocity and phase composition on the mass transfer of α - amyloglucosidase in polyethylene glycol (PEG) - sodium sulfate aqueous two-phase system. The results suggested that only cationic surfactants can be employed for extraction of enzyme for operating pH < pI. Further, the use of higher surfactant concentration and higher dispersed phase velocity resulted in an increase in the mass transfer coefficient. In the present work, the effect of following variables on mass transfer coefficient and dispersed phase holdup has been investigated:

a) various cationic surfactants

b) dispersed phase velocity

c) scaleup

2. MATERIALS AND METHODS

Materials

In the present work a sodium sulfate- PEG - water (buffered) aqueous two phase system was used. The PEG used had moleculer weight of 4000. Sodium sulfate, glacial acetic acid, and di-tetraborate were of analytical grade (Loba Chemicals, Co., Bombay, India). Water was glass double distilled. *p*- Nitrophenyl- α - D - galactopyranoside (pNPG) was procured from Sigma Chemical Co. Hexadecyl trimethyl ammonium chloride (HTAC), dodecyl trimethyl ammonium chloride (DTAC) and hexadecyl pyridinium chloride (HPC) were of commerecial grade (HICO Chemicals, Bombay, India).

Solution preparation and Analysis

Predetermined weighed quantities of PEG and sodium sulfate were added to a weighed quantity of buffer. The entire mixture was stirred for about 4 h for equilibration. The phases were then allowed to settle overnight, and then each phase was carefully removed and filtered. The pH was adjusted to 4.3 with the help of potassium dihydrogen orthophosphate (0.1 kmol/m^3). The equilibrium diagram for this system has been reported.[9] The system composition (in % wt/wt) and physical properties are given in Tables 1 and 2, respectively. β - Galactosidase (BG) was used as solute. BG activity was analysed using a previously described method.[10] The method involved catalytic decomposition of pNPG to *p*- nitrophenol, which was analysed using a Perkin-Elmer 3B (UV/Vis) spectrophotometer by measuring absorbance at 400 nm. The extent of decomposition of pNPG (after 2 h of reaction) is directly proportional to the enzyme activity. A calibration chart was prepared using known BG activity. The partition coefficient was calculated as $m = C_u / C_l$.

3.   EXPERIMENTAL

The experimental setup consisted of two parts (Figure 1). The extraction column (C) used was a perspex column 30 mm diameter and 1000 mm tall fitted with an expansion zone of size 150 x 150 x 150 mm at the top. The CGA generator used was the same as described by Save *et al.*[4] For scale-up studies, columns of 65 and 100 mm diameter fitted with an expansion zone of proportionate size were used. A peristaltic pump was incorporated between the CGA generator and column for pumping CGA.

TABLE 1 : STARTING COMPOSITIONS FOR GENERATING AQUEOUS TWO-PHASE SYSTEMS

SYSTEM No.	COMPOSITION		
	PEG (%)	SODIUM SULFATE (%)	BUFFER (%)
1	13.6	6.5	79.7
2	15.0	8.0	77.0
3	17.8	7.8	74.3
4	14.4	10.8	74.3
5	20.0	8.1	71.9

TABLE 2 : PHYSICAL PROPERTIES OF THE PHASE SYSTEM EMPLOYED

SYSTEM No.	VISCOSITY	m Pa. s	DENSITY	kg/m³	pH
	PRP	SRP	PRP	SRP	
	μ_d	μ_c	ρ_d	ρ_c	
1	14.16	1.25	1071	1115	4.3
2	20.33	1.35	1072	1138	4.3
3	20.37	1.36	1073	1150	4.3
4	24.77	1.44	1082	1062	4.3
5	28.37	1.50	1083	1167	4.3

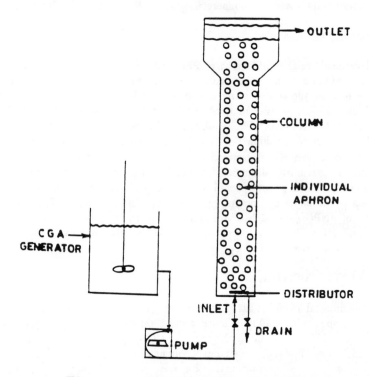

Figure 1.   Experimental setup.

An aqueous two-phase system of desired composition was prepared according to the procedure given earlier. In all experiments the salt rich phase (SRP, heavy) formed the continuous phase, while PEG rich phase (PRP, light) converted to CGA formed the dispersed phase. The column was operated in the semi-batch mode. The light phase was converted to CGA having 65% gas holdup by adding suitable quantity of desired surfactant and stirring at 6000 rpm for 15 s. After this the speed of agitation was lowered to 1000 rpm long enough to keep CGA suspended to the end of the experiment. These CGA were then continuously pumped at the bottom of the column where the CGA separated into indivual aphrons of PEG rich phase which travelled through the dispersion and coalesced at the top. The coalesced layer was removed through the overflow outlets. A multiorifice CGA distributor was employed as suggested by Save *et al.*[3]

In the present work, transfer of enzyme was investigated wherein enzyme was dissolved in PRP at predecided activity level 30 min before converting it into CGA. BG was found to be stable at concentrations of surfactant employed for much longer time than the duration of the experiment. The flow of CGA was monitored through the precalibrated peristaltic pump. The flow of CGA was allowed for 600 s after its appearnce at the base of the column. This time ensured the steady state with respect to the dispersion, assuring a constant interface level and PRP flowrate from the top outlet. The enzyme at the inlet and outlet was measured and its reproducibility was checked by repeating experiments. The performance of the contactor was evaluated on the basis of the overall mass transfer coefficient $K_D a$ estimated by taking the mass balance for BG over a differential height *dh* of the column:

$$ LdCp = K_D a \, (C_p - mC) S dh $$

which on integration and substitution of the following boundry condition gives,

$$ Cp = C_i \qquad \text{at} \qquad h = 0 $$
$$ Cp = C_o \qquad \text{at} \qquad h = H_D $$

$$ K_D a = \frac{L}{V} \ln \frac{C_i - mC}{C_o - mC} $$

The partition coefficients were measured for both BG at the various surfactant concentrations employed.[3]

4.   RESULTS AND DISCUSSION

Effect of surfactant type and its concentration

Figures 2a and 2b show the effect of various cationic surfactants, namely, HTAC, DTAC and HPC on enzyme mass transfer coefficients. It shows that the performance of HTAC and HPC yield mass transfer coefficients of similar magnitude while DTAC results in comparatively higher mass transfer coefficients. This can be explained on the basis of surfactant behaviour in ATPS as follows[11]:

FIGURE 2A: EFFECT OF DISPERSED PHASE VELOCITY, PHASE COMPOSITION AND
TYPE OF SURFACTANT ON ENZYME MASS TRANSFER COEFFICIENT:
SURFACTANT CONCENTRATION 0.12 kg/m³

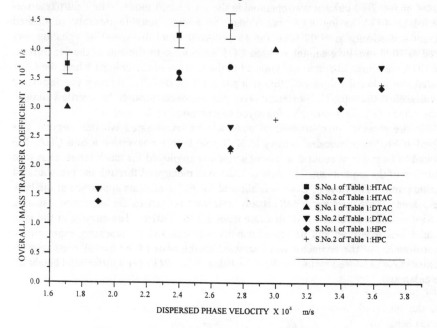

FIGURE 2B: EFFECT OF DISPERSED PHASE VELOCITY, PHASE
COMPOSITION AND SURFACTANT TYPE ON MASS TRANSFER
COEFFICIENT: SURFACTANT CONCENTRATION 0.33 kg/m³

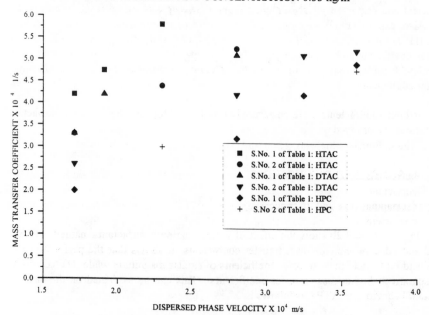

Surfactants present in the PRP offer resistance for enzyme mass transfer from PRP to SRP. It is known that the cationic surfactants prefer PEG rich phase increasingly proportional to their alkyl chain length. Hence, the resistance offered by HTAC and HPC is higher as compared to DTAC owing to their larger alkyl chain length. This results in lower mass transfer coefficients with HTAC and HPC as compared to DTAC.

The increase in mass transfer coefficient observed with increase in surfactant concentration is similar to one observed and explained by Save *et al.*[3]

Effect of dispersed phase velocity

Figures 2a and 2b also show the variation of mass transfer coefficient ($K_D a$) with dispersed phase velocity (U_d). The dependence of mass transfer coefficient on U_d can be described by

$$K_D a \quad \alpha \quad U_d$$

This linear dependence is explained in detail by Save *et al.*[3] which assumes that for a given CGA system, K_D is constant. Further, a is also proportional to the number of aphrons per unit volume, which in the absence of coalescence increases linearly with U_d. Similar behaviour was observed for systems 3, 4 and 5.

Thus, the observed linear dependence of $K_D a$ on U_d agrees with the theoretically expected behavior.

Effect of scale-up

In order to investigate the effect of scale-up on dispersed phase mass transfer coefficient and dispersed phase holdup, column diameter was varied from 30 mm to 100 mm. Also, experiments were carried out at various dispersion height to column diameter ratios (H/D) for each column diameter. Figures 3 and 4 show the effect of H/D on mass transfer coefficient and dispersed phase holdup respectively at various dispersed phase velocities. It is clear that scale-up has no effect on dispersed phase holdup as well as mass transfer coefficient.

5.0   CONCLUSION

The following conclusions are drawn from the above discussion:

i)   mass transfer coefficient increases almost linearly with the alkyl chain length of the surfactant,

ii)   scaleup has no effect on dispersed phase holdup as well as enzyme mass transfer coefficient.

FIGURE 3: EFFECT OF SCALE UP ON DISPERSED PHASE HOLDUP:
S.No. 2 OF TABLE 1

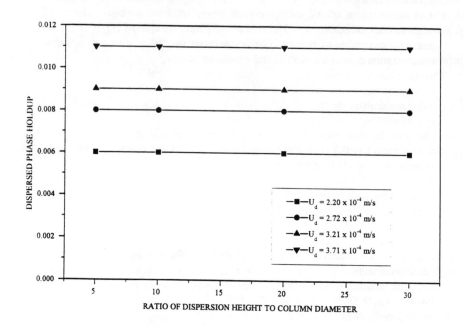

FIGURE 4: EFFECT OF SCALE UP ON ENZYME MASS TRANSFER COEFFICIENT:
S.No. 2 OF TABLE 1

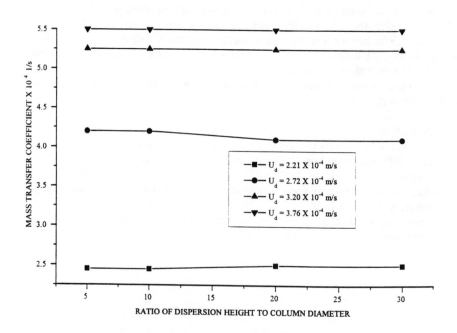

NOMENCLATURE

a	interfacial area m^2/m^3
C_i	inlet concentration of enzyme in PEG rich phase, mg/ml
C_l	concentration of enzyme in salt rich phase, mg/ml
C_o	outlet concentration of enzyme in PEG rich phase, mg/ml
C_u	concentration of enzyme in PEG rich phase, mg/ml
C	average concentration of enzyme in salt rich phase, mg/ml
C_p	average concentration of enzyme in PEG rich phase, mg/ml
D	column diameter, m
d_b	aphron diameter, m
H	height of the dispersion, m
K_D	overall mass transfer coefficient, m/s
L	volumetric flowrate of dispersed phase, m^3/s
m	partition coefficient
S	cross section area of the column, m^2
V	volume of the dispersion, m^3
U_d	superficial dispersed phase velocity, m/s

subscripts

c	continuous phase
d	dispersed phase
l	lower phase (SRP)
u	upper phase (PRP)

REFERENCES

1.H. Walter, D.E. Brooks and D. Fisher (eds.), `Partitioning in Aqueous Two-Phase Systems: Theory, Methods, Uses and Applications to Biotechnology', Academic Press, New York, U.S.A., 1985.

2.J.B.Joshi, S.B. Sawant, K.S.M.S. Raghav Rao, T.A. Patil, K.M.Rostami and S.K. Sikdar, Bioseparation, 1990, 1, 311.

3. S.V.Save, V.G. Pangarkar and S. Vasantkumar, Biotech.Bioeng., 1993, 41, 71.

4. S.V.Save, Ph.D.(Tech.) Thesis, University of Bombay, 1993.

5. S.V.Save and V.G. Pangarkar, Chem.Eng.Comm., 1994, in press.

6. S.V.Save, V.G. Pangarkar and S. Vasantkumar, Sepn. Tech., 1994, in press.

7. S.V.Save and V.G. Pangarkar, J. Chem. Tech. Biotech., 1994, accepted for publication.

8. F.Sebba, `Foams and Biliquid Foams: Aphrons', John Wiley and sons, London, 1989.

9. S.P.Pathak, S. Sudha, S.B. Sawant, and J.B. Joshi, Biochem.J., 1991, 46, B31-B34.9

10. J.R.Ford, A.J. Nunley and Y.J.Li, Anal. Biochem. , 1973, 54, 120.

11. S.V.Save and V.G. Pangarkar, Bioseparation, 1994, accepted for publication.

Modelling the Partition Behaviour of Proteins in Aqueous Two-phase Systems

Anette S. Schmidt and Juan A. Asenjo

BIOTECHNOLOGY AND BIOCHEMICAL ENGINEERING GROUP,
DEPARTMENT OF FOOD SCIENCE AND TECHNOLOGY, THE UNIVERSITY OF
READING, PO BOX 226, READING RG6 2AP, UK

The effect of protein hydrophobicity and charge in aqueous two-phase systems (ATPS) has been investigated. PEG/phosphate, PEG/sulphate, PEG/citrate and PEG/dextran systems in the presence of low (0.6 %w/w) and high level (8.8 %w/w) of NaCl have been evaluated using 12 proteins. The surface hydrophobicity of the proteins was measured by ammonium sulphate precipitation as the inverse of their solubility (1/m*). The hydrophobicity values correlated well with the partition coefficients obtained in the three PEG/salt systems at high concentration of added NaCl. The PEG/citrate system has a higher hydrophobic resolution (H) to exploit differences in the protein's hydrophobicity. The charge of proteins evaluated by electrophoretic titration curves has much less effect on partition than hydrophobicity in PEG/salt systems.

1 INTRODUCTION

ATPS have important potential in down-stream processing as a large-scale continuous operation for the separation of proteins[1,2]. However, the wider implementation in industry has been restrained in part by a lack of models to predict the partition behaviour of proteins. Fundamental theories derived from classical polymer solution thermodynamics are being developed[3], based mostly on PEG/dextran systems. The value of the partition coefficient, K, defined as the ratio of the protein concentration in the top phase to that in the bottom phase, relies on the physico-chemical properties of the target protein and contaminants (e.g. hydrophobicity, charge, molecular weight) and their interactions with those of the chosen system (e.g. composition, ionic strength, addition of specific salt ions, pH).

This paper describes a part of our work on a model to correlate the physico-chemical properties of a protein to its partition coefficient in ATPS using the so-called "modified group contribution" approach based on the simple correlation[1,4]:

$$K = K_{hphob} \cdot K_{el} \cdot K_{size} \cdot K_{sol} \cdot K_{aff} \tag{1}$$

where: K_{hphob}, K_{el}, K_{size}, K_{sol}, and K_{aff} denote the contribution to the overall partition coefficient of hydrophobicity, electrostatic forces, size, solubility, and affinity, respectively. The effect of protein hydrophobicity and charge on their partitioning in a number of selected ATPS including PEG 4000/salt and PEG 4000/dextran systems are discussed in this paper.

2 MATERIALS AND METHODS

2.1. Materials

PEG with a nominal molecular weight of 4,000 (Da) was obtained from Fluka. Dextran T500 was purchased from Pharmacia Diagnostic, Sweden. All other chemicals were analytical grade. α-Amylase, Amyloglucosidase and Subtilisin were obtained from Boehringer Mannheim, Germany. Thaumatin was a gift from Four-F Nutrition, Northallerton, England. α-Chymotrypsinogen A, Bovine Serum Albumin, Conalbumin, Invertase, α-Lactalbumin, Lysozyme, Ovalbumin, and Soybean Trypsin Inhibitor were all purchased from Sigma, UK.

2.2. Preparation of phase systems

Phase systems were prepared from stock solutions of PEG (50 %w/w), sodium/potassium phosphate (40 %w/w), magnesium sulphate (23.3 %w/w), sodium citrate (28 %w/w), dextran (25 %w/w), and NaCl (25 %w/w), stored at 4 °C. The proteins were added at a final concentration of 1 g/l. All partition experiments were carried out at 20 °C and pH 7. Top and bottom phase were assayed for protein concentration.

2.3. Precipitation

All proteins were dissolved in sodium phosphate buffer (0.05 M, pH 7) at a constant concentration of 2 g/l. Solid ammonium sulphate was added slowly to a weighed protein solution while mixing and left to equilibrate for 15 minutes at 24 °C. After centrifugation (14,000 rpm, 20 min) the supernatant was assayed for protein concentration.

2.4. Electrophoretic titration curves

The titration curves were obtained with the PhastSystem (Pharmacia LKB Biotechnology)[5] using PhastGel I.E.F. 3-9 (linear pH gradient from 3 to 9).

2.5. Protein assay

Protein concentration was measured by a modified Lowry assay[6] using a bicinchoninic acid (BCA) kit at 37 °C, supplied by Pierce & Warringer, England, using standard curves for each protein. Blank systems without protein were used as reference.

3 RESULTS AND DISCUSSION

PEG/salt (phosphate, sulphate, citrate) and PEG/dextran at three levels of added NaCl (0, 0.6, 8.8 %w/w) were selected based on their ability to illustrate differences in physico-chemical properties of proteins[7-9]. Proteins were chosen with a wide range of values of hydrophobicity, molecular weight, and isoelectric point (pI) as given in **Table 1.**

3.1. Hydrophobicity

In PEG/salts systems at high NaCl (>7.5 %w/w), hydrophobic proteins such as α-Amylase[7] and Thaumatin[8] partition strongly to the top phase. Although reverse phase chromatography and hydrophobic interaction chromatography also exploit hydrophobicity[10], the obtained hydrophobicity did not correlate well with K in ATPS[4,9]. On the other hand, hydrophobicity measured by precipitation correlated well with the partition in PEG/phosphate systems at high concentration of NaCl[4,9]. To quantify the protein surface hydrophobicity, the point m* at which the protein begins to precipitate was found as described previously[4,10-12]. The maximum solubility of the protein (its hydrophilicity) was defined by this point and hence the surface hydrophobicity <u>in solution</u> as 1/m*[4,9]. The 12 proteins could then be divided into 2 groups[12] with respect to their hydrophobicity, 1/m*. Charge and molecular weight has no significant effect in the measurement of m* and 1/m*[12] as observed by the results in **Table 1**.

Table 1 Protein Molecular weight, isoelectric point and hydrophobicity (1/m*).

Proteins	Molecular Weight [Da]	pI	1/m* [kg/mol]
α-Amylase (Amy)	50,000	4.9	0.484
Amyloglucosidase (Amg)	97,000	3.5	0.321
Bovine Serum Albumin (BSA)	67,000	4.7	0.312
α-Chymotrypsinogen A (Chy)	23,600	8.9	0.544
Conalbumin (Con)	77,000	5.5	0.345
Invertase (Inv)	270,000	3.4	0.339
α-Lactalbumin (Lac)	17,400	5.1	0.353
Lysozyme (Lys)	12,000	11.0	0.418
Ovalbumin (Ova)	43,000	4.9	0.358
Subtilisin (Sub)	27,500	8.4	0.462
Thaumatin (Tha)	22,200	11.5-12.5	0.473
Trypsin Inhibitor (TrI)	24,500	4.6	0.503

The model for predicting the effect of protein hydrophobicity on the partition behaviour in ATPS[12] based on that of Eiteman and Gainer[3] for small molecules and peptides is shown in equation (2):

$$\log K_{hphob} = H \cdot \log (1/m^*) - H \cdot \log (P_0) \qquad (2)$$

where log 1/m* and log P_0 represent the solute's hydrophobicity and the intrinsic hydrophobicity of the given ATPS, respectively. The hydrophobic resolution (H) describes the ability of the system to discriminate between two solutes of different hydrophobicity. The values of log 1/m* were plotted against the partition coefficient (log K) in the 4 different systems at the 3 different levels of NaCl[12], as shown for PEG/phosphate in **Figure 1**.

The ATPSs were classified by the constants in equation (2) as given in **Table 2**. An extremely good correlation was obtained in all PEG/salt systems at 8.8 %w/w NaCl concentration. At lower NaCl and in PEG/dextran systems the linear fit was weaker due to the effect of other physico-chemical properties like charge and molecular weight. A remarkably good correlation was also achieved in the PEG/citrate system at low NaCl. The resolution was generally higher in PEG/citrate systems, indicating that they are better able to separate proteins by their hydrophobicity than other systems. For all PEG/salt systems the resolution increased as NaCl concentration increased. Surprisingly, the resolution in PEG/dextran systems was comparable to PEG/salt systems at low NaCl. When the hydrophobicity of the protein equals the hydrophobicity of the system (log 1/m*

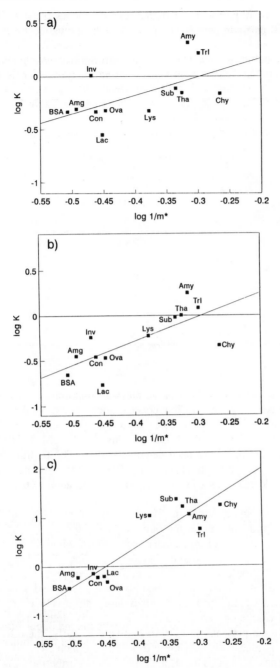

Figure 1 The relationship between partition coefficient (log K); in PEG(13 %w/w)/ phosphate (10.7 %w/w) systems with **a)** 0 %w/w (r = 0.59), **b)** 0.6 %w/w (r = 0.74), and **c)** 8.8 %w/w NaCl (r = 0.92); and the hydrophobicity (log 1/m*). Abbreviations are given in **Table 1**.

= log P_0) the protein will partition equally between the phases (K = 1). If log 1/m* > log P_0, proteins will partition preferably to the top and *vice versa*.

Table 2 The hydrophobic resolution (H) and the intrinsic hydrophobicity (log P_0) of the systems with varying NaCl concentration at pH 7 and the correlation coefficient (r).

Systems:	NaCl [%w/w]	H	log P_0	r
PEG/Phosphate	0	1.72	-0.30	0.59
	0.6	2.68	-0.29	0.74
	8.8	7.91	-0.45	0.92
PEG/Sulphate	0	1.13	0.05	0.24
	0.6	1.98	-0.06	0.43
	8.8	8.51	-0.44	0.92
PEG/Citrate	0	2.06	-0.31	0.81
	0.6	3.78	-0.31	0.86
	8.8	11.47	-0.42	0.93
PEG/Dextran	0	0.88	-0.34	0.40
	0.6	2.76	-0.33	0.62
	8.8	2.49	-0.46	0.6

3.2. Charge

In addition to hydrophobicity, charge plays an important role in the partition of proteins. In PEG/dextran systems, at low concentration of added salts and by changing the pH, the effect of the electrostatic potential across the interface of the two phases, caused by small differences in the affinity of ions for the phases, and the sign and magnitude of net charge on K has been demonstrated[1,13]. The partition of proteins has been correlated with the partition of salts in PEG/dextran systems[13]. For negatively charged proteins, K decreases as a function of NaCl (up to about 0.6 %w/w NaCl), and *vice versa*[1,13].

In PEG/salt systems a change in pH and charge of proteins can affect K[4,7]. Due to the high concentration of salt in the bottom phase, the pH of the top phase will determine partitioning due to electrostatic effects. Electrophoretic titration curves for proteins have proven useful for charge evaluation to predict ion exchange behaviour[14,15]. The electrophoretic titration curve reveals the mobility of the protein and hence its charge throughout a pH gradient[15]. The protein's mobility is closely related to and used as its net charge[15]. As pH increases the protein becomes more negatively charged. In order to isolate the effect of charge on partition, log (K/K_{hphob}) was plotted using K_{hphob} from the correlations given in the previous section, in **Figure 2**. Clearly, the effect of charge on K in these systems is much less than the effect of hydrophobicity, also the correlation coefficients are not nearly as good as for hydrophobicity. Furthermore, in the absence of NaCl, the value of K decreases with charge; with 0.6 %w/w NaCl K is almost constant with charge, and with 8.8 %w/w NaCl it shows an increase. The fact that a better correlation is not obtained with log (K/K_{hphob}) must reflect the effect of other factors in equation (1), such as molecular weight and possible affinity for the PEG. These factors

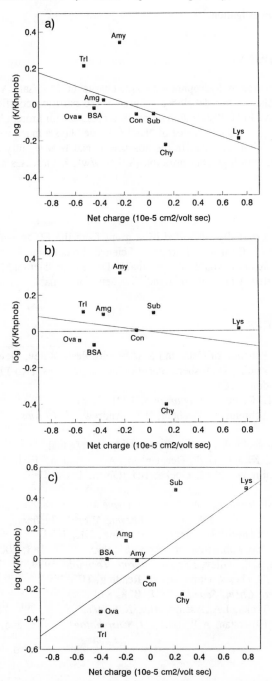

Figure 2 The relationship between partition coefficient (log K/K_{hphob}); in PEG/phosphate systems with **a)** 0 %w/w (r = 0.58), **b)** 0.6 %w/w (r = 0.26), and **c)** 8.8 %w/w NaCl (r = 0.73); and the protein net charge. Abbreviations are given in **Table 1**.

are presently under investigation.

4 CONCLUSIONS

The effect of protein hydrophobicity and charge in PEG/salt ATPS has been investigated. A correlation was developed to predict the partition coefficient based on protein hydrophobicity. In all PEG/salt systems, K increases with protein hydrophobicity and this effect is stronger in the presence of NaCl. Charge plays a much weaker role than hydrophobicity in PEG/salt systems. In the absence of NaCl, K slightly decreases with increased charge, and with high concentration (8.8 %w/w), K increases as a function of charge.

Acknowledgement

This work was partially supported by a grant from the EC to whom thanks are due. Financial support from the European Science Foundation is also gratefully acknowledged. The authors would like to thank Dr. Folke Tjerneld and Dr. Göte Johansson, Biochemistry, University of Lund, Sweden, for valuable discussions.

5 REFERENCES

1. P.Å. Albertsson, 'Partition of Cells and Macromolecules', Wiley, New York, 1986.
2. H. Walter, D.E. Brooks, D. Fisher, 'Partitioning in Aqueous Two-Phase Systems', Academic Press, London, 1985.
3. M.A. Eiteman, J.L. Gainer, *Bioseparation*, 1991, **2**, 31.
4. J.A. Asenjo, A.S. Schmidt, F. Hachem, B.A. Andrews, *J. Chromatogr.*, 1994, **668**, 47.
5. Pharmacia LKB Biotechnology. PhastSystem™ Users Manual, 1991.
6. P.K. Smith, R.I. Krohn, G.T. Hermanson, A.K. Mallia, F.H. Gartner, M.D. Provenzano, E.K. Fujimoto, N.M. Goeke, B.J. Olson, D.C. Klenk, *Anal. Biochem.*, 1985, **150**, 76.
7. A.S. Schmidt, A.M. Ventom, J.A. Asenjo, *Enzyme Microb. Technol.*, 1994, **16**, 131.
8. O. Cascone, B.A. Andrews, J.A. Asenjo, *Enzyme Microb. Technol.*, 1991, **13**, 629.
9. F.M. Hachem, PhD Thesis, University of Reading, UK, 1992.
10. W. Melander, C. Horvath, *Arch. Biochem. Biophys.*, 1977, **183**, 200.
11. T.M. Przybycien, J.E. Bailey, *Enzyme Microb. Technol.*, 1989, **11**, 264.
12. A.S. Schmidt, PhD Thesis, University of Reading, UK, 1994.
13. G. Johansson, *Acta Chem. Scand.*, 1974, **B28**, 873.
14. E. Watanabe, S. Tsoka, J.A. Asenjo, *Ann. N.Y. Acad. Sci.*, 1994, **721**, 348
15. L.A. Haff, L.G. Fägerstan, A.R. Barry, *J. Chromatogr.*, 1983, **266**, 409.

Rapid and Effective Method of Myeloperoxidase (MPO) Recovery

V. N. Senchenko and S. V. Yarotsky[a]

ENZYMES STEREOCHEMISTRY LABORATORY, ENGELHARDT INSTITUTE OF
MOLECULAR BIOLOGY, RUSSIAN ACADEMY OF SCIENCES, MOSCOW
117984, RUSSIA

[a]NATIONAL RESEARCH CENTRE FOR ANTIBIOTICS, MOSCOW 113105,
RUSSIA

1 INTRODUCTION

Myeloperoxidase [donor: H_2O_2 oxidoreductase, EC 1.11.1.7.] is
a component of the oxygen–dependent system of polymorphonuclear
neutrophils.[1] In the presence of hydrogen peroxide and halo-
gen ions MPO has the ability to produce hypohalogenite ions.
These are responsible for rapid oxidative degradation of a
wide variety of biological objects including bacteria, fungi,
viruses etc.[2-3]

MPO is a heme–containing glycoprotein with a molecular
mass in the region of 120–160 kDa and consists of two heavy
and two light subunits.[4] MPO has been prepared from several
sources, including normal human leukocytes.[4-10] Previously
described MPO isolation methods included cation–exchange,
gel–filtration or affinity chromatography and sometimes
hydrophobic interaction in various order. As a rule, enzyme
yield wasn't more than 40% of initial activity. Obviously,
MPO properties are suggestive of both scientific and pharmaceutical significance.[10-13] However, MPO of animal origin
is potentially biologically hazardous in causing side effects
by antigen–antibody reactions.[10] MPO of human origin is
extremely difficult to obtain in large quantities due to the
limited source. Under these circumstances, an object of the
present study is to develop a method for human MPO recovery
with high yield.

2   RESULTS AND DISCUSSION

2.1 Extraction and primary fractionation

Polled buffy coats from 1 litre of normal human blood (or with elevated bilirubin content) were twice washed in hypo-osmotic solution to lyse contaminating erythrocytes.[7] White cells were suspended in 5 vol. of 0,05 M potassium phosphate buffer, pH 7,0, containing 0,5% cetyltrimethylammonium bromide (CETAB),[4] then 2 g. glass beads (size : 75-100 microns) were added. Cell disruption was carried out by a motor-driven homogenizer at $-20^{o}C$, 2000 rev/min, 5 min. After centrifugation (30.000xg, 30 min) this procedure was repeated with pellet and the green supernatants were combined. The initial fractionation step involved ammonium sulfate precipitation of MPO from solution after extraction. The extract was adjusted with solid $(NH_4)_2SO_4$ to 50% saturation, stirred for 1 h. with cooling and separated (20.000xg, 20 min.) to remove precipitate. Then MPO was precipitated at 70% saturation with $(NH_4)_2SO_4$, pellet was collected by centrifugation and dissolved in 0,05 M potassium phosphate buffer, pH 7,0.

2.2 Hydrophobic interaction chromatography

The solution from the previous step was saturated by $(NH_4)_2SO_4$ up to 30% and applied to a column (10x150 mm) with

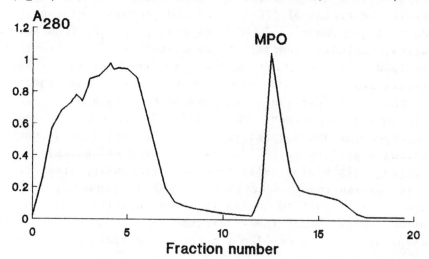

Figure 1 *Purification of MPO by hydrophobic interaction chromatography*

Butyl-Toyopearl 650 M ("TosoHaas"). The column was first equilibrated with 0,05 M potassium phosphate buffer, pH 7.0, containing 30% of $(NH_4)_2SO_4$. The column was washed with several volumes of the same buffer. MPO was eluted with 20% solution of $(NH_4)_2SO_4$ in 0,05 M potassium phosphate buffer, pH 7,0 (flow rate 100 ml/h). The active fractions were combined. This step allowed rapid and efficient MPO separation; most concomitant greenish-brown admixtures were retained in the column when MPO was eluted (Figure 1). After each operation the column was washed out according to the manufacturer's recommendations.

2.3 Affinity chromatography

It is well known that MPO is able to bind with concanavalin A due to its carbohydrate chains. Con A – Sepharose 4B was prepared using cyanogen bromide-activated Sepharose 4B ("LKB-Pharmacia") according to the usual procedure and equilibrated with 0,1 M sodium acetate, pH 6,0, containing 0,1 M NaCl and 1 mM $MgCl_2$. The addition of 0,1 vol. of 1,0 M sodium acetate, pH 6,0, 1M NaCl and 10 mM $MgCl_2$ was allowed to settle the fraction with MPO activity and applied to the equilibrated column (10x50 mm). The column was washed with several column volumes of the equilibrating buffer. MPO was further recovered by elution with 0,1 M sodium acetate, pH 4,7, 0,1 M NaCl and 0,5 M α-methyl-D-glucopyranoside (or α-methyl-D-mannopyranoside).

Except where noted, the procedures were carried out at room temperature.

The final preparation (about $7,0x10^4$ units, 2,7 mg of protein) was free from CETAB. The enzyme can be transferred into suitable buffer solution by dialysis and frozen or freeze-dried. Both frozen and freeze-dried preparations can be stored for a long time without loss of activity. Typical results of the MPO purification procedure are summarized in Table 1.

Table 1 *Purification of human MPO from 1 litre of blood*

Purification step	Total protein mg	Total activity (units)	Specific activity (units/mg)	Yield %	R_z
CETAB extract	78,9	87000	1000	100	0,1
$(NH_4)_2SO_4$	48,4	82000	1700	95	0,3
Butyl-Toyopearl 650 M	6,8	73000	11000	84	0,5
Con A-Sepharose	2,7	69600	25800	80	0,8

At each step of the purification, fractions were assayed for both protein concentration[14] and MPO activity using guaiacol as substrate.[15] The purity of the final enzyme was determined by native and SDS-PAGE electrophoresis[7,16] and UV-visible spectra.[4,7] The method provides a possibility for preparation of highly purified human MPO ($R_z=A_{430}/A_{280}=0,8$)[7] with maximal yield (80% of total activity). The variations in total blood formula don't influence the procedure results. The leukocyte fractions remaining after isolation of the traditional blood products are also suitable for MPO preparation.

Thus, the MPO purification procedure involves only three steps: 1) ammonium sulphate fractionation, 2) hydrophobic interaction, 3) affinity chromatography, without any additional manipulations. A similar sequence of steps with several differences allows the recovery MPO from porcine blood leukocytes with high efficiency. This may be of special interest due to the resemblance of many corresponding human and porcine proteins, for example, insulins of both origins.

REFERENCES

1. S.J.Klebanoff, *J.Bacteriol.*, 1968, **95**, 2131.
2. R.A.Clark and S.J.Klebanoff, *J.Clin.Invest.*, 1979, **64**, 913.
3. J.M.Albrich, C.A.McCarthy and J.K Hurst, *Proc.Natl. Acad.Sci.USA*, 1981, **78**, 210.
4. Y. Morita, H.Iwamoto, S.Aibara, T.Kobayashi and

E.Hasegawa, *J.Biochem.*, 1986, **99**, 761.

5. H.Iwamoto, Y.Morita, T.Kobayashi and E.Hasegawa, *J.Biochem.*, 1988, **103**, 688.

6. M.R.Andersen, C.L.Atkin and H.J.Eyre, *Arch.Biochem. Biophys.*, 1982, **214**, 273.

7. R.L.Olsen and C.Little, *Biochem.J.*, 1983, **209**, 781.

8. R.L.Olsen and C.Little, *Biochem.J.*, 1984, **222**, 701.

9. B.E.Svensson, K.Domeij, S.Linvall and G.Rydell, *Biochem.J.*, 1987, **242**, 673.

10. E.Hasegawa and T.Kobayashi, *US Patent*, **4306025**, 1981.

11. E.W.Odell and A.W.Segal, *Biochim.Biophys.Acta*, 1988, **971**, 266.

12. C.E.Cooper and Odell, *FEBS Lett.*, 1992, **314**, 58.

13. R.A.Cuperus, H.Hoogland, R.Never and A.O.Muijsers, *Biochim.Biophys.Acta*, 1987, **912**, 124.

14. M.M.Bradford, *Anal.Biochem.*, 1976, **72**, 248.

15. S.O.Pember, R.Shapira and J.M.Kinkade, *Arch.Biochem. Biophys.*, 1983, **221**, 391.

16. U.K.Laemmli, *Nature*, 1970, **277**, 680.

Simple High Level Purification of Two Related Proteins, Interleukin-1β and Interleukin-1 Receptor Antagonist, Produced by Recombinant *E. coli*

O. M. P. Singh,[a] K. P. Ray,[b] M. A. Ward,[c] D. J. MacKay,[d]
B. S. Baines,[e] A. R. Bernard,[f] and M. P. Weir[a]

[a]DEPARTMENTS OF BIOMOLECULAR STRUCTURE, [b]CELLULAR AND
MOLECULAR SCIENCES, [c]STRUCTURAL CHEMISTRY, [d]INTERNATIONAL
REGULATORY AFFAIRS, AND [e]NATURAL PRODUCTS DISCOVERY, GLAXO
RESEARCH AND DEVELOPMENT LTD., GREENFORD ROAD, GREENFORD,
MIDDLESEX UB0 4HE, UK

[f]GLAXO INSTITUTE OF MOLECULAR BIOLOGY SA, 14 CHEMIN DES AULX,
CASE POSTALE 674, 1224-PLAN-LES-OUATES, GENEVA, SWITZERLAND

INTRODUCTION

The interleukin 1 group of proteins, produced by activated macrophages /monocytes, have a major role in mediating immune and inflammatory responses. Interleukin-1α and β (IL-1α and β) are the two agonists, able to elicit the full spectrum of cellular activities such as modulation of prostaglandin E_2 , interleukin-8, interleukin-2 receptor α subunit and intercellular adhesion molecule (ICAM-1) production. The interleukin-1 receptor antagonist (IL-1ra), by competing in the type I & II receptor occupancy by both interleukin-1α and β, exerts a powerful control in maintaining the *status quo* [1].

In order to study the role of these molecules in disease, demands of pure protein requirement are such that recombinant technology is the only logical way to obtain sufficient material. Consequently, we have expressed both of these proteins in *E.coli*. Both of these proteins have been cloned and expressed in *E. coli* before [2,3]. Our expression system for IL-1β has been described [4] and the procedure for expressing the IL-1ra was essentially similar to that used for the expression of interleukin-5 [5] except IL-1ra gene replaced the interleukin-5 gene. The IL-1β was produced in *E.coli* entirely as soluble protein recoverable from cytoplasmic extract whereas the IL-1ra protein was produced both as soluble and insoluble inclusion bodies in about equal proportion. The expression level for both proteins was near 5% of cell protein.

Purification strategy: In order to effect rapid and high level recovery of these proteins the purification strategy utilised conventional chromatography namely ion-exchange, hydrophobic interaction and gel permeation. The number of steps required to achieve high level purity was kept to a minimum so as to avoid losses, consequently dialysis and precipitation of the proteins were not considered. In the case of the receptor antagonist purification of only the soluble protein was attempted.

Methods: Fermentation conditions for both expression systems were as reported previously [4,5]. Essentially, the IL-1β expression was induced by the addition of 1mM IPTG to the media in 1L shake flasks and harvested once the cell density resulted in the OD_{550nm} of the media reaching 2.5. The cells were harvested by centrifugation of the media at 10,000g for 20min at 4°C. The IL-1ra protein expression was induced by raising the fermentation temperature from 30 to 42°C and the cells were harvested as before. Lysis of interleukin-1β containing cells was performed in 25mM Tris HCl buffer, pH 7.4, containing 5mM benzamidine, 1mM EDTA, 1mM DTT and 0.1mM sodium azide. Interleukin-1 receptor protein containing cells were lysed in 20mM Tris HCl buffer, pH 8.0 containing 2mM EDTA, 5mM DTT and 5mM benzamidine. In both cases cell lysis was effected by sonication of the 10% (wet weight) slurry using the Branson continuous flow sonifier (model 450) and the cellular debris was removed by centrifugation at 10,000g for 20min at 4°C .

After clarification, 2ml of the extract containing IL-1β was loaded onto a monoQ column (HR5/5; Pharmacia UK) equilibrated with 10mM Tris HCl buffer, pH 7.4, containing 1mM EDTA. Interleukin-1β protein did not bind to this matrix and eluted following the first protein peak after 50% of the column volume wash. The fractions containing the protein were identified from SDS-page gels which were run using the phast gel system (Pharmacia, UK) and silver stained [6]. All other chromatography steps were similarly analysed. The pH of the pooled sample containing protein was adjusted to 5.0 by the addition of acetic acid. This was then loaded onto a monoS column (HR5/5; Pharmacia UK) which had been equilibrated with 20mM Tris-acetate buffer pH 5.0, containing 1mM EDTA. Elution of the protein was effected by running a shallow NaCl gradient (0-200mM) in 15 column volumes.

The extract from the IL-1ra was loaded onto a monoQ column (HR5/5, Pharmacia, UK) with the same buffer containing EDTA only. After removing the unbound material with a 5ml buffer wash, the protein was eluted with a 0-300mM NaCl gradient over 30 column volumes followed by a further wash with NaCl containing buffer. The antagonist protein eluted with 80mM NaCl. The pooled material was diluted to 50% with 2M $(NH_4)_2\,SO_4$ in the same buffer as before and loaded onto a phenyl-Sepharose column (HR5/5, Pharmacia, UK). After washing the column with 5ml of buffer containing 1M $(NH_4)_2\,SO_4$ the protein was eluted with reverse gradient of $(NH_4)_2SO_4$ from 1.0 to 0M. The major protein peak contained all of the antagonist protein. The pooled material was applied to a 100ml Superdex G70 column equilibrated with Tris HCl buffer containing only 200mM NaCl.

Large scale production of both proteins involved using the Biopilot system and column sizes were increased appropriately to accept 500g wet weight cell extract. The cells were cultured in 50L fermenters and grown to OD_{550nm} of 10-20.

Protein and biological function analysis: Edman sequencing of both proteins was carried out using an Applied biosystems model 476A instrument following the manufacturers instructions . Buffer components were removed from samples for mass spectroscopy by reverse phase chromatography using ProRPC column (HR5/5, Pharmacia, UK) where the sample was rapidly eluted with 0-100% gradient of 0.1% CH_3COOH in CH_3CN, lyophilised and then dissolved in

50%CH_3CN containing 0.1%CH_3COOH. This was then infused into an electrospray
ionisation source fitted with a Finnigan MAT TSQ-700 triple quadrupole mass
spectrophotometer. Immunoassay of both proteins was performed using the Cistron
Kit for IL-1β and the quantakine kit (British Biotechnology, UK) for IL-1ra. Isoelectric focusing (IEF) was performed using pH 5-8 range Phast gels (Pharmacia,
UK) which were fixed with 20% trichloroacetic acid and stained (6).
        Prior to biological assay the endotoxin level in the protein sample was
reduced by chromatography using detoxigel (Pierce Warriner, UK) following
manufacturers instructions. The biological activity of IL-1β was determined as
described [7]. Essentially the protein was incubated with EL4 cells (C57/BL/6
mouse thymoma cell line) which produce IL-2 in response. The IL-2 was measured
from the resulting conditioned media by measuring enhancement in the [3]H-
thymidine uptake by the CTLL cells [cytotoxic T cell line; 8]. The biological
activity of the IL-1ra was examined from its ability to antagonise the induction of
interleukin-8 in the pulmonary epithelial cell line, A549 [9] and the IL-2 receptor α
chain by the HSB.2 cells [7] by IL-1α or β.

        Results & Discussion: The two step procedure described for the
isolation of the IL-1β utilised only two chromatography steps namely anion and
cation -exchange. As described the interleukin-1β protein did not bind to the
MonoQ matrix. However under these conditions a considerable amount of the
E.coli proteins bound to the matrix and were removed by washing the column with
buffer containing 0.5M NaCl. After adjusting the pH to 5.0 the post monoQ
material was applied to the monoS column. Under these conditions the protein
bound to the matrix and it could also be recovered by applying a very shallow NaCl
gradient. The protein appeared to elute in three distinct peaks (See figure 1)
followed by some *E.coli* components. SDS-page analysis of these showed that all
three contained polypeptide of about 17.5Kda in size.

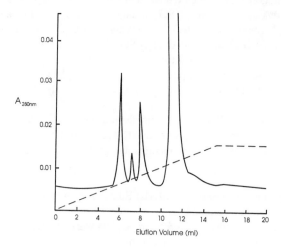

Figure 1. Elution profile of the cation exchange chromatography for the purification of IL-1β. The dashed line represents the NaCl gradient (0-0.2M) followed by isocratic elution with 0.2M NaCl in buffer as described in text. The first three u.v. absorbing peaks contain IL-1β.

Immunoassay confirmed that this material was indeed interleukin-1β. IEF analysis of the three samples indicated that this chromatography step had resulted in the resolution of three distinct protein species. The pI value of the major species was near 6.9 to 7.0, the next species pI value was near 6.7 to 6.8 and the minor species separated as a diffuse band. Edman sequencing of the two species confirmed the presence of initiating methionine with the species with pI of 6.7-6.8 whereas the major protein band was entirely free of initial methionine. The biological assay confirmed that this protein displayed all the functions attributed to interleukin-1β (specific activity 2-4x 10^8u/mg of protein), however the methionine plus species was consistently about five fold weaker in its receptor affinity. This was also reflected in the biological assays (specific activity 4-6x10^7u/mg of protein) in agreement with previous reports [10] and also confirming the view that the N-terminal region of this protein may be an important determinant in receptor binding as indicated from the various mutational and epitope mapping studies [11,12].

Previous reports on the purification of interleukin-1β [13,14] have reported multistep procedures yielding high purity material showing all the biological properties expected. However, the protein so isolated was found to be a mixture containing at least three species with two major ones resulting from incomplete processing of the initiating methionine. Further separation using chromatofocusing appeared to resolve this mixture [15].

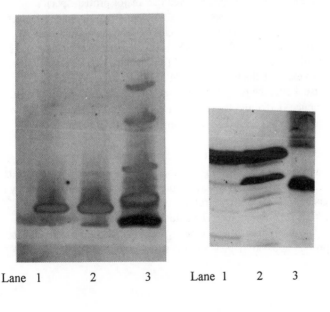

Lane 1   2   3   Lane 1   2   3

figure 2a   figure 2b

Figure 2. SDS-page separation of the pooled IL-1β after chromatography using the monoS matrix. Figure 2a Lane 1 and 2 show doubling protein loading and lane 3 shows the separation of low molecular weight protein standards (Pharmacia, UK). Iso-electric focusing profile of the pooled IL-1β fraction is seen in figure 2b, lane 1, standard IL-1β protein in lane 2 as purified by another group (see ref.13) and lane 3 shows the separation of IEF standards used to calibrate the pH range of the gel.

A later report [16] described the use of two ion exchange steps followed by gel permeation to resolve the two major species described and also the desAla form. Examination of the published theoretical titration curve of the protein appears to indicate differences in the ionisation of native and met plus species only in the pH range of 6-9 and that at pH 5.0 both species behave similarly [15]. However the results from our monoS chromatography appear not to agree with this, since the proteins must be differently ionised to allow this separation. The different ionisation properties of the methionine and alanine at the N-terminus of interleukin-1β must impose sufficient difference in the property of these protein to allow their separation. ^{1}H-nmr spectra of both species appear identical thus there would appear to be no gross differences in the structural features [10] but local effects cannot be ruled out. Figure 2 shows the results from SDS-polyacrylamide gel electrophoresis our pooled N-terminus methionine processed protein and the comparison of the IEF separation pattern with protein purified by another group [13].

The isolation of interleukin-1 receptor antagonist using ion exchange, hydrophobic interaction and gel permeation chromatography (see figure 3) resulted in protein which was over 95% pure as judged by both SDS-page and iso-electric focussing with a pI value of 5.4 to 5.5. A minor contaminant with pI of 5.2 was also noted. Mass spectroscopic analysis showed that major protein species contained unprocessed methionine (mass of 17324Da) and that the minor species was fully processed (mass of 17257Da). Ten cycles of Edman degradation confirmed that the sequence was MRPSGRKSSK, no secondary sequences were detected thus also confirming the high purity of this protein.

In order to examine the biological role of these molecules in any inflammatory disease model the protein should be free of bacterial endotoxins as these themselves induce IL-1α and IL-1β from activated macrophages [17]. The detoxigel chromatography reduced the level from 5-10ng/ml to below detection (0.1ng/ml).

Biological function assays of IL-1β indicated that this protein could induce IL-2 by the EL4 cells , interleukin-8 by the A549 cells and IL-2 receptor α chain by the HSB.2 cells. All of these assays could be specifically antagonised by the IL-1ra protein thus confirming that the proteins we have produced using recombinant *E.coli* are fully functional. The natural IL-1ra isolated from urine is glycosylated [18], it would appear that the biological function of the recombinant unglycosylated protein does not depend on the sugar moieties.

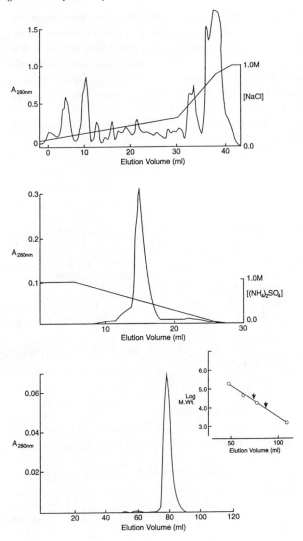

Figure 3. Chromatograms of the elution pattern from mono-Q , phenyl-Superose and Superdex G75 matrices. IL-1ra protein was eluted from the mono Q column with approximately 100-120mM NaCl (top figure), elution of the phenyl-Superose matrix with a reverse gradient, 1.0 to 0M $(NH_4)_2SO_4$ resulted in the protein eluting in the main protein peak (see middle figure) and finally the gel permeation chromatography elution profile was consistent with the protein having a molecular mass of near 17.5KDa. Inset shows calibration curve obtained when a standard mixture of proteins is separated under identical conditions. The arrows mark the position of boundary for the IL-1ra protein elution from the gel permeation column.

In summary we report the development of procedures for the isolation of two related cytokine proteins, IL-1β and IL-1ra. The methods allow direct scale up of the isolation procedure and indeed we have been able to purify 500-600mg of high quality protein from 500g of cells with below detection level contamination by bacterial endotoxins. Full biological spectrum of activities is displayed by these proteins. Thus, using the procedures we have described how large quantities of these proteins can be produced to understand the diverse roles of IL-1 in disease and in health and allow structural studies to be undertaken to see whether they may shed light towards means of intervention in disease.

References :

1. C.A. Dinarello & S.M. Wolff, N Engl J Med., 1993, **328** . 106-113
2. P.E. Auron, A.C. Webb, L.J. Rosenwasser, S.F. Mucci, A. Rich, S.M. Wolff & C.A. Dinarello, Proc. Natl. Acad. Sci. U.S.A., 1994, **81**, 7907-7911
3. S.P. Eisenberg, R.J. Evans, W.P. Arend, E. Verderber, M.T. Brewer, C.H. Hannum & R.C. Thompson, Nature, 1990, **343**, 341-346
4. D.C. Humber , M. Allsopp, B.A. Coomber, L.G. D'Urso, C.W. Dykes, A.N. Hobden, M. Mitchell & S.A. Noble, Nucleosides & Nucleotides, 1987, **6**, 413-414
5. A.E .Proudfoot, D. Fattah, E.H. Kawashima , A. Bernard & P.T. Wingfield, Biochem. J., 1990, **270**, 357-361
6. J.H. Morrissey, Anal. Biochem., 1981, **117**, 307-310
7. N. Smithers, PhD Thesis, 1990, Brunel University, London
8. S. Gillis & K.A. Smith, Nature, 1977, **266**, 154-156
9. T.J. Standiform , S.L. Kunkel, M.A. Basha , S.W. Chensue, J.P. Lynch III, G.B. Toews, J. Westwick & R.M. Strieter, J.Clin. Invest.,1990, **86**,1945-1953
10. P.T. Wingfield , P. Graber , N.R. Movva , A.M. Gronenborn & H.R. MacDonald, FEBS Letts., 1987, **215**, 160-164
11. F. Guinet, J-D .Guitton, N. Gault, F. Folliard, N. Touchet, J-M. Cherel, A.Crespo, A. Destourbe, P. Bertrand, P. Denefle, J-F. Mayaux, A. Bousseau, M. Duchesne, B. Terlain & T. Cartwright, Eur.J. Biochem., 1993, **211**, 583-590
12. E. Labriola-Topkins, C. Chandran, K.L. Kaffka, V.S. Biondi, B.J. Graves, M. Hatada, V.S. Madison, J. Karas, P.L. Killian & G. Ju, Proc. Natl. Acad. Sci. U.S.A., 1991, **88**, 11182-11186
13. P.T. Wingfield, M. Payton, J. Tavernier, M. Barnes, A. Shaw, K. Rose, M.G. Simona, S. Demczuk, K. Williamson & J-M. Dayer, Eur. J. Biochem., 1986, **160**, 491-497
14. C.A. Meyers, K.O. Johanson, L.M. Miles, P.J. McDevitt, P.L. Simon, R.L. Webb, M-J. Chen, B.P. Holskin, J.S. Lillquist & P.R. Young, J. Biol. Chem., 1987, **262**, 11176-11181
15. P.T. Wingfield, P. Graber, K. Rose, M.G. Simona & G.J. Hughes, J. Chromatography, 1987, **387**, 291-300

16. Y. Kikumoto, Y-M. Hong, T. Nishida, S. Nakai, Y. Masui & Y. Hirai, Biochem. Biophys. Res. Communs., 1987, **147**, 315-321

17. C. A. Dinarello, Blood , 1991, **77**, 1627-1652

18. G.J. Mazzei, P.L. Seckinger , J-M. Deyer & A.R. Shaw, Eur. J. Immunol., 1990, **20**, 683-689

Acknowledgements: We would like to thank Ms Sarah Witham, Mrs Nicki Searle and Dr Nicholas Smithers for the biological assay support. Dr Allan shaw (currently at Merck Sharpe & Dohme Research Laboratories, West Point, USA) for initiating the expression work on the IL-1ra, Dr Richard Dennis for guidance with the mass spectroscopic analysis, Dr Gerardo Turcatti (Glaxo Institute of Molecular Biology, Geneva) and Ms Shila Patel for their sterling efforts to sequence our protein samples. Mr Pierre Graber and Dr Paul Wingfield (Glaxo Institute of Molecular Biology, Geneva) are thanked for supplying a sample of purified IL-1β.

Novel Boronophthalide-based Polymers for the Selective Separation of Sterols

C. R. Smith, M. J. Whitcombe, and E. N. Vulfson

BBSRC INSTITUTE OF FOOD RESEARCH, READING LABORATORY, EARLEY GATE, WHITEKNIGHTS ROAD, READING RG6 2EF, UK

ABSTRACT

A new polymerizable derivative of boronophthalide has been prepared and its use as a functional monomer in the molecular imprinting of sterols has been demonstrated. Cooperativity in binding has been shown by a comparison of polymers imprinted with androst-5-en-3β,17β-diol, androst-5-en-3β-ol and androst-5-en-17β-ol.

1. INTRODUCTION

The application of principles of molecular recognition to the preparation of highly specific adsorbents and chromatographic materials has attracted much attention in recent years. The synthesis of specific adsorbents, using a template-mediated approach, is termed **Molecular Imprinting**[1] and the products are known as **Imprinted Polymers**. Materials of this type have been investigated as HPLC stationary phases with the aim of performing highly selective chiral separations.[2-4] In chromatographic procedures however, the binding groups within the recognition site should be capable of undergoing a very rapid and reversible interaction with the ligand. The interaction between boronic acids and hydroxyl groups (formation of boronate esters[5,6]) evidently satisfies the above criteria and polymers containing aryl boronic acids have been successfully used for the separation of monosaccharides and their derivatives. For example, by incorporating a monosaccharide diester of 4-vinylphenylboronic acid (1) into the polymer, it has been possible to create chiral cavities possessing two boronic acid groupings, with the ability to resolve racemates of the original template molecule.[5,6]

(1)

Following our interest in the preparation of adsorbents specific to sterols,[7] we initially investigated the usefulness of 4-vinyl phenylboronic acid (1) in the preparation of polymers specific to various mono-hydroxy steroids. The boronic acid was synthesized according to the method of Kamogawa and Shiraki[8] from 4-bromostyrene by the preparation of the Grignard reagent and its subsequent treatment with trimethyl borate. However it was found that this methodology, although perfectly applicable to compounds possessing a diol functionality, did not generally provide the specificity required for the separation of molecules containing non-adjacent hydroxyl groups. No difference between the binding abilities of sterol-imprinted and non-imprinted polymers was observed.

2. BORONOPHTHALIDE-BASED POLYMERS

Boronophthalide (5) is known to be a remarkably stable boronic acid derivative.[9] This compound, being more suitable for interactions involving only one alcohol functionality, was therefore an attractive alternative to arylboronic acids. It has an added advantage over arylboronic acids of being relatively stable to oxidation and in the presence of strong acids and bases.

Grignard formation with 2-bromotoluene (2), followed by reaction with trimethyl borate gave 2-methylphenylboronic acid (3). This reacted with bromine under the influence of light to give (4) which was subjected to alkaline hydrolysis to yield the required boronophthalide (5).[8] Ester formation of boronophthalide with various sterols was verified using NMR spectroscopy. The addition of a drying agent (4Å molecular sieves) to solutions of boronophthalide and sterol (1:1 molar ratio) resulted in almost complete conversion to the ester.

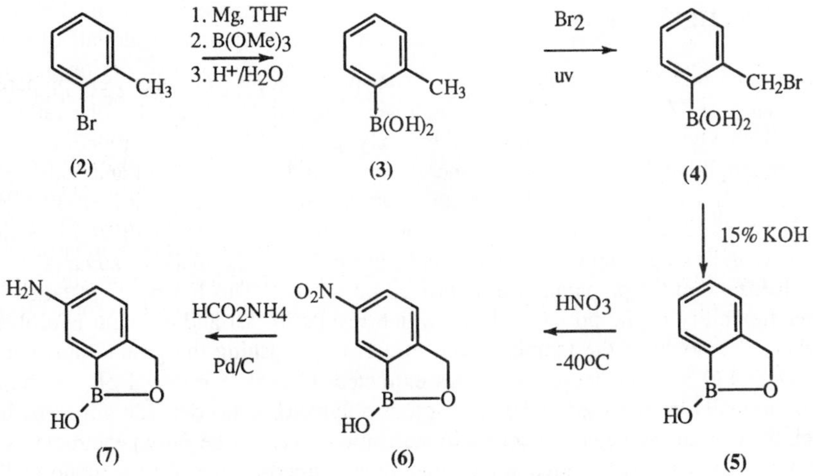

Scheme 1. Synthesis of boronophthalide and its amino derivative.

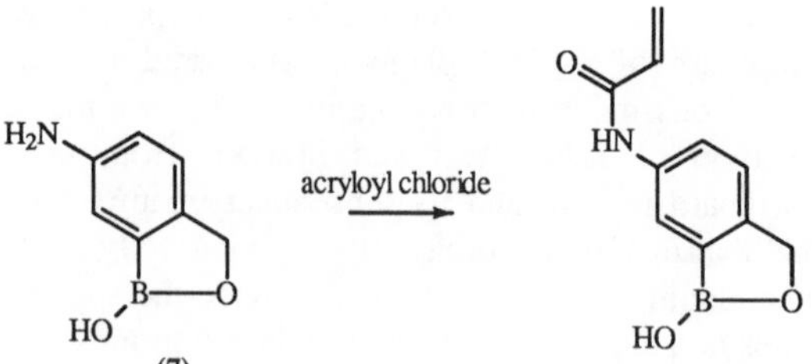

Scheme 2. Synthesis of the polymerizable boronophthalide derivative **(8)**

A polymerizable boronophthalide was required. It was known that boronophthalide could be nitrated without deboronation or oxidation of the methylene group,[10] so it was decided that the acrylamide **(8)** would be the primary target (c.f. acrylamides of arylboronic acids[11]). Nitration of **(5)** and reduction of 5-nitroboronophthalide **(6)** using 10% palladium on charcoal and ammonium formate, gave 5-aminoboronophthalide **(7)** which was subsequently treated with acryloyl chloride (**Scheme 2**) to give the required 5-(acrylamido)boronophthalide **(8)**. Full details of the synthesis will be described elsewhere.[12] Attempts to react **(8)** with sterols **(9)**, **(10)** and **(11)** in chloroform resulted in suspensions containing much undissolved material. On addition of 4Å molecular sieves to drive the equilibrium however, the solutions became clear. NMR spectroscopy showed the complete absence of starting materials in these clear solutions and a quantitative yield of the product ester, as shown in **Figure 1**, for the formation of the 2:1 adduct between **(8)** and androst-5-en-3β,17β-diol, **(11)**. Ester formation was confirmed by the characteristic 0.8 PPM downfield shift of the methine protons on both the 3 and 17-carbon atoms of **(11)**.

3. SYNTHESIS OF POLYMERS

Following promising results obtained by NMR spectroscopy, polymers were prepared, in the presence of 4Å molecular sieves, which were imprinted with a 1:1 molar ratio of the polymerizable boronophthalide to steroid hydroxyl groups. The template monomers chosen were boronophthalide esters of the isomeric sterols: androst-5-en-3β-ol and androst-5-en-17β-ol, **(9)** and **(10)** respectively, and the corresponding diol, androst-5-en-3β,17β-diol **(11)**. Ethylene glycol dimethacrylate (95 mole% with respect to the template) was chosen as the crosslinking agent as studies indicate that it has marked advantages over divinylbenzene in the preparation of imprinted polymers.[13] This has been attributed to the greater flexibility of the polymer chains which results in fast and efficient hydrolysis of templates. Removal of the templates was effected by washing the ground polymer with 99:1 THF:H₂O until no more sterol was extracted. Complete removal of the templates cannot normally be expected with macroporous polymers, as the dense, highly crosslinked nuclei, formed during heterogeneous polymerization, may not be fully penetrable. A non-imprinted polymer was prepared at the same molar composition as for the mono-hydroxy-imprinted polymers, but using methanol in place of sterol. Polymer compositions are summarized in **Table 1**. The solvent for all polymerizations was chloroform.

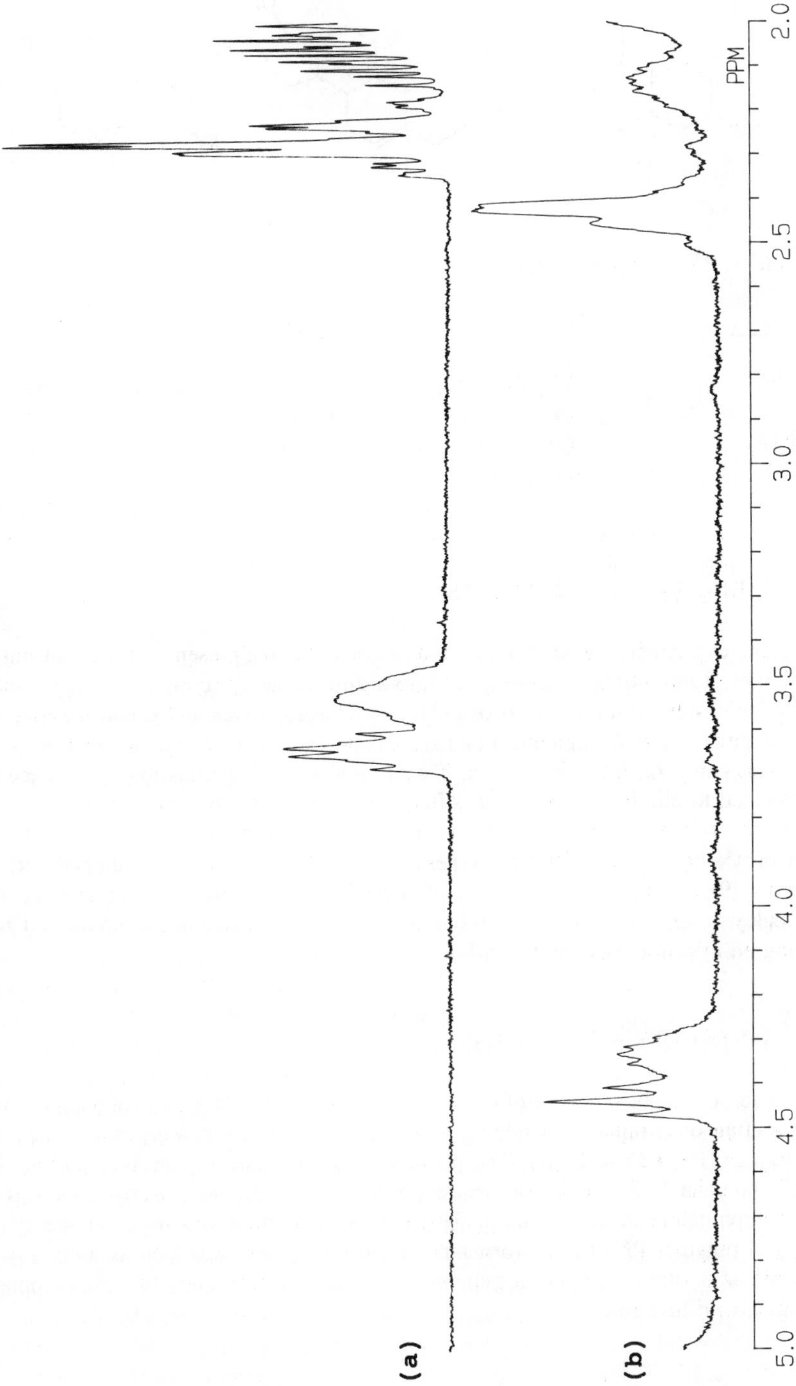

Figure 1. Partial ^{1}H NMR spectra of: (a). (11) in CDCl$_3$ and (b). 2:1 complex of (8) and (11) in CDCl$_3$ after 16 hours.

(9)   (10)   (11)

Table 1. Polymer preparations.

Polymer	Template	% Loading	Yield %
PN	MeOH	5	61
P3	(9)	5	76
P17	(10)	5	87
PD	(11)	5	58

4. BINDING EXPERIMENTS

Binding experiments were performed in batch, and in the presence of 4Å molecular sieves. Polymer (20mg-40mg) was weighed into a 2mL capacity screw top vial. A 2mL aliquot of a 2mM solution of (9), (10) or (11),in ethyl acetate was added and the vial shaken at room temperature. At high molecular sieve concentrations and long incubation times, some transesterification from the solvent became noticeable.[14] Incubation times were therefore shortened to eliminate this effect. After separation of the polymer by centrifugation, the concentration of sterol in the supernatant was then determined by GC analysis. This was performed on a Hewlett-Packard series 5890A gas chromatograph equipped with FID. For sterols (9) and (10), 1μL samples of trimethylsilyl derivatives, prepared according to Sweeley *et. al.*[15] were used. Solutions of (11) were analysed as the un-derivatized sterol, using an injection volume of 10μL.

5. RESULTS AND DISCUSSION

In order to establish the usefulness of the polymerizable boronophthalide (8) for the imprinting of templates containing a single hydroxyl group, we initially prepared three polymers (P3, P17 and PD). The polymerization mixture for the preparation of P3 and P17 contained 5 mol% of androst-5-en-3β-ol (9) and androst-5-en-17β-ol (10). Boronophthalide methyl esters at the same concentration was used for the synthesis of control polymer PN. The performance of these polymers was then assessed using 2 mM solutions of the corresponding ligands in ethyl acetate and the results obtained are summarized in **Table 2**.

Table 2. Binding of **(9)** and **(10)** to polymers **P3**, **P17** and **PN**.

Polymer	Re-binding of Template molecule	Binding of Template Molecule to **PN**	Specific Binding
P3	58 μmol/g	19 μmol/g	39 μmol/g
P17	65 μmol/g	34 μmol/g	31 μmol/g

The control polymer **PN** showed a relatively low uptake of both ligands (typically about 10-15%) even when molecular sieves were added to the reaction mixture to promote the formation of the ester bond (**Table 2**). As expected, the binding to the imprinted polymers was significantly higher and the specific uptake of **P3** and **P17** were calculated to be 39 and 31 μmol/g respectively.

Encouraged by these results we prepared a polymer imprinted with androst-5-en-3β,17β-diol **(11)**. The formation of covalent linkages between the hydroxyl groups of this template and 2 molecules of **(8)**, which was independently confirmed by NMR, should lead to precise positioning of the boronophthalide groups in the recognition sites of **PD**. This in turn should result in a certain degree of cooperativity in the interaction between boronophthalide residues and the hydroxyl groups of **(11)**. Indeed, this was found to be the case: **PD** showed a much better uptake of androst-5-en-3β,17β-diol as compared to both non-imprinted (**PN**) and mono-hydroxy sterol imprinted (**P3** and **P17**) polymers (**Table 3**). The cooperative interactions of functional groups in the recognition sites of imprinted polymers are currently being investigated in our group, with a view to enhancing the specificity of these materials.

Table 3. Concentration of **(11)** following incubation of a 2mM solution in ethyl acetate with polymers **PN**, **P3**, **P17** and **PD**, (30 and 50mg/ml) for one hour in the presence of 4Å molecular sieves. (Transesterification was negligible under these reaction conditions.)

	Concentration of **(11)** detected by gc	
Polymer	Polymer concentration = 30mg/ml	50mg/ml
PN	1.96mM	1.80mM
P3	1.50mM	1.47mM
P17	1.63mM	1.48mM
PD	1.31mM	1.14mM

6. CONCLUSIONS

(i) A new polymerizable boronophthalide derivative has been prepared and its usefulness for imprinting of templates containing individual or spatially separated hydroxyl groups has been established.

(ii) Cooperativity in the interaction between boronophthalide residues and the hydroxyl groups of androst-5-en-3β,17β-diol (11) has been demonstrated.

7. ACKNOWLEDGEMENTS

The authors would like to thank Eric Needs for the GC analysis, Esther Rodriguez for assistance with sample preparation and BBSRC (formerly AFRC) for financial support.

8. REFERENCES

1. K. Mosbach, Trends Biochem. Sci., 1994, 19, 9.
2. G. Wulff and J. Haarer, Makromol. Chem., 1991, 192, 1329.
3. B. Sellergren, Makromol. Chem., 1989, 190, 2703.
4. L. Fischer, R. Müller, B. Ekberg and K. Mosbach, J. Am. Chem. Soc., 1991, 113, 9358.
5. G. Wulff, W. Vesper, R. Grobe-Einsler and A. Sarhan, Makromol. Chem., 1977, 178, 2799.
6. G. Wulff and W. Vesper, J. Chromatogr., 1978, 167, 171.
7. C. Alexander, C.R. Smith, E.N. Vulfson and M.J. Whitcombe, *These Proceedings*, page 22.
8. H. Kamogawa and S. Shiraki, Macromolecules, 1991, 24, 4224.
9. H.R. Snyder, A.J. Reedy and W.J. Lennarz, J. Am. Chem. Soc., 1958, 80, 835.
10. W.J. Lennarz and H.R. Snyder, J. Am. Chem. Soc., 1959, 82, 2172.
11. G. Wulff and H-G. Poll, Makromol. Chem., 1987, 188, 741.
12. C.R. Smith, M.J. Whitcombe and E.N. Vulfson, *in preparation*.
13. G. Wulff, J. Vietmeier and H-G. Poll, Makromol. Chem., 1987, 188, 731.
14. D.P. Roelofsen, J.A. Hagendoorn and H. van Bekkum, Chem. Ind., 1966, 1622.
15. C.C. Sweeley, R. Bentley, M. Marita and W.W. Wells, J. Am. Chem. Soc., 1963, 85, 2497.

The Effects of Salt Concentration and pH on α-Chymotrypsin Activity During Reversed Micellar Extraction

G. Spirovska and J. B. Chaudhuri

SCHOOL OF CHEMICAL ENGINEERING, UNIVERSITY OF BATH, CLAVERTON DOWN, BATH BA2 7AY, UK

1. INTRODUCTION

The partition and solubilisation of proteins in microemulsion droplets has been applied to selective protein purification and the biosynthesis of water-insoluble compounds[1, 2]. The commercial exploitation of this technique is limited by protein losses in interfacial precipitates and the subsequent loss of enzyme activity. We are investigating the nature of the microemulsion environment and partitioning process on the loss of protein mass and specific activity. In this paper we present our initial studies on the enzyme α-chymotrypsin. This is a medium sized enzyme of molecular weight 25 kD, and a pI of 8.5.

2. MATERIALS AND METHODS

2.1 Materials

Sodium di-2-ethylhexyl sulfosuccinate (AOT) approx. 99% pure; α-chymotrypsin (from bovine pancreas); sodium chloride, potassium chloride and anhydrous calcium chloride were obtained from Sigma Chemical Co. Ltd., Poole Dorset. Glutaryl-L-phenylalanine 4-nitroanilide (GPNA) and methylsulfoxide (DMSO) were obtained from Fluka Chemicals. All the other salts required for buffer (sodium phosphate, acetate, citrate) preparation were obtained from BDH Chemicals, Merck Ltd, Poole, Dorset.

2.2 Experimental Methods

2.2.1 Forward Transfer. The protein solutions were prepared in suitable buffers at the pH required and at concentrations of 1mg/mL with the ionic strength of the solutions adjusted by addition of KCl, CaCl$_2$ or NaCl. The aqueous protein solutions were mixed by inversion in tightly stoppered 25mL vials in a rotating mixer

(Stuart Scientific Co., UK) with an equal volume of 50mM AOT in isooctane for 20min at 80rpm. All transfer experiments were carried out at 20±2°C.

     2.2.2 *Backward Transfer.* The micellar phase containing the solubilized protein from the forward transfer step was added to an equal volume of aqueous solution (buffer at required pH with 1M KCl) and mixed by inversion for 1 hour. For the experiments where the influence of the salt concentration on the solubilisation was examined the backward transfer was performed when using 50mM phosphate buffer pH 7 with 0.1M KCl.

2.3 Analytical Methods

     2.3.1 *Protein Concentration.* Protein concentration in both aqueous and organic phases was measured by absorbance at 280nm (Cecil 3000, Cecil Instruments Limited, Cambridge, UK). An extinction coefficient of 50400 $M^{-1}cm^{-1}$ was used to determine the protein concentration in solution (both in water and in the micelle).

     2.3.2 *Measurement of α-chymotrypsin Activity.* The activity of α-chymotrypsin was determined by using GPNA (glutaryl-phenyl-alanyl-4-nitroanilide) as a substrate[3]. The release of p-nitroanilide was followed at 410nm. Its molar extinction coefficient was 8800 M^{-1} cm^{-1} [3]. The specific activity of the enzyme retained in the aqueous phase after forward transfer was measured as was the specific activity of the protein recovered after the backward transfer. The specific activity is expressed relative to the activity in the feed solution.

     2.3.3 *Water Content in the Organic Phase.* The water content of the organic phase was measured by the Karl Fischer technique (Mettler DL37 Coulometer). These measurements were carried out in triplicate with an average deviation of ±2%.

3. RESULTS AND DISCUSSION

3.1 The Effect of Salt Type on Micelle Size

The size of a reverse micelle is expressed by means of the w_0 value, which is the molar ratio of water to surfactant. Water content may be controlled by the salt concentration in the feed solution. Solutions of varying salt concentration for three different salts were contacted with a 50mM AOT/isooctane solution. The phases were contacted for 12h which was found to be sufficient to reach constant amounts of water in the organic phase. The equilibrium values of w_0 were measured as a function of the initial M^+Cl^- concentration ($M^+ = K^+$, Na^+, Ca^{2+}) in the original aqueous solution. Results from this experiment are shown in **Figure 1**.

The salt type and concentration have a strong influence on w_0. The results show that for a given concentration of MCl w_0 increases when the radius of the cation decreases; this is in agreement with similar recent studies[2].

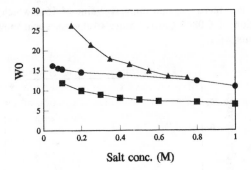

Figure 1.   *Effect of salt type of micelle size.*   ($\bullet = Ca^{2+}$, $\blacksquare = K^{+}$, $\blacktriangle = Na^{+}$)

Increasing the surfactant concentration at constant salt concentration (0.1M $CaCl_2$ added to 20mM acetate buffer pH 5) does not show any influence on w_0 (**Fig 2**). This implies that increasing the surfactant concentration yields an increase in the number of micelles of a constant size.

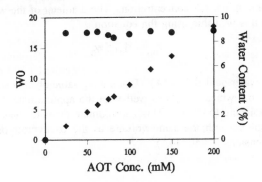

Figure 2.   *Dependence of water uptake on surfactant concentration.* ($\bullet = w_0$, $\blacklozenge = $ *water content*)

3.2 Protein Transfer as a Function of Micelle Size

α-chymotrypsin (1mg/ml) was transferred into a reverse micelle solution of 50mM AOT in isooctane. Transfer was made from aqueous solutions of different initial $CaCl_2$. **Figure 3** shows the solubilisation of α-chymotrypsin as a function of the micelle water content (size).

The results show that there is a minimum w_0 value of 15.6 which corresponds to an initial concentration of 0.08M CaCl$_2$ before α-chymotrypsin may be solubilised in significant amounts.

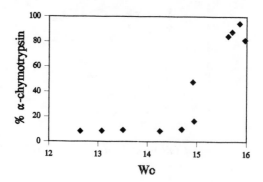

Figure 3.   *Dependence of α-chymotrypsin solubilisation on w_0*

The effect of increasing salt concentration acts to reduce the micelle size. α-chymotrypsin may only be solubilised above a limiting w_0 and thus protein transfer is most significant at low salt concentration. The diameter of the water pool may be calculated from this w_0 value using the equation[6]

$$r_{wp} = 1.75w_0 \qquad\qquad (1)$$

where r_{wp} is the water pool radius (Å). Using a w_0 value of 15.6 the diameter of the water pool is 54.6Å which correlates well with the approximate diameter of the α-chymotrypsin molecule (53.8 Å). This approximate diameter was calculated as the diameter of a sphere with the same volume as the α-chymotrypsin molecule (an ellipsoid of dimensions 51×40×40Å).

3.3 Forward Transfer of α-chymotrypsin

α-chymotrypsin (1mg/mL) was transferred into a reverse micelle solution of 50mM AOT in isooctane. Transfer was made from aqueous solutions of 20mM acetate buffer solutions with 0.08M CaCl$_2$, the pH was adjusted by 5M NaOH addition. The results are shown in **Figure 4.**

In common with other reverse micelle studies[4, 5] it was found that significant solubilisation occurred at pH values below the isoelectric point (pH 8.5). Above the isoelectric point, and at low pH values there was little solubilisation.

At these extremes of pH interfacial precipitation was observed. By mass balancing the protein concentration in the aqueous and organic phase this precipitation was quantified **(Fig. 4)**. A good correlation between protein solubilisation and precipitation is observed.

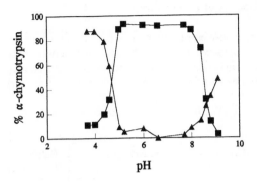

Figure 4. *Forward transfer of α-chymotrypsin as a function of pH. (■ = solubilisation, ▲ = precipitation)*

The results shown in **Figures 1** and **3** may be summarised in **Figure 5**, which shows the effect of the feed solution salt (NaCl) concentration on protein solubilisation. **Figure 5** also shows the specific activity of the protein that remains in the feed solution expressed relative to the initial activity. The retained specific activity increases with salt concentration.

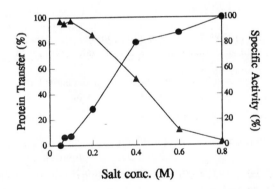

Figure 5. *Influence of salt concentration on protein transfer and retained activity. (▲ = protein transferred to organic phase, ● = relative specific activity of protein retained in feed solution)*

The explanation for this is most likely that at the lower salt concentrations most of the protein is transferred and comes in to contact with the surfactant and the organic solvent. At higher salt concentrations little protein is transferred, and a smaller proportion of the protein comes into contact with the organic phase.

3.4 Backward Transfer of α-chymotrypsin

Back transfer studies were carried out on microemulsions containing maximal amounts of protein, ie those transferred into the micelles between pH 5-8.5. Activity measurements were made on the protein recovered in the aqueous phase (**Table 1**). The results are expressed relative to the initial activity in the feed solution.

Table 1. *Protein and Activity Recovery During Back Transfer*

[NaCl] (M)	Recovery (%)	Relative Specific Activity (%)
0.05	22.8	35
0.07	22.6	34
0.1	19.9	18

The amount of protein recovered from the micelle is low compared to other proteins, for example, lysozyme[4]. There is significant activity loss during the transfer process. It is not possible to say whether the protein loss occurs during the forward or back transfer step, or as a result of both.

4. CONCLUSIONS

1. w_0 is very sensitive to salt concentration and salt type. The smaller the size of the cations the greater the amount of water solubilised.

2. α-chymotrypsin solubilisation is strongly dependent on the micelle size as measured by w_0. Significant solubilisation occurs at w_0 values greater than 15.7.

3. α-chymotrypsin is best transferred into the reverse micellar phase at pH values between 5 and 8.5. The reduced protein solubilisation outside this range is accompanied by significant losses of protein in the form of interfacial precipitation.

4. As a result of the relationship between micelle size and salt concentration, α-chymotrypsin is most greatly solubilised into reverse micelles at low salt concentrations.

5.. The retained specific activity of the solubilised protein remaining in the feed phase is minimal at low ionic strength. Retained specific activity increases with salt concentration as less protein is transferred and thus less protein interacts with the organic phase interface.

6   The back transfer of α-chymotrypsin is not very efficient with a maximum recovery of 22% under the conditions studied here. The retained activity of the recovered protein is low with 35% the greatest value.

5. ACKNOWLEDGEMENTS

We are grateful to the Separation Processes Committee of the SERC for funding this work under Grant No GR/H47135.

6. REFERENCES

1.   B.D.Kelley, D.I.C. Wang, and T.A. Hatton, <u>Biotechnol Bioeng</u>, 1993, <u>42</u>, 1199

2.   G.Marcozzi, N.Correa, P.L.Luisi and M.Caselli, <u>Biotechnol Bioeng</u>, 1991, <u>36</u>, 1239.

3.   B.F.Erlanger, F.Edel, and A.G.Cooper, <u>Arch Biochem Biophys,</u> 1966, <u>115</u>, 206.

4.   T.Kinugasa, K.Watanabe and H.Takeuchi, <u>Ind Eng Chem Res</u>, 1992, <u>31</u>, 1827.

5.   S.F.Matzke, A.L.Creagh, C.A.Haynes, J.M.Prausnitz and H.W.Blanch, <u>Biotechnol Bioeng</u>, 1992, <u>40</u>, 91.

6.   P.L.Luisi, M.Giomini, M.P.Pileni, B.H.Robinson, <u>Biochem. Biophys. Acta</u>, 1988, <u>947</u>, 209

Laboratory Scale Resolution of Short Chain Chiral Carboxylic Acid

L. P. Szabo, D. Kallo,[a] and L. Szotyory[b]

RESEARCH INSTITUTE FOR TECHNICAL CHEMISTRY OF THE HUNGARIAN ACADEMY OF SCIENCES, 8201 VESZPRÉM, PO BOX 125, HUNGARY

[a]CENTRAL RESEARCH INSTITUTE FOR CHEMISTRY OF THE HUNGARIAN ACADEMY OF SCIENCES, 1525 BUDAPEST PO BOX 17, HUNGARY

[b]VESZPRÉM UNIVERSITY, 8200 VESZPRÉM EGYETEM U. 4, HUNGARY

1 ABSTRACT

Material of crystalline character, zeolite A was used as a support and onto the surface via ethylene-diamine spacer L-leucine, L-glutamic acid, L-lysine and L-phenylalanine were chemically bonded to prepare chiral stationary phases. The laboratory scale resolution of (+/-)-3-bromo-2-methyl-propionic acid (DL-BMPA) on these stationary phases was studied. The new packings proved to be efficient for the laboratory scale resolvation of DL-BMPA.

2 INTRODUCTION

As a result of the intensive interest in asymmetrical (chiral) molecules in several scientific disciplines, there is a growing demand not only for analysis to determine enantiomeric purity but for the preparative scale resolution as well. Accompanying this development there has been a growing understanding of the mechanism by which enantio-selectivity occurs.

In general terms, to distinguish the handedness of a solute, or to differentiate such on a chiral phase a usually three attachment points are necessary. The model, originally proposed by Dalgliesh[1] and later emphasized by Pirkle[2-5] and Davankov[6] means that three simultaneous interactions must occur with either solute enantiomer to differentiate it by preferential interactions, such as structural differences, size, charge distribution, hydrogen bonding, hydrophobic or other interactions may contribute to the stability differences between diastereomeric solvates.

Direct resolution on a chemically bonded chiral stationary phase appears very promising, because it needs the less additives. Complexing species could create problem for preparative applications, because inevitably these additives must be separated from the enantiomeric solutes.

The number of silanol groups potentially available for covalent chemical bonding is about 300 mmol/100 g adsorbent for a silica of surface area of 450 m^2/g.

Article[7] on the preparation of nitrogen-containing chiral stationary phases from silica bead give on the basis of nitrogen analysis only 20-60 mmol optically active species content pro 100 g dry adsorbent.

Comparing the previous data, only one tenth or fifth of the active sites were used for derivatization . It is partially owing to the subsequent derivatization steps having lower than 100 % conversion. The result is that the used silica seems to have a low (40-90 m^2/g) active specific surface.

The commercial silica gels have about 5 hydroxyl groups per 100 $Å^2$ of surface, it means that the surface demand of one -OH group is 20 $Å^2$. At the preparation of stationary phases [10,12] the authors determined 55-400 $Å^2$ spatial requirement depending on the structure of organic molecules. It appears again that about the tenth of functional hydroxyl groups keep their potential for derivatization.

It was stated that a support of 30-60 m^2/g specific surface with wide pores (80 Å) and uniform pore size distribution is advantageous. Naturally the pore size of support must be adjusted to the size of chiral selector. Otherwise the pores could be blocked, or the chiral molecules inside the pores could "associate" with one another, and poor resolution is resulted.

We presumed that above requirements can be fulfilled when a zeolite with appropriate crystal size and active sites to prepare chiral stationary phase is chosen as a support. Commercial zeolite A molecular sieve meets these requirements despite its pore size of 4 Å which is too small for penetration of chiral molecules to be separated, its crystal size of 1-4 μm ensures the desired specific surface area of some square meters per gram. Chiral selector can chemically be bonded to active sites of the surface positions, which are of well defined co-ordinations, determined by crystal structure.

In order to eliminate steric hindrance, or "association" of chiral molecules of organic layer, spacer of short chain lenght and selectors of small spatial requirement were bonded onto the support surface.

In the development of new phases, Pirkle et al.[2-5,8] used the concept of reciprocity, i.e. if a chiral stationary agent X will separate the enantiomers of Y it means that a stationary chiral agent Y should be able to separate the enantiomer of X. This concept led us to the experiments to prepare chiral stationary phases on the basis of literature [7-12] for the resolution of short chain carboxylic acid without derivatization.

3 EXPERIMENTAL

The solvents methanol (MeOH), acetonitrile (MeCN), water, were purchased from MERCK (MERCK, Darmstadt) and were of HPLC grade. The racemates of DL-3-bromo-2-methyl-propionic acid (DL-BMPA) were presented by FÜZFÖ-NIKE (Hung. Chem. Works) and trifluoroacetic-acid, L-leucine, L-glutamic acid, L-lysine, L-phenylalanine, ethylene diamine, dicyclohexyl carbodiimide (DCC) were obtained from Aldrich (Aldrich-Chemie) as well as other reagents used for the preparation of the new stationary phases.

Chromatography was performed with a MERCK-HITACHI L-4500 Diode Array Detector, a L-6200A Intelligent pump, a Rheodyne Model 7125 injector equipped with a 20 μl sample loop, an EPSON LQ-100 printer. The run control, data-acquisition and processing was performed by DAD 6500 system manager program. To the determination of D/L-BMPA enantio ratio MERCK ChiraSpher (250 mm x 4 mm) chiral stationary phase under isocratic conditions was used. Baseline separation was achieved at H_2O:MeCN=50:50 eluent of pH=2 (trifluoroacetic acid) (Figure 1).

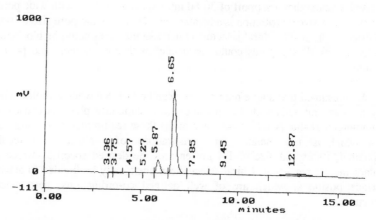

```
Method Title   : Method File
Column Type    : ChiraSpher 5um
Operator Name  : D-6500 DAD System Manager
Solvent A      : MeCN:H2O:TFAc
Solvent B      :
Solvent C      :
Comments : Sample: DL-BMPA 1g/1,
           Flow rate: 0.5 ml/min,detection: UV 210nm
           Eluent: MeCN:H2O=50:50
```

Figure 1 *Typical chromatogram of DL-BMPA resolution*

3.1 Preparation of Stationary Phase[13] for Laboratory Scale Resolution

From zeolite A (manufactured by Ajka Timföld, Hungary, composition: 32% SiO_2, 28% Al_2O_3, 20% Na_2O, 20% H_2O) 50 g was refluxed for half an hour in ammonium chloride, 1mol/l solution. After decantation it was repeated in the presence of fresh ammonium chloride solution, until 40% of Na^+ was exchanged. The drying was carried out at room temperature and was completed at 105 °C . The thermal decomposition in quartz tube, at gentle air stream, with programmed heating (10 °C/min until 200 °C, 1 h 200 °C, 10 °C/min till 450 °C, 1 h 450 °C) was realised to produce H^+-zeolite. The zeolite under dry N_2 was poured into 100 g of $SOCl_2$. After an hour reflux it was filtered and washed with 3 x 100 ml chloroform. In the presence of 80 g ethylene diamine (spacer) it was refluxed for an hour and then was filtered and washed with 3 x 100 ml ethyl acetate. Onto the surface -NH_2 groups of modified zeolite, N-benzoyled L-amino acids in the presence of DCC were chemically bonded (Table 1). The reagent excess was removed with exhaustive washing in ethyl acetate, chloroform, hexane, chloroform, ethyl acetate, ethanol, MeOH:H_2O=60:40, this later was used for the storage of prepared chiral phases, too.

Table 1 *List of prepared stationary phases*

Adsorbent	Chiral Selector
A	L-leucine
B	L-glutamic acid
C	L-lysine
D	L-phenyl alanine

3.2 Laboratory Scale Resolution of DL-BMPA

The resolving capability of the prepared chiral stationary phases were investigated in batch scale experiments. In each case 5g prepared adsorbent was conditioned for 30 min in double distilled water of pH=2 (trifluoroacetic acid). After suction, it was perfectly mixed in 10 ml of stock solution of DL-BMPA (1 mg/ml having L/D ratio of 2.3). The excess solution of DL-BMPA was sucked. The DL-BMPA adsorbed onto the surface was step by step partially chromatographed off with 5 ml portions of different eluents (MeCN, MeOH, H_2O pH=2). The filtrate containing D- and L-BMPA in different ratios were analysed on the ChiraSpher column.

The regeneration of adsorbents after each experiment was carried out in aqueous-methanol and within 20 cycles the results were reproducible, proving there was no damage of organic chiral coverage.

4 RESULTS AND DISCUSSION

On the basis of N analysis the surface coverage were for adsorbent A: 27.8 mmol/100 g, for adsorbent B: 24.9 mmol/100 g, for adsorbent C: 40 mmol/100 g and for adsorbent D: 35 mmol/100 g dry adsorbent.

These datas prove even if the zeolite has a low specific surface, about 30 m^2/g, which was used for the derivatization, the chemically bonded selector amount is in each case in the range, usually known from the literature. These give reason for utilizing the adsorbents for the laboratory scale resolution of DL-BMPA.

The measure of enantiodependent interaction was evaluated by the ratio of L/D-BMPA, as shown in Figure 2-5 for the adsorbents A-D, respectively.

The first points of the diagrams show that the L/D=2.3 enantio ratio of the starting stock solution was increasing to 2.6-2.9 owing to the different enantioactivities of the adsorbents A-D. The D antipode was enriched on the surface of adsorbents, hence the L content of filtrate of the solution was growing.

The elution tendencies of a certain eluent on adsorbent A is differing from that of on adsorbent B, C and D. It is due to differences in the interaction strenghts occurring between the different selectors and DL-BMPA. The main selecting powers are multiple H-bondings, offered by the =N-H and -CO-OH groups being present in all the selectors and BMPA as well. The suspected principal of optical resolution by diastereomeric-solute complexes is involving at least two hydrogen bonds. The additional functional groups of the selectors ($-CO-C_6H_5$, $-NH_2$, -COOH) and analyte ($-Br-CH_2$, -COOH) are strengthening or loosening the interactions (steric effects) (Figure 6).

The atomic distances between the stereogenic centre and the functional groups of the selector are very similar to that of analyte. There are probably strict limitations for complexation of the solute. The conformational freedom is low and can be predicted, that the prepared stationary phases are capable for the resolution of enantiomers having similar structure to BMPA.

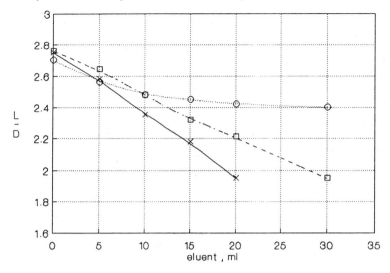

Figure 2 *L/D-BMPA enantioratio versus eluent volume on adsorbent A, with eluents: x:H₂O pH=2, o :MeOH, □ :MeCN*

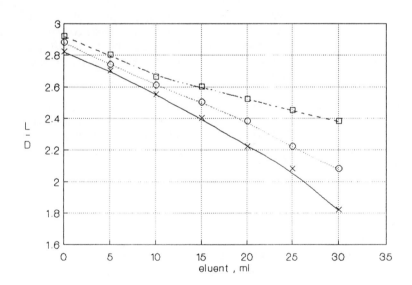

Figure 3 *L/D-BMPA enantioratio versus eluent volume on adsorbent B, with eluents: x:H₂O pH=2, o :MeOH, □ :MeCN*

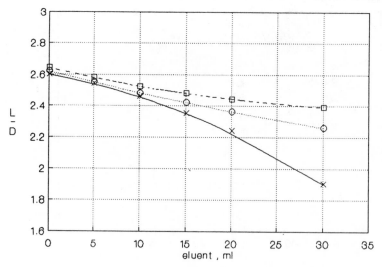

Figure 4 *L/D-BMPA enantioratio versus eluent volume on adsorbent C, with eluents: x:H₂O pH=2, o :MeOH, □ :MeCN*

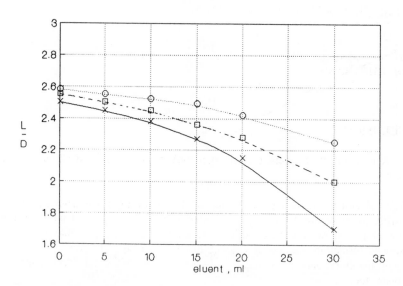

Figure 5 *L/D-BMPA enantioratio versus eluent volume on adsorbent D, with eluents: x:H₂O pH=2, o :MeOH, □ :MeCN*

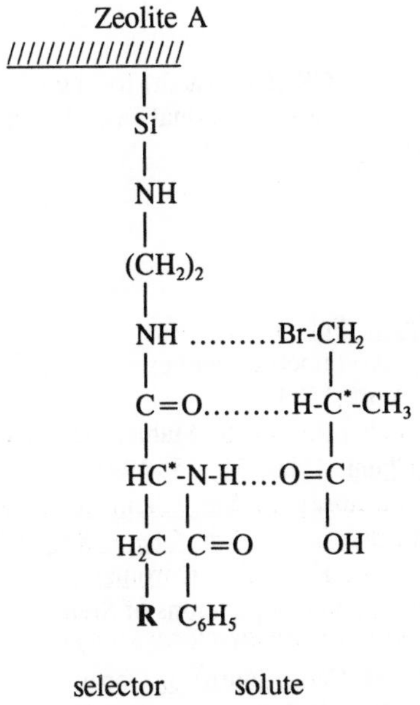

Zeolite A

/////////////////

|
Si
|
NH
|
$(CH_2)_2$
|
NH ....Br-CH$_2$
| |
C=O.........H-C*-CH$_3$
| |
HC*-N-H....O=C
| | |
H$_2$C C=O   OH
| |
R C$_6$H$_5$

selector   solute

R:-CH(CH$_3$)$_2$; -CH$_2$-COOH; -(CH$_2$)$_3$-NH$_2$; -C$_6$H$_5$

Figure 6 *Illustration of possible interactions between the chiral selector and DL-BMPA solute*

5 CONCLUSION

The zeolite A was suitable to prepare chiral stationary phases of desired chiral organic layer content.

The chiral selector L-amino acids can be regarded as chiral propionic acids as well as the DL-BMPA to be resolved on it. These materials were chosen to ensure an ideal situation for chiral interactions (Dalgliesh: three point attachment) for diastereomeric transient solvate formation, because the atomic distances and the location of functional groups around the chiral carbon atom are nearly the same on the chiral selector and the BMPA solvate too.

The BMPA enantiomers were selectively retained on adsorbents A-D. In batch experiments the differences in L/D-BMPA ratio are differing from each other according to the type of amino acid selectors. The results are promising for laboratory scale up resolution of DL-BMPA.

6 Acknowledgement

The authors are grateful to MERCK (Darmstadt) for the ChiraSpher column provided and for financial support of the Hungarian National Foundation for Fundamental Researches through grant number T 007401.

References

1. C. E. Dalgliesh, J. Chem. Soc. 1952, 3940
2. W. Pirkle and J. Finn, "Asymmetric Synthesis, Analytical Methods", J. D. Morrison, Academic Press New York,1983.
3. W. H. Pirkle, T. C. Pochapsky, G. S. Mahler, D. E. Corey, D. S. Reno, D.M. Alessi, J. Org. Chem. 1986, 51, 25, 4991
4. W. H. Pirkle, T. C. Pochapsky, J. Am. Chem. Soc. 1986, 108, 2, 352
5. W. H. Pirkle, T. C. Pochapsky, J. Am. Chem. Soc. 1986, 108, 5627
6. V. A. Davankov and A. A. Kurganov, Chromatographia 1983, 17, 686.
7. R. W. Souter "Chromatographic Separations of Stereoisomers" CRC Press, Inc., 1985 Chapter 3, p.135.
8. W. H. Pirkle and J. E. McCune, Journ. of Chrom. 1989, 471, 271
9. L. Ladányi, I. Sztruhár, P. Slégel, G. Vereczkey-Donáth, Chromatographia 1987, 24, 477
10. O.-E. Brust, I. Sebastian, I. Halász, Journ. of Chrom. 1973, 83, 15.
11. H. Deuel, G. Huber, R. Iberg, Helv. Chim. Acta, 1950, 157, 1950
12. H. Engelhardt, D. Mathes, Journ. of Chrom. 1977, 147, 311
13. Hung. Pat. P 9401125, 1994

Utilization of Temperature-induced Phase Separation for Purification of Biomolecules

Folke Tjerneld,[a] Patricia A. Alred,[a] and J. Milton Harris[b]

[a]DEPARTMENT OF BIOCHEMISTRY, UNIVERSITY OF LUND, PO BOX 124, S-221 00 LUND, SWEDEN

[b]DEPARTMENT OF CHEMISTRY, UNIVERSITY OF ALABAMA IN HUNTSVILLE, HUNTSVILLE, AL 35899, USA

1 INTRODUCTION

There are a number of polymers which in water solution exhibit reversed solubility. The polymers have a lower critical solution temperature (LCST). Above this critical temperature the polymers are no longer soluble and phase separation occurs. This temperature is called the cloud point of the system. Examples of thermo-separating polymers are ethylene oxide/propylene oxide random copolymers and hydrophobically modified cellulose derivatives.[1,2] The random copolymer Ucon 50-HB-5100 has a composition of 50% ethylene oxide (EO), 50% propylene oxide (PO) and has a cloud point of 50°C. Its molecular weight is 4000. A water phase and a liquid, concentrated Ucon phase are formed at temperatures above the cloud point of Ucon.[2] Factors determining the cloud point are EO/PO ratio, molecular weight and salt concentration. Non-ionic surfactants, like Triton X-114, also have a LCST in water, and this property has been used for isolation of membrane proteins.[3]

Aqueous two-phase systems are widely used for separation and purification of biomolecules.[4,5] These systems are formed in water solutions of two incompatible polymers, such as poly(ethylene glycol) (PEG) and dextran. PEG is enriched in the upper phase and dextran in the lower phase. Both phases contain 80 to 95% water. Cells, cell particles and biomolecules, such as proteins, DNA and RNA, will distribute between the upper and lower phases as a result of several factors, including type and molecular weight of polymers, salt, pH, size and surface properties.[4-6]

A new type of aqueous two-phase system, composed of an ethylene oxide/propylene oxide random copolymer as upper phase and either dextran or hydroxypropyl starch as lower phase polymer has been developed.[7-9] Partitioning in aqueous two-phase systems can in this way be combined with temperature-induced phase separation. A biomolecule can first be partitioned into a phase containing a thermo-separating polymer. This phase can then be removed and the temperature increased above the cloud point of the polymer. This will result in a new phase separation between a water phase and a polymer phase. The biomolecule can be separated from the polymer by partitioning to the water phase and the polymer can be reused. Both protein[7-9] and steroid[10,11] purification with this technique have been studied.

2 PROTEIN PURIFICATION

A protein purification method has been developed where Ucon/dextran or Ucon/hydroxypropyl starch aqueous polymer two-phase systems are used in a first step.[7] The composition of the system is selected so that the target protein is partitioned to the Ucon phase and the cell debris to the dextran or starch phase. The Ucon phase is removed and isolated in a separate container. The cloud point of Ucon is lowered to 40°C by addition of salt. The temperature is increased above the cloud point of Ucon which leads to the formation of a new two-phase system with an upper water phase and a lower Ucon phase. The high polymer concentration (>40%) in the lower phase leads to exclusion of the protein from this phase, and the protein is partitioned totally to the water phase. The polymer can be recycled back to the first extraction step. The target protein is obtained in the water phase free of polymer.

2.1 Purification of Intracellular Enzymes

The method has been used for purification of intracellular enzymes from baker´s yeast.[7] Purification of 3-phosphoglycerate kinase (3-PGK) from yeast homogenate is shown in Table 1. The phase system used was Ucon 50-HB-5100 (Union Carbide, New York, NY) and dextran T500 (Pharmacia, Uppsala, Sweden) with 20% added homogenate. The extraction process began with a primary Ucon/dextran two-phase system at room temperature. 3-PGK had a partition coefficient of 0.56, while K for total protein was 0.16. After separation, the upper phase was removed and the temperature increased to 40°C. In the new two-phase system, the enzyme partitioned completely into upper water phase, leaving the lower Ucon phase free of contamination. This Ucon phase was recovered. The polymer could be recycled and used in a second extraction of the original dextran lower phase. Total recovery of 3-PGK from both extractions in the combined upper water phases obtained by temperature-induced phase separation at 40°C was 56.8%, with a purification factor of 5.2.

Table 1 *Purification of 3-phosphoglycerate kinase from yeast homogenate using recycled Ucon to perform a second extraction. System is 6% Ucon 50-HB-5100, 3% dextran T500, 0.01M TEA-HCl buffer, pH 8.0, 0.2M sodium sulfate, and 20% yeast homogenate. K is the partition coefficient, C_t/C_b, and G the distribution ratio, $G = K \times (V_t/V_b)$.*

Step		K (22°C)	G (22°C)	Purification Factor[a] (40°C)	% Units Recovered in Water Phase
1st extraction	3-PGK	0.56	0.85	6.6	36.6
	Protein	0.16	0.45	-	-
2nd extraction	3-PGK	0.41	0.66	3.8	20.2
	Protein	0.20	0.44	-	-
Total for both extractions[b]				5.2	56.8

a) The specific activity of 3-phosphoglycerate kinase in the homogenate was 0.853 units mg^{-1} protein, equivalent to a purification factor of 1.
b) Calculated for combined water phases.

2.2 Temperature-Induced Phase Separation at Ambient Temperature

In order to lower the temperature of phase separation an ethylene oxide/propylene oxide random copolymer composed of 20% EO and 80% PO ($EO_{20}PO_{80}$) was synthesized.[8] This polymer has a cloud point of $18^{\circ}C$ and has been used to perform temperature-induced phase separation close to room temperature. The same process as above was performed using a two-phase system composed of $EO_{20}PO_{80}$ and dextran T500 (Tables 2a and 2b). The primary two-phase partitioning was performed at $4^{\circ}C$ and the temperature-induced phase separation at $24^{\circ}C$.[8] Partition of glucose-6-phosphate dehydrogenase (G6PDH), hexokinase and 3-phosphoglycerate kinase from yeast homogenate was determined. K values in the original $EO_{20}PO_{80}$/dextran system at $4^{\circ}C$ were 0.18 for G6PDH, 0.53 for hexokinase, 0.98 for 3-PGK and 0.05 for total protein content. The $EO_{20}PO_{80}$ polymer is hydrophobic and the partitioning in the $EO_{20}PO_{80}$/dextran system reflects the hydrophobicity of the proteins. 3-PGK is the most hydrophobic of the proteins studied. The majority of the water-soluble proteins are hydrophilic and therefore the total protein is strongly partitioned to the bottom phase. In order to increase enzyme recovery, the volume of upper phase was increased. After temperature-induced phase separation the total yields in the upper water phase at $24^{\circ}C$ were 50% for G6PDH, 67% for hexokinase and 72% for 3-PGK (Table 2b). The very low partition coefficient for total protein meant that fairly good purification factors were obtained: 5.3, 15.3 and 15.4 for the three enzymes, respectively.

Table 2a *Partition of glucose-6-phosphate dehydrogenase, hexokinase and 3-phosphoglycerate kinase from yeast homogenate. Primary phase system: 8.5% $EO_{20}PO_{80}$, 2.0% dextran T500, 0.02 M sodium phosphate buffer, pH 7.0, and 20% yeast homogenate. K and G values at $24^{\circ}C$ are not given as there was no detectable enzyme activity or protein in lower phase at this temperature. Volume ratio (V_t/V_b) at $4^{\circ}C$ was 4.0.*

Sample	$K(4^{\circ}C)$	$G(4^{\circ}C)$
Protein	0.05	0.20
G6PDH	0.18	0.75
Hexokinase	0.53	2.21
3-PGK	0.98	4.07

Table 2b *Purification of glucose-6-phosphate dehydrogenase, hexokinase and 3-phosphoglycerate kinase from yeast homogenate. Primary phase system: 8.5% $EO_{20}PO_{80}$, 2.0% dextran T500, 0.02M sodium phosphate buffer, pH 7.0, and 20% yeast homogenate. Y= % yield of enzymes. PF = purification factor at $24^{\circ}C^a$.*

Sample	G6PDH		Hexokinase		3-PGK	
	PF	Y	PF	Y	PF	Y
Raw homogenate	1	-	1	-	1	-
After centrifugation	2.4	71	2.1	64	3.2	96
Upper phase - 4°C	4.2	52	13.0	69	15.7	80
Upper phase - 24°C	5.3	50	15.3	67	15.4	72

[a]*The specific activity of an enzyme (units per mg protein) in the homogenate was equivalent to a purification factor of 1.*

2.3 Purification Scheme Using Temperature-Induced Phase Separation

A scheme for enzyme purification using temperature-induced phase separation is shown in Figure 1. This purification process allows for recycling of copolymer solution, and recovery of enzyme in a water/buffer solution at 24°C. Original $EO_{20}PO_{80}$/dextran system is formed at 4°C. In this system, it has been shown that most of total protein will partition to lower dextran phase, and upper copolymer phase can be removed and placed at 24°C. The upper water phase formed after temperature increase contains the target enzyme, and is virtually free of copolymer. The low temperature (24°C) at which this copolymer phase separates from water eliminates any heat denaturation of sensitive enzymes and also does not require addition of salt in order to obtain phase separation at a low temperature.

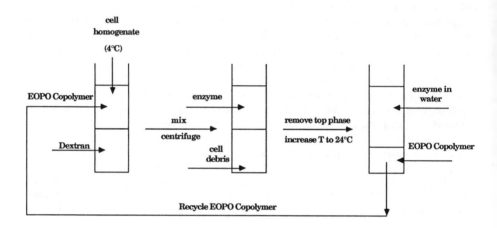

Figure 1 *Enzyme purification scheme using aqueous two-phase partitioning at 4°C and temperature-induced phase separation at 24°C, with recycling of the $EO_{20}PO_{80}$ copolymer.*

3 AFFINITY PARTITIONING

Temperature-induced phase separation has been utilized where the affinity ligand Procion Yellow HE-3G was covalently bound to Ucon for purification of enzymes.[9] In this technique an initial affinity partitioning step is performed in a Ucon/dextran aqueous two-phase system containing Ucon-ligand and cell extract. This step is similar to earlier work where PEG with bound affinity ligands has been used for affinity partitioning.[12] The Ucon phase is then isolated in a separate container. After temperature increase above the cloud point of Ucon the enzyme can be recovered in the water phase and the Ucon-ligand in the Ucon phase. Ucon-ligand plus Ucon can be recycled for a renewed extraction.

3.1 Affinity Partitioning Studied with a Pure Enzyme

Table 3 shows the results of affinity partitioning and temperature-induced phase separation with pure glucose-6-phosphate dehydrogenase. G6PDH was partitioned in a

Ucon/dextran phase system with 0.2% Ucon-Procion Yellow HE-3G.[9] In this phase system G6PDH was partitioned to the top Ucon-rich phase (K=4.5). After phase separation had occurred at 22°C, the Ucon-containing phase was removed. Sodium sulfate was added to this Ucon-phase to a concentration of 0.2M. This solution was placed in a water bath at 40°C for 15 minutes in order to induce phase separation. By the salt addition and temperature increase the enzyme was dissociated from the affinity ligand. In the new phase system G6PDH was totally partitioned to the upper water-salt phase. The enzyme was recovered to 88% in the water phase. The Ucon-Procion Yellow partitioned to the lower, Ucon-rich phase (K_L=0.06) and could be recovered to 77% along with Ucon. There was no enzyme present in the lower Ucon and Ucon-Procion Yellow phase at 40°C.

Two partitionings are achieved by performing temperature-induced phase separation on the phase which contains the Ucon-ligand-enzyme complex. The enzyme is partitioned 100% to the water phase due to the steric exclusion from the concentrated polymer phase. The Ucon with bound ligand is partitioned to the Ucon phase because of the thermodynamic force which favours phase separation between Ucon and water at this temperature.

Table 3 *Affinity partition of glucose-6-phosphate dehydrogenase with recovery of Ucon 50-HB-5100-Procion Yellow HE-3G. System is 5.1% Ucon 50-HB-5100, 7% dextran T500, 0.2% Ucon-Procion Yellow HE-3G, and 0.01M sodium phosphate buffer, pH 7.0. The amount of glucose-6-phosphate dehydrogenase was 34 units. K values at 40°C are for the partition between the water and Ucon phase formed by the increase in temperature.*

$K_E(22°)^a$	$K_E(40°)$	$K_L(40°)^b$	% G6PDH recovered in water phase	% Ucon-PrY recovered in Ucon phase
4.51	>100	0.06	88.0%	77.1%

a) K_E is the partition coefficient for G6PDH.
b) K_L is the partition coefficient for Ucon-Procion Yellow HE-3G.

3.2 Affinity Purification of Enzyme from Yeast Extract

Glucose-6-phosphate dehydrogenase was purified from yeast extract in a Ucon/dextran aqueous two-phase system using 0.2% Ucon-Procion Yellow HE-3G (Table 4).[9] In the initial phase system the enzyme was extracted by the Ucon-ligand to the top phase (K=12). The bulk proteins were partitioned to the bottom phase (K=0.32). The upper phase was isolated in a separate container. Sodium sulfate and sodium chloride were added, both at 0.2M, and the temperature was raised to 40°C. In the new two-phase system formed at 40°C the enzyme was recovered in the water phase with a yield of 79% and a purification factor of 4.2. The partition coefficient for the enzyme in the water/Ucon phase system was >100. Ucon-Procion Yellow was recovered in the lower Ucon phase with a yield of 85%. No protein could be detected in this Ucon phase.

Table 4 *Purification of glucose-6-phosphate dehydrogenase from yeast extract. System is 6.3% Ucon 50-HB-5100, 9% dextran T40, 0.2% Ucon-Procion Yellow HE-3G, 0.02M sodium phosphate buffer, pH 7.0, and 5.7% yeast extract. K values at $40^\circ C$ are for the partition between the water and Ucon phase formed by the increase in temperature. K is the partition coefficient, C_t/C_b.*

Sample	K $(22^\circ C)$	K $(40^\circ C)$	Purification Factor[a] $(40^\circ C)$	% Recovered at $40^\circ C$
G6PDH	12.4	>100	4.2	78.8[b]
Protein	0.32	>100	-	-
Ucon-PrY	24.6	0.32	-	84.6[c]

a) *The specific activity of G6PDH in the PEG precipitated homogenate was 0.475 units mg^{-1} protein, equivalent to a purification factor of 1.*
b) *Recovered in water phase at $40^\circ C$.*
c) *Recovered in Ucon phase at $40^\circ C$.*

4 CONCLUSIONS

A new purification technique for biomolecules has been developed by combination of temperature-induced phase separation and partitioning in aqueous two-phase systems. The technique is based on the use of a thermo-separating polymer as one of the phase-forming polymers in an aqueous two-phase system. The thermo-separating polymer can be removed from the protein by a moderate temperature increase. All proteins so far studied have been excluded from the polymer phase formed above the cloud point. The target protein is obtained in a water/buffer solution after only two purification steps. A polymer with attached affinity ligand can be separated from the protein solution by raising the temperature above the cloud point of the polymer. The temperature of phase separation can be controlled by the salt concentration and by the polymer hydrophobicity. A commercially available polymer (Ucon 50-HB-5100 from Union Carbide) with a cloud point of $50^\circ C$ has been successfully used. With this polymer it was possible to use a covalently bound affinity ligand for enzyme purification. The use of affinity ligand increased the specificity of the purification technique. An ethylene oxide-propylene oxide random copolymer with a cloud point of $18^\circ C$ has been synthesized and used for protein purification. With polymers which have low cloud point it is possible to perform temperature-induced phase separation at temperatures close to room temperature. With the use of a hydrophobic polymer with low cloud point fairly high purification factors were achieved. This was due to the partitioning of hydrophilic proteins to the lower phase in the primary phase system. More hydrophobic proteins could be partitioned to the phase containing the hydrophobic thermo-separating polymer. It was possible to purify proteins according to surface hydrophobicity with the use of the described combination of aqueous two-phase partitioning and temperature-induced phase separation.

Temperature-induced phase separation makes it possible to achieve important simplifications when aqueous two-phase systems are used in bioseparations. The polymer can easily be removed from the target protein. The polymer can be reused for repeated extractions. The purified target protein is obtained in a water solution which facilitates integration with other purification techniques.

Acknowledgements

This research is supported by grants from the Swedish National Board for Industrial and Technical Development (NUTEK) and from the Swedish Research Council for Engineering Sciences (TFR). Berol Nobel AB, Stenungsund, Sweden, is thanked for the synthesis of the 20% ethylene oxide/80% propylene oxide random copolymer.

References

1. S. Saeki, N. Kuwahara, M. Nakata and M. Kaneko, *Polymer*, 1976, **17**, 685.
2. H.-O. Johansson, G. Karlström and F. Tjerneld, *Macromolecules*, 1993, **26**, 4478.
3. C. Bordier, *J. Biol. Chem.*, 1981, **25**, 1604.
4. P.-Å. Albertsson, 'Partition of Cell Particles and Macromolecules', 3rd ed., Wiley, New York, 1986.
5. H. Walter, D.E. Brooks and D. Fisher, 'Partitioning in Aqueous Two-Phase Systems: Theory, Methods, Uses and Applications to Biotechnology', Academic Press, Orlando, Florida, 1985.
6. F. Tjerneld, in: 'Poly(Ethylene Glycol) Chemistry: Biotechnical and Biomedical Applications', J.M. Harris, ed., Plenum Press, New York, 1992, Chapter 6, p. 85.
7. P.A. Harris, G. Karlström, and F. Tjerneld, *Bioseparation*, 1991, **2**, 237.
8. P.A. Alred, A. Kozlowski, J.M. Harris and F. Tjerneld, *J. Chromatogr.*, 1994, **659**, 289.
9. P.A. Alred, F. Tjerneld, A. Kozlowski and J.M. Harris, *Bioseparation*, 1992, **2**, 363.
10. P.A. Alred, F. Tjerneld and R.F. Modlin, *J. Chromatogr.*, 1993, **628**, 205.
11. R.F. Modlin, P.A. Alred and F. Tjerneld, *J. Chromatogr.*, 1994, **668**, 229.
12. G. Johansson, in: 'Methods in Enzymology', W.B. Jakoby, ed., Academic Press, New York, 1984, Vol. 104, p. 356.

The Rapid Monitoring of Biological Particles to Facilitate Their Recovery and Purification

S. Tsoka, I. Holwill, M. Hoare, C. Lewis,[a] J. Brookman,[b] and K. Gull[b]

THE ADVANCED CENTRE FOR BIOCHEMICAL ENGINEERING, DEPARTMENT OF CHEMICAL AND BIOCHEMICAL ENGINEERING, UNIVERSITY COLLEGE LONDON, TORRINGTON PLACE, LONDON WC1E 7JE, UK

[a]BRITISH BIOTECHNOLOGY LTD., WATLINGTON ROAD, COWLEY, OXFORD OX4 5LY, UK

[b]SCHOOL OF BIOLOGICAL SCIENCES, DEPARTMENT OF BIOCHEMISTRY AND MOLECULAR BIOLOGY, UNIVERSITY OF MANCHESTER, STOPFORD BUILDING, OXFORD ROAD, MANCHESTER M13 9PT, UK

ABSTRACT

Virus-like particles (VLPs) expressed intracellularly by the yeast *S.cerevisiae* have set the framework for the development of a wide range of biologicals, particularly as carriers for viral antigens. The use of photon correlation spectroscopy (PCS) for the rapid evaluation of the concentration and purity of the VLPs that will aid the complex purification strategy is discussed. Particular emphasis is placed on the recovery of the VLPs from similarly sized and larger cell debris by centrifugal particle classification.

PCS offers rapid measurement, but the deconvolution of the complex light scattering signal does not lead to a unique particle size distribution. Data is given for the use of PCS to detect relatively small quantities of contaminating debris in the presence of the particles.

To increase the detection efficiency of PCS, use may be made of the specific interaction of antibodies with the VLPs. Preliminary data is given for the development of such a PCS-based immunoassay to evaluate VLP concentration.

1. INTRODUCTION

Virus-like particles are non-infectious protein aggregates, of approximately 60 nm in diameter, produced by the yeast microorganism *S. cerevisiae* and are the intermediate structures by which the yeast transposon (Ty) moves from one chromosome to the other[1, 2, 3]. Using genetic technology, a foreign gene can be inserted in the gene that encodes for VLP formation. The resulting fusion proteins assemble into VLPs that display the foreign protein on their surface in multiple copies and possess a significant potential for forming the basis of novel biological products including vaccines, diagnostics, research agents and therapeutics[4, 5, 6]. Due to their intracellular formation, the task of recovering VLPs from cell homogenate is particularly difficult and a large number of purification stages are needed. Successful industrial production of VLPs strongly depends on the efficiency of evaluating their concentration during the production and purification stages. Therefore, a rapid means of assaying VLPs is

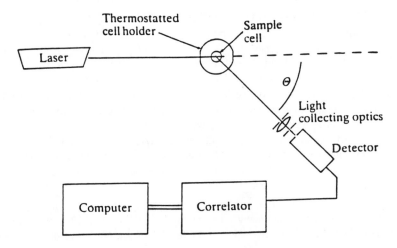

Figure 1: *Photon correlation spectroscopy: diagram showing the component parts of a PCS system. Θ is the scattering angle. ($\Theta=90^o$ in our experiments).*

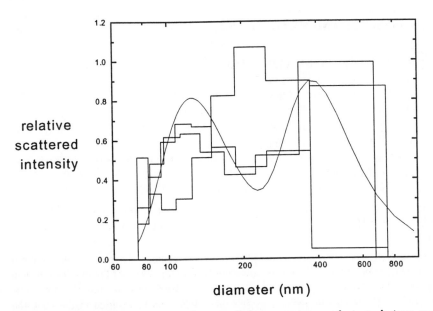

diameter (nm)

Figure 2: *The performance of a dynamic light scattering analysis technique may be assessed to some extent by simulation of the scattered light signal from a known size distribution. Here, a histogram method is tested, where simulated data from the true distribution (——) is passed to the algorithm with differing degrees of noise added. In the two cases of lower noise (----, 0.05%) and (·····, 0.1%), the distribution is not accurately reconstructed but does reflect the nature of the distribution, i.e. bimodal. When the noise is sufficiently high (- -, 0.5%), the program returns a "smoothed" version of the real distribution.*

needed, so that such monitoring may be achieved on-line.

Optical methods, such as photon correlation spectroscopy (PCS), offer very rapid responses and are able to measure in the submicron range, thus being a promising technique to analyse VLPs. In PCS (**Fig.1**) light from a continuous, visible laser beam is directed through an ensemble of macromolecules or particles in suspension and moving under Brownian motion. Some of the laser light is scattered by the particles and this scattered light is collected by a lens and detected by a photon detector which generates an electric signal proportional to the light intensity detected[7]. Due to the Brownian motion of the particles in suspension, the intensity scattered from the sample fluctuates on a time scale related to the time taken for a particle to diffuse a distance comparable with the light wavelength. From the intensity fluctuations a diffusion coefficient is calculated which can be related to particle size.

The problem of inverting light scattering data into the required distribution is ill-posed, that is for one set of data many solutions can be fitted to within the experimental noise in the collected data[8]. To overcome this effect, prior knowledge of the range and likely components of the size distribution may be used to aid the deconvolution, or alternatively the raw data may be used to detect the difference between a measured unknown and a previously measured control[9]. It should be noted however that improvement of the data quality must be the first step to a reliable measurement.

2. MATERIALS AND METHODS

VLPs 5620/12 were used (VLP-producing strain BJ2168/pMA5620). Frozen VLP stock solutions (0.84mg/ml) were donated from British Biotechnology Limited (Cowley, Oxford, UK). The particles were stored in phosphate buffered saline at -70°C. The purity of the sample was quoted as >90% by densitometry.

The anti-TyG6 antibody that was used was of the IgA class. It was purified at the University of Manchester by affinity chromatography using a protein A sepharose column and was concentrated to 0.6 mg/ml with a Centricon system. The antibody was stored in phosphate buffered saline at -70°C.

For the PCS-based immunoassay, VLP and antibody solutions were thawed to room temperature before use. 20 μL aliquots of the VLP solution were diluted with 130 μL of water. 0-56 μL of antibody were added into the diluted VLP solutions. Each antibody-VLP solution was pipetted into a 150 μl volume flow cell (Hellma England Ltd.) and was placed in the cell holder of the PCS instrument (Malvern 4700, Malvern Instruments, Worcester, UK). The temperature was 20.1°C and the data collection time was 100 s. Scattering of the laser light by the sample was measured at a 90° angle and converted by the computer software to size distributions based on intensity, weight and number of particles present in the sample.

Computer simulations and deconvolution were carried out using an IBM P.C. programmed in Zortech C++ language. The noise was simulated using a Gaussian random distribution and added in equal amounts across the correlogram generated. The deconvolution program used was a histogram method[10].

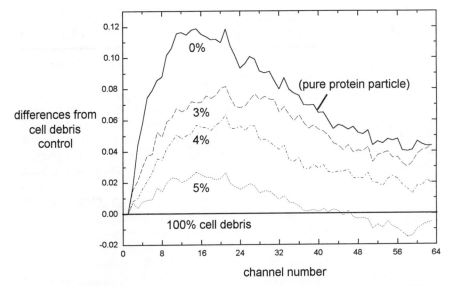

Figure 3: *PCS measurement of a mixture of clarified yeast homogenate and VLPs to assess the level of removal of cellular debris. The debris consists of cell wall fragments and cell contents. The percentages in the figure represent the volume fraction of cell debris to protein. The control is cell debris in the absence of the VLPs.*

3. RESULTS AND DISCUSSION

PCS for rapid analysis

Simulation experiments have indicated that for rapid data analysis where noise in the data is high, the solutions are seen to not necessarily reproduce the true particle distribution. An example is shown in **figure 2**, where different levels of noise are imposed onto simulated data that would be measured given a sample mixture of particle size distribution shown as the true distribution in the figure. At lower noise levels the bimodal nature of the sample is detected, but the amount in each peak does not accurately reflect the true sample. At higher noise levels only an average of the true distribution is returned as the solution.

An alternative approach to detect reliably product particle in homogenate, is presented here where the difference in correlograms due to a homogenate plus a product and homogenate alone are subtracted. The difference in this case will provide a measure of the amount of product present. The noise in the correlogram will be the determining factor in distinguishing between curves and, due to the subtraction process, will be larger than the correlogram noise by a factor of $\sqrt{2}$. However there is no instability due to algorithm performance in estimating the size distribution. The key consideration is the stability of the correlogram for the homogenate alone, so that in the situation where the measurement is to be applied, the appropriate baseline is subtracted.

Effecting a change in the signal proportional to VLP concentration by reagent

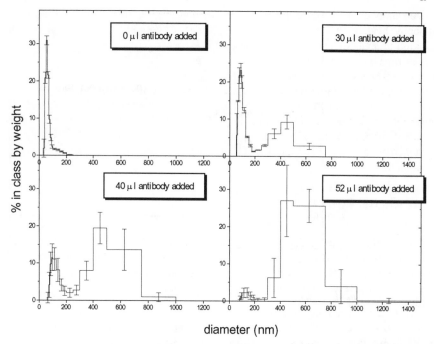

diameter (nm)

Figure 4: Size distribution for different levels of antibody addition to a VLP suspension. VLP solution with no antibody present shows one peak at a mean size of 67 nm on a weight basis. As antibody is added, a second peak gradually appears, and the mean sizes for the lower and the upper size peak increase with antibody addition. Both antibody coating the particles and coated VLPs cross-linking are proposed as the mechanism of the immunoprecipitation reaction. Error bars relate to the standard deviation for five measurements.

addition should further sensitise the method.

PCS-based immunoassay for VLPs

The underlying principle here is that if a size change is effected by binding antibodies on the VLP surface, this change could be detected and would be similar regardless of background. In that case, VLP concentration could be evaluated by relating it to size increase of the particles.

The objective in this paper is to study the effect of antibody concentration and size change on PCS signal with a view to relate VLP size change to concentration. The set of experiments presented here is of a preliminary nature with a view to confirming that the antibody-VLP interaction results in a size change which is detectable using PCS.

Figure 4 shows size distributions for different levels of antibody addition to a suspension of virus-like particles. The size distributions for the VLPs are in accordance with Burns *et al* [11] using electron micrographs. The polydispersity as shown by PCS is an intrinsic property of the VLPs and the mean size of 67 nm is in good agreement.

As antibody was added to the VLP solution, rate experiments confirmed that antibody binding to the VLPs was rapid compared with the experiment time. The addition of antibody leads to the appearance of a second larger peak and the mean size values for both the smaller peak and the larger peak increase with antibody addition. The two forms of size change suggest two mechanisms of antibody-VLP interaction, where the antibody molecules coat the VLP surface and also cause particles to cross-link to form agglomerates. The research is now being extended to explore the measurement of the antibody related size change for the rapid assay of VLPs in the presence of cell debris as a precursor to the study of recovery of VLPs from cell homogenate by centrifugal particle classification, by selective flocculation and by precipitation.

ACKNOWLEDGEMENT

UCL is the Science and Engineering Research Council's Interdisciplinary Research Centre for Biochemical Engineering and the Council's support to the participating UCL departments is gratefully acknowledged. The financial support of the Alexander S. Onassis Public Benefit Foundation is gratefully acknowledged, as is the support of British Biotechnology Ltd.

REFERENCES

1. J.R. Cameron, E.Y. Loh, R.W. Davis, Cell, 1979, 16, p. 739.

2. D.J. Garfinkel, J.D. Boeke, G.R. Fink, Cell, 1985, 42, p. 507.

3. J.D. Boeke, D.J. Garfinkel, C.A. Styles, G.R. Fink, Cell, 1985, 40, p. 491.

4. S.M. Kingsman, A.J. Kingsman, Vaccine, 1988, 6, p. 304.

5. A.J. Kingsman, S.E. Adams, N.R. Burns, S.M. Kingsman, Trends in Biotechnology, 1991, 9, p. 303.

6. S.E. Adams, K.M. Dawson, K. Gull, S.M. Kingsman, A.J. Kingsman, Nature, 1987, 329, p. 68.

7. R.J.G. Carr, R.G.W. Brown, J.G. Rarity, D.J. Clarke, "Biosensors: Fundamentals and Applications", Turner, A.P.F., Karube, I., Wilson, G.S.(ed), Oxford Science Publications, 1987, p.679.

8. J.G. McWhirter, E.R. Pike, J. Phys., 1978, A11, p. 1729.

9. I.L. Holwill, G.B. Davies, N.J. Titchener-Hooker, G. Parry, M. Hoare, IChemE Research Event , 1994, 1, p. 114.

10. G.C. Fletcher, P.S. Ramsay, Optica Acta, 1983, 30(8), p. 1183.

11. N.R. Burns, H.R. Saibil, N.S. White, J.F. Pardon, P.A. Timmins, S.M.H. Richardson, B.M. Richards, S.E. Adams, S.M. Kingsman, A.J. Kingsman, The EMBO Journal, 1992, 11, p. 1155.

Physico-chemical Database Development for Baculovirus-produced Proteins: The Rational Design of Large Scale Protein Purification

R. E. Turner,[a] B. S. Baines,[b] and J. A. Asenjo[a]

[a]BIOTECHNOLOGY AND BIOCHEMICAL ENGINEERING GROUP, DEPARTMENT OF FOOD SCIENCE AND TECHNOLOGY, UNIVERSITY OF READING, PO BOX 226, READING RG6 2AP, UK

[b]GLAXO GROUP RESEARCH LIMITED, GREENFORD, MIDDLESEX, UK

1 ABSTRACT

The complex protein components of product streams from baculovirus infected insect cells have been characterised on the basis of their physico-chemical properties. These include the main protein contaminants, expressed as titration curves, molecular weight and relative hydrophobicities. The subsequent data-base is being interfaced with heuristic 'expert' information set up in a computer based system. Once completed the expert system will choose the most appropriate high resolution purification steps based on physico-chemical data to obtain maximum yield (and purity) of the target protein with the minimum number of stages.

2 INTRODUCTION

Today, scale-up and design of an effective industrial protein purification process represent one of the major and most delaying problems in bringing a new protein (and process) to the market. This is to a large extent because there is a significant lack of expertise in choosing an optimal purification sequence with a minimum number of steps and maximum yield. Also there is a lack of understanding of efficient ways of eliminating trace contaminants to obtain extremely high levels of purity that is necessary for these proteins. This is mainly due to a lack of organized knowledge and data on physical, chemical and molecular properties of target proteins and contaminants, particularly those important for their behaviour in separation operations.

The database of physico-chemical properties of intracellular and extracellular protein streams of baculovirus infected insect cells is at present being interfaced to a 'hybrid' expert system[1]. Such a system will then calculate the separation coefficients for the different properties and for all the contaminants present (figure 1). The rationale used, has already been discussed[2], it includes a deviation factor (DF) for each property between the product protein and each of the contaminant proteins. It also possesses efficiency factors for separation techniques (e.g. ion-exchange is very efficient in exploiting small differences in charge whereas the separation of proteins by gel filtration is not). The total concentration of protein is included to find the "separation selection coefficient " (proteins present in large concentrations should be separated first) and a cost element is included to give the "economic separation coefficient." This correlation is then used to choose the most efficient techniques to separate the target protein from the main protein contaminants. Careful consideration is given to the effect of efficiency, protein concentration and cost factors on the final calculation of the separation selection coefficient. The result is a sequence of operations with the highest confidence factors and minimum number of high resolution purification operations.

The aim of this paper is to show the development of a database on physico-chemical properties of main protein contaminants from baculovirus infected cells that is being implemented in the expert system.

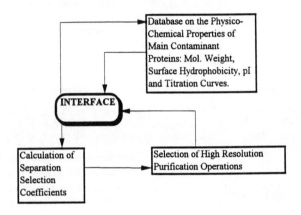

Figure 1

3 METHODS

3.1 Preparative Isoelectric Focusing

Both intra and extracellular product streams were concentrated using 3K MWCO membranes before isoelectric focusing (IEF). The use of solubilising agents was minimised due to their effect on certain physico-chemical properties. Protease inhibitor 'cocktails' were present in all the protein samples. Focusing was initially carried out in a pH gradient of 3 - 10. Narrow gradients were also used to produce numerous aqueous fractions with one to two concentrated proteins. All preparative IEF was conducted on a LKB 2117 Multiphor at 5^0C, 8 watts (constant power), using Ultrodex^{TM} granulated gel medium containing Pharmalytes^{TM}. Focusing times ranged from 12 to 18 hours.

3.2 Charge Density as a Function of pH

Each protein fraction was applied to a IEF gel across a prefocused pH gradient of 3 - 9. After an average of 50 volt hours (Vh), a 'titration curve' (TC) of one or more of the proteins was created. Once molecular weight had been established using a combination of SDS-PAGE, native PAGE and gel filtration, the charge density (number of charges per gram of protein) of each protein as a function of pH was determined. All analytical electrophoretic techniques were done using a Phast system (Pharmacia).

3.3 Hydrophobicity

The surface hydrophobicity of the proteins of interest was estimated using hydrophobic interaction chromatography. The surface hydrophobicity was correlated with the concentration of salt needed to elute the protein from a hydrophobic matrix. All measurements were made with an FPLC (Pharmacia) in 1ml columns containing Phenyl Superose (low substitution) (Pharmacia). The salt used was ammonium sulphate.

3.4 Isoelectric point vs Molecular Weight

In the initial characterisation of insect cell fermentation streams the technique of two-dimensional electrophoresis was used to relate isoelectric point with molecular weight. The method used was based on that developed by O'Farrel[3]. For both first and second dimensions the Bio-Rad Mini Protean II was used in conjunction with Bio-Rad internal 2-D markers to standardise each gel.

4 RESULTS AND DISCUSSION

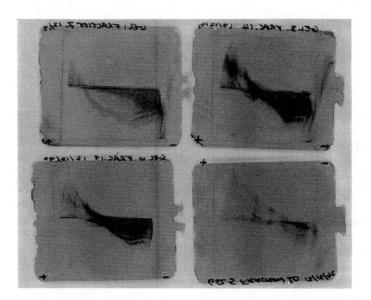

Figure 2: Titration curve analysis (pH 3-9) of intracellular insect cell proteins.

The principle physico-chemical characteristics that have been studied are:
- Charge density as a function of pH (figure 2)
- Hydrophobicity
- Molecular weight
- Isoelectric point

From two-dimensional polyacrylamide electrophoresis (2-D PAGE) using silver staining it is possible to say that from mechanical lysis of Baculovirus infected insect cells approximately 114 individual proteins are liberated and for an extracellular product stream (i.e., conditioned media containing fetal calf serum), approximately 19 proteins are present. Obviously the method of detection will dictate the amount of information accumulated from such a procedure. The term 'major contaminant' in this work is used to represent proteins that are present from the mg/g range for cell lysate and mg/l for extracellular proteins.

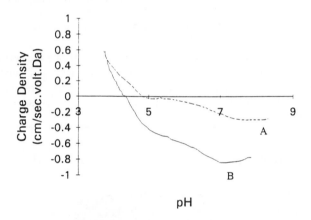

Figure 3: Charge density vs. pH for two extracellular insect cell proteins.

For each of the major protein contaminants, a titration curve has been created (figure 3). Figure 3 represents the titration curve of two major proteins isolated from baculovirus infected insect cell conditioned medium. From TC results it is possible to predict ion-exchange separations[4]. For example at pH 4.0 the charge densities in Figure 3 are similar and thus cation exchange chromatography at pH 4.0 would not be expected to resolve these two proteins (figure 4). Where as at pH 7.0 the charge density difference is quite large. One might predict from this information that anion exchange at pH 7.0 would resolve both proteins, protein A being eluted before protein B (figure 5). The limits of valid correlation between charge density and ion exchange separation are being investigated using both the insect cell database as well as with model proteins. In particular we are studying the numerical correlation between these two parameters and whether it is net charge per protein, charge density (net charge / molecular weight) or charge /surface area that determines behaviour in an ion exchanger.

The margin to which small scale HIC can be used in the determination of surface hydrophobicity is presently being assessed. At present, all work has been conducted in a

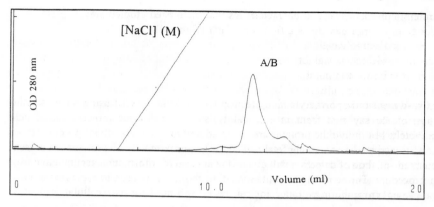

Figure 4: Fractionation of extracellular proteins (represented in figure 3) by cation exchange chromatography. Column 5mm x 5 cm. Mono S matrix (Pharmacia). Sample volume 500 µl. Buffer A, 20 mM Citric acid pH 4.0. Buffer B, as A + 1 M NaCl. Flow rate 1 ml.min^{-1}.

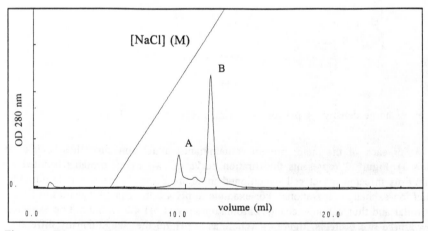

Figure 5: Fractionation of extracellular proteins (represented in figure 3) by anion exchange chromatography. Column 5 mm x 5 cm. Mono Q matrix (Pharmacia). Sample volume 500 µl. Buffer A, 20 mM Tris-HCl pH 7.0. Buffer B, as A + 1 M NaCl. Flow rate 1 ml.min^{-1}.

narrow pH range and at a fixed temperature. The influence of temperature, pH, column size and choice of matrix in the determination of surface hydrophobicity is being studied. This is providing more accurate information than just indicating general hydrophobic trends present in the protein streams. The use of HIC as a high resolution step has already been investigated and was shown to achieve only half the resolving power of ion-exchange chromatography [4].

Both molecular weight and isoelectric point have been determined individually and together in two-dimensional analysis (figure 6). Molecular weight of each protein contaminant is on its own an unimportant constant, other than in crude first step product stream fractionation i.e., ultrafiltration and potential polishing steps such as gel filtration. Molecular weight is important however in our determination of charge density for the prediction of ion-exchange behaviour. It may also allow correlations between surface hydrophobicity and hydrophobic interaction separations.

All contaminants which can be detected by silver stained 2-D PAGE gels have been characterised sufficiently to provide an adequate database for the expert system. Obviously the database can never be totally complete, what it can do is provide sufficient information to isolate a target protein from all it's major contaminants and thus most of the minor ones, by suggesting the most effective purification scheme. If the target is 'surrounded' by low concentration contaminants for example in isoelectric point and molecular weight, then further characterisation would be merited or the choice could then be made as to the expression vector or the possibility to introduce properties in the protein to facilitate separation.

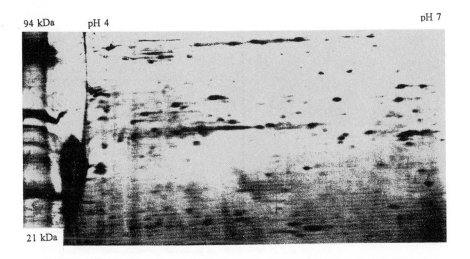

Figure 6: Two dimensional gel electrophoresis of intracellular proteins from cultured insect cells infected with baculovirus.

5 CONCLUSIONS

The rational design of protein purification processes clearly show that properly developed expert systems can be a vital tool to assist with solving the knowledge-intensive and heuristic-based problem of process synthesis in biotechnology. The interfacing of the prototype expert system with a database of the physicochemical and thermodynamic properties of main protein contaminants in specific production streams is an important improvement. This will allow the selection of high resolution purification operations on a much more rational basis, resulting in a improved process selection and thus process design.

References

1. Leser, E. W. and Asenjo, J. A. , 'Building an Expert System to Assist the Rational Design of Large Scale Protein Purification Processes'. Separations for Biotechnology III, University of Reading, UK, 1994
2. Asenjo, J. A. and Maugeri, F., Frontiers in Bioprocessing II, ACS Books, Washington, 1991
3. O'Farrel, P. H., J.Biol. Chem., 250 (1975) 4007-4021
4. Watanabe, E., Toska, S., Asenjo, J. A., Ann. N. Y. Acad. Sci. (in press)

Foam Separation for Enzyme Recovery: Maintenance of Activity

J. Varley and S. K. Ball

BIOTECHNOLOGY AND BIOCHEMICAL ENGINEERING GROUP, DEPARTMENT OF FOOD SCIENCE AND TECHNOLOGY, UNIVERSITY OF READING, PO BOX 226, READING RG1 2AP, UK

1 INTRODUCTION

The preferential adsorption of surface active species, *eg* proteins, at a gas-liquid interface can be exploited as a separation technique. Foam separation (illustrated in Figure 1), which is based on this phenomenon, has been discussed in the literature as a potential protein purification method[1-3,8-13] It is mechanically simple, can be operated continuously, is easily scaleable and available at low cost[1-3]. In the foam separation process, gas is bubbled through a solution containing a range of components with differing surface activities. At the surface of the solution, a foam will form (the stability of the foam will depend on both the operating conditions and the biological characteristics of the solution). The composition of the foam is expected to differ from the composition of the solution: the foam being richer in the more surface active component.

Proteins tend to adsorb to gas-liquid interfaces, increasing their stability. Stabilisation of the gas-liquid interface takes place in several stages: i) diffusion of proteins through the bulk liquid, ii) adsorption at the gas-liquid interface; this may be reversible or irreversible, and iii) molecular rearrangement may occur. It is generally accepted[4,5] that two types of film can form at the air-water interface: i) a dilute film: molecules unfold, ii) a concentrated film, consisting of both native and denatured molecules with possible molecular aggregation. The protein film formed at the gas-liquid interface may be a monolayer or multi-layer film[6,7]. The type of film formed will depend on protein properties, protein structure and protein concentration in the bulk liquid and at the gas-liquid interface.

There are several possible objectives for a foam separation process: i) to maximise protein enrichment ratio (protein enrichment ratio = protein concentration in foam/protein concentration in initial solution), ii) to maximise protein recovery (protein recovery = (mass of protein in foam/initial mass of protein) x 100 %), iii) to maximise partition of one component from a multi-component mixture. Retention of activity for enzymes, through maintenance of native structure, during the foam separation process will also be important.

Previous research on foam separation has focused on measurements of enrichments and recoveries for single proteins from aqueous solutions for a range of process conditions[8-12]. Only limited results have been reported for separation of proteins from multi-component mixtures[3,12,14]. There are many reports in the literature that foaming causes protein denaturation; however several studies also indicate that, with careful choice of operating conditions, conformational changes at an air-water interface can be avoided[5,15].

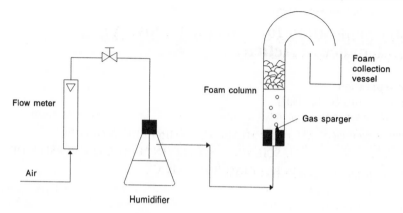

Figure 1 Foam Separation Apparatus

The investigation reported here examines the effect of initial protein concentration and gas flowrate on protein enrichment and recovery and percentage of enzyme activity retained on foaming, for a range of well characterised enzymes.

2 MATERIALS

Trypsin 1 (T0134) from porcine pancreas, pepsin (P6887) from porcine stomach mucosa, trypsin 2 (T8253) from bovine pancreas, lysozyme from chicken egg white (L6876) and catalase from bovine liver (C10), *Micrococcus lysodeikticus* (M-3770) BAPNA (B4875), bovine hemoglobin (H2625), ammonium sulphate and trizma base were supplied by Sigma Chemical Co. All other chemicals were AnalaR grade and supplied by BDH. The known properties of these enzymes are: pI = 10.7, 8, 10.4, 7, and 3 for trypsin 1, trypsin 2, lysozyme, catalase and pepsin respectively; molecular weights 23,000, 14,500, 240,000, 35,000 , for trypsin, lysozyme, catalase and pepsin respectively. Trypsin and pepsin are flexible proteins, whilst lysozyme is globular and rigid. The number of disulphide bridges per molecule at the pI is 5, 4, 3 for trypsin, lysozyme and pepsin respectively.

3 METHODS

Protein concentrations and enzyme activity assays

Protein concentration was determined using a spectrophotometer measuring absorbance at 280 nm and published extinction coefficients. Lysozyme activity was measured using *Micrococcus lysodeikticus* as a substrate, according to the method of Canfield (1963)[16]. The assay for trypsin is based on the hydrolysis of BAPNA by trypsin which releases 4-nitroaniline which can be measured by reading the increase in absorbance at 395 nm[17]. Pepsin activity was measured using the method described by Anson (1938)[18] and modified by Ryle (1984)[19]. Catalase activity was measured using the standard method detailed by the suppliers (Sigma Chemical Co.).

Surface tension measurement

The surface tension of each enzyme solution was measured using the Wilhelmy plate method, with a platinum plate and a K12C tensiometer (Kruss Ltd, Royston, UK).

Foam separation

The foam separation apparatus is shown in Figure 1. In all experiments, foam separation was carried out as a batch process with solutions of single proteins. The protein solution under consideration was poured into a vertical, glass column (column height = 0.27 m, column diameter = 0.026 m). Pre humidified air, at a pre-set flowrate, was continuously injected at the base of the column through a sintered glass sparger (pore size 16 to 40 µm). Foam was collected from the column exit and allowed to collapse. When no further foam was formed, any foam and solution remaining in the column was collected as the retentate. The volumes, total protein content and specific activity of the initial solution, foam and retentate were measured for each experiment. The protein enrichment ratio, protein recovery and retention of initial enzyme activity were then calculated for the foam fraction. The protein enrichment ratio and recovery are defined above. The percentage of initial enzyme activity retained after foaming = (specific enzyme activity in the foamate/specific enzyme activity in the initial solution) x 100). All experiments were carried out at room temperature, atmospheric pressure and humidity.

Results from two series of experiments are reported in this paper. In the first series, the enzymes pepsin and trypsin 1 were considered. The enzymes were buffered as follows: pH 3 and 4.8 citric acid/disodium hydrogen orthophosphate; pH 10.7 sodium carbonate/sodium bicarbonate buffer. 50 ml of enzyme solution was used in each experiment and the gas flowrate was set at 200 cm^3/min, unless otherwise stated. In the second series of experiments, the enzymes trypsin 2, catalase and lysozyme were considered. The enzymes were diluted with 0.1 M NaCl and the required pH achieved by addition of NaOH or HCl. 100 ml of enzyme solution was used in each experiment and the gas flowrate was set at 200 cm^3/min, unless otherwise stated.

4 RESULTS AND DISCUSSION

Surface tension

Figure 2 shows that the surface tension decreases with increasing enzyme concentration. The surface tension should approach a constant level at the critical micelle concentration (CMC) for each protein. For the concentration range considered here, the CMC is not reached for all enzymes. For trypsin 1 and catalase, the CMC at the pI occurs at 0.1 mg/ml and 0.5 mg/ml respectively. The pH has a pronounced effect on the surface tension concentration profile, as illustrated in Figure 2 for trypsin 1 and pepsin. The surface tension is lowest at the pI of the enzyme. It is widely reported in the literature that foam separation will be most effective when operated at concentrations below the CMC. It is expected that a saturated monolayer of protein is formed at this concentration and that further increases in interfacial concentration will be small[8]. However, it has also been shown that multi-layer adsorption, which can occur at concentrations above or below the CMC, can significantly increase interfacial concentrations [20,21].

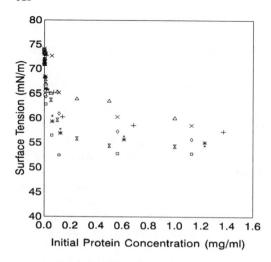

Figure 2 Dependence of Surface Tension on Initial Protein Concentration.
□ trypsin 1 pH = 10.7, x trypsin 1 pH = 4.8, ◊ trypsin 1 pH = 3, ▫ pepsin pH = 10.7,
+ pepsin pH = 4.8, ✳ pepsin pH = 3, △ lysozyme pH = 10.4, ✗ catalase pH =7.

Protein enrichment ratio

As shown in Figure 3, the protein enrichment ratio decreases as the initial protein concentration increases. At low concentrations, effective enrichment ratios are achieved. However, as the initial protein concentration is increased, the enrichment ratio approaches 1 and the effectiveness of the concentration obviously decreases.

There is a lower limit to the protein concentration at which foam separation can be operated. A minimum concentration is required to generate a stable foam; this will depend on the protein structure and the surface properties of the interfacial protein film. It must be remembered that the foaming process considered here is a batch process. Therefore, the protein concentration in the liquid will be depleted, as protein is carried into the foam, until conditions are reached when a stable foam can no longer be formed.

Protein recovery

Although enrichment decreases with increasing initial protein concentration, protein recovery increases (Figure 4). This is a result of increases in foam volume produced with increasing initial protein concentration (Figure 5).

Enzyme activity

Figure 6 shows clearly that the percentage of initial enzyme activity retained after foam separation increases with increasing initial protein concentration in all cases, and that for certain conditions, high levels of activity can be maintained after foam separation.

The surface concentration of protein at the bubble surface will be changing with protein concentration in the bulk liquid. For lysozyme, Graham and Phillips[20,21] have shown that for an air water interface, over the entire initial protein concentration range considered here, surface concentrations are well above those found for monolayer protein coverage

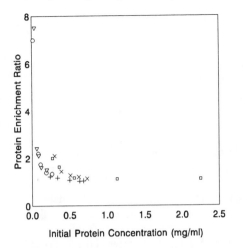

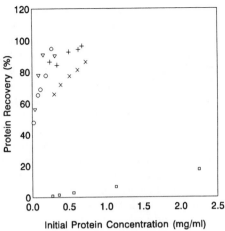

Figure 3 Dependence of Protein Enrichment Ratio on Initial Protein Concentration ▽ trypsin 1, + trypsin 2, O pepsin, □ lysozyme and x catalase (pH = pI for all proteins).

Figure 4 Dependence of Protein Recovery on Initial Protein Concentration. ▽ trypsin 1, + trypsin 2, O pepsin, □ lysozyme and x catalase (pH =pI for all proteins).

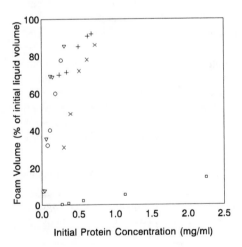

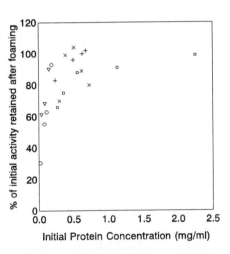

Figure 5 Dependence of Foam Volume Produced on Initial Protein Concentration. ▽ trypsin 1,+ trypsin 2, O pepsin, □ lysozyme and x catalase (pH = pI for all proteins).

Figure 6 Dependence of Percentage of Initial Enzyme Activity Retained on Initial Protein Concentration. ▽ trypsin 1, + trypsin 2, O pepsin, □ lysozyme and x catalase (pH = pI for all proteins).

(3.0 mg/m^2 for lysozyme). Also, at these concentrations, multi-layers will exist in the protein film giving surface coverages of 5-10 mg/m^2 (diffusion in the bulk liquid phase is not limiting, over the initial protein concentration range considered here[20]). Obviously, if the bubble surfaces are to be saturated in the bulk liquid phase, it is essential that the interfacial area available is not in excess of that just required for saturation conditions. For all the conditions considered here, order of magnitude calculations, to determine the interfacial area available for protein adsorption in the bulk liquid of a foam column, indicate that the protein present is always far in excess of that needed to saturate the interface (saturation values for monolayer coverage are typically 1-2 mg/m^2 [4,5]). It is therefore feasible, that, under all conditions, the gas-liquid interface (bubble surface) is saturated *ie* a concentrated protein film is formed at the interface. Such conditions favour maintenance of native structure and maintenance of activity, as discussed above.

In order to explain the variation in retention of enzyme activity with initial protein concentration, the effect of bulk concentration on foam volume collected and rates of liquid drainage in the foam must be considered. The foam phase consists of liquid associated with the interfacial protein film (*ie* at the gas-liquid interfaces in the foam) and the interstitial liquid between the foam cells. The interstitial liquid will have a concentration equal to that of the bulk liquid (assuming the interfacial concentration has reached saturation, whilst in the bulk liquid). The properties of the interstitial liquid will influence rates of liquid drainage in the foam phase. As the concentration in the interstitial liquid increases, liquid drainage will be retarded thus leading to higher foam flowrates[22] and foam volumes (see Figure 5). As the foam flowrate increases, the residence time of the protein in the foam is reduced, providing a shorter period of time for rearrangement of protein structure, which may lead to losses in specific activity. It is therefore, proposed that retention of activity decreases with decreasing concentration as a result of increased residence time of the foam phase in the column.

This theory can be tested by considering results observed for the variation of retention of activity with gas flowrate. The retention of enzyme activity increases as the gas flowrate increases (see Table 1). If the interface is saturated at all concentrations, as suggested above, then the major effect of changing the gas flowrate is to change the foam flowrate and hence residence time of the protein films in the foam phase. At low gas flowrates, the residence time is sufficient for considerable inactivation to take place. The extent of this inactivation decreases as the residence time in the column increases.

Table 1 Influence of gas flowrate on percentage of initial activity retained after foaming.

Enzyme	Initial protein concentration (mg/ml)	Gas flowrate (cm^3/min)	Protein enrichment	Protein recovery (%)	Percentage of initial activity retained (%)
Lysozyme	2.2	60	1.2	2	87
		100	1.2	3	86
		200	1.1	17	99
		400	1.0	33	98
Pepsin	0.2	100	2.2	50	74
		200	1.3	78	92
		300	1.6	78	90
Trypsin 1	0.1	100	2.2	66	66
		200	1.6	80	70
		300	1.6	80	90

6 CONCLUSIONS

Foam separation columns for recovery of enzymes can be designed to provide high levels of retained activity in the foamate. Providing that a saturated protein film is formed at the bubble interface, conditions which favour retention of activity in the foam are those which reduce the residence time in the foam phase. Any inactivation will be a function of the residence time and enzyme structure. It is likely that conditions which favour retention of enzyme activity conflict with those that favour high enrichment ratios. Therefore, it is important to decide at the outset of the design stage of a foam separation system, whether the objective is to optimise the process in terms of protein enrichment, recovery or retention of enzyme activity. This will obviously depend on the final product requirements.

7 ACKNOWLEDGEMENTS

This work was supported by the AFRC. The authors would also like to acknowledge the contributions made to the experimental work by Marie Fouilhe and Laurent Panzani.

REFERENCES

1. F. Uraizee and G. Narsimhan, Enzyme Microb Technol 1990, 12, 222.
2. F. Uraizee and G. Narsimhan, Enzyme Microb Technol 1990, 12, 315.
3. P. Sarkar, P. Bhattacharya, R. N. and M. Mukherjea, Biotech Bioeng, 1987, 29, 934.
4. M. T. A. Evans, J. Mitchell, P. R. Mussellwhite and L. Irons, "Surface Chemistry of Biological Systems", Plenum Press, Ed M. Blank, 1970, p 1.
5. L. K. James and L. G. Augenstein, "Advances in Enzymology", Interscience Publishers, Ed F. F. Nord, 1966, Vol 28, p1.
6. J. R. Hunter, P. K. Kilpatrick, R. G. Carbonell, J Colloid Interface Sci, 1990, 137, 2, 462.
7. T.A. Horbett and J. L Brash, American Chemical Society 1987, 1.
8. R. W. Schnepf and E. L. Gaden, Biochem Microb Technol Eng 1959, 1, 1.
9. L. Brown, G. Narsimhan and P. C. Wankat, Biotech Bioeng, 1990, 36, 947.
10. R. D. Gehle and K. Schugerl, Appl Microbiol Biotechnol 1984, 20, 133.
11. S. E. Charm, "Adsorptive Bubble Separation Techniques", Academic Press, Ed R. Lemlich, 1972, Chapter 9, p 157.
12. S. E. Charm, J. Morningstar, C. M. Matteo and B. Paltiel, Anal Biochem, 1966, 15, 498.
13. M. London, M. Cohen, and P. B. Hudson, Arch Biochem Biophys, 1953, 75, 1746.
14. A. Thomas and M. A. Winkler, "Topics in Enzyme and Fermentation Biotechnology", Ellis Horwood, Ed A. Wiseman, 1977, Vol 1, Chapter 3, p43.
15. D. C. Clark, L. J. Smith, and D. R. Wilson, J Colloid Interface Sci, 1988, 121, 1, 136.
16. R. E. Canfield, J Biol Chem, 1963, 238, 2698.
17. J. Withka, P. Moncuse, A. Baziotis, R. Maskiewicz, J Chromatorgr, 1987, 398, 175.
18. A. P. Ryle, "Methods of Enzymatic Analysis", Ed Bergmeyer, 1984, p223.
19. M. Anson, J Gen Physiol, 1938, 22, 79.
20. D. E. Graham and M. C. Phillips, J Colloid Interface Sci, 1979, 70, 3, 403.
21. D. E. Graham and M. C. Phillips, J Colloid Interface Sci, 1979, 70, 3, 415.
22. B. Haryono, MPhil Thesis, Reading University, submitted February 1994.

Salting Out Phenomena in Aqueous Two-phase Partition, Hydrophobic Interaction Chromatography, and Fractional Precipitation

Rayduen Wang, Karl Ottomar, Jonathan Huddleston, and Andrew Lyddiatt

BIOCHEMICAL RECOVERY GROUP, BBSRC CENTRE FOR BIOCHEMICAL ENGINEERING, SCHOOL OF CHEMICAL ENGINEERING, UNIVERSITY OF BIRMINGHAM, EDGBASTON, BIRMINGHAM B15 2TT, UK

1. INTRODUCTION.

Protein purification utilising systems composed of mixtures of polymer and salt involves the distribution of solutes between two phases one of which is rich in salt and the other rich in PEG (poly(ethylene glycol)) (1). It is attractive to consider the mechanism of partition in these systems to be one involving a balance of solubilities derived from the salting out effect of high concentrations of salt in the lower (salt rich) phase and the excluded volume of PEG in the upper (PEG rich) phase (2). In this way the similarity to other biotechnically important processes such as precipitation and hydrophobic interaction chromatography (HIC) is emphasised. Here some aspects of the behaviour of proteins in each of these biotechnically important purification procedures is examined. Specifically the partitioning behaviour of a number of pure proteins is compared during HIC and PEG-salt aqueous two-phase extraction. The partition of intracellular yeast protein is re-examined (3) and some aspects of the salting out behaviour of BSA are presented. The strong dependency of these separation processes on the solution behaviour of proteins is thereby emphasised.

2. METHODS.

Hydrophobic Interaction Chromatography of pure proteins.

HIC was performed on an LKB HPLC system equipped with a Bio-Rad PEG 300-10 column (Bio-Rad RSL, Belgium), 150mm long and 4.6mm in diameter packed with 10μm diameter silica particles of 30nm pore size and derivatised with low molecular weight PEG. Isochratic retention times of 20μl injections of commercial preparations of proteins at a concentration of 1.5 to 3 mg/ml were measured for various salt concentrations in the running buffer at a flow rate of 0.2 ml/minute. Salt concentration was varied by changing the mixture fraction of two limit buffers. The limiting buffer concentrations were 2% w/w phosphate and 30% w/w phosphate having a constant weight ratio of 18:7 K_2HPO_4 : KH_2PO_4. Under these conditions the pH varied between 7.15 and 7.48 with increasing salt concentration as a result of the change in the activity coefficient of the mono-basic salt over this concentration range. Retention was expressed as the distribution coefficient D (4,5) where,

$$D = \frac{t_e - t_0}{t_0} \qquad [1]$$

and t_e is the retention time during hydrophobic interaction chromatography and t_0 is the hold up retention time of unretained solute.

Partitioning of purified proteins.

The same commercially obtained proteins were partitioned in systems composed of two different molecular weights of PEG (1450 and 1000 Daltons, Sigma UK), and at four different tie line lengths as indicated in Figures 1 and 2 precise details of which may be found in reference 3. Potassium phosphate was of the same weight ratio as used in the HIC experiments. All experimental systems for partitioning were selected to have a volume ratio of unity, since volume ratio has been shown to have a pronounced effect on the partition coefficient of proteins at longer tie line lengths in these systems (2). From well mixed stock solutions of the selected systems aliquots of known weight, having a volume of either 10 or 5 ml were withdrawn. Accurately weighed, freeze dried protein samples were then added to a final overall concentration of 3 mg/ml. Concentrations in each phase and partition coefficients were determined for the equilibrated and separated systems.

Proteins used in correlating HIC and partition.

The following proteins, obtained from Sigma Chemicals, Fancy Road, Poole, Dorset were used to in the studies correlating HIC and partitioning; BSA, ovalbumin, cytochrome *c*, lysozyme, ribonuclease A, α chymotrypsin, myoglobin and α amylase.

Partitioning of yeast homogenate.

Yeast homogenate was prepared by wet-milling of baker's yeast (40% w/v) (Dyno-mill type KDL 0.6L) in 20 mM phosphate buffer pH 7.0 at a throughput of 13 Lh^{-1} and an agitator tip speed of 10.5 ms^{-1}. Aqueous two phase systems of differing tie line lengths and PEG molecular weights were prepared by direct addition of PEG and phosphate to 40% w/v yeast homogenate as detailed in reference 3.

Precipitation studies.

Precipitation behaviour of BSA was studied in PEG and phosphate solutions (0.05L) at protein concentrations from 1 to 8 mg/ml in presence or absence of 0.1 moles/kg potassium thiocyanate. PEG 1450 or di-potassium hydrogen orthophosphate were added incrementally and following equilibration the solution was sampled for determination of soluble protein following centrifugation.

Determination of protein concentrations.

The concentration of pure proteins was determined by U.V. spectrophotometric absorbance at 280nm and total protein in yeast homogenates by the method of Bradford (5).

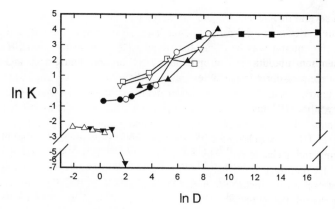

<u>Figure 1.</u> Relationship between the partition coefficient and the distribution coefficient. Partition coefficients determined at four tie line lengths in a PEG 1000 - potassium phosphate system. Distribution coefficient determined as the slope of lnD in HIC and extrapolated to the salt concentration of the lower phase. Symbols: (O) BSA, (∇) ovalbumin, (Δ) cytochrome c, ($\square$) lysozyme, ($\bullet$) ribonuclease A, ($\cdot$) α chymotrypsin, ($\blacklozenge$) myoglobin, ($\blacksquare$) α amylase.

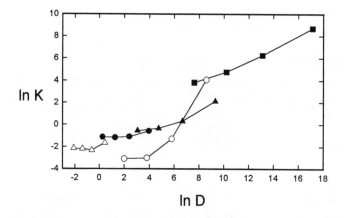

<u>Figure 2.</u> Relationship between the partition coefficient and the distribution coefficient. Partition coefficient determined at four tie line lengths in a PEG 1450 - potassium phosphate system. Distribution coefficient determined from the slope of lnD in HIC extrapolated to the salt concentration of the lower phase of the systems. For interpretation of symbols see Figure 1.

3. RESULTS AND DISCUSSION.

The partition coefficient of several pure proteins was determined in PEG-phosphate aqueous two-phase systems using PEG of nominal molecular weights (Sigma) of 1000 Da and 1450 Da at several tie line lengths. Isochratic retention times were also determined by HIC in phosphate buffer on a PEG-silica column. The salt concentration required to cause interactions on the column was less than that required for phase partition. In order to compare the two sets of data, retention on the HIC column was extrapolated to the concentration of the phosphate in the lower phase of the two phase system and the results are shown in Figures 1 and 2. There are no proteins in the sample whose behaviour is markedly different in HIC and aqueous two-phase partition. Proteins predisposed to unusual partitioning behaviour such as cytochrome *c* and α amylase are also highlighted by their behaviour in HIC. Cytochrome *c* displays a strong preference for the salt phase during partitioning and is characterised by a weak interaction during HIC even at high salt concentrations. By contrast α amylase shows strong PEG phase preference during partition and interaction is quickly promoted by moderate concentrations of salt during HIC. This emphasises the expected important role of salting out in determining the distribution of the proteins between the two mildly hydrophobic phases involved despite their differing physical nature. One may expect that the application of HIC in the preliminary design of partitioning steps for protein purification could usefully simplify the search for suitable systems by indicating the likely phase preference of target molecules relative to contaminants.

Partition of high concentrations of yeast protein (up to 40 mg/ml in one or other phase) in a series of PEG-phosphate aqueous two-phase systems covering a range of nominal molecular weights from 700 Da to 3350 Da and a range of tie line lengths from 20-50 % w/w are shown in Figures 3 and 4. Figure 3 illustrates the importance of the

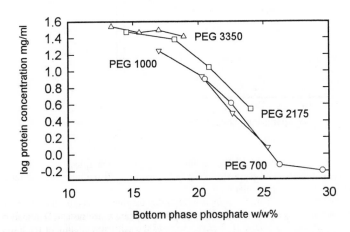

Figure 3. Log total protein concentration in the bottom phase of PEG - potassium phosphate systems as a function of phosphate concentration in the lower phase. 40% w/v yeast homogenate partitioned in systems of differing molecular weight and tie line length. Symbols denote data for systems composed of PEG 3350 Da. (Δ), PEG 2175 Da. (□), PEG 1000 Da. (∇) and PEG 700 Da. (O).

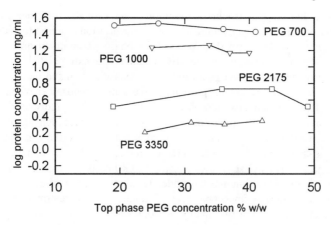

Figure 4. Log total protein concentration in the top phase of PEG - potassium phosphate systems as a function of PEG concentration in the top phase. Symbols and experimental details as for Figure 3.

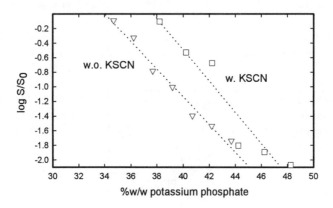

Figure 5. Precipitation of BSA by addition of di-potassium hydrogen orthophosphate in presence or absence of 0.1 M/kg potassium thiocyanate expressed as log S/S_0 : where S_0 is the initial concentration and S the final after addition of salt to a particular level.

salt concentration in determining the concentration of protein present in the lower phase of these systems which, since they are intended to model technologically useful purification systems, may be close to saturation - see (7). Regardless of the PEG molecular weight, the concentration of protein in the bottom phase is controlled solely by the salting out strength of that phase. In contrast, in the same systems the concentration of total protein in the top phase is largely unaffected by the PEG concentration in the phase and is controlled almost wholly by the polymer molecular weight. This suggests some simplification of current heuristic partitioning design procedures may be possible and that, taken in conjunction with the HIC data, the salting

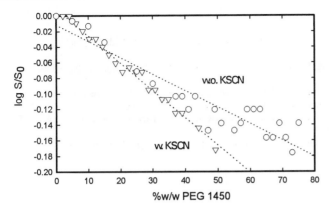

Figure 6. Precipitation of BSA by addition of PEG 1450 in presence or absence of 0.1 M/kg KSCN expressed as log S/S_0 .

out strength of the lower phase may be one of the most important factors determining observed partitioning behaviour.

Neutral salts are frequently added to aqueous two-phase systems to adjust the partitioning of proteins. This is frequently ascribed to electrostatic effects (1). The behaviour of BSA during precipitation with PEG and phosphate, with and without the addition of potassium thiocyanate, is shown in Figures 5 and 6. Figure 5 shows that the solubility of BSA in phosphate is considerably increased by the presence of the neutral salt. Figure 6 shows that there may be some decrease in solubility of BSA in PEG solutions containing neutral salt, but this is less pronounced. In itself this might be sufficient to account for the decreased partition coefficients often observed upon addition of neutral salt. Whilst not precluding the existence of electrostatic determinants of partitioning behaviour, this data suggests that these effects may be difficult to disentangle from changes in solubility behaviour brought about by neutral salt addition and from changes in intrinsic solubility due to pH change.

4. CONCLUSIONS.

The importance of salting out in PEG-salt partition has been demonstrated. This is a common feature of several generic purification processes including precipitation and HIC as well as phase partition and may be susceptible to common theoretical treatment (2,3). HIC may provide a rapid method of use in the simplification of the design of partitioning recovery steps. Salting out and exclusion are aspects of protein solution behaviour which may be of some significance in determining and modulating the activity (or its loss) of proteins within living cells and during purification procedures. Rationalisation in terms of molecular structural features of proteins should be an important current goal in biochemical recovery studies. It is likely that partitioning will have a strong role to play in the elucidation of these important features of protein structure.

5. REFERENCES.

1. J.G. Huddleston, A. Veide, K. Köhler, J. Flanagan, S-O. Enfors and A. Lyddiatt., <u>Tibtech.</u>, 1991, 9, 381.
2. J.G. Huddleston, R. Wang, J.A. Flanagan, S.O'Brien and A. Lyddiatt., <u>J. Chrom. Biomedical Applications</u>, 1994, In press.
3. J.G. Huddleston K.W. Ottomar, D. Ngonyani and A. Lyddiatt., <u>Enz. Microb. Technol.</u>, 1991, 13, 24.
4. R. Wang. MSc Examination Thesis, University of Birmingham. (1993).
5. C.J.O.R. Morris and P. Morris, 'Separation Methods in Biochemistry', Pitman, London, 1976 p35.
6. M.M. Bradford., <u>Anal. Biochem.</u>, 1976, 72, 248.
7. A.S. Schmidt, A.M. Ventom, and J.A. Asenjo., <u>Enz. Microb. Technol.</u>, 1994, 16, 131.

A Study of the Effects of Biomass Morphology Upon Partition in the System Propanol/Water/n-Decanol in the Presence of *Aureobasidium pullulans*

L. R. Weatherley,[a] E. A. Doherty-Speirs,[b] J. A. S. Goodwin,[b] and J. C. Slaughter[b]

[a]DEPARTMENT OF CHEMICAL ENGINEERING, THE QUEEN'S UNIVERSITY OF BELFAST, BELFAST BT9 5AG, UK

[b]DEPARTMENT OF CHEMICAL AND PROCESS ENGINEERING, HERIOT WATT UNIVERSITY, EDINBURGH EH14 4AS, UK

1. INTRODUCTION

The manufacture of products such as antibiotics, vitamins, amino acids, organic acids and alcohols by fermentation with microorganisms forms an important sector of the bioprocess industries. Extraction and purification of such products can be difficult and expensive as they are usually present in the final broth at low concentration, with whole cells, cell fragments, medium components such as salts and excess sugars, metabolic intermediates, and the products of side reactions. Direct product recovery from unfiltered fermentation broths by extraction, in principle results in a simpler process having fewer operations and with potential of improved process efficiency. Similar advantages would also accrue in the context of extractive fermentation processes.

The aim of this study was to determine the effects of biomass morphology upon partition of iso-propanol from broth phase into the solvent n-decanol. The broth system studied was a dimorphic fungus Aureobasidium pullulans which is an organism which can be present as both filamentous and a unicellular form, the proportion of each being controlled by growth conditions. In addition, the organism produces an extracellular polymer known as pullulan, which was of interest as it was thought that the presence of microbial polymers would have an effect on interfacial mass transfer coefficients. The amount of pullulan produced could also be controlled by the growth conditions.

Published data on whole broth extraction systems is confined largely to mass transfer studies (1,2) and to the application of specific contactor designs (3,4) . There is little systematic data published concerning partition in these or in similar systems.

A general finding relevant to partition in biological systems was reported by Walter et al (5) who found a linear relationship between the log of the partition coefficient and the surface charge of protein and viruses. It was also stated that the partition coefficient is exponentially related to both the net charge and the size of the molecule.

Another factor influencing the extraction behaviour of whole fermentation broths is the presence of biosurfactants which represent a diverse group of compounds although no systematic study of these is presented here. The major features of the fermentation system <u>Aureobasidium pullulans</u> are well documented. The effect of pH during the fermentation was investigated by Heald <u>et al</u> (6) and showed that pH exerts an important influence upon morphology, the percentage of unicells in the culture and the amount of pullulan both increasing with pH. It was noted that although the amount of unicells in the broth increased, the level of total biomass changed very little. A further observation was that pH only influenced morphology during the initial growth phase.

Other unpublished work shows that pH has an effect on the morphology of broths in continuous cultures with only yeast forms found at pH 6.0, only filamentous forms at pH 2.5, and an optimum yield of pullulan at pH 4.5. The yield of pullulan is also dependent on the morphological balance of the culture, which is affected by pH, stirring rate, temperature, and dissolved oxygen concentration.

In summary it can be stated that pullulan production in batch fermentation is initiated when ammonium is exhausted and glucose is in excess.

2. MATERIALS AND METHODS

Fermentation

The fermentation broth used in this study was produced using a 2 litre LH mini-fermenter, with a working volume of 1.6 litres. This was a batch fermenter with a direct drive agitator and two 6-blade impellers. Air was pumped into the broth through the agitator shaft, directly below the lower impeller, to aid gas dispersion. An online filter (0.45μm pore size, Whatman) was used to sterilise air. The temperature of the broth was measured using a resistance thermometer, and controlled to a pre-set value using a heating element in the broth. The pH was monitored using a probe in the broth (Ingold), and controlled at a pre-set value by the automatic addition of 0.5M sodium hydroxide and sulphuric acid. The inoculum was added aseptically to the broth, by gravity feed. The fermenter was sterilised by autoclaving while full.

<u>Aureobasidium pullulans</u> was maintained at 4°C on potato dextrose agar (PDA) slopes. These were routinely subcultured to PDA plates and used in inoculum preparation. The culture was checked for contamination by routine subculture of fermentation broths and stock cultures onto Nutrient Agar plates. The following medium was used in the preparation of inoculum cultures for fermentation, and as a control for the fermentation studies (known as APM medium). All chemicals were GPR grade (BDH) except the yeast extract which was microbiological grade (Oxoid). The pH was adjusted to pH 4.5. In the fermentation studies, the pH and ammonium sulphate concentrations were varied.

The medium contained the following in g/l:

Sucrose	30
Monopotassium phosphate	5.0
Sodium chloride	1.0
Ammonium sulphate	0.6
Yeast extract	0.4
Magnesium sulphate heptahydrate	0.2

Analysis

(a) Fungal biomass

The culture broth (10ml) was filtered through a pre-weighed and pre-dried nylon mesh filter (45 micrometer pore size, Lockertex). This removed the filamentous form of the organism, but not the yeast form. The filter was then washed with 20ml of distilled water, and dried to a constant weight in an oven, at $100^{\circ}C$.

(b) Yeast biomass

The filtrate from (a) above (10ml) was passed through a pre-dried and pre-weighed glass fibre filter (Whatman GF/C grade). The filter was then washed with 20ml of distilled water, and dried to a constant weight in an oven, at $100^{\circ}C$.

The pullulan dry weight and the cell activity were also determined routinely.

Iso-propanol in water

Iso-propanol in the aqueous phase was measured using GC analysis. The 1µl samples were injected into a carbowax column using a Hamilton syringe fitted with a Chaney adapter. The GC was set up with the injection temperature at $120^{\circ}C$, the oven temperature $80^{\circ}C$, and the detector temperature at $200^{\circ}C$. Samples were run against an external standard of propanol in distilled water. Fermentation broth samples were pre-filtered to remove biomass prior to injection in the GC.

Iso-propanol in decanol

Iso-propanol in decanol was measured using the same column but using different temperature settings. The injection temperature was set at $200^{\circ}C$, and the detector temperature set at $300^{\circ}C$. The oven was set at $80^{\circ}C$ for one minute, and then increased at a rate of $30^{\circ}C/min$ until it reached $150^{\circ}C$, at which it was held at for 3 minutes. This removed the decanol in the sample preventing overlap in further samples. Samples were run against an external standard of propanol in decanol.

Aureobasidium pullulans was grown for 48hrs in shake flask culture. (values in brackets show percentage of morphology present), see Table 1.

Table 1 Effect of broth pH on cell growth

	pH3	**pH4.5**	**pH6**
Filaments (g/l)	3.79 (85%)	2.36 (54%)	1.55 (46%)
Yeast (g/l)	0.69 (15%)	2.04 (46%)	1.83 (54%)
Pullulan (g/l)	0.30	3.38	3.36
Pullulan yield (g/g yeast)	0.43	1.66	1.83

Partition Coefficient Measurement

Known amounts of aqueous phase (1-10 g) were added to weighed amounts of decanol (approximately 10g). A range of amounts of decanol (1-10g) was also added to weighed amounts of aqueous phase (approximately 10g). The two phases were shaken together in a glass sample vial, and left for 48hr for equilibration to occur. Both phases were then sampled using disposable plastic pipettes, with great care taken to ensure there was no entrainment. The propanol concentration in each phase was estimated by GC analysis. The partition coefficient was found by taking the gradient of the line of aqueous concentration vs solvent concentration. Since both carrier phases were pre-saturated it was assumed that any effects due to mutual solubility could be neglected.

3. RESULTS AND DISCUSSION

(A) Effect of pH

Initial experiments were carried out in shake flask culture to determine the effect of pH on the morphological balance of the culture. Standard APM media was used (100ml in a 500ml conical flask). Each flask was inoculated with 5ml of a 48hr culture, grown at pH 4.5. The experiments were carried out in triplicate. The flasks were incubated at 28°C for 48hrs, on an orbital shaker at 300rpm. The flasks were then analysed for filamentous fungus concentration, yeast concentration, and pullulan concentration. The results can be seen in **TABLE 1** show that the proportion of yeast cells in the broth can be controlled by pH thus confirming earlier work.

(B) Propanol Utilisation Experiment

It was essential to eliminate the possibility of propanol utilisation by the organism in order to ensure a valid material balance during the partition experiments.

Standard APM shake flasks were grown, at pH 3.0, 4.5, and 6.0. After 48hrs of growth 0.5 ml of propanol was added to the flask, the foam bung was covered with foils to prevent evaporation, and the flasks incubated for a further 48hrs.

Table 2 Partitioning of pure propanol solutions in decanol.

Initial propanol conc (wt%)	Partition Coefficient D
1	0.66
2.5	0.68
5.0	0.71
5.0 (pH 5 buffer)	0.70
5.0 (pH 7 buffer)	0.71
10.0	0.81
20.0	0.89

Each experiment was carried out in triplicate, with water/propanol flasks incubated as a control. Samples were taken from the flasks, and the propanol concentration was measured by GC. It was found that there was no reduction in the propanol concentration, and concluded that under these conditions <u>Aureobasidium</u> <u>pullulans</u> was unable to utilise propanol.

(C) Partitioning Experiments

Each partition coefficient was calculated from at least ten sets of equilibrium concentrations and was checked for reproducibility. The data for distribution of iso-propanol between n-decanol and water are summarised in **TABLE 2**. The value of the partition coefficient is dependent on the initial concentration of the propanol solution used, with partition increasing with respect to concentration, possibly due to dimerisation of the propanol in the solvent phase.

Partition coefficients were measured in whole broths, yeast broths (where filaments have been filtered off), and in broth filtrates (where yeast cells were removed by centrifugation). Broths containing varying amounts of total cells, of filamentous forms, and of yeast were also studied. The partition coefficients are summarised in **TABLE 3**. These data show that the partition coefficients in the broth experiments were typically higher compared with those in the pure system at the corresponding starting concentrations. Of particular note is the behaviour exhibited by Broth 1 where the partition coefficient of the yeast fraction broth is higher than that of either the whole broth or of the filtrate. Broth 1 filtrate showed the lowest partition coefficient. A sample of Broth 1 (unfiltered) was stored for 3 months at 4 $^{\circ}$C, and the partition coefficient measured again. The value was 0.54, considerably lower than any previous broth value and thus showing a significant ageing effect.

In an attempt to quantify the effect of cells further, broths were produced, and cells washed and concentrated, and resuspended in both buffer and broth filtrate and the partition coefficients measured. This allowed a range of yeast and mycelial concentrations, and effect of pullulan to be studied. By resuspending some cells in buffer the effect of any surface active agents in the broth could also be quantified.

Table 3 Partitioning in broth systems

Broth System		Initial Propanol (wt%)	D
Broth 1	Whole broth	10.0	1.0
	(After 3 mths)	1.0	0.54
	Yeast fraction	10.0	1.14
	Filtrate	10.0	0.94
Broth 2	Whole broth	10.0	0.96
	Yeast fraction	10.0	0.93
Broth 3	Whole broth	1.0	0.69

Broth 1	Filaments	2.36g/l	**Broth 2**	Dry wts	**Broth 3**	Filaments	2.02g/l
	yeast	2.04g/l		unknown		yeast	1.29g/l
	pullulan	0.3g/l				pullulan	0.75g/l

Independent material balances on propanol were conducted for every set of partition data as a means of validating each extract phase concentration.

The data in **TABLE 4** show that the effect of pullulan upon partition is minimal. When cells were resuspended in buffer, the partition coefficient was only slightly higher than that for the same cell concentration in filtrate. The partition values for the "yeast-only" samples were higher than those for mycelia only samples, and the whole broth sample. All the values obtained are higher than those for pure systems under identical conditions.

Propanol can partition into decanol to a higher degree in the presence of A pullulan broths. This effect seems to be most apparent with the yeast cells. High concentrations of filamentous cells seem to depress the improved partitioning, although at concentrations up to 2 g/l there is no reduction in the partition coefficient compared to the pure system.

The effect is not solely due to the presence of cells however, as the partition coefficient for the filtrate is also higher than for a pure system. The presence of pullulan has no significant effect upon partition. The highest partition coefficient was obtained in the sample containing cells and buffer. This shows that the effect is not due to the production of soluble surface active agents during fermentation, as the cells used in the buffer samples had been washed. It is possible that the propanol itself can wash surfactants from the cell surface or make the cells more permeable, causing cell contents to leak out. This is possible, as the sample which showed the best partitioning (Y4) also had the lowest surface tension.

Table 4 Partitioning in Broth 5

Sample	D
filtrate	0.78
M1	0.71
M2	0.79
Y1	0.87
Y2	0.92
Y3	0.94
Y4	1.08
1 wt% solution	0.66
1 wt% in pH 5 buffer	0.70

Dry wt Whole Broth	g/l
Filaments	3.8
Yeast	0.15
Pullulan	0.84
M1 Filaments in filtrate	2.0
M2 Filaments in buffer	2.1
Y1 Yeast in filtrate	1.5
Y2 Yeast in filtrate	1.8
Y3 Yeast in buffer	1.7
Y4 Yeast in buffer (all filtrates contain pullulan)	1.2

From previous propanol utilisation experiments, mass balance calculations, it is known that propanol is not being taken up by the cells, nor by evaporation. There is no clear explanation for the higher partition coefficients found in the Aureobasidium broths. The yeast cells increase the partition of propanol into decanol, although the reason for this is not understood. For more conclusive evidence, experiments with very high cell concentrations (more than 5 g/l) would have to be carried out.

5. CONCLUSIONS

The partition coefficients for propanol in water (1-20 wt%) into decanol were 0.66-0.89. Aureobasidium pullulans cultures were unable to metabolise or actively take up propanol in shake flask culture. Rather than reducing propanol partitioning, the presence of A. pullulans cells, and the yeast form in particular, actually improved the partitioning of propanol into decanol. This was also true for whole broths, and for washed cells in buffer.

6. REFERENCES

1   Crabbe P.G., Tse C.W., Munro P.A., Biotechnology & Bioengineering, 1986, vol.28, 939.
2   L.R.Weatherley, G.Allen, J.A.S.Goodwin & M.B. Haig, Proceedings of The International Solvent Extraction Conference, eds. D H Logsdail & M J Slater, Elsevier Applied Science, London, 1993.
3   Anderson D.W., Lau E.F. Chem Eng Prog, 1955, 51, 507.
4   Katinger H, Wibbelt F, Scherfler H. vt-Verfahrenstechnik, 1981, No.15, 179.
5   Walter.H, Brooks. D E, Fischer. D. "Partitioning in aqueous two-phase systems". Academic Press, New York, 1985.
6   Heald P J, Kristianson B. (1985). Biotechnology and Bioengineering, 1985, 27, 1516.

7. ACKNOWLEDGEMENT

The authors acknowledge the financial support of SERC Grant No. GR/HI4175.

Spray Drying as the Final Step in Down-stream Processing of Xylanases of *Trichoderma reesei*

L. Werner, W. Stöllnberger, F. Latzko, and W. Hampel

INSTITUTE OF BIOCHEMICAL TECHNOLOGY AND MICROBIOLOGY, UNIVERSITY OF TECHNOLOGY, GETREIDEMARKT 9/172, A-1060 VIENNA, AUSTRIA

SYNOPSIS

Spray drying processes are widely used for large scale preservation of biological materials; moreover there is an increasing demand of drying catalytic active proteins. In this paper evidence is given that spray drying is a reasonable method for the recovery and preservation of Xylanases of *Trichoderma reesei*. Several concentrated enzyme solutions containing 10% (w/v) KCl, sucrose, $MgSO_4$ or K_2SO_4 as carrier substances were dried in a laboratory spray dryer at an air flow of 0.7 Nm^3/h, inlet air temperatures of 110-120°C and outlet air temperatures of 64-75°C. The best results were obtained by using KCl as carrier at an inlet air temperature of 110°C. The dried preparation showed a yield in enzyme activity of 54%.

1 INTRODUCTION

The industrial use of enzymes is often restricted by inadequate long-term stability, high viscosity and poor degree of purity of liquid preparations, which are obtained after removal of biomass. The unfavorable handling of sticky liquids initiated investigations on new downstream techniques for the preservation and stabilization of enzymes. Therefore it has been attempted to preserve enzymes by removing the adhering water. In general freeze-drying was used as very careful method yielding preparations with high activity and low water content. Despite the advantages, the time consuming and discontinuous procedure causes a high cost of operation.

In contrast spray drying does not show any of these effects and proved itself as a technique widely used for drying of bioproteins as yeast, whey or slaughterhouse blood (Masters and Vestergaard, 1978). Unfortunately the thermal stress on the enzyme protein during this procedure is rather high, which seems to be the reason that there is almost no industrial application of spray drying in enzyme technology except for thermostable proteases and amylases (Herlow, 1974). Besides thermal instability and the low protein content of enzyme solutions produced by cultivating microorganisms may cause severe problems. Because of the increasing demand for enzymatic preparations to be used in paper manufacturing, spray drying techniques have been tested and adapted for cellulases (Himmel et al., 1986). Even solutions of yeast-lytic enzymes, which exhibit maximal thermal stability at 30-50 °C and show a low protein content (1 g/l) can be dried at inlet air temperatures of 110°C-120°C. A yield in enzymatic activity of 43 to 57% has been shown, if substances acting as carriers or stabilizers were added (Werner et al., 1993).

In this paper we show that spray drying may be used for the preservation of xylanases of *Trichoderma reesei*, if carriers are added and process parameters are optimal, although protein concentration in the liquid is at a very low level (0.4 g/l) and the enzyme shows only moderate thermal stability (50 °C, Tenkanen et al., 1992).

2 MATERIALS AND METHODS

The crude enzyme solution was obtained by cultivating *Trichoderma reesei* RUT-C30 in media containing xylan and xylitol as carbon source. Biomass and solids were removed by centrifugation (Centrifuge Universal Junior III, drum insert, Heraeus-Christ GmbH, Osterode, GFR), separation (Separator LWA205; Westfalia A.G.; Oelde, GFR) and membrane filtration (cellulose acetate membrane, 0.45 µm; Sartorius AG, Göttingen, GFR). The membrane filtration was performed at a pressure of 2 bar and a temperature of 4°C. The protein solution was concentrated by crossflow-ultrafiltration at 4°C (Minisette system, membrane-cassette Alpha MW 10 kD; Filtron Tech.Corp., Clinton, MA, USA). The retentate-flow was set at 3 l/min and the membrane pressure at 1.8 bar. After ultrafiltration the membrane cassette was washed twice with 100 ml citrate buffer (0.05 M, pH=5). Spray drying was performed on a laboratory-system (Mod. 190, Büchi A.G., Flawil, Switzerland) by adjusting air-flow at 0.7 Nm^3/h, aspirator setting to 7 Nm^3/h and delivery rate of the protein solution to 250 ml/h. Inlet air temperature was varied between 110 and 120°C; the resulting temperature of the outlet air was in the range 64-78°C. A volume of 100 to 500 ml of enzyme solution was dried during one trial; substances such as KCl, K_2SO_4, $MgSO_4$ and sucrose (10% final concentration) were added as carrier and stabilizer, respectively, and the isolated powder was stored at 4°C. For enzyme activity determination 0.1g powder was dissolved in 10 ml citrate buffer (0.05 M; pH=5) and dialyzed against 500 ml of the same buffer for at least 10 hours.

Assay of enzyme activity: The standard reaction mixture consisted of 0.5 ml of suitably diluted enzyme and 1 ml substrate (10 mg/ml, xylan [Lenzing A.G., Lenzing, Austria] suspended in citrate buffer (0.05 M; pH=5). The solution was incubated at 50°C for 15 min and submitted to the determination of reducing sugar (Miller, 1959). One unit was defined as that amount of enzyme liberating 1µmole of glucose per min under these conditions.

Protein was estimated according to Bradford (1976).

3 RESULTS

Recovery of xylanase

Even after centrifugation and separation some of the autolysed mycelium and the unconsumed insoluble carbon source remains in the culture liquid. For removing the solids and the stable foam, additional membrane filtration is essential. This step can be performed without great loss in activity. The clear yellow solution can be concentrated easily by crossflow-ultrafiltration yielding protein concentrations of 4 g/l. The filtration rate of the enzyme solution is about 0.5 h/l. Apparently 80% of the total activity of the crude liquid are found in the retentate solution and 0.4% in the filtrate (**table 1**). Cleaning the membranes with buffer shows no effect; the washing solutions contain only 0.1% of the total activity.

Table 1   Recovery of Xylanase

	V [ml]	v.act [U/l]	t.act [U]	η [%]
culture filtrate	8300	134 000	1 112 400	100.0
membrane filtrate	8230	134 000	1 103 000	99.2
ultrafil. permeate	7230	620	4 500	0.4
ultrafil. concentrate	990	910 500	910 500	81.5

V [ml] ... volume of enzyme solution
v.act [U/l] ... volumetric activity of the enzyme solution
t.act [U] ... total activity
η [%] ... yield in enzyme activity

Optimization of Spray drying parameters

The main parameters to be manipulated in laboratory models of spray dryers include inlet air temperature, delivery rate of the sample, aspirator capacity and air flow at the nozzle (= spray flow).

Aspirator capacity at 7 Nm^3/h and a spray flow 0.7 Nm^3/h minimize product losses from particle precipitation. The influence of inlet (T_1) and outlet temperature (T_2) on total activity has been studied in several experiments. As can be seen in **table 2** the yield of activity (η) increases with decreasing inlet air temperature, the recovery of solid substance (RC) shows a contrary effect. At inlet air temperatures below 115°C there is an imminent danger of forming deposits in the column and the cyclone of the drying system. The water content (WC) of the solid product depends on both temperature of the inlet air and delivery rate of the enzyme solution. In order to minimize the outlet air temperature and water content of the product, the delivery rate has been set at 220 ml/h.

Table 2 Influence of Operating Temperatures on Yield of Xylanase Activity

$T_1[°C]$	$T_2[°C]$	RC[%]	WC [%]	$\eta[\%]$
120	75	90	1.6	34.6
115	71	93	1.6	49.1
110	64	79	1.8	54.0

The influence of the type of carriers has been studied in further experiments with KCl, K_2SO_4, $MgSO_4$ and sucrose as carrier substances (**table 3**). Inlet air temperature was set at 115°C, sample delivery rate at 250 ml/h. The carrier concentration was 10% in the sample liquid. As can be seen in **table 3**, the water content of the dried preparations is different. The hygroscopy of some carrier substances may cause problems. In the case of sucrose and magnesium sulphate the fine powder aggregated into large lumps. With the exception of those using sucrose as carrier, the enzyme preparations showed no loss of activity after storing at 4°C for one year.

Table 3 Percentage of recovered enzyme activity with different carrier substances

Carrier	η [%]	WC [%]
KCl	49.1	1.6
Sucrose	9.6	2.4
KCl-Sucrose 1:1	17.8	2.2
$MgSO_4$	58.1	14.5
K_2SO_4	48.9	1.2

Spray Drying of a whole batch

The recovery from the fermentation broth was achieved as already described. An amount of 50g potassium chloride was added to a half of the concentrated solution (500 ml). Drying was achieved at a heating block temperature of 115°C and a sample feed rate of 250 ml/h.

Outlet air temperature was influenced by changes in process parameters. Consequently a profile of inlet and outlet air temperature allows control of the drying process. **Figure 1** shows a profile of inlet and outlet air temperature. The monitoring of temperature was started two minutes after switching on the heating and at t=0.25 h distilled water was pumped into the system. At t=0.7 h the parameters were constant and the sample was delivered to the heating block. After 2 hours the whole solution was dried.

The recovery of solid substance was 91 % and that of enzymatic activity 52 %. The specific activity of the dried enzyme preparation was 5318 U/g. <u>The total</u> <u>recovery referring to culture broth was</u> <u>43%.</u>

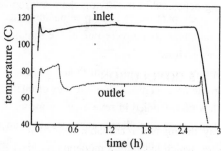

Figure 1 Time course of air temperature

4 DISCUSSION

Spray drying procedures have been used predominantly in industrial enzyme technology for the preservation of heat stable amylases and proteases (Neubeck, 1980) and inclusion of enzymes in surfactants for dustless preparations (Herlow, 1974). In recent years this technique has been used for the preparation of cellulolytic (Himmel et al., 1986), pectinolytic (Mikhailova et al., 1989) and lipolytic enzymes (Kerzel and Mersmann, 1993).

In general drying was achieved by inlet air temperatures of 120-180°C and settings of air and liquid flow rate which resulted in outlet air temperatures of 40-80°C. But high temperatures and the resulting thermal stress are extremely detrimental to non-thermostable enzymes and will give only poor activity yields. Nevertheless, the results presented in this paper show that inlet air temperatures below 120°C can be used in spray drying of non-thermostable enzymes. Moreover, a considerable influence on the enzyme activity recovered can be attributed to the type of substance added to the enzyme solution as stabilizer or carrier. Addition of polyols (sorbitol), disaccharides (sucrose, lactose), polysaccharides (starch, cellulose, galactomannan, carboxymethylcellulose), polymers (polyvinylpyrrolidone) and their derivates, as well as a broad variety of soluble and insoluble salts (KCl, $MgSO_4$, $Ca_3(PO_4)_2$) was tested and studied by several authors (Yoo and Lee, 1993; Yamada and Soga, 1992; Neubeck, 1980; Gölker, 1987; Mikhailova et al., 1989; Werner et al., 1993). Beside its stabilizing effect on the enzyme protein the physical properties of the additive will influence its applicability when used together with low inlet air temperatures for spray drying. Additives forming crystalline powders should be preferred as there are no heavy wall deposits and the product can be easily collected from the cyclone. In contrast polyols and some sugars form honey-like droplets, most of them being deposited in the drying chamber from which recovery is difficult.

5 CONCLUSION

Spray drying is a reasonable method for the preservation of Xylanase of *Trichoderma reesei*, which has only a moderate thermostability (50°C, Tenkanen et al, 1992). For that reason losses in enzyme activity of 40% and more are explainable, even in the presence of stabilizers. The kind of stabilizer added influences considerably the maximum recovery of the enzyme. Consequently the search for new and better carrier substances should be promoted urgently. At the moment an industrial application of spray drying is charged by the high loss in activity.

6 REFERENCES

1. M. Bradford, Anal. Biochem., 1976, 73, 248-254
2. C. Gölker, Biotech.Forum, 1987, 4, 158-166
3. A. Herlow, Brit. 1,344,253, 1974
4. M. Himmel, K. Oh, M. Tucker, C. Rivard, K. Grohmann, Biotechnol. Bioeng. Symp., 1986, 17, 413-423
5. P. Kerzel, A. Mersmann, BioTec, 1993, 5 (1), 44-47
6. K. Masters, I. Vestergaard, Process Biochem., 1978, 13 (1), 3-6
7. R.V. Mikhailova, A.G. Lobanok, L.I. Sapunova, Ser. Biyal. Navuk, 1989, 6, 62-65
8. G.L. Miller, Anal.Chem., 1959, 31, 426-28
9. C.E. Neubeck, U.S. 4,233,405, 1980
10. M. Tenkanen, J. Puls, K. Poutanen, Enzyme Microb. Technol., 1992, 14 (7), 566-74
11. L. Werner, F. Latzko, W. Hampel, Biotechnol. Tech. 1993, 7 (9), 663-666
12. N. Yamada, Y. Shoga, Europ.Pat.Appl. EP 501.375, 1992
13. B. Yoo, C.M. Lee, J.Agric.Food Chem., 1993, 41, 190-192.

Extraction of Penicillin G with Sulfoxide Extractants

Lei Wen and Li Zhou

DEPARTMENT OF CHEMICAL ENGINEERING, TSINGHUA UNIVERSITY, BEIJING 100084, PEOPLE'S REPUBLIC OF CHINA

1 INTRODUCTION

Penicillin G is an important and widely—used antibiotic, its traditional extraction process with n—butylacetate from filtered fermentation broth has the disadvantages of high power consumption for recovery of n—butyl acetate from the raffinate of extraction process and a lower recovery yield of penicillin G (the typical recovery yield for two stages of solvent extraction is about 92–96%, the recovery yield of back—extraction is about 95–97%). To overcome these shortcomings, Schugerl and others [1-8] have been researching the reactive extraction of penicillin systematically and some new extraction systems containing amine extractants were adopted. Li Zhou et al [9] carried out similar research work. Yan Zhifa [10] researched TBP extraction system. These new extraction systems are superior to the single n—butyl acetate system for penicillin extraction. The recovery yield of penicillin G during the extraction process may be increased, but owing to a little difficultly in the back-extraction process, in addition to the toxicity of amine and phosphorus—based extractants, these new extractants are restricted in practice. Thus, in our experiments the sulfoxide extractants petroleum sulfoxide (PSO) and di—isooctyl sulfoxide (DISO), were adopted for penicillin extraction.

The extraction equilibrium and extraction rate were researched by using an imitation feed (model media) and filtered fermentation broth of penicillin G. The mathematic models of extraction equilibrium were developed.

Then, the bench—scale extraction experiments are carried out in the cascade of centrifugal extractors.

2 EXPERIMENTAL

2.1 Extraction System

The adopted extractant, petroleum sulfoxide, is a mixture of sulfoxide (80%) and non—sulfoxide (20%), produced from diesel oil with high content of sulfur. The structure of sulfoxide is as follows:

$$\text{(structure)} - C_{12}H_{25}$$
$$S$$
$$\overset{\|}{O}$$

Di–isooctyl sulfoxide is synthesized by ourselves and its structure is:

$$CH_3-(CH_2)_4-\underset{\underset{C_2H_5}{|}}{CH}-\underset{\underset{O}{\|}}{S}-\underset{\underset{C_2H_5}{|}}{CH}-(CH_2)_4-CH_3$$

both of them are nontoxic in reality.

The physical parameters of them are tabulated in Table 1.

Table 1   *Physical properties of* PSO *and* DISO

physical properties Extractants	molecular weight	density, d_4^{25}	viscosity, mN / m	dioptre n_D^{25}	boiling point, ℃	solubility in H₂O	toxicity
PSO	250	0.95	14.4 (28℃)	1.539	300	0.179g / 100g (25℃)	non
DISO	274	0.8995	24.09 (25℃)	/	/	low	non

The diluent is sulphonated kerosene, and a certain percent of n–octanol is added to prevent the formation of a third phase.

The aqueous feed is an imitation feed and filtered fermentation broth of penicillin G, and the later is the practical fermentation broth produced in the pharmaceutical factory.

2.2 Experimental Installation

The extraction equilibrium experiments were carried out in specially made test tubes, and centrifugal extractor with rotary drum ofφ20mm is used for bench–scale experiments.

The aqueous and organic phase were pumped by metering pump, and the countercurrent extraction is realized.

2.3 Experimental Results

2.3.1 Petroleum sulfoxide–n–octanol–sulfonated kerosene System.   At first, the experiments of extraction performance comparison of PSO with n–butyl acetate were carried out, the experimental results are plotted in Fig.1.

It can be seen from Fig.1 that the extraction performance of PSO is superior to that of n–butyl acetate for penicillin extraction.

Since penicillin G is easier to degrade at low pH value than at high pH value, and the extraction of penicillin G can be carried out at higher pH value with PSO than that with n–butyl acetate. So the degradation rate of penicillin G decreases during the extraction process of PSO.

For PSO system the extraction equilibrium data at various equilibrium pH values in aqueous feed and various extractant concentrations in organic phase were determined. Experimental results at ambient temperature are illustrated in Fig.2 and 3.

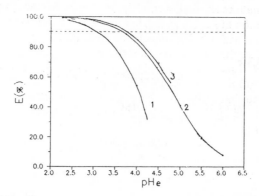

Figure.1   Comparison of extraction performance of PSO, DISO with n–butyl actate for penicillin G
1–n–butyl acetate, 2–30% PSO–5% n–octanol–65% sulphonated kerosene(V / V), 3–30% DISO–5% n–octanol–65% sulphonated kerosene(V / V).

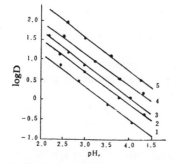

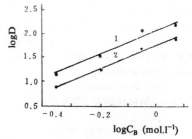

Figure.2   Relationship between extraction distribution coefficient of penicillin G and equilibrium pH value in aqueous feed
Cp = 0.03 mol / l
[PSO]: 1–5%, 2–10%, 3–15% 4–20%, 5–30%(V / V)
A / O = 2 / 1

Figure.3   Relationship between extraction distribution coefficient of penicillin G and PSO concentration in organic phase
Cp = 0.032 mol / l
pHe: 1–2.5, 2–3.0,
A / O = 2 / 1

(Cp: Initial concentration of penicillin G in aqueous feed)

It can be seen from Fig.2 that the slopes of curves between lg D and pHe are about −1, the slopes of curves shown in Fig.3 are equal to 2. The experiments discovered also that the effect of penicillin G initial concentration on its distribution coefficient is not notable because the penicillin concentration in the organic phase is far from its saturation concentration.

The extraction of penicillin G was carried out also from the fermentation broth, and the experimental results show that the extraction yield is a little lower than that from the model media.

The extraction rate of penicillin G for PSO system was also examined, experimental results show that the extraction equilibrium can be reached in less than 10s.

The back–extraction of penicillin G from this extraction system is easy and the back–extraction yield is near to 100% under the condition of O / A = 2 / 1 by using 0.15 mol / l NaHCO₃ as the back–extraction reagent.

2.3.2 Di–isooctyl   sulfoxide   (DISO)–n–Octanol–Sulphonated   Kerosene System. The extraction performance of DISO system for penicillin G is very similar to but is superior to that of PSO system, and the extracted amounts of miscellaneous acids and pigments are lower than that of the PSO system, The experimental results are plotted in Figs. 4, 5.

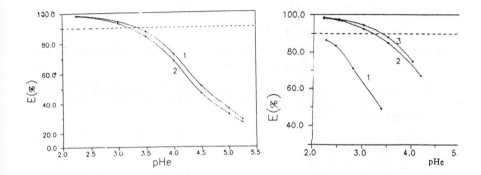

Fig.4   *Degree of extraction of penicillin G as a function of the pH value in real fermentation broth*
1. *30%DISO–5%n–Octanol –Kerosene (V / V); Cp=0.031mol / l.*
2. *5%DISO–n–butyl   acetate (V / V): (O / A=1 / 3, Aqueous phase: Fermentation broth+0.025%PPB)*

Fig.5   *Extraction of penicillin G with DISO–kerosene system*
1. *10%DISO–5%n–Octanol –Kerosene, 2. 25%DISO– 5%n–Octanol–Kerosene, (V / V)*
3. *30%DISO–5%n–Octanol –Kerosene (V / V), (Cp=0.033mol / l, A / O=1 / 2, model media)*

3 EXTRACTION DISTRIBUTION MODELS

Based on the extraction equilibrium experiments, the extraction reaction is supposed as follows:

$$HP \overset{ka}{\rightleftharpoons} P^- + H^+ \tag{1}$$

$$nP^- + nH^+ + qB \overset{K}{\rightleftharpoons} (HP)_n \cdot B_q \tag{2}$$

Where, HP represent penicillin G, B represents extractant.

From the definition of equilibrium constant and equation (1) and (2), we have:

$$k_a = \frac{[H^+][P^-]}{[HP]} \tag{3}$$

$$K = \frac{[(HP)_n.B_q]}{[P^-]^n.[H^+]^n[B]^q} \tag{4}$$

the extraction distribution coefficient

$$D = \frac{n[(HP)_n B_q]}{([P^-]+[HP])} \tag{5}$$

Combine equations (6), (4), (3) , and substitute $k_a = 10^{-2.75}$

$$K = \frac{D}{n} \frac{(1+[H^+]/k_a)^n}{([P^-]+[HP])^{n-1}[H^+)^n[B]^q} \tag{7}$$

$$LOG(D) = (LOG(K) + LOG(n)) - LOG([P^-] + [HP]) + n[LOG([P^-] $$
$$+ [HP]) - PH_e - LOG(1 + 10^{2.75 - pH_e})] + qLOG[B] \tag{8}$$

Regressing the extraction equilibrium data the following extraction distribution models were developed separately for PSO and DISO extraction systems:
For PSO extraction system;

$$LOG(D) = 6.8786 + 0.3025LOG(Ce) - 1.3025[pHe + LOG(1 + 10^{(2.75 - pH_e)})]$$
$$+ 2.1228LOG[B]_e \tag{9}$$

where, Ce is the equilibrium concentration of penicillin G in aqueous phase, $Ce = ([P^-] + ([HP])$
$[B]_e$ is the equilibrium concentration of PSO in the organic phase.
Under the conditions: the initial concentration of penicillin G in aqueous feed is $0.03 \sim 0.05$ mol / l, the initial concentration of PSO in organic phase is $0.6 \sim 1.5$ mol / l, and pHe < 4, equation (1) can be simplified as follows:

$$LOG(D) = 5.7256 - 1.3025pHe - 1.3025LOG[1 + 10^{(2.75 - pHe)}]$$
$$+ 2.1228LOG[B]_o \tag{10}$$

where, $[B]_o$ represents the initial concentration of PSO in organic phase.
For DISO extraction system, the following model was developed;

$$LOG(D) = 7.24 + 0.148LOG(Ce) - 1.148[pHe + LOG(1 + 10^{(2.75 - pHe)})]$$
$$+ 1.881LOG[B]_e \tag{11}$$

Thus, it was deduced that the extract structure of penicillin G is (HA) $\cdot$ B$_2$, and it was examined by infrared spectrum analysis.

4 BENCH–SCALE EXTRACTION EXPERIMENTS

Bench–scale extraction experiments were carried out with the DISO extraction system by using a cascade of centrifugal extractors with ϕ20mm rotary drum (extraction: 3 sets, back–extraction: 2 sets).
According to the extraction equilibrium data, the experimental conditions are

determined as follows:

Extraction step; aqueous feed: filtered fermentation broth of penicillin G, 0.25% of PPB was added for de−emulsifying, and the acidity was regulated by adding H_2SO_4. Organic phase: 30% DISO−5% n−octanol−kerosene

    flow ratio: A / O = 3 / 1

    Back−extraction step; back−extraction reagent: 0.30mol / 1 $NaHCO_3$

    flow ratio: A / O = 1 / 3

The experimental results are tabulated in the following table 1.

Comparing the tabulated data it can be seen that the total recovery yield of Penicillin G in the extraction process with DISO extraction system is higher than that in the traditional n−butylacetate extraction system.

In this paper the emulsification during the extraction process was researched. The experiments show that emulsifying is result from the reaction of denatured protein with organic phase.

The methods for de−emulsifying were researched. The experimental results show that ultrafiltration of filtrered fermentation broth of penicillin G is an effective method for eliminating protein, for example, emulsification during the extraction of penicillin G can be eliminated completely when PAN flat or tubular membrane module with interception of molecular weight 30,000 is adopted for protein ultrafiltration before extraction

Table 2   *Bench−scale extraction experimental results* [*]

No.	1	2	3	4
Extraction system	30% DISO −5% n−Octanol −Kerosene	30% DISO −5% n−Octanol −Kerosene	30% DISO −5% n−Octanol −Kerosene	n−Butylacetate
Concentration of penicillin G (mol / l)	0.021	0.013	0.038	0.022
pHe (extraction)	2.2–2.5	2.4	~3	1.8–2.2
pHe (back− extraction)	7.0	7.3	/	7.25
Extraction yield (%)	95.4	94.5	97.8	93.4
Back−extraction yield (%)	98.6	99.5	/	96.9
Total recovery yield (%)	94.06	94.03	/	90.5

  [*] : phase ratio in extraction: A / O = 3 / 1. in back−extraction: A / O = 1 / 3

5 CONCLUSIONS

(1) Based on the study on extraction equilibrium of penicillin G extraction dis-

tribution models are developed for PSO and DISO extraction system separately.

(2) The extraction performances of new extraction systems, PSO and DISO for penicillin G are superior to that of n–butyl acetate, the recovery yield of penicillin G may increase by 3%.

(3) Ultrafiltration of the filtered fermentation broth can eliminate the protein effectively, decrease the emulsifying trend, and a new extraction flowsheet with ultra filtration step was presented.

REFERENCES

1. M.Reschke and K.Schugerl, Chem, Eng. J., 1984, 28, B1.
2. M.Reschke and K.Schugerl, Chem, Eng. J., 1984, 28, B11.
3. M.Reschke and K.Schugerl, Chem, Eng. J., 1984, 29, B25.
4. M.Reschke and K.Schugerl, Chem, Eng. J., 1985, 31, B19.
5. M.Reschke and K.Schugerl, Chem, Eng. J., 1985, 32, B1.
6. Z.Likidis and K.Schugerl, Biotech. and Bioeng., 1985, 30, 1032.
7. Z.Likidis and K.Schugerl, Biotech. Eng., 1985, 3, 79.
8. Z.Likidis and K.Schugerl, Chem. Eng. Sci., 1985, 43, 27.
9. Li zhou et al., Research Report for Extraction of Penicillin G with Amine Extractants, 1990.
10. Yan Zhifa, Yu Shuqiu and Chen Jiayong, Eng. Chemistry & Metallurgy (China), 1992, 13, 51.

Purification and Characterization of Recombinant MPB70

A. Whelan, W. P. Russell, F. Nawaz, D. G. Newell, and
R. G. Hewinson

CENTRAL VETERINARY LABORATORY, NEW HAW, ADDLESTONE, SURREY,
UK

1   INTRODUCTION

MPB70 is an immunogenic secreted antigen of *Mycobacterium bovis*.[1]   This protein contains species specific epitopes[2] and is a candidate antigen for the development of blood based veterinary diagnostic tests for *M.bovis* infection.[3] Due to the difficulties in growing *M.bovis* and subsequently purifying the antigen, a recombinant form of MPB70 has been produced in *Escherichia coli*.[4]   This protein is secreted into the periplasmic region of the cell[4] following cleavage of the signal peptide.   In this way MPB70 can be separated from contaminating cytoplasmic proteins.   In this manuscript a simple technique is described for the purification of recombinant MPB70 from the periplasmic fraction of *E.coli* using a two stage electrophoretic technique.   The purified recombinant protein has been characterised and its immunological activity assessed.

2   METHODS

2.1   Osmotic Shock Procedure.   Recombinant *E.coli*, expressing MPB70, was cultured overnight at 37°C in 4 litres of Luria-Bertani broth containing 0.1% (w/v) glucose and 2ml ampicillin (100mg/ml).   The culture was harvested by centrifugation at 7000g, 15 min at 4°C and resuspended in 400ml ice-cold 20% (w/v) sucrose, 10mM Tris/HCl (pH 7.5). Twelve millilitres of EDTA solution (0.5M, pH 8) was added and incubated for 10 min on ice.   The cell suspension was centrifuged for 10 min at 11,500g, 4°C.   The supernatant fluid was removed and the pellet rapidly resuspended by vigorous agitation in 50ml ice cold distilled water.   The mixture was incubated for 10 min on ice, centrifuged for 10 min at 11,500g, 4°C, and the supernatent (periplasmic fraction) harvested.

2.2 Preparative Isoelectric Focusing. Carrier ampholytes (1ml), pH range 3-10, were added to the periplasmic fraction. This mixture was loaded onto the Bio-Rad Rotofor isoelectric focusing (IEF) cell and run for 5 hours at constant power (12W). The fractions were then harvested and those fractions containing recombinant MPB70 identified using analytical SDS-PAGE followed by western blotting[4] with a monoclonal antibody specific for MPB70, SB10.[2] The fractions containing MPB70 were pooled and concentrated to 1ml using Centricon 10 (Amicon) centrifugal concentrating tubes.

2.3 Preparative SDS-PAGE. The Bio-Rad Model 491 Prep-Cell was used for preparative SDS-PAGE. The discontinuous buffer system of Laemmli[5] was used for continuous elution gel electrophoresis. An acrylamide concentration of 15% (15%T) was used to separate the polypeptides. The volume of the separating gel, cast in the 28mm internal diameter casting tube, was 40ml, the volume of the stacking gel (5%T) was 6ml. The MPB70 fractions obtained following preparative IEF were mixed 1:1 (v/v) with sample buffer (62.5mM Tris-HCl pH6.8, 10% (v/v) glycerol, 10% (w/v) SDS, 0.05% (w/v) bromo-phenol blue and 2% (v/v) beta-mercaptoethanol) and incubated at 95°C for 5 min. The sample (6mg total protein) was then loaded onto the Bio-Rad Model 491 PrepCell. The PrepCell was run at constant current (40mA) for 10 hours. Running buffer was circulated through the elution chamber at 1ml/min; 5ml fractions were collected. The fractions containing MPB70 were identified by western blotting as before. These fractions were pooled and concentrated to 1ml as before followed by dialysis in phosphate buffered saline (PBS, pH7.2).

2.4 Amino Acid Sequencing. The N-terminal amino acid sequence was obtained for purified recombinant MPB70 by automated Edman degradation from 100µg of protein.

2.5 Analytical 2-D PAGE. Purified recombinant MPB70 (40µg) was separated by analytical 2-D PAGE (Millipore 2-D Investigator) according to manufacturers instructions. Separated polypeptides were stained using a silver stain kit (Bio-Rad). Antibody activity was assessed by western blotting using sera from a cow naturally infected with *M.bovis* or using the monoclonal antibody SB10.

2.6 T-Cell Proliferation Assay. Peripheral blood mononuclear cells were separated from fresh heparinized blood obtained from two *M.bovis* infected (CG165 and CG145) and two uninfected (CF160 and CF56) cattle by density gradient centrifugation. 15ml of blood was layered onto 15ml Histopaque 1.077 (Sigma) in a 50ml Accuspin tube and centrifuged at 1000g for 30 min at room temperature.
The mononuclear cells at the interface were harvested and washed three times with Hanks Balanced Salt Solution (containing 1% heparin) by centrifugation at 800g for 10 min. The final cell pellet was resuspended in 1ml of complete RPMI 1640 medium.

Viable cells were enumerated by 1% erythrosin dye exclusion and adjusted to 2×10^6 cells/ml.

96 well, flat-bottomed, microtitre plates (Nunc) were used for the proliferative assays. Solutions of native and recombinant forms of MPB70 (100µl) were dispensed in triplicate at an initial concentration of 20 and 10 µg/ml and doubling dilutions using complete RPMI were performed in the microtitre plate to 9 and 4 ng/ml respectively. Cells from each animal were dispensed into the wells at 2×10^6 cells/ml in 100µl volumes. Complete RPMI media (100µl) was added to the control unstimulated wells. The total culture volume was 200µl for all assays. The cells were incubated for 5 days at 37°C in a humidified atmosphere of 5% CO_2 and 95% air followed by an 18 hour pulse of 0.5mCi ^{3}[H]thymidine (specific activity : 25 Ci/ml). The cells were harvested on glass fibre filters using a cell culture harvester (Innotech). The filters were dried and the incorporation of ^{3}[H]thymidine determined using a Beta plate counter (Packard Matrix 96). Two separate experiments were performed and the results expressed as the difference between the mean cpm with antigen and the mean cpm without antigen.

3 RESULTS

3.1 Purification of Recombinant MPB70. Following IEF and preparative SDS-PAGE purification, recombinant MPB70 was purified from the periplasmic proteins (Figure 1, lane 1) to a single band as observed on a silver stained SDS-PAGE gel (Figure 1, lane 3). The yield of purified recombinant MPB70 was 0.5mg from 2.5g (wet weight) of *E.coli* cells, this was equivalent to 0.5mg of purified protein recovered from 30mg of periplasmic proteins. Western blotting of the purified sample against the MPB70 specific monoclonal antibody SB10 confirmed the single protein band to be recombinant MPB70 (data not shown).

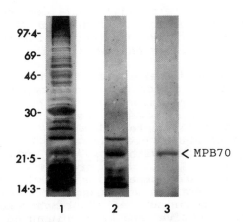

Figure (1). *Silver stain of the periplasmic sample (1), Rotofor sample (2), and PrepCell sample (3).*

3.2 Physical Characterisation of Recombinant MPB70.
Using N-terminal sequencing, the first 17 amino acids of the
purified recombinant MPB70 were shown to be identical to
those of the native protein[6] (Figure 2).

1				5					
Gly	Asp	Leu	Val	Gly	Pro	Gly	Val	Ala	Glu

10				15		
Tyr	Ala	Ala	Ala	Asn	Pro	Thr

Figure (2). *N-terminal amino acid sequence of recombinant
MPB70.*

Following separation by analytical 2-D PAGE, isomeric
forms of recombinant MPB70, which differed slightly in
isoelectric point, were observed (Figure 3a).

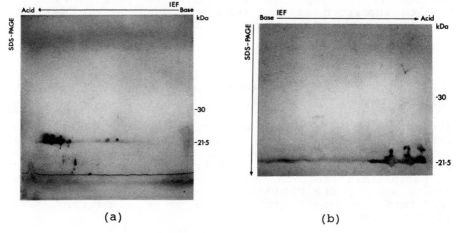

                    (a)   (b)

Figure (3). *Silver stain (3a) and western blot (3b) of
purified recombinant MPB70 separated on analytical 2-D PAGE gels incubated with sera from an
M.bovis infected cow.*

3.3 Immunological Characterisation of Recombinant MPB70
The B-cell reactivity of purified recombinant MPB70 was
investigated by separation of the purified protein by
analytical 2-D PAGE followed by western blotting using sera
from a cow naturally infected with M.bovis. All three
isomeric forms of MPB70 were recognised by the sera from the
cow naturally infected with M.bovis (Figure 3b). In order
to investigate the T cell proliferative response to
recombinant MPB70 and compare its activity with the native
protein, the polypeptides were used for T cell proliferative
assays. The comparison of T lymphocyte proliferative
responses using native and recombinant MPB70 for cells
harvested from M.bovis infected and uninfected cattle showed
that recombinant MPB70 elicited a comparable proliferative
response to that observed for the native antigen (Figure 4).

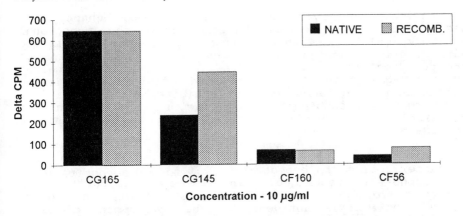

Figure (4). *Comparative bovine T lymphocyte proliferative response to native and recombinant MPB70. CG165 and CG145, M.bovis infected; CF160 and CF56 uninfected.*

4   DISCUSSION

The intention of this study was to develop a technique for the bulk purification of an immunologically active form of the *M.bovis* antigen, MPB70. In the past the two stage electrophoretic technique utilized in this study has proved to be a simple two step method for the bulk purification of proteins[7] and therefore this technique was utilized to develop a purification strategy for MPB70.

A crude protein mixture containing recombinant MPB70 was obtained from the periplasmic region of an overnight culture of *E.coli* cells by osmotic shock. This enabled separation of recombinant MPB70 from the cytoplasmic proteins which may constitute up to 80% of all *E.coli* proteins. Recombinant MPB70 was then purified to homogeneity using a two stage electrophoretic technique. Purified protein (0.5mg) was recovered from 30mg of *E.coli* periplasmic proteins. The further assessment of the purity of recombinant MPB70 by analytical 2-D PAGE showed the existence of isomeric forms of the same molecular weight but differing in isoelectric point. The existence of 3 isomeric forms of native MPB70 has previously been reported.[8]

To ensure that recombinant MPB70 was cleaved by the *E.coli* endopeptidase at the same site as by the *M.bovis* endopeptidase, the N-terminal sequence of the first 17 amino acids of purified recombinant MPB70 was determined. The sequence of recombinant MPB70 was found to be identical to that of the native protein confirming the cleavage position.

Antibody reactivity of the purified recombinant MPB70 was confirmed by western blotting with the monoclonal antibody, SB10, and with sera from a cow naturally infected with *M.bovis*. Recognition of all three isomeric forms of the purified protein was observed following probing against SB10 and against the *M.bovis* infected cattle sera confirming that each of the isomeric forms exhibited B cell activity. Furthermore, recombinant MPB70 was shown to elicit a comparable T-cell proliferative response to that observed for the native antigen.

This report has described the purification and characterisation of recombinant MPB70 by use of a two stage electrophoretic technique. The use of this purification strategy for bulk purification of MPB70 and other recombinant *M.bovis* antigens for the development of blood based diagnostic tests for *M.bovis* infection is under investigation.

5   REFERENCES

1.   Fifis et al., Scan. J. Immunol., 1989, 29, 91.

2.   Wood et al., J. Gen. Microbiol., 1988, 134, 2599.

3.   Harboe et al., J. Clin. Microbiol., 1990, 28, 913.

4.   Hewinson and Russell, J. Gen Microbiol., 1992, 139, 1253.

5.   Laemmli et al., Nature, 1970, 227, 680.

6.   Radford et al., J. Gen Microbiol., 1990, 136, 265.

7.   Hochstrasser et al., Applied and Theoretical Electrophoresis, 1990, 1, 265.

8.   Harboe et al ., Infect. Immunol., 1986, 52, 293.

Polymeric Adsorbents for Cholesterol Prepared by Molecular Imprinting

M. J. Whitcombe, M. E. Rodriguez, and E. N. Vulfson

BBSRC INSTITUTE OF FOOD RESEARCH, READING LABORATORY, EARLEY GATE, WHITEKNIGHTS ROAD, READING RG6 2EF, UK

ABSTRACT

A new method of molecular imprinting, based on 4-vinylphenyl carbonate esters is described. Imprinted polymers using cholesterol as the template were prepared with surface areas of 26 and 440 m^2g^{-1}. Both polymers show specific binding of cholesterol. Rapid binding kinetics were shown by the high surface area polymer but at the expense of a large non-specific contribution to the overall binding.

1. INTRODUCTION

The technique of molecular imprinting is still a developing technology, with initial reports of several new applications for imprinted polymers published in 1993 (see accompanying paper,[1] and references therein). Most of these recent advances have centred around polymers synthesized by the non-covalent approach to molecular imprinting developed by Mosbach and co-workers.[2-5] This methodology relies on the self-assembly of polar monomers around a highly functionalized template molecule during the preparation of a densely cross-linked polymer. The resulting polymers (imprinted with a single enantiomer) have also been shown to resolve racemic mixtures on an analytical scale, when used as HPLC stationary phases.[6-8] The same is true of polymers prepared by the covalent imprinting method of Wulff et al.[9]

There is however a major drawback to these "conventional" approaches to the preparation of imprinted polymers for the purposes of separation, which we have sought to address. It is still not clear whether imprinting methodologies can be applied efficiently to a template with a single functionality, for example a hydroxyl group. In order to investigate this problem we have chosen sterols as a suitable model system, and cholesterol (1) as an initial target with particular relevance to the food industry. An added benefit of this research would be the development of cost-effective methods for the selective removal of cholesterol from foodstuffs, using recyclable polymeric adsorbents. The main purpose of this investigation however, is to extend the general applicability of imprinting methods. This is also the aim of an accompanying paper,[10] which describes another novel approach to this problem.

In this communication we report a new imprinting strategy for compounds with single (or spatially separated) hydroxyl group(s), based on (4-vinyl)phenyl carbonate esters. Using cholesteryl (4-vinyl)phenyl carbonate **(2)** as a **covalent** template monomer, polymers specific to cholesterol have been prepared and subjected to preliminary characterization. The reversibility of **non-covalent** binding of cholesterol to the imprinted polymers was demonstrated and the binding properties were related to the internal surface area of the polymers.

2. SYNTHESIS OF IMPRINTED POLYMERS

Full details of the synthesis of monomer **(2)** and a more comprehensive range of polymers will be presented elsewhere,[11] but a brief description of polymer syntheses is given below:

2.1 Polymerizations

For the preparation of 5g of imprinted polymer, monomer **(2)**, (0.620g, 5mole%), ethylene glycol dimethacrylate, (4.380g, 4.168mL, 95mole%) and azo-*bis*-isobutyronitrile, (0.0745g, 1mole% with respect to double bonds) were placed in a Quickfit test tube. Solvent, either hexane (polymer **I**), or a 3:1 mixture of propan-2-ol and toluene (polymer **II**), was added at the ratio of 2mL per gram of monomer mixture. The tube was sealed with an adaptor fitted with a PTFE vacuum stopcock and degassed on a vacuum line with several freeze-thaw cycles. The degassed polymerization mixture was placed in a water bath at 65°C for 24 hours. The polymer was recovered by breaking-up the bulk with a spatula and washing the particles of polymer onto a sintered glass funnel with methanol. The air-dried polymer was ground and extracted with methanol to remove any low molecular weight impurities before template removal.

Table 1. Details of imprinted polymer preparations.

Polymer	Solvent	Yield %	BET[*] surface area, m^2g^{-1}
I	hexane	93	26
II	propan-2-ol:toluene 3:1	96	440

[*]Determined from multipoint nitrogen adsorption isotherms.

Table 2. Imprinted polymers, following template cleavage.

Polymer	Parent polymer	Yield of polymer %	Yield of theoretically available cholesterol %
III	I	78	73
IV	II	81	48

2.2 Template removal

The standard method of template removal is by hydrolysis with methanolic NaOH. For example, polymer **I**, (2.0g) was heated to reflux for 6 hours with 1M NaOH in methanol. At the end of this time the reaction mixture was poured into dilute HCl. The polymer was obtained by filtration, washed with water and methanol. A subsequent methanol extraction of the polymer in a soxhlet apparatus was followed by vacuum drying to yield polymer **III**. The washings from the hydrolysis work-up were extracted with diethyl ether after the addition of brine. The combined organic extracts were dried and evaporated to yield cholesterol, a recrystallized sample being identical by IR, NMR and Mpt. to the authentic substance. A similar hydrolysis of polymer **II**, yielded polymer **IV**.

Details of two polymer preparations are set out in **Table 1**, above and similarly, following template cleavage, in **Table 2**.

3. BINDING EXPERIMENTS

The uptake of cholesterol by imprinted polymers **III** and **IV** , from a solution in hexane, was used as a model system to investigate the polymer binding characteristics. The general method of measurement is described below.

3.1 Measurement of kinetics

Polymer (9-20mg, depending on the individual experiment) was weighed into a 2mL capacity screw top vial. Cholesterol in hexane (2-4mM, 1-2mL) was added and the vial shaken for a pre-determined time interval. At the end of this period the solution was rapidly filtered through a compressed cotton wool plug in the tip of a syringe barrel, into an HPLC sample vial. In other experiments, further solvent, or a second, more highly concentrated, solution of cholesterol was added and the mixture shaken for a further pre-determined time interval before filtration and analysis. The concentration of each individual sample was then determined by HPLC. The HPLC analysis was performed using Gilson 303 pumps equipped with an ACS light-scattering mass detector and a Shimadzu SIL-9A autosampler. Samples were analysed on a 25cm, 5μ Spherisorb column (Anachem), at room temperature, using a flow rate of 1.5ml/min. Elution was with a linear gradient from 20% ethyl acetate:hexane to 100% ethyl acetate over 8 minutes.

The results for single kinetic runs, and for a study of the reversibility of binding to polymer **IV**, are given in **Figures 1** and **2** respectively.

4. RESULTS AND DISCUSSION

As mentioned above, the major problem associated with the imprinting of a hydroxy-steroid was the lack of suitable approaches. The use of cholesteryl acrylate and dicholesteryl 4-vinylphenylboronate as templates appeared to be unsatisfactory. In the first instance the templates could not be removed from the polymer, even under drastic conditions of hydrolysis (refluxing 5M NaOH in methanol). Similar problems were encountered by Byström *et al.*[12] The removal of cholesterol from 4-vinylphenylboronic acid-containing polymers presented no difficulties, but these materials showed practically no specificity when compared to non-imprinted controls. This was probably due to the 2:1 complex of cholesterol and boronic acid giving rise to poorly defined binding sites.

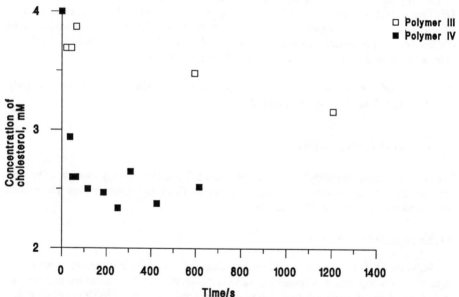

Figure 1. A comparison of the kinetics of binding of cholesterol to polymers **III** and **IV**, (20mg of polymer + 2mL of 4mM cholesterol in hexane).

To overcome these problems we have developed a new approach to the imprinting of alcohols based on the use of (4-vinyl)phenyl carbonates. This polymerizable functionality has a number of advantages compared to the alternatives. In particular it is readily and efficiently hydrolysed. The template is covalently linked to the polymer during the polymerization step, giving well defined binding cavities. The loss of CO_2 during template cleavage creates a phenolic hydroxyl, positioned so as to facilitate non-covalent binding. As a method of producing imprinted polymers for non-covalent binding, the covalent nature of the imprinting step ensures that the ability of the polymerization solvent to stabilize monomer-template interactions is not a major concern.

We have prepared a number of copolymers of ethylene glycol dimethacrylate and (**2**) under different reaction conditions and embarked on a thorough characterization of each of the products. It should be noted that the morphology of polymers of this type is generally dependent on the solvent in which the polymerization is carried out. Its role is to act as the "porogen", creating a network of interconnecting pores throughout the rigid polymer matrix. The most convenient measure of the porosity is the internal surface area (**Table 1**) which can be determined from nitrogen adsorption isotherms and calculated using the BET equation.[13] Two polymers, **I** and **II**, prepared using hexane and propan-2-ol:toluene respectively, and their hydrolysed counterparts, **III** and **IV**, (**Table 1** and **2**) were selected as representative examples of high and low surface area materials, for functional studies.

Initially we investigated the uptake of cholesterol, from a 2mM solution in hexane, by polymers **III** and **IV**. It was found that both polymers showed a significant specific binding of cholesterol of approximately 35μmole and 50μmole of cholesterol per gram of polymer respectively, (calculated after subtraction of some non-specific binding obtained with the corresponding non-imprinted analogues). It was also found that the amount of cholesterol specifically bound to the polymers at equilibrium was only marginally dependent on their surface areas (**Table 1**), whereas the non-specific binding increased significantly with increasing surface area. Non-specific binding was probably due to carboxyl groups formed by hydrolysis of ethylene glycol dimethacrylate. Alternative methods of template removal, involving milder hydrolysis conditions, or not involving hydrolysis at all, are currently being investigated.

In order to investigate further the effect of the surface area on the binding properties of imprinted polymers, the kinetics of cholesterol uptake was studied in some detail (**Figure 1**). Polymer **IV**, derived from the high surface area material, equilibrated extremely rapidly with free cholesterol, and maximum uptake was seen after approximately 100s, whereas for polymer **III**, equilibrium was not achieved even after 20 minutes. Since the chemical composition of the polymers was essentially the same, the marked difference in the rate of cholesterol uptake by **III** and **IV** should be attributed to the surface area.

The rapid kinetics seen in the case of polymer **IV** allowed us to study the reversibility of binding in a series of experiments varying the concentrations of both polymer and cholesterol, either by dilution, or by the addition of a second, more concentrated solution of cholesterol. These results are depicted in **Figure 2**. In each case, perturbation of the concentration of cholesterol and polymer caused a new equilibrium concentration to be rapidly established. The newly measured concentrations were consistent with those determined for unperturbed samples, prepared with the same ratio of polymer and cholesterol as the final mixtures.

Clearly the rapid and reversible interaction of ligands with imprinted polymers is essential for their use as stationary phases in chromatographic separations. This however may not be a crucial factor for large scale bulk separations. The above findings, that the surface area of polymers determines the kinetics of interaction, but probably not the overall number of binding sites, are currently under detailed investigation in our group.

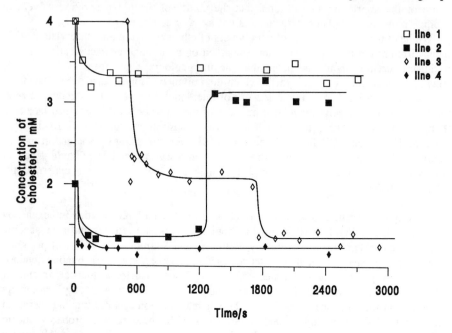

Figure 2. Reversibility study for the binding of cholesterol to polymer **IV**.

Key: Line 1: (□) 9mg polymer + 2mL 4mM cholesterol.
      Line 2: (■) (10mg polymer + 2mL 2mM cholesterol) or (9mg polymer + 1.8mL 2mM cholesterol, with the addition of 0.2mL 22mM cholesterol after 1200s)
      Line 3: (◊) (2mL 4mM cholesterol, 20mg polymer added after 500s) or (1mL 4mM cholesterol, 10mg polymer added after 500s, with the addition of 1mL of hexane after 1700s).
      Line 4: (♦) 10mg polymer + 2mL 2mM cholesterol.

5. CONCLUSIONS

(i) A novel approach to imprinting templates with a single (or multiple, spatially separated) hydroxyl group(s) has been developed;

(ii) This approach relies on a covalently-bound template to create a non-covalent binding site, therefore complex formation between polar monomers and template functionality cannot be influenced by the polymerization solvent.

(iii) Imprinted polymers capable of the specific binding of cholesterol have been prepared using this methodology;

(iv) The kinetics of cholesterol adsorption, but not the number of specific binding sites, appeared to be markedly dependent on the internal surface area of the polymers. Non-specific binding however increased with increasing surface area, with the consequence that faster exchange of ligands resulted in lower overall specificity.

6. ACKNOWLEDGEMENTS

We would like to thank the Co-responsibility Fund and the AFRC for financial support and Professor R. Burch of the Chemistry Department, University of Reading, for his assistance in providing the surface area measurements.

7. REFERENCES

1. C. Alexander, C.R. Smith, E.N. Vulfson and M.J. Whitcombe, *These Proceedings*, page 22.
2. R. Müller, L.I. Andersson, and K. Mosbach, Makromol. Chem., Rapid Commun., 1993, 14, 637.
3. E. Hedborg, F. Winquist, I. Lundström, L.I. Andersson and K. Mosbach, Sensors and Actuators A, 1993, 37-38, 796.
4. G. Vlatakis, L.I. Andersson, R. Müller and K. Mosbach, Nature, 1993, 361, 645.
5. K. Mosbach, Trends Biochem. Sci., 1994, 19, 9.
6. B. Sellergren, Makromol. Chem., 1989, 190, 2703.
7. K.J. Shea, D.A. Spivak and B. Sellergren, J. Am. Chem. Soc., 1993, 115, 3368.
8. L. Fischer, R. Müller, B. Ekberg and K. Mosbach, J. Am. Chem. Soc., 1991, 113, 9358.
9. G. Wulff and J. Haarer, Makromol. Chem., 1991, 192, 1329.
10. C.R. Smith, M.J. Whitcombe and E.N. Vulfson, *These proceedings*, page 482.
11. M.J. Whitcombe, M.E. Rodriguez and E.N. Vulfson, *in preparation*.
12. S.E. Byström, A. Börje and B. Akermark, J. Am. Chem. Soc., 1993, 115, 2081.
13. S. Brunauer, P.M. Emmett and E. Teller, J. Am. Chem. Soc., 1938, 60, 309.

Computer Aided Desk-top Scale-up and Optimization of Chromatographic Processes

D. J. Wiblin

BIOSEP, BIOTECHNOLOGY SERVICES, AEA TECHNOLOGY, 353 HARWELL, OXON OXII ORA, UK

ABSTRACT

The significance of the use of modelling to aid chromatographic process design can be gauged merely by assessing the extent to which academic establishments and some keen industrial companies employ their resources. However, in most instances the work fails to become commercially viable due to several reasons: the models lack generality; extensive validation is not attempted; the programs are too difficult to use; or the funds for development are exhausted before the benefits are realised. Here we will present a new Microsoft® Windows™ compatible computer package to aid the chromatographer in obtaining accurate performance predictions at large scale. This package is called Simulus, and is currently being used by numerous BIOSEP member companies. Experimental data from several key areas will be presented to demonstrate the validity of Simulus in the prediction at a number of scales of operation. Performance of Simulus will be assessed in the prediction of gel-filtration and affinity chromatography separations, together with the new technique of fluidised bed adsorption and more familiar processes including ion exchange. In each case data from small scale experiments were analysed using Simulus giving characteristic adsorption parameters. Using these parameters, predictions were made of the expected performance of a scaled-up process. These predictions compared favourably with experimental results from the scaled-up process.

1 INTRODUCTION

Background

The use of adsorption and chromatographic techniques throughout the chemical and pharmaceutical industry is extensive. However, a full process optimisation of high value products, which are often proteinaceous, is often restricted by the value of the products involved. Furthermore, the commercial environment necessitating 'first-to-the-market' competitiveness, often restricts research programmes to process validation and not to optimising the operating efficiency in the full-scale purification facility.

Previous work[1-5] has shown that mathematical equations which model the diffusion, adsorption and convection of adsorbate molecules within a process can be solved to give accurate predictions of the purity and yield of the final product. Noble *et al*[6] demonstrated a set of programs which could be used to assess the effect of operational parameters, such as flow rate or column aspect ratio, on the performance of a process. Moreover, the paper showed that the simple equations applied could model complex equilibria existing in ion exchange columns.

Although the theoretical credibility of computer aided design packages is unquestioned, its use will not be widespread until the following criteria are met:

- generic programs are written with diverse capabilities;

- extensive experimental validation work is undertaken; and,
- simple, user-friendly packages are compiled.

To this end the Simulus design package was developed[7,8] and tested on real biological systems. This design package is incorporated into a proprietary design report available to members of BIOSEP. This paper, however, will seek to address some of the criteria highlighted.

Scope of the Work

The two types of models which BIOSEP employ are based on the dynamic equilibria between molecules in the adsorbed and desorbed states and the rates of diffusion across particle boundary layers and through the pores.

One major limitation to modelling elution chromatography is the ability to characterise the sharp peak which often results. With existing models, the column is normally broken down into a number of discrete cells; however, to model the movement of a sharp peak as it moves down the column requires a highly concentrated discretisation localised to the position of the peak. With the latest version of our programs this could only be achieved by increasing the total column discretisation to such an extent that the program's operation would be laboriously slow. This paper will present the finite volume discretisation for 'moving grid' modelling of gel-filtration, which gives rise to a significant decrease in run-time. Experimental validation of this technique will also be given.

In the second stage of the work to develop the programs into a commercially viable tool for the chromatographer, we have produced a Windows™ compatible version of our suite of adsorption programs, which is called Simulus. Validation will also be given in several new key areas of purification.

2 MODELLING TECHNIQUES USED IN SIMULUS

Existing Models

The basic models and equations which Simulus uses were transferred directly from the original suite of well documented programs[3,4,7,9,10], whose worthiness for modelling a number of systems is now well established. The validation of Simulus, however, is ongoing and, since several new purification techniques are now commercially available, further validation of their capability to model such new adsorption processes is sought. This paper highlights Simulus's capacity to characterise some of these processes.

In essence the models assume that the limitation to mass transfer is due to diffusion, and not to adsorption kinetics. Hence it is assumed that the time taken to reach equilibrium between adsorbed and desorbed phases is very small. The equilibrium constant is defined by K_d and the maximum capacity of the adsorbate molecule for the system is given by Q_m.

Moving Grid Strategy

The importance of the models' capability to model sharp elution peaks is paramount for a detailed and accurate estimation of a process's efficiency. The original models, in their simplicity, were proficient at modelling these types of experimental data, but required a large discretised grid. Furthermore, the computational processing power required made the feasibility of using such models impractical. To approach this problem a new model was included into Simulus, called the 'moving grid'.

In the simplest form of elution chromatography, size exclusion chromatography, or gel filtration, a sample is loaded onto the top of a column and the mixture is separated by diffusion effects only. Here the need for modelling adsorption is removed and the resulting equations which can model the process are given below.

When considering a gel-filtration column the convection of adsorbate along the length of the column can be described by (1).

$$\frac{\partial c_b}{\partial t} = D_a \frac{\partial^2 c_b}{\partial z^2} - v \frac{\partial c_b}{\partial z} - \frac{3k_f V_S}{R V_L}(c_b - c_i) \qquad (1)$$

Here the last term represents the quantity of adsorbate which diffuses across the boundary layer. The adsorbate concentration at the surface of the particle is given by (2);

$$\left.\frac{\partial c_i}{\partial r}\right|_{r=R} = \frac{k_f}{D_e \varepsilon_i}(c_b - c_i) \tag{2}$$

The diffusion of adsorbate into the pores of the particle is described by (3),

$$\frac{\partial c_i}{\partial t} = D_e \left(\frac{\partial^2 c_i}{\partial r^2} + \frac{2}{r} \frac{\partial c_i}{\partial r} \right) \tag{3}$$

assuming the boundary condition holds that the rate of mass transfer at the centre of the particle is zero (4).

$$\left.\frac{\partial c_i}{\partial r}\right|_{r=0} = 0 \tag{4}$$

The simultaneous solution to these differential equations cannot be undertaken analytically, but instead relies upon numerical solution techniques. To solve numerically, however, requires appropriate column and particle discretisation which was where the original models arrived at problems.

The solution to the moving grid discretisation system can be achieved using the finite volume method[11]. The finite volume discretisation method essentially evaluates the flux of material across each grid boundary. However, if the grid moves then this simply increases or decreases the flux across the boundary accordingly. This is a complex solution[12] and beyond the scope of this paper. The solution employs instantaneous re-gridding which is carried out at each time step along the integration to maintain 98% of the loaded mass within the confines of the grid (**Figure 1**). Increasing this 'captured' percentage only results in elongation of the grid, and a corresponding reduction in resolution.

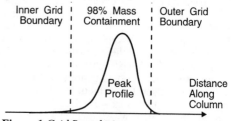

Figure 1 *Grid Boundaries*

The remaining grid is divided evenly between the inner and outer boundaries (**Figure 2**). This movable, expanding grid allows maximum resolution of the peak with a minimum of processing power. The corresponding scenario with the fixed grid (also shown) cannot adequately characterise the diffusive mass transfer due to numerical deficiencies in the discretisation.

3 EXPERIMENTAL VALIDATION

Amongst many others, previous validation work includes the purification of aspartic acid using Duolite A162[9], the adsorption of an intracellular enzyme leucine dehydrogenase onto DEAE Spherodex[13] and the uptake of α-amylase by QMA Spherosil[13]. The three processes modelled below attempt to reinforce the validation of Simulus.

The methods employed throughout the modelling of the data were first elucidated by Noble *et al*[6].

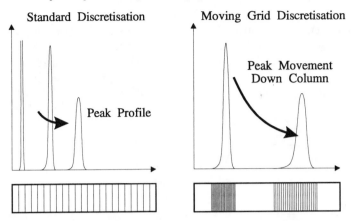

Figure 2 *Comparison of Standard and Moving Grid Discretisation*

Expanded Bed Adsorption

The first process is the purification of α-amylase using Expanded Bed Adsorption (EBA™). The unclarified mixture was prepared by dissolving 2 mg.ml^{-1} of a crude, bacterial source α-amylase into Tris/HCL buffer at pH 9. 5 g.l^{-1} of dried yeast was added to simulate a fermentation broth. The clarified solution was loaded at 10 ml.min^{-1} onto a 10 x 1.6 cm column packed with Pharmacia Streamline DEAE media. The unclarified 'broth' was loaded at 100 ml.min^{-1} onto an 55 x 5 cm expanded bed (also using Streamline DEAE).

The results from the packed bed study were analysed using Simulus to obtain the parameters (**Table 1**) which characterise the adsorption process. The confidence intervals are statistical values that Simulus gives to describe the effect of varying the parameter. Tight confidence intervals for one parameter implies that the simulation results are heavily dependant on that parameter. **Figure 3a** shows a comparison between the experimental curve and the curve 'fitted' by Simulus.

The characteristic parameters were then taken and used to predict the output profile of the expanded bed study. The comparison between the real and predicted results are given in **Figure 3b**. This prediction represents approximately a 20-fold scale-up in capacity.

Affinity Fluidised Bed Adsorption

The second of the two processes is the purification of a monoclonal antibody using affinity fluidised bed adsorption. The approach for this modelling study was the same as for the expanded bed analysis. In the initial packed bed experiments an ideal system was selected. A solution of 20 mM trisodium phosphate containing 0.27 mg.ml^{-1} of the antibody was loaded at 0.49 ml.min^{-1} onto a 5 x 0.5 cm Prosep-G packed bed. The resulting frontal curve was analysed to obtain parameters which characterise the adsorption kinetics (**Table 2**). For the analysis the value of the maximum capacity, Q_m, was fixed at 13.5 mg.ml^{-1} as this was deduced previously from batch experiments. Simulus could not fit a value for the liquid film diffusion coefficient, k_f, which indicates that pore diffusion is the mass transfer rate limiting step and that liquid film diffusion is rapid. The comparison of the experimental and the fitted curve is shown below in **Figure 4a**.

A similar solution, at 0.275 mg.ml^{-1} was then applied to a fluidised bed process with the same type of media, Prosep-G. The bed was fluidised to 16.2 x 1.0 cm at a constant flow rate of 1.95 ml.min^{-1}. The range of particle sizes for Prosep-G (63-150 µm) ensured that bed stability was observed, as the larger particles tend to congregate at the bottom of the column, whereas the smaller, lighter particles are carried to the top. As little axial motion was detected, the same model was used as with the packed bed but with an increased bed voidage; from 0.33 to 0.78. The results from the experiment were compared to predictions given by Simulus (**Figure 4b**).

Table 1 *Estimated Parameters from Simulus for α-amylase Adsorbing onto Streamline.*

Parameter	Fitted Value	Lower (5%) ci[†]	Upper (95%) ci[†]
Q_m	0.598 mg.ml^{-1}	0.581	0.616
K_d	5.70x10^{-4} mg.ml^{-1}	5.15x10^{-4}	6.23x10^{-4}
D_e	5.50x10^{-12} m^2.s^{-1}	4.95x10^{-12}	6.15x10^{-12}
k_f	2.67x10^{-6} m.s^{-1}	1.53x10^{-6}	4.75x10^{-6}

[†]ci - confidence interval

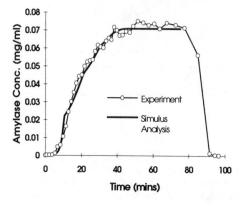

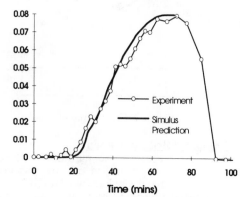

Figure 3a *Comparison Between the Adsorption of α-amylase onto Streamline DEAE using a Packed Bed and the Results Fitted by Simulus.*

Figure 3b *Comparison Between the Adsorption of α-amylase onto Streamline DEAE using EBA Technology and the Prediction Given by Simulus.*

Table 2 *Estimated Parameters from Simulus for IgG Adsorbing onto Prosep-G.*

Parameter	Fitted Value	Lower (5%) ci	Upper (95%) ci
Q_m	13.5 mg.ml^{-1}	n/a[†]	n/a[†]
K_d	0.049 mg.ml^{-1}	0.047	0.052
D_e	3.97x10^{-11} m^2.s^{-1}	3.77x10^{-11}	4.30x10^{-11}
k_f	3.81x10^{-5} m.s^{-1}	Not fitted	Not fitted

[†]n/a - not applicable

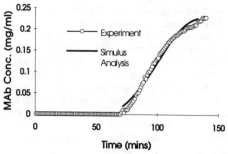

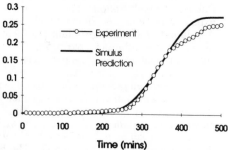

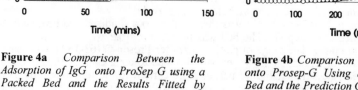

Figure 4a *Comparison Between the Adsorption of IgG onto ProSep G using a Packed Bed and the Results Fitted by Simulus.*

Figure 4b *Comparison of Adsorption of IgG onto Prosep-G Using an Affinity Fluidised Bed and the Prediction Given by Simulus.*

Gel Filtration

The final model system seeks to validate the use of the moving grid strategy for gel filtration. In the initial studies the selected system was pure α-amylase dissolved into Tris/HCl buffer at pH 7.2. This solution was loaded onto various columns packed with Sephacryl S100HR. To alleviate non-specific binding, 50 mM sodium chloride was also added to the loading and elution buffers. The analysis of the mass transfer processes was undertaken using the smallest sized column at 11.2 x 1.6 cm where the linear flow rate was 30 cm.hr^{-1} (**Figure 5a**). The fitted parameters are given in **Table 3**. Once again the value for k_f is not fitted as pore diffusion is the rate limiting diffusion step. It is interesting to note that the value for D_e is less than that for α-amylase diffusing into Streamline DEAE. Here, F_D is the distribution coefficient which governs the selectivity of the α-amylase/Sephacryl system. This parameter is well documented[14]. The value of the effective intraparticle porosity, ε_i, is also evaluated by Simulus, as a large molecule will diffuse into a lesser proportion of the matrix than a small molecule. This is inherent in all size exclusion systems.

Table 3 *Fitted Parameters from Simulus for α-amylase Loading onto Sephacryl S100HR.*

Parameter	Fitted Value	Lower (5%) ci	Upper (95%) ci
ε_i	0.382	0.379	0.385
F_D	0.510	0.498	0.521
D_e	2.42×10^{-12} m^2.s^{-1}	1.08×10^{-12}	3.85×10^{-12}
k_f	5.3×10^{-5} m.s^{-1}	not fitted	not fitted

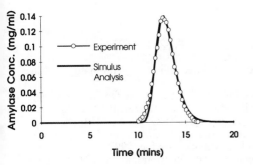 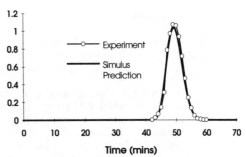

Figure 5a *Comparison Between the Elution Profiles for α-amylase and the Results Fitted by Simulus for a 22.5 ml Column.*

Figure 5b *Comparison Between the Elution Profiles for α-amylase and the Prediction Given by Simulus for a 465.2 ml Column.*

The elution profiles of the remaining experiments were then compared to the predictions by Simulus. Accurate correlations were found between the simulations and reality. One such comparison is given in **Figure 5b**. In this experiment the amylase solution was loaded at 60 cm.hr^{-1} onto a 87.7 x 2.6 cm column (465.2 ml), corresponding to an increase in α-amylase throughput of 4000 %.

In further validation work an industrial gel filtration process was modelled which gave equally convincing results.

4 THE SIMULUS PACKAGE

A Windows™ programming language, Microsoft® Visual Basic™, was used to create a suite of highly interactive user-friendly Windows compatible interfaces to help the chromatographer to rapidly create, edit and execute basic input files[15]. This progression to a user-friendly interface has been instrumental in addressing one of the most difficult problems for chromatographic modellers in that the benefit of the work is not fully realised unless the computer programs are readily accessible by chromatographers.

5 CONCLUSION

It is apparent that the inclusion of the moving grid strategy is of great benefit in the modelling of separation processes. The accuracy of the validation with gel filtration, the simplest form of elution chromatography, strengthens this claim.

The Windows™ compatibility demonstrates the versatility of Simulus, a true desktop package, which is no more difficult to utilise than conventional, well established spreadsheets or word processors.

With the incorporation of the moving grid discretisation strategy into the current BIOSEP adsorptive programs and the improvements in operational ease, Simulus's overall level of predictive performance was enhanced.

PARAMETRIC DEFINITIONS

c_b Bulk phase concentration (kg.m^{-3})
c_i Particulate phase concentration (kg.m^{-3})
D_a Axial dispersion coefficient (m.s^{-2})
D_e Pore diffusion coefficient (m.s^{-2})
F_D Distribution coefficient (usually K_D)
K_d Dissociation constant (m.s^{-1})
k_f Liquid film diffusion coefficient (m.s^{-1})
Q_m Maximum capacity (mg.ml^{-1})

r Radial distance (m)
R Particle radius (m)
t Time (s)
v Superficial bulk fluid velocity (m.s^{-1})
V_L Liquid phase volume (m^3)
V_S Solid phase volume (m^3)
z Axial distance (m)

REFERENCES

1   A. C. Liapis and D. W. T. Rippin, Chem. Eng. Sci., 1978, 32, 619-627.
2   H. A. Chase, J. Chromatog., 1984, 297, 179-202.
3   G. H. Cowan, I. S. Gosling, J.F.Laws and W. P. Sweetenham, J. Chromatog., 1986, 363, 37.
4   G. H. Cowan, I. S. Gosling, and W. P. Sweetenham, in *Separations for Biotechnology*, M. S. Verrall and M. J. Hudson (eds.), Ellis Horwood, Chichester, 1987, 152.
5   B. J. Horstmann and H. A. Chase, Chem. Eng. Res. Des., 1989, 67, 243-254.
6   J. B. Noble, G. H. Cowan, W. P. Sweetenham and H. A. Chase, in *Ion Exchange Advances*, M. J. Slater (ed.), Elsevier Applied Science, London, 1992, 214.
7   G. H. Cowan, J. B. Noble and W. P. Sweetenham with contributions from H. A. Chase and B. J. Horstmann, *DR1, Prediction of the Performance of Batch Tank Adsorbers and Fixed Bed Adsorption Columns*, BIOSEP, 1988.
8   D. J. Wiblin, *Simulus Theoretical and Technical Reference*, BIOSEP, 1994.
9   G. H. Cowan in *Adsorption: Science and Technology*, A. E. Rodrigues, M. D. LeVan and D. Tondeur (eds.), NATO ASI Series E, Applied Sciences, 517-537.
10 G. H. Cowan, I. S. Gosling and W. P. Sweetenham, J. Chromatog., 1989, 484, 187-210.
11 J. H. Furziger, *Numerical Methods for Engineering Applications*, Wiley & Son, 1981.
12 R. G. Myhill, *A Mathematical Solution for Gel Filtration*, Private Communication, November 1992.
13 J. B. Noble, *PhD Thesis*, Imperial College, London, 1991.
14 L. Hagel, H. Lundström, T. Andersson, and H. Lindblom, J. Chromatog., 1989, 476, 329.
15 D. J. Wiblin, *Simulus User Guide*, BIOSEP, 1994.

ACKNOWLEDGEMENTS

Acknowledgement is made to BIOSEP, 353 Harwell, Oxon, OX11 0RA, U.K., for permission to publish the work presented in this paper. Copyright in this paper remains the property of AEA Technology, 353 Harwell, Oxon, OX11 0RA, U.K.
Windows™ and Visual Basic™ are trademarks of the Microsoft® Corporation.

Identification of Enzyme Activity Loss Caused by Gas Bubbles

K. I. T. Wright and Z. F. Cui

DEPARTMENT OF CHEMICAL ENGINEERING, UNIVERSITY OF EDINBURGH, KINGS BUILDINGS, EDINBURGH EH9 3JL, UK

1 INTRODUCTION

The use of air sparging to create a gas-liquid two-phase crossflow to overcome concentration polarisation has proven to be an effective technique for enhancing ultrafiltration and microfiltration[3,4,7]. We have recently reported flux enhancements of up to 250% and 90% respectively for the ultrafiltration of dyed dextran and bovine serum albumin solutions using this technique[3,4]. However, the deleterious effects of air addition to shear sensitive biological fluids, such as protein solutions, could severely limit the wider application of this technique. There is therefore a need for the further identification and quantification of the damage caused by bubbles.

The effects of hydrodynamic shear and the influence of the air-liquid interfaces upon protein solutions have been reported[6,8-10]. However similar correlations between bubble mediated damage and protein activity have not been so extensively covered.

In this work we report experimental results from a study on the effects of bubble damage on proteins in solution. The location of the site of major damage was identified, and the influence of sparge tube diameter, protein concentration and gas flowrate were examined. Two enzyme systems, Lactate dehydrogenase (LDH) and Yeast alcohol dehydrogenase (YADH) were used as sample proteins, using enzyme activity as a measure of bubble mediated damage.

2 MATERIALS AND METHODS

2.1 Enzymes and Analytical Technique

The two enzymes, Lactate dehydrogenase (Type II, EC 1.1.1.27, product no. L2500) from rabbit muscle and Yeast alcohol dehydrogenase (product no. A7100) were purchased from Sigma. The LDH arrived as a stock solution with an activity of 860 Units/mg of protein with 11 mg of protein per ml. Freeze dried YADH was reconstituted as per instructions to give a stock solution with an activity of 3400 Units/ml (10 mg/ml).

LDH catalyses the conversion of pyruvate to lactate, while simultaneously converting β-Nicotinamide Adenine Dinucleotide (NADH) to its oxidised form, NAD. YADH uses the reverse reaction, NAD to NADH, to aid the conversion of ethanol to acetaldehyde. The reduced form of this cofactor (NADH) absorbs light at a wavelength of 340nm, while its oxidised form (NAD) does not. By using a microplate reader (Dynatech MR5000) to follow the rate of oxidation or reduction of this cofactor in the respective reactions, an assessment of enzyme activity can be made.

Reaction rates determined using these methods gave correlation coefficients of greater than 0.99, with the standard deviation for each set of triplicate assays being less than 7% in all of the samples tested. The protocols for these two enzyme assays have been reported elsewhere[11].

2.2 Bubble Column Apparatus

Two vertical bubble columns were constructed from transparent perspex tube with an internal diameter of 12.7mm. The longest column (1m) was water jacketed for temperature control, with four sample ports spaced evenly along the column enabling the location of the principle site of protein damage. The holdup of this column was 90ml. The shorter column (0.4m) allowed a closer examination of the damage occurring at the stage of bubble burst using a reduced hold up volume of 20 ml.

Stainless steel sparge tubes, with different internal diameters of 0.8mm, 1.2mm, 1.5mm and 2.0mm respectively, were set in a rubber stopper, sealing the base of the column. Compressed air was supplied to the sparger via a rotameter and a non-return valve.

2.3 Experimental Conditions

Four concentrations of each enzyme were examined, ranging from 0.011 mg/ml to 0.044 mg/ml (activities of 9.46 and 37.84 U/ml) for LDH and 0.01 and 0.05 mg/ml (3.4 to 17 U/ml) for YADH. With the short column, three gas flow rates were examined, 20, 50 and 100 cc/min, which gave superficial gas velocities of 2.6, 6.6 and 13.2 mm/s, respectively. A higher gas flowrate of 150cc/min (18.8mm/s superficial gas velocity) was used in the larger column. Protein concentrations were determined using the Lowry assay method (Sigma kit no. 690-A) and a Jenway spectrophotometer (model no. 6105).

2.4 Experimental Procedure

The column and sparge tube assembly were cleaned thoroughly and rinsed with ultra-pure water and the appropriate buffer solution prior to the start of each experiment. Test solutions were prepared from stock enzyme and buffer solutions. The bubble column was rinsed with 5 ml of the test solution, which was discarded before the actual experiment.

Long column: 90ml of test solution was added to the column. At time t = 0, air at the desired flowrate was injected. Samples were taken in triplicate from each of the four sample ports at regular intervals during the experiment and stored on a microtitre plate for analysis at the end of the experiment. In this way, the activity changes at each of the different locations, the bottom port where the bubbles were generated, the lower and upper ports where the bubbles were rising, and the top port where bubbles burst, could be monitored.

Short column: 20ml of test solution was added to the column. Three 100 μl samples were taken from the column and stored on a microtitre plate. At time t = 0, air at the desired flowrate was injected. At the end of the experiment, which generally lasted for 10 min, the air was stopped, the tube sealed with parafilm and inverted several times to resuspend any precipitate formed on the walls of the tube. Three samples were then taken from this final mixture and stored on the sample plate. The apparatus was then dismantled and cleaned for the next test.

Samples were stored until the end of the series of tests, with little variation in the activity of the stored samples detected within a 3 hour period. Dilutions of the more concentrated solutions were made to allow the analysis of samples within the range of the assay techniques. Results are expressed as percentage losses in specific activity, based on the initially recorded enzyme activity.

3 RESULTS AND DISCUSSION

3.1 Location of Major Activity Loss

Results obtained from the 1m bubble column are presented in Figure 1. The rate of activity loss was found to be different at each location. The greatest damage was found to occur at the top of the column, and hence it was concluded that the major cause of damage to proteins was from the bursting of bubbles. This is in agreement with those results reported for the damage of mammalian cells in sparged vessels[2], where foam formation and bubble rupture have been identified as the principle mechanisms of cellular damage.

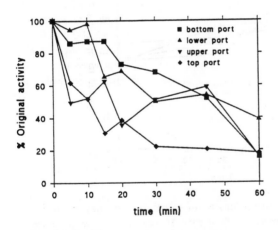

Figure 1. LDH activity changes in the 1m bubble column
(air flow rate 150 ml/min, initial LDH activity 9.46 U/ml)

It has also been reported[1,5] that when bubbles burst, small droplets, originating from either the rupture of the bubble film or the fluid jet resulting from the collapse of the bubble cavity, are propelled to the gas phase above the liquid. This was confirmed in these experiments by the progressive formation of protein precipitates situated on the tube wall above the air-liquid interface. Evaporation from these droplets, while in the air phase or entrained on the tube wall, would result in a loss in the activity of the contained protein via dehydration.

Protein losses during experiments in which the larger column was used were typically 10% . Protein flocs, formed at the air-liquid interface, were fully dispersed throughout the column by the end of the experimental period. This gradual dispersion of flocs was thought to be a result of bubble mediated mixing and the increased density of the aggregates.

3.2 Protein Precipitates and Foaming

In the experiments using the short column, the precipitate formed during bubbling was resuspended into the bulk solution prior to the measurement of enzyme activity, with a minimum recovery of 85% of the original protein. By measuring the activity and concentration of samples from the column, prior to the resuspension of wall bound precipitate, the contribution of these aggregates upon enzyme activity was examined. It was found that a majority of the activity losses observed in these sparged experiments was attributable to the formation of protein precipitates, suggesting that bubble mediated precipitate formation was the major cause of enzyme inactivation. A good agreement between the relative percentage losses in the measured activity after the resuspension of the precipitate and the level of proteins in the bulk solution, prior to resuspension, was obtained (Figure 2) indicating little change in the specific activity of the enzyme retained in solution.

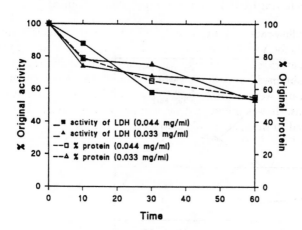

**Figure 2. Protein loss from solution and enzyme activity
(short column, gas flow rate 50 cc/min, 0.8 mm sparge)**

Increasing the diameter of the bubble column could be expected to reduce the loss of active protein by decreasing the influence of the tube wall deposits upon the bulk solution. However, this would not eliminate precipitate formation completely, as micro-drops would still be propelled into the gas phase during bubble burst. Whether protein precipitate formed in this way can be resuspended without a loss of activity requires further investigation.

Protein loss through foaming was thought to be another factor contributing to the observed activity loss. The characteristics of the foam layer developed by each of the two enzyme solutions tended to differ, with LDH creating an unstable, thin layered foam and YADH developing a more stable foam. This may explain why YADH is more sensitive to bubbling, as shown in Figures 4 and 5.

3.3 Effects of Gas Flow Rate and Sparger Size

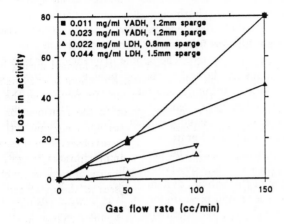

Figure 3. Effect of gas flowrate (YADH-30min bubbling, large column; LDH-10min bubbling, short column)

The effects of gas flow rate upon the activity of both enzyme systems is presented in Figure 3. The results for YADH were obtained after 30 minutes of sparging in the long column, with the averaged specific activity for the whole column reported. The results for LDH were obtained after 10 minutes sparging of the short column.

Higher gas flow rates were shown to lead to a greater loss in the specific activity of the enzyme solutions. It was found that, at similar gas flow rates, the degree of damage was also related to the sparge tube diameter (Figure 4).

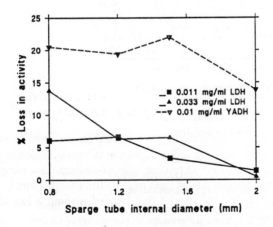

Figure 4. Activity loss after 10min bubbling (short column, gas flow rate 50 cc/min)

These results can be explained by the influence of the bubble size and frequency on the protein losses observed within the column. In the short column experiments, all of the gas flow rates examined gave a dispersed, non-slugging bubble flow. In the larger column, at 150 cc/min, the dispersed bubbles produced at the sparge tube were observed to coalesce to form slugs within the first 30 cm of the column.

In dispersed flow, the bubble size depends upon surface tension, viscosity of the solution, sparger size and the gas flowrate. At similar conditions, the bigger the sparger, the bigger the bubbles generated. The bubble frequency, based on a specific bubble size, depends upon the gas flowrate. Smaller bubbles are known to have a greater energy associated with their burst[2], and tends to cause foaming as it is more difficult for small bubbles to disengage from the gas liquid interface. Consequently, the combination of a high gas flow rate and a smaller sparge size would contribute to a higher degree of damage to entrained proteins.

The results obtained using sparge tube sizes of less than 1.5 mm showed little difference in the degree of activity loss observed (Figure 4). This phenomenon may be related to the influence of the bubble dynamics associated with the size of the sparge tube, e.g. surface tension and pressure forces. It is thought that these factors are either too small, or vary little, over the range of sparge sizes up to an internal diameter of 1.5 mm.

3.4 Effects of Protein Concentration

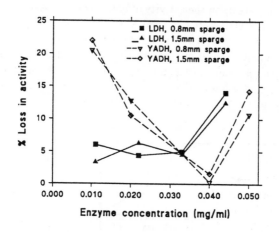

Figure 5. Influence of enzyme concentration
(short column, 10min bubbling, gas flow rate 50 cc/min)

As shown in Figure 5, a marked difference between the behaviour of the two enzyme solutions was observed at concentrations below 0.04 mg/ml, with a decrease in activity loss observed for increasing concentrations of YADH, while little difference in activity loss was observed over a similar range for LDH. These results may be explained by the previously observed differences in the stability of the proteins inter- action with the air-liquid interface. YADH forms a stable foam layer which is capable of entraining enzyme throughout the experiment, with a low level of resuspension in the bulk fluid. At low concentrations, the loss of protein to the foam layer and wall deposition would remove a higher proportion of the total protein from solution. As the bulk concentration of YADH is increased, the proportion of enzyme associated with the foam layer and wall deposition would be lower, reducing the overall effect on the percentage activity loss of the solution. The instability of the foam generated with LDH, suggests that the reentrainment of LDH into the bulk solution occurs readily, restricting the losses of protein to deposition upon the walls of the tube.

At concentrations above 0.04 mg/ml there was an abnormal increase in the loss of specific enzyme activity for both enzymes. Possible explanations for this behaviour may be attributed to the formation of protein aggregates in the solution, or the inhibiting effect between enzyme molecules themselves.

4 CONCLUSIONS

It is concluded that the principle damaging effect from gas bubbles occurs at the surface of the solution where bubbles burst. Preliminary experimental results suggest that enzyme activity loss is mainly caused by the bubble mediated loss of proteins from solution, through the formation of protein precipitates associated with bursting bubbles and the foam layer.

Enzyme activity loss was found to be dependent upon the gas flowrate, sparger size and enzyme concentration. A higher gas flowrate, a smaller sparge and a lower concentration of enzyme were found to result in a higher proportional loss in enzyme activity. The characteristics of the protein and the relative affinity of the proteins for the air-liquid interface were again found to be of importance.

Acknowledgments

This work is sponsored by the Science and Engineering Research Council (Grant GR/J/46388) and carried out in the Chemical Engineering Department, Edinburgh University. The authors would like to thank Dr C. L. Pritchard, Head of Department, for his encouragement and support.

References

1. Blanchard, D. C. and Syzdek, L. D., *J. Geophys. Res.*, **77**, 5087.
2. Cherry, R. S., Hulle, C. T. , *Biotechnol. Prog.*, 1992, **8**: 11.
3. Cui, Z.F., IN: Patterson, R. (Ed.), *Effective membrane processes- New perspectives*, Mech. Eng. Publications Ltd., London, 1993, 237-246.
4. Cui, Z. F. & Wright, K. I. T., *J. Membrane Sci.* 1994,to be published.
5. Garcia-Briones, M. and Chalmers, J. J., *Ann. N. Y. Acad. Sci.*, **665**, 219.
6. Harrington, T. J.,Gainer, J. L., Kirwan, D. J., *Enzyme Microb. Technol.* 1991, **13**: 610.
7. Lee, C. K., Chang, W. G., Ju, Y. H., *Biotechnol. Bioeng.* 1993, **41**: 525.
8. Narendranathan, T. J., Dunnill, P., *Biotechnol. Bioeng.* 1982, **24**: 2103.
9. Thomas, C.R., Nienow, A.W., Dunnill, P., *Biotechnol. Bioeng.* 1979, **21**: 2263.
10. Virkar, P.D., Narendranathan, T.J., Hoare, M., Dunnill,P., *Biotechnol.Bioeng.* 1981, **23**: 425.
11. Wright, K. I. T, Carless, L. and Cui, Z. F., *The 1994 IChemE Research Event.* Vol.1, 88-90.

Direct Recovery of Secondary Metabolites from Fermentation Broth

S. V. Yarotsky, Yu. V. Zhdanovitch, L. F. Yakhontova,
M. S. Bulychova, L. I. Nasonova, S. M. Navashin, M. A. Sinitsin,
and S. A. Kobzieva

NATIONAL RESEARCH CENTRE FOR ANTIBIOTICS, MOSCOW 113105,
RUSSIA

1 INTRODUCTION

Isolation of many exocellular secondary metabolites from fermentation broth often begins with the major problem of preliminary cell and insoluble media components separation followed by adsorption onto a suitable matrix with further elution of target compounds. Traditional methods of filtration or centrifugal separation present some disadvantages, such as the use of complicated equipment, process duration and loss of product.

First attempts at direct recovery of metabolites from fermentation broth involved antibiotics which could be isolated by adsorption on a resin matrix.[1] There are examples of the application of such procedure for preparation of streptomycin,[2] sisomycin, gentamycin, geneticin (antibiotic G-52),[3] etc.

Isolation of streptomycin[4] was performed by dynamic discontinuous adsorption using an appropriate fluidized-bed column. This process brings some problems:
- high pressure drop of column filling;
- loss of resin by elutriation with simultaneous loss of adsorbed product;
- sticking together of adsorbent particles with local channel formation which reduces mass transfer;
- necessity of special broth pretreatment (heating and multiple dilution with water) to reduce viscosity.

Recovery of some aminoglycoside antibiotics[3,5,6] was

carried out in a batch operation by agitation of broth with solid adsorbent until equilibrium is achieved. Then adsorbent is separated on vibrating filter. In this procedure:
- mechanical agitation leads to crumbling of adsorbent particles with subsequent loss of resin and product during filtration;
- the separated adsorbent contains large quantities of mycelia and insoluble media components. Therefore it is additionally necessary to rinse off the adsorbent and employ large size (1.0–1,6 mm) resin particles.

2 RESULTS AND DISCUSSION

We have developed an alternative procedure for secondary metabolites recovery from fermentation broth consisting in their direct adsorption on a solid adsorbent without preliminary separation of non–soluble admixtures. This method is practically free from the disadvantages described above. For this purpose we use a system which permits at simultaneous saturating of the adsorbent by target compounds and washing out of mycelia and other "mechanical" components. The main part of this system is a column of special design. A typical size of pilot–scale column is 100x1000 mm. There are a 10 partitions inside the column with equal distance from each other. The partitions supplied with opposite directed blades for flow turbulization. Column also equipped with pulsation generator driven by compressed air. This construction gives a possibility for achieving fluidized-bed conditions with high level of mass-exchange and low pressure drop. There is an opportunity for discontinuous and recurrent mode.

Typical conditions of operation are as follows:
adsorbent (for example, IRC–50(NH_4^+), quantity up to 4 L;
flow rate:

– static mode	50 L/hour
– dynamic mode	up to 15 L/hour
maximal total broth volume	up to 250 L
pulsation frequency	120–140 puls/m
pulsation amplitude	60–80 mm

antibiotic (concentration in the broth, mg/L)	extraction efficacy, %
gentamycin (1500)	97–98
sisomycin (800)	95–97
kanamycin (4000)	95–97
streptomycin(12000)	94–96

If necessary, accompanying related admixtures (minor components) as well as pigments can be separated, for example, by
gradient elution. From our experience there is a possibility
of practically total (92–95%) and fast (4–6hours) recovery of
different products, in particular aminoglycoside antibiotics,
directly from fermentation broth avoiding loss of adsorbent
and product.

The desorption of target product and adsorbent regeneration are carried out in the same system. For example, in the
case of streptomycin isolation, the process does not include
the traditional step of fermentation broth filtration; for
gentamycin recovery effluent decolorization and demineraliza-
tion steps are also avoided.

As a general result we have developed a truly one–step
method for preparation of purified compounds with 10–12%
yields higher than traditional approaches. The same results
were investigated during isolation and purification of heparin and some other products. From our experience there are
no limitations in process scaling up.

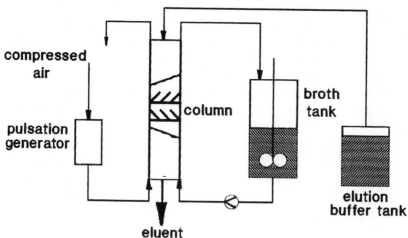

Figure 1 *Outline of system for direct metabolites recovery*

The method gives a possibility for:
- simplifying biosynthetic product isolation process;
- increasing total product yield;
- diminishing product and adsorbent losses;
- environmental issues improvements;
- decrease of time and utilities consumption.

REFERENCES

1. C.R.Bartels, G.Kleiman, J.N.Korzun, et al., *Chem.Eng. Prom.*, 1958, **54**, 49.
2. I.D.Boyko, E.S.Bylinkina, V.F.Terekhova, M.G.Nechaeva, *USSR Med.Prom.*, 1962, 18.
3. *US Patent*, **3997403**, 1976.
4. *USSR Patent*, **310470**, 1968.
5. *USSR Patent*, **1271078**, 1984.
6. B.N.Laskorin, L.I.Vodolasov, et al., *Zh.Prikl.Khim.*, 1990, **63**, 139.

New Technology for Extraction and Concentration of Gibberellins from Filtered Fermentation Broth

Li Zhou, Lei Wen, Zhang Xianhong, Xu Ping, and Gao Huating

DEPARTMENT OF CHEMICAL ENGINEERING, TSINGHUA UNIVERSITY, BEIJING, PEOPLE'S REPUBLIC OF CHINA

1 INTRODUCTION

Gibberellins are important hormones for plant growth and are widely used in the growth of rice, wheat, fruits and vegetables. The current production flowsheet of gibberellins is as follows:

Fermentation broth → Acidification and filtration → Back adjustment of pH value → Concentration by evaporation → Extraction by ethyl acetate → Concentration of ethyl acetate extract and crystallization of gibberellins → Scrubbing and drying of gibberellin crystals.

The main disadvantage in this production process is the low recovery yield in the evaporation concentration step accompanied by the high energy consumption. The extraction process was researched in order to replace the evaporation concentration process.

In a previous paper [1] the extraction of gibberellins by various extractants was examined, these extractants include neutral, amine extractants and synergistic extraction systems.

Based on the study of extraction equilibrium and extraction technological conditions, a new solvent extraction flowsheet for gibberellins was developed.

In the new process a solvent extraction cycle was adopted for the concentration of gibberellins in filtered fermentation broth instead of the evaporation process. Thus both the extraction and concentration processes for gibberellins were combined in a single solvent extraction process, and the advantages of high recovery yield and low energy consumption were obtained comparing with the current technological process.

2 EXPERIMENTAL

2.1 Experimental System

Aqueous feed: filtrate of fermentation broth

Concentration of gibberellins: $\sim 1000u$ / ml (850u / mg), pH≈ 2

Organic phase: phosphorous based extractant(for example, TBP is one choice and it is in C.P. grade)–sulphonated kerosene (in different volume ratio).

2.2 Experimental Installation

The cascade experiments are carried out in specially made test tubes with volume of 25ml, and the annular centrifugal extractor with ϕ20mm rotor are used for bench–scale experiments. The layout of this extractor is shown in Fig.1.

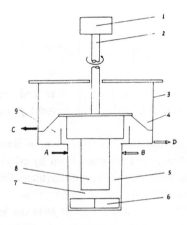

1–motor   2–shaft   3–casing   4–collecting annulus for heavy phase
5–annular gap   6–baffles   7–gap   8–rotary cylinder
9–collecting annulus for light phase
A,B–entrances of heavy and light phases
C,D–exit of heavy and light phases

Figure.1   *Layout of centrifugal extractor*

This extractor possesses the specification of very low solvent hold–up and very short retention time in the extractor. It is useful for extraction systems which emulsify readily and for the treatment of sensitive products.

Both the phases are pumped in countercurrent ways by metering pumps.

2.3 Experimental Results

2.3.1Experimental Results of Extraction Cascade   Based on the extraction and back–extraction equilibrium data under the following flow ratios: A / O = 5 for extraction, O / A = 3 for back–extraction, and according to the design index: extraction yield = 99%, back–extraction yield = 99%, the theoretical computation shows that three stages for extraction and three stages for back–extraction are needed. Figure 2 illustrates the three stage countercurrent extraction flowsheet.

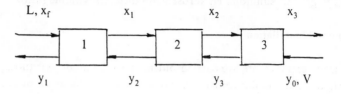

Figure.2   *Three stages countercurrent extraction flowsheet*

solving the following matrix:

$$
\begin{bmatrix}
-(1+\varepsilon) & \varepsilon & 0 \\
1 & -(1+\varepsilon) & \varepsilon \\
0 & 1 & -(1+\varepsilon)
\end{bmatrix}
\begin{bmatrix}
x_1 \\
x_2 \\
x_3
\end{bmatrix}
=
\begin{bmatrix}
-x_f \\
0 \\
-\dfrac{V}{L}y_0
\end{bmatrix}
$$

where ε represents extraction factor
and for $y_0 = 0$, $x_f = 1.5644$ mg / ml, $L / V = 5 / 1$,
The stagewise concentration of gibberellins are obtained:
   $x_1 = 0.323$ mg / ml, $x_2 = 0.065$ mg / ml, $x_3 = 0.0112$ mg / ml,
   $y_1 = 7.766$ mg / ml, $y_2 = 1.559$ mg / ml, $y_3 = 0.269$ mg / ml,
The experimental results for the extraction cascade are listed in Table 1.

Table 1   *Experimental results of extaction cascade*

x, mg / ml, in aqueous phase		y, mg / ml, in organic phase	
x_1	0.3345	y_1	7.45
x_2	0.0691	y_2	1.503
x_3	0.0194	y_3	0.2595

It can be seen from the tabulated data that they are almost consistent with the computed data, but owing to the different analytical methods for measuring the gibberellin concentration in aqueous and organic phases, there is a small difference between the calculated extraction yield in terms of the gibberellin concentration in the aqueous phase and in the organic phase, 98.78% of extraction yield is obtained calculated based on the aqueous phase concentration and 95.2% based on the organic phase concentration.

In Table 2 the experimental results of the back–extraction cascade are tabulated.

Table 2   *Experimental results of back–extraction cascade*

y, mg / ml, in organic phase		x, mg / ml, in aqueous phase	
y_1	1.2575	x_1	22.5381
y_2	0.0210	x_2	3.6629
y_3	0.0401	x_3	0.5329

It can be calculated from the listed data that 99.5% back–extraction yield is obtained.

The total recovery yield in the extraction cycle is 94.7~98.3%.

2.3.2 Results of Extraction Bench–scale Experiments

According to the experimental results for the extraction cascade, the following experimental conditions were chosen for bench–scale extraction by using centrifugal extractor with ϕ20mm rotor.

Extraction

Filtrate of fermentation broth: pH ≈ 2
Flow ratio (A / O): 4~5:1
Operation temp: ambient
Number of centrifugal extractor in bench–scale cascade: 4

Back–extraction

Back–extraction reagent: saturated $NaHCO_3$ solution
Flow ratio (O / A): 3:1
Operation temp: ambient
Number of centrifugal extractor in bench–scale cascade: 4
The experimental results are listed in Table 3.
Seven groups of experimental data are listed. Although there are some differences in every batch of fermentation broth, the total recovery yield of giberellins are almost the same except for the data of No.4.

Comparing these results with those for evaporative concentration, the recovery yield of gibberellins during the concentration step increases by a factor of about 15%.

The obtained concentrated feed is clear and transparent, it is easy to treat in further steps.

Table 3   *Bench–scale extraction experimental results*

No.		1	2	3	4	5	6	7
Filtered fermentation broth	u / ml	960 clear	980 muddy	930 clear	1000 muddy	1200 clear	940 clear	790
	pH	2.0	2.07	2.24	2.10	2.22	2.59	2.64
Throughput, L		30.0	27.12	20.0	25.0	22.15	21.31	20.5
Extn. flowratio, A / O		4:1	4:1	4:1	4:1	4:1	4:1	4:1
Extn. yield, %		98.12	98.20	98.14	94.88	99.11	99.17	98.84
Back–extraction ratio O / A		3:1	3:1	3:1	3:1	3:1	3:1	3:1
Back–extraction yield, %		98.08	99.0	99.05	97.15	98.86	99.32	98.94
Total recovery yield, %		96.23	97.22	97.24	92.17	97.98	98.19	97379
Conc. feed (u / ml)		22500	19900	27200	—	11300	11200	13700
Inspissation times		23.4	20.3	29.2	—	9.4	11.9	17.3

Using the following treatment steps, i.e., extraction of gibberellins by ethyl acetate → concentration of extract and crystallization of gibberellins → scrubbing and drying of gibberellin crystals, the product of the required purity is obtained.

In addition the hydrodynamic and mass transfer experiments were also carried out in a single industrial scale centrifugal extractor with φ230mm rotor. The experimental results show that its operation performance is satisfactory and the mass–transfer efficiency is also high enough, there is no obvious scale–up effect from φ20mm to this industrial scale extractor.

2.4 Emulsion breakage

For developing this process, the de–emulsifying reagent was added into the extraction system and the de–emulsifying experiments were carried out systematically, the experimental results show that 1231 ($C_{12}H_{34}BrN$) or 1681 ($C_{16}H_{42}BrN$) are the most suitable of the 12 reagents tested for elimination of emulsification during the extraction process, and the amount added to the filtered fermentation broth is only 0.075%(w / w).

3 CONCLUSIONS

Based on the extraction equilibrium and technological research a new technological process for extraction of gibberellins was developed as follows:

Fermentation broth→ Acidication and filtration→ Back adjustment of pH value→ Extraction and concentration by extraction cycle→ Extraction by ethyl acetate → Concentraction of ethyl acetate and crystallization of gibberellins→ Scrubbing and drying of gibberellin crystals.

Where, the concentration of gibberellins with a extraction cycle instead of the evaporation concentration possesses remarkable superiority, the recovery yield of gibberellins is high, and the energy consumption is low. All of these are very attractive.

REFERENCES

1   Li Zhou et al., Proceedings of ISEC'93, Edited by D.H.Logsdail and M.J.Slater,Vol.2, 1056. (1993), Elsevier Applied Science.

Author Index

Subject Index